Coherent Optical Interactions in Semiconductors

NATO ASI Series

Advanced Science Institutes Series

A series presenting the results of activities sponsored by the NATO Science Committee, which aims at the dissemination of advanced scientific and technological knowledge, with a view to strengthening links between scientific communities.

The series is published by an international board of publishers in conjunction with the NATO Scientific Affairs Division

A	**Life Sciences**	Plenum Publishing Corporation
B	**Physics**	New York and London
C	**Mathematical and Physical Sciences**	Kluwer Academic Publishers
D	**Behavioral and Social Sciences**	Dordrecht, Boston, and London
E	**Applied Sciences**	
F	**Computer and Systems Sciences**	Springer-Verlag
G	**Ecological Sciences**	Berlin, Heidelberg, New York, London,
H	**Cell Biology**	Paris, Tokyo, Hong Kong, and Barcelona
I	**Global Environmental Change**	

Recent Volumes in this Series

Volume 325 —Statistical Mechanics, Protein Structure, and Protein Substrate Interactions
edited by Sebastian Doniach

Volume 326 —Linking the Gaseous and Condensed Phases of Matter: The Behavior of Slow Electrons
edited by Loucas G. Christophorou, Eugen Illenberger, and Werner F. Schmidt

Volume 327 —Laser Interactions with Atoms, Solids, and Plasmas
edited by Richard M. More

Volume 328 —Quantum Field Theory and String Theory
edited by Laurent Baulieu, Vladimir Dotsenko, Vladimir Kazakov, and Paul Windey

Volume 329 —Nonlinear Coherent Structures in Physics and Biology
edited by K. H. Spatschek and F. G. Mertens

Volume 330 —Coherent Optical Interactions in Semiconductors
edited by R. T. Phillips

Volume 331 —Hamiltonian Mechanics: Integrability and Chaotic Behavior
edited by John Seimenis

Volume 332 —Deterministic Chaos in General Relativity
edited by David Hobill, Adrian Burd, and Alan Coley

Series B: Physics

Coherent Optical Interactions in Semiconductors

Edited by

R. T. Phillips

University of Cambridge
Cambridge, United Kingdom

Plenum Press
New York and London
Published in cooperation with NATO Scientific Affairs Division

Proceedings of a NATO Advanced Research Workshop on
Coherent Optical Interactions in Semiconductors,
held August 11–14, 1993,
in Cambridge, United Kingdom

NATO-PCO-DATA BASE

The electronic index to the NATO ASI Series provides full bibliographical references (with keywords and/or abstracts) to more than 30,000 contributions from international scientists published in all sections of the NATO ASI Series. Access to the NATO-PCO-DATA BASE is possible in two ways:

—via online FILE 128 (NATO-PCO-DATA BASE) hosted by ESRIN, Via Galileo Galilei, I-00044 Frascati, Italy

—via CD-ROM "NATO Science and Technology Disk" with user-friendly retrieval software in English, French, and German (©WTV GmbH and DATAWARE Technologies, Inc. 1989). The CD-ROM also contains the AGARD Aerospace Database.

The CD-ROM can be ordered through any member of the Board of Publishers or through NATO-PCO, Overijse, Belgium.

Library of Congress Cataloging-in-Publication Information
On file

ISBN 0-306-44737-1

A Division of Plenum Publishing Corporation
233 Spring Street, New York, N.Y. 10013

Printed in the United States of America

PREFACE

The NATO Advanced Research Workshop on Coherent Optical Processes in Semiconductors was held in Cambridge, England on August 11-14, 1993. The idea of holding this Workshop grew from the recent upsurge in activity on coherent transient effects in semiconductors. The development of this field reflects advances in both light sources and the quality of semiconductor structures, such that tunable optical pulses are now routinely available whose duration is shorter than the dephasing time for excitonic states in quantum wells. It was therefore no surprise to the organisers that as the programme developed, there emerged a heavy emphasis on time-resolved four-wave mixing, particularly in quantum wells. Nevertheless, other issues concerned with coherent effects ensured that several papers on related problems contributed some variety. The topics discussed at the workshop centred on what is a rather new field of study, and benefited enormously by having participants representing many of the principal groups working in this area.

Several themes emerged through the invited contributions at the Workshop. One important development has been the careful examination of the two-level model of excitonic effects; a model which has been remarkably successful despite the expected complexities arising from the semiconductor bandstructure. Indeed, modest extensions to the two level model have been able to offer a useful account for some of the complicated polarisation dependence of four-wave mixing signals from GaAs quantum wells. This work clearly is leading to an improved understanding of excitons in confined systems. To proceed much further in the theoretical description demands the greater rigour of many-body models, and these extend impressively the developing insight into transient effects in the rather complicated environment of a semiconductor.

The problem of experimental distinction between quantum beating and polarisation interference was another theme, and significant progress was reported in this area. Both time-domain and frequency-domain techniques are now being pursued, and the results - particularly those related to coupling of excitons in GaAs quantum wells - have led to some interesting surprises. Other topics which emerged in various guises in different papers include spin-related effects, the dynamics of free carriers, coherent phonons and vertical transport. Time-domain studies with resolution of 100fs and below also prompted stimulating discussion of limits to Markovian behaviour, and the dynamics of mixing processes in quantum wells. This volume also includes the contributed papers, many of which expand upon the topics reviewed in the invited papers.

In addition to the editor, the Workshop's organising committee comprised Izo Abram, Richard Harley and Ernst Göbel, and to these people especially I would like to express my gratitude. We were fortunate to receive financial assistance from Coherent (UK) Ltd, Melles Griot-Photon Control, Oxford Instruments and Spectra Physics Lasers.

I hope that this collection of papers proves to be useful in stimulating work in this exciting field; if it does it will be in large measure due to the effort and care of the numerous contributors.

Richard Phillips
Cavendish Laboratory, Cambridge
January 1994

CONTENTS

OPTICAL DEPHASING OF EXCITONS IN III-V SEMICONDUCTORS

[1]J. Kuhl, E.J. Mayer, G. Smith and R. Eccleston,
[2]D. Bennhardt, P. Thomas, K. Bott and O. Heller

[1]Max-Planck-Institut für Festkörperforschung
Heisenbergstr. 1, 70506 Stuttgart 80, Germany
[2]Philipps University, Physics Dept.
Renthof 5, 35032 Marburg, Germany

1. INTRODUCTION

Rapid progress in the development of mode-locked laser systems during the past decade has boosted the time resolution attainable in nonlinear optical spectroscopy well below 100 fs. Ti:sapphire lasers which can directly generate pulses as short as 12 fs [1] mark the most recent milestone of this evolution. This new generation of fs solid state lasers surpasses the older colliding-pulse mode-locked (CPM) dye laser [2] by far with respect to output power and stability, and most importantly, by the broad tunability of the fs output between 680-1000 nm. Combining these lasers with frequency converters like harmonic generators [3] or optical parametric oscillators [4] extends the tunability range to the ultraviolet, visible, near and middle infrared regime.

Therefore, semiconductor physicists now have powerful experimental tools at their disposal to study ultrafast dynamics of nonequilibrium carrier distributions which limit the speed of ultrafast electroectronic devices needed for future communication systems. Various new short-pulse-spectroscopy techniques such as degenerate-four-wave mixing (DFWM) [5], photon echo (PE) [6] or quantum beat (QB) experiments [7] allow studies of these dynamics on a time scale where the elementary relaxation steps, such as scattering by phonons, defects or other quasi-particles, which lead to the decay of the initially prepared coherent optical state, become directly observable.

Particular attention has been given to the nonlinear optical properties and dynamics of excitons which are the dominating optical features in the region close to the band edge of a semiconductor. Many studies have revealed remarkable variations of the optical nonlinearity and the relaxation dynamics with the dimension of the excitonic system.

Coherent Optical Interactions in Semiconductors
Edited by R.T. Phillips, Plenum Press, New York, 1994

In this paper, we will summarize the present state of information about exciton dephasing in III-V semiconductors and present our recent theoretical and experimental studies on the dependence of the created optical coherence and its subsequent decay on the polarization of the excitation pulses [8,9]. The polarization dependence of coherent optical experiments has been almost completely disregarded in the past. Systematic analysis demonstrates the important influence of coherent coupling between excitonic states caused by exciton-exciton interaction or the presence of disorder in most samples. Phenomenological theoretical models for these coupling processes provide a consistent explanation of many discrepancies and surprising results reported previously, including the change of signal decay rates [10,11] and switching from a PE like signal to free polarization decay [11] with polarization configuration of DFWM-experiments.

The paper is organized as follows: After briefly explaining the basic concept of optical phase coherence of excitons (Chap. 2) we will illustrate the principles of DFWM with ultrashort laser pulses in Chap. 3. Then we present a short summary on the present state of knowledge on dephasing in III-V semiconductors and discuss the dominant dephasing mechanisms (Chap. 4). The bulk of the paper (Chap. 5-7) deals with recent theoretical and experimental studies of the polarization dependence of time-integrated (TI) and time-resolved (TR) DFWM on excitons in GaAs QW's.

2. EXCITONIC DEPHASING

A consistent a priori description of an optically excited semiconductor requires a many-body theory due to the large number of strongly interacting states in a crystal. Although such theories have been developed during the past decade [12], their widespread application is still hampered because of their complexity and the inherently time-consuming numerical calculations.

The excitation of excitons in a semiconductor can be discussed, to a good approximation, in the frame of a simple two-level model providing at least a qualitatively correct description of the relaxation dynamics. If more than one exciton state is involved, this model has to be extended to a corresponding multi-level system. The energetic positions of the levels are taken from the experiment and their dynamics are characterized by introducing phenomenological relaxation times.

The time evolution of the polarization and population of simple non-interacting homogeneously broadened two-level systems after excitation by a short laser pulse is described by the optical Bloch equations formulated in the density matrix formalism [13] (see Chap.5). The dynamics of such a system after turning off the excitation are completely determined by the population lifetime T_1 and the dephasing time T_2. T_1 and T_2 account for the decay of the population created in the excited state and the decay of the macroscopic polarization P originating from the phase relaxation of the individual coherently excited dipoles, respectively. Since population relaxation implies also the decay of P, the relation

$$\frac{1}{T_2} = \frac{1}{2T_1} + \frac{1}{T'}, \tag{1}$$

holds where T' represents pure dephasing and T_2 corresponds to the homogeneous linewidth Γ_h of the optical transition $\Gamma_h = \frac{h}{\pi} \cdot T_2$ which is very often covered in the frequency domain by inhomogeneous broadening characterized by the inhomogeneous linewidth Γ_{inh}.

Equation (1) comprises the well-known relation $T_2/2 \leq T_1$ which sets an upper limit of T_2 since energy relaxation always implements loss of the phase coherence. The dependence of T_1 on T_2 has been disregarded in the past. As demonstrated recently, T_1 increases with decreasing T_2 for 2D excitons if T_1 is dominated by radiative recombination [14].

For extension to a three- (or more) level system, one starts with two-level systems and introduces a phenomenological coupling between either the ground or excited levels. The additional degree of freedom in the three-level system is accounted for by a spectral relaxation time T_3. The relation of the different relaxation times is then extended to

$$\frac{1}{T_2} = \frac{1}{2T_1} + \frac{1}{T'} + \frac{1}{T_3} \quad . \tag{2}$$

Optical excitation of a two-level system by a short laser pulse prepares a coherent superposition of the two states. After the pulse is turned off the phase coherence decays due to scattering processes with the phase coherence time T_2. Simultaneous coherent excitation of both excited states in a 3-level system leads to an interference between the distinct resonances and a concomitant periodic modulation of the signal with a frequency corresponding to the level splitting. If the two resonances share a common level the modulation originates from real QB's. Beating caused by superposition of the radiation fields of two independent oscillators on the detector is called polarization interference (PI). QB spectroscopy which is now a well established technique for investigations of semiconductors will be discussed in other contributions to this book.

3. DEGENERATE-FOUR-WAVE MIXING

Typical dephasing times of free-exciton transitions in III-V compounds range from 0.1 to 10 ps. This loss in phase coherence can be directly monitored in the time domain in a transient DFWM experiment with pico- or subpicosecond laser pulses [5]. The most versatile and experimentally simplest version is the two-pulse self-diffraction configuration where two laser pulses with frequency ω and wavevectors $\vec{k}_1$ and $\vec{k}_2$ impinge noncollinearly onto the sample with a variable time delay which is shorter than or comparable to T_2. The first pulse sets up a macroscopic coherent polarization of the optical transition since all excited dipoles are forced by the optical field of the pulse to oscillate in phase. The second delayed pulse probes the coherent part of the oscillation left behind from the first pulse via interference of the material's polarization field with the optical field of the pulse. This interference leads to the formation of a spatially periodic, optically-induced absorption grating. The subsequent diffraction of part of the second pulse off this grating results in the emission of signal beams in the phase-matching directions $\vec{k'}_3 = \vec{k}_2 + (\vec{k}_1 - \vec{k}_2) = \vec{k}_1$ and $\vec{k}_3 = \vec{k}_2 - (\vec{k}_1 - \vec{k}_2) = 2\vec{k}_2 - \vec{k}_1$. (It should be mentioned that for two noncollinearly propagating beams a small phase mismatch $\Delta\vec{k} = (2\vec{k}_2 - \vec{k}_1) - \vec{k}_3 \neq 0$ is unavoidable, implying a slightly reduced amplitude of the diffracted beam).

The source of the signal is the 3rd order polarization created by the nonlinear interaction of three exciting optical pulses incident onto the sample at t_1, t_2 and t_3.

This polarization is given by the following expression:

$$P_i^{(3)}(t, t_1, t_2, t_3) = \sum_{jkl} \int_{-\infty}^{t} \int_{-\infty}^{t'} \int_{-\infty}^{t''} dt'''dt''dt' \tag{3}$$

$$\chi_{ijkl}^{(3)}\ (t, t' - t_1, t'' - t_2, t''' - t_3)\ E_j(t' - t_1) \cdot E_k(t'' - t_2)\ E_l(t''' - t_3)$$

Here the $\chi_{ijkl}^{(3)}$ are the contributing components of the 3rd order susceptibility tensor which involve the polarization decay dynamics.

In a two-pulse experiment, the exciting electric field is given by the superposition of the two time-delayed laser pulses

$$\vec{E}\ (t) = \vec{e_1}\ E_1(t)cos(\vec{k_1}\vec{r}\ - \omega t) + \vec{e_2} \cdot E_2(t - \tau)cos\left[\vec{k}_2\,\vec{r} - \omega(t - \tau)\right] \tag{4}$$

with $\vec{e_i}$ and E_i being the polarization vector and the envelope of the laser pulse, respectively, and τ the time delay between the exciting pulses.

Calculation of the polarization for a system of inhomogeneously broadened two-level systems requires an additional integration to average over the spatially varying magnitude of the dipole moments.

$$P_i^{(3)inh} = \int_{o}^{\infty} h(\omega_o) \cdot P_i^{(3)hom}\ (\omega_o)d\omega_o \tag{5}$$

with $h(\omega_o)$ the exciton distribution function determined by the inhomogeneous lineshape.

The rapidly oscillating polarization $P_{2\vec{k}_2 - \vec{k}_1}^{(3)}(t, \tau)$ leads to the emission of the signal

$$I(t, \tau)\ \propto\ |P^3\ (t, \tau)|^2 \tag{6}$$

3.1. Time-Integrated DFWM Signal

The dephasing can be detected if the total energy $I(\tau)$ diffracted into the direction $\vec{k_3}$ is recorded by a slow (time-integrating) detector as a function of the delay between the two exciting pulses.

$$I(\tau)\ \propto \int_{-\infty}^{+\infty} |P(t, \tau)|^2\ dt \tag{7}$$

For pulses much shorter than the dephasing time, the decay of the signal intensity is quite well approximated by an exponential function of the delay between the two pulses

$$I(\tau)\ \sim\ exp(-\frac{\tau}{T_D}) \tag{8}$$

with $T_D = T_2/2$ and $T_D = T_2/4$ for homogeneously and strongly inhomogeneously broadened transitions, respectively. The two-pulse experiment itself cannot differentiate between homogeneous and inhomogeneous broadening unless the dephasing is so clearly resolved that accurate comparison with spectral data is possible.

Difficulties in the evaluation of the experimental data arise if the dephasing which occurs during excitation cannot be neglected or if the homogeneous and inhomogeneous contributions to the linewidth are almost comparable. In this case, one has to solve the optical Bloch equations for a two-level system, taking into account the finite pulse duration, and fitting the theoretical DFWM signal shape to experimental data. In

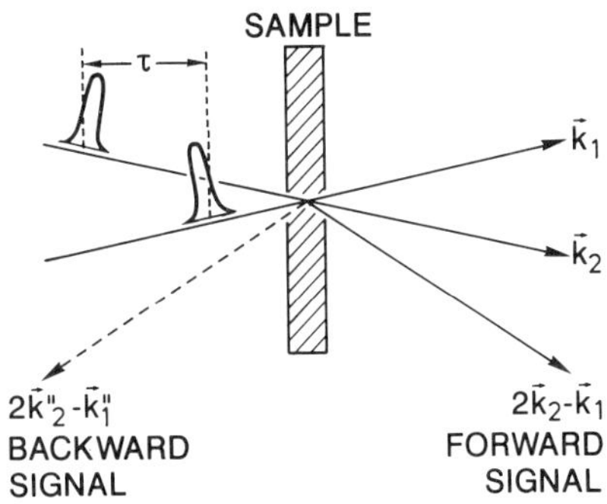

Fig. 1 Geometry of exciting beams and generated signal for forward and backward reflected DFWM.

the standard experimental configuration (see Fig.1), which is utilized by most groups working in this field, the excitation pulses are incident from one side onto the sample and the signal, diffracted in the forward direction, is monitored behind the sample. This geometry necessitates the use of optically thin samples and usually implies polishing and etching of the substrate to eliminate the influence of its optical properties. Substrate removal usually results in creation of remarkable strain within the sample accompanied by a corresponding change of the optical properties. These problems are prevented by the backward reflection geometry [15] which utilizes the small signal diffracted back into the half-space from where the laser pulses impinge onto the sample. This configuration permits DFWM experiments on the original as-grown sample without etching and polishing. In thin layers ($d \leq 30$nm), the intensities diffracted in the forward and backward direction, respectively, are almost equal because of the partial breakdown of momentum conservation: in such thin layers, only the wavevector components parallel to the layer have to be conserved. This configuration is particularly suitable to study excitations in thin layers, such as single QW's, without being influenced by the optical properties of absorbing substrates.

Weiner *et al.* [16] have shown that three pulse grating experiments which have been traditionally applied to study orientational relaxation, excited state energy relaxation or spatial diffusion can be utilized for dephasing measurements as well. In this geometry, see Fig. 2, the optically-induced grating formed by pulses 1 and 2, which are temporally separated by the variable delay τ, is detected by a third probe pulse with wavevector $\vec{k}_3$ incident onto the grating after a fixed delay T such that T is large compared to the pulse duration and the initial fast energy relaxation, but small compared to the interband relaxation which erases the grating. This probe pulse is scattered off the grating into the directions $\vec{k}_4=\vec{k}_3 \pm(\vec{k}_1 - \vec{k}_2)$. Compared to two-pulse scattering, this three-pulse experiment offers several advantages for dephasing measurements including resolution below the pulse width, perfect phase matching, insensitivity to rapid energy relaxation and clear differentiation between homogeneous and inhomogeneous broadening. In addition, measurements of the signal intensity in dependence of the delay between the second and third pulse yield a simultaneous determination of the energy relaxation time T_1. The phase-conjugated three-pulse geometry (see Fig.2), where pulse 3 is incident onto the sample from the direction opposite to the direction of pulse 2, provides an enhanced signal-to-noise ratio over the forward configuration [17].

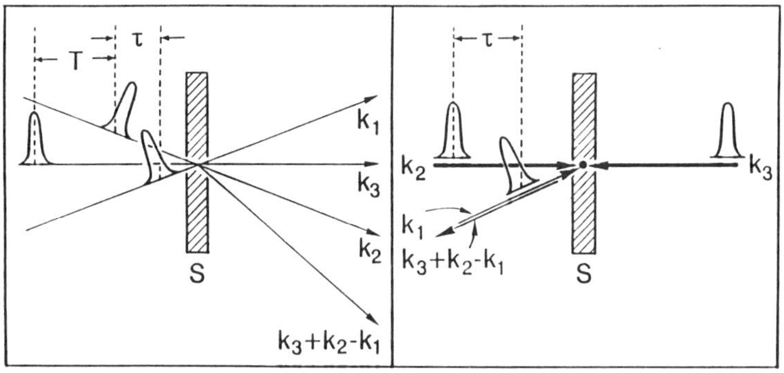

Fig. 2 Experimental geometries for dephasing measurements with 3-pulse DFWM. Left: Forward diffraction geometry; right: Phase conjugation geometry.

3.2. Time-Resolved DFWM

The time-integrated signal provides no information about whether the nonlinear signal has been emitted as a photon echo or a free-polarization decay, as expected for homogeneously and inhomogeneously broadened transitions, respectively. This important information, which permits a separate determination of homogeneous and inhomogeneous parts of the overall linewidth, becomes attainable if the FWM signal $I(t,\tau)$ in Eq. (7) is time-resolved. Time-resolution of 10-20 ps is accomplished if the slow photodetector is replaced by a streak camera. This technique has been applied by Noll, *et al.* [18] to resolve the stimulated photon echo due to localized excitons in CdSSe which have dephasing times of several 100 ps. Free excitons in GaAs and other III-V compounds, however, have much shorter ($< 10ps$) dephasing times. Time-resolution in this regime requires cross-correlation techniques, e.g. up-conversion of the DFWM signal in a nonlinear optical crystal by a synchronous femtosecond laser pulse [19]. A typical experimental configuration for up-conversion is shown in Fig. 3. The sample is excited sequentially by two laser pulses with wavevectors $\vec{k}_1$ and $\vec{k}_2$ delayed by τ and the FWM signal emitted into the direction $2\vec{k}_2 - \vec{k}_1$ is mixed with a third reference pulse in a thin ($< 1mm$) nonlinear optical crystal. The intensity of the second harmonic signal for a fixed delay τ between the exciting pulses is recorded by a second photomultiplier as a function of the real time separation t between the reference pulse and the two excitation pulses.

For a homogeneously broadened transition, the maximum signal occurs at $t = \tau$ and decays with $T_2/2$ (free polarization decay). In contrast, for an inhomogeneous line, the

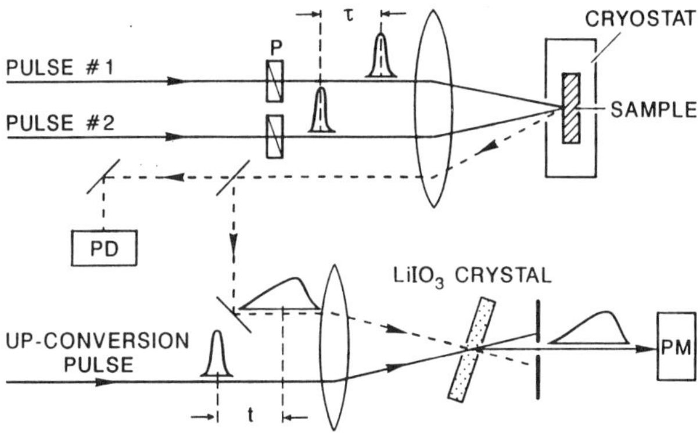

Fig. 3 Experimental set-up for time-resolved DFWM.

FWM signal is emitted as a spontaneous photon echo at $t = 2\tau$. If the inhomogeneously broadened exciton transition can be viewed as an ensemble of noninteracting two-level systems, the temporal width of the spontaneous echo corresponds to the inverse of the inhomogeneous linewidth and thus permits a precise determination of this quantity. Furthermore, TR-DFWM allows an unambiguous distinction between quantum beats and polarization interference of independent quantum mechanical systems [21,22].

4. DEPHASING TIMES AND MECHANISMS FOR EXCITONS IN III-V SEMICONDUCTORS

4.1 Dephasing Processes of Excitons in GaAs

Phase destroying events can be classified into two different groups:

1. Interactions of excitons with the imperfections of the sample and
2. Collisions of excitons with other quasi-particles such as optical or acoustic phonons, excitons and free electrons or holes.

The first group includes scattering of excitons by impurities, the roughness of the GaAs/$As_xGa_{1-x}As$ interfaces or by fluctuations of the Al concentration within the $Al_xGa_{1-x}As$ barrier material. The latter contribution is expected to increase with decreasing well width because of the increasing extension of the exciton wavefunction into the barriers. The contributions originating from the crystal defects may vary considerably from sample to sample, depending on its quality. A well-defined quantitative variation of the magnitude and distribution of these crystal imperfections is still well beyond the potentials of MBE technology. Thus, a systematic study of the dependence of the dephasing rate on deviations of the crystal structure from the ideal order is presently impossible. Recently we have shown that sample disorder effects a coupling of excitonic states which complicates the interpretation of DFWM experiments. Under such conditions, the extraction of T_2 from the signal necessitates special polarization conditions (see Chaps. 5.2 and 7).

On the other hand, the modification of T_2 by the presence of optical and acoustic phonons, and free carriers or excitons can be easily explored on one and the same sample by temperature and excitation density dependent dephasing measurements, respectively (see Chaps. 4.3 and 4.4.).

4.2. Intrinsic T_2 of Excitons in 3D and 2D GaAs

The majority of published dephasing studies in III-V semiconductors concern the hh-exciton transition in bulk GaAs and GaAs/(Al,Ga)As QW's [22]. At very low temperatures (< 1.8K) T_2 is determined by interactions of the hh-exciton with the crystal imperfections and thus may vary considerably from sample to sample. For high quality material, T_2 amounts to 7ps for 3D GaAs [23] and between 10-25 ps [24,25] for the best QW's. An unambiguous determination of the variation of T_2 with QW thickness L_z is still missing. Intuitively, an increase of T_2 with decreasing L_z is expected since the restriction of the exciton mobility for the direction perpendicular to the layers implies a reduced interaction with the crystal environment. However, surface roughness or alloy fluctuations in the barriers tend to establish the opposite effect since the extension of the exciton wavefunction into the barrier increases with decreasing L_z. Honold *et al.* [25] observed an almost constant T_2 for thicker QW's (15nm< L_z < 30nm) and a drop by about a factor of 2 between 15 and 13 nm. This rapid drop is even more pronounced for lh excitons which are shown in the same paper to have dephasing times only about 20-30% shorter than those of the hh excitons for L_z >15 nm. The enhanced dephasing of lh excitons for L_z < 13nm is caused by a Fano-type resonance of the lowest lh exciton state with the e/hh continuum states leading to a rapid decay of the lh exciton. It should be noticed, however, that the data of ref. [25] have to be considered with care since the experiments were performed with cross-polarized (CP) laser pulses. As shown in Chap. 5.2, the signal decay in this geometry is sensitive to inhomogeneity of sample disorder and does not always reflect the proper dephasing time. Only slight differences in T_2 for lh and hh excitons in thicker QW's, where the lh exciton state is located well below the band edge of e/hh continuum transition, have been also reported in Ref. [26].

4.3 Influence of Disorder

The presence of strong inhomogeneous broadening leads to a remarkable change of T_2 with exciton energy across the inhomogeneous line [11,27,28]. On the low energy side, T_2 values of several tens of picoseconds have been observed and attributed to excitons localized by QW width fluctuations. Dephasing of these localized excitons is mainly due to phonon-assisted migration [28]. At about the center of the linear absorption line, the dephasing time begins to decrease rapidly since the exciton energy passes the mobility edge separating localized states below line center from quasi-delocalized states above. Increased localization of the exciton associated with the presence of strong disorder in the sample corresponds to an increased admixture of large k-vector excitons to the wavefunction resulting in a quasi-continuous transition from a two-level system in momentum space in the absence of disorder to a noninteracting 2-level system in real space for strong disorder.

A comparison of $T_2 = 530$ fs, measured at 5K for hh excitons in a 20nm (In,Ga)As/(In,Al)As QW [29] with the more than ten times larger values in GaAs QW's demonstrates the strong influence of chemical disorder (alloy fluctuations) present in the ternary QW material, in addition to structural disorder. A theoretical model for dephasing of optical excitations in a disordered semiconductor including also the influence of exciton/phonon interactions has been presented in Ref. [30]. Experimental studies showed that elastic disorder scattering is an essential dephasing mechanism in $Al_xGa_{1-x}As$ in contrast to the situation in $CdS_{1-x}Se_x$, where the contribution of disorder to dephasing is negligible. The different behavior could be explained by the microscopically different range of the disorder potential as compared to the exciton Bohr radius.

4.4. Influence of Exciton/Phonon Interaction

The low temperature variation of the homogeneous linewidth $\Gamma_h = h/\pi T_2$ in a high quality sample T_s can be fairly well described by a linear function of sample temperature T_s

$$\Gamma_h(T_s) = \Gamma_h(0) \; + \; \beta_{ph} \cdot T_s \tag{9}$$

and can be attributed to anti-Stokes scattering with acoustic phonons [31]. $\Gamma_h(0)$ summarizes all contributions caused by collisions with defects. It should be noted, however, that the relation between T_2 and sample temperature may change drastically in a strongly disordered sample.[30]

If the temperature is increased beyond 15 and 30K for the 3D and 2D sample, respectively, the growth of Γ_h with T_s becomes much faster owing to the onset of exciton ionization at these temperatures and the concomitant appearance of an additional highly effective dephasing process: scattering of excitons by free electrons and holes. Typical parameters for $\Gamma(0)$ and β_{ph} are presented in table 1. Evidently, the dephasing efficiency of exciton/acoustic phonon collisions in QW's is smaller than in bulk GaAs and is reduced with decreasing well width in contrast to theoretical predictions [32]. This discrepancy is likely caused by the lack of proper phonon wavefunctions in 2D. At still higher temperatures, the interaction of excitons with optical phonons gains increasing importance which finally leads to ionization and a concomitant loss of exciton phase coherence within 300 fs at room temperature [33].

Table 1. Homogeneous linewidth and temperature coefficient for the line broadening

L_z(nm)	194	27.7	13.5	12
$\beta_{ph}(\mu eV/K)$	17±1	±1	5±1	2.5± 0.5
$\Gamma(0)$(meV)	0.18 ± 0.02	0.40 ± 0.02	0.48 ± 0.02	0.11 ± 0.01

4.5. Enhanced Exciton Dephasing by Exciton/Exciton and Exciton/Free-Carrier Scattering

The rate at which the phase coherence is lost appears to become faster at higher excitation densities due to the increase of exciton/exciton collisions. The broadening of the homogeneous linewidth can be well described by a linear density dependence

$$\Gamma_h(N) \;=\; \Gamma_h(0) \;+\; \beta_{xx}\;(a_B^{nD})^n\; E_B^{nD} \times N \tag{10}$$

with n=2 and n=3 for the 2D and the 3D exciton, respectively. N is the 2 or 3-dimensional density of excitons, a_B^{nD} and E_B^{nD} are the exciton Bohr radius and exciton binding energy for the 2D(3D) system. β_{xx} is a dimensionless parameter for the line broadening [22,24,34].

A very similar relation holds for the influence of collisions between the exciton and free carriers (electrons or holes) on Γ_h. In this case N represents the free-carrier density and β_{xx} in Eq. (10) has to be replaced $\beta_{xe,h}$. Experimentally determined values of β_{xx} and $\beta_{xe,h}$ are shown in table 2. The data demonstrate exciton/free-carrier collisions to be almost 10× more efficient than exciton/exciton collisions in destroying the phase of 2D as well as 3D excitons. This result is explained by the long range Coulomb field of free carriers. The surprisingly strong exciton/exciton interaction, which is detectable at densities as small as $2 \times 10^9 \mathrm{cm}^{-2}$ or $2 \times 10^{14} \mathrm{cm}^{-3}$ for 2D and 3D samples, respectively, is caused by dipole/dipole interaction, phase-space filling and exchange effects.

Table 2. Broadening parameters for the homogeneous linewidth for exciton/exciton β_{xx} and exciton/free-carrier scattering $\beta_{xe,h}$

L_z(nm)	a_B(nm)	E_B(meV)	$\Gamma(0)$(meV)	β_{xx}	$\beta_{xe,h}$
194	13.5	4.2	0.18 ± 0.02	10 ± 1	100 ± 5
12	9.5	8.5	0.11 ± 0.01	1.5 ± 0.3	11.5 ± 2.0

A direct comparison of the collisional broadening of the 2D and 3D exciton transitions, caused by the interaction with additional free carriers or incoherent excitons, becomes available if the particle distance is used as the parameter for the excitation density. The inter-particle distance r_b normalized to the respective exciton Bohr radius, can be easily calculated from the particle density [35]

$$\begin{aligned} 3\mathrm{D} : \mathrm{r_b} &= (4\pi a_B^3 N/3)^{-1/3} \\ 2D : r_b &= (\pi a_B^2 N)^{-1/2} . \end{aligned} \tag{11}$$

Figure 4 depicts the line broadening of 2D and 3D excitons due to exciton/exciton (X-X) and exciton/free-carrier (X-eh) collisions versus r_b. It is evident that 2D excitons are much more efficient in their interactions with free carriers or other excitons as compared with their 3D counterparts at the same normalized inter-particle distance. The line broadening of the 2D exciton transition measured at an inter-particle distance of 5-10 Bohr radii amounts to about 2-4 times the value found at the same distance for the 3D exciton for both exciton/exciton as well as exciton/free-carrier collisions. This result can be understood as a consequence of the considerably weaker screening of the Coulomb interaction in 2D as compared with 3D [36]. The relative unimportance of screening in 2D systems leads to an increasing interaction of the colliding excitons via repulsive forces which originate from the Pauli exclusion principle for identical fermions [36,37].

5. THEORY OF POLARIZATION-DEPENDENT DEGENERATE-FOUR-WAVE-MIXING

The optical dipole transitions at the band edge of a GaAs/(Al,Ga)As QW for light propagating perpendicular to the layers can be described by two independent, degenerate three-level systems representing the $J = \frac{1}{2}$ and $J = \frac{3}{2}$ valence and the $J = \frac{1}{2}$ conduction electrons (see Fig.5). σ^+ and σ^- represent the transitions allowed for clockwise and counter-clockwise circularly polarized light. Transitions from the $m_J = \pm\frac{3}{2}$ valence bandstates are hh-exciton transitions and those from $m_J = \pm\frac{1}{2}$ are lh-exciton transitions. In the two-pulse DFWM experiment, two short optical pulses with electric fields $\vec{E}_1(\omega, \vec{k}_1, t - t_1)$ and $\vec{E}_2(\omega, \vec{k}_2, t - t_2)$ impinge on the sample at different times t_1 and t_2. In order to describe the different orientations of the linearly polarized electric field vectors $\vec{E}_1$ and $\vec{E}_2$, we define an $(\vec{x}, \vec{y})$ coordinate system in the plane of the QW

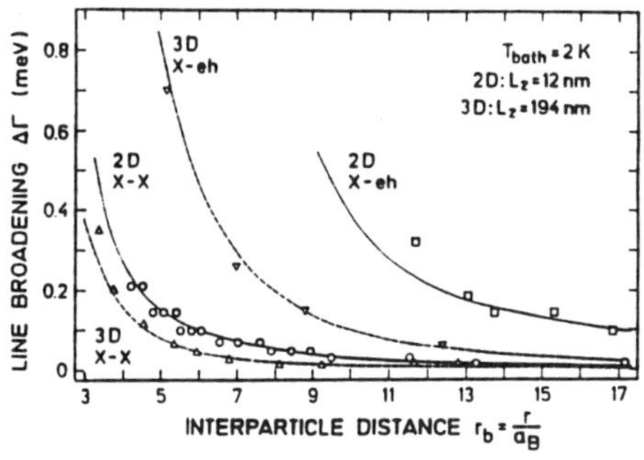

Fig. 4. Line broadening of 2D (circles and squares) and 3D excitons (triangles) subjected to collisions with excitons (x-x) or free carriers (x-e,h) as a function of normalized inter-particle distance.

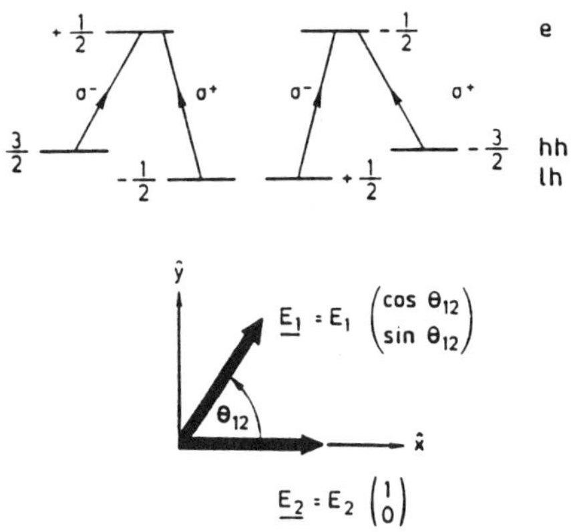

Fig. 5. hh and lh exciton circularly polarized transitions in a GaAs/(Al,Ga)As quantum well (upper part) and schematic diagram illustrating polarization notation used in the text.

(see Fig.5) with $\vec{E}_2$ oriented along $\vec{x}$ and $\vec{E}_1$ at some angle θ_{12} to $\vec{E}_2$. We neglect any $\vec{z}$ component of the light fields because in the experiments, the propagation directions of the pulses within the sample deviate by less than 1° from the growth direction. For $\vec{E}_1$ parallel to $\vec{E}_2$, the two circularly polarized components of $\vec{E}_1$ are in phase with those of $\vec{E}_2$. Choosing $\theta_{12} \neq$ introduces a phase shift between the corresponding circular components of $\vec{E}_1$ and $\vec{E}_2$, causing strong variations of the DFWM signal.

To calculate the total 3rd-order nonlinear polarization $\vec{P}^{(3)}_{2\vec{k}_2-\vec{k}_1}$ for the pair of uncoupled 3-level systems in Fig. 5, we start with the optical Bloch equations for an N×M level system using the notation of Ref. [38]

$$
\begin{aligned}
\partial_t p_{mn} + i(\epsilon^c_n - \epsilon^v_m)P_{mn} + \gamma^{cv}_{nm}P_{mn} &= i\sum_{m'} v_{mm'}R_{m'n}(t) - i\sum_{n'} R_{mn'}(t)c_{n'n} \\
\partial_t v_{mm'} + (i\epsilon^v_{m'} - i\epsilon^v_m + \gamma^{vv}_{m'm})(v_{mm'} - \delta_{mm'}) &= i\sum_{n} (p_{mn}R^*_{m'n}(t) - R_{mn}(t)p^*_{m'n}) \\
\partial_t c_{n'n} + i(\epsilon^c_n - \epsilon^c_{n'})c_{n'n} + \gamma^{cc}_{nn'}c_{n'n} &= -i\sum_{m} (R^*_{mn'}(t)p_{mn} - p^*_{mn'}R_{mn}(t)) \, .
\end{aligned}
\tag{12}
$$

We will confine the following discussion to homogeneously broadened transitions and assume δ-like excitation pulses since we are mainly interested in the qualitative dependence of the signal intensity, its polarization and temporal shape on the polarization of the incident beams. With the exception of the pulse overlap region, these signal characteristics do not drastically change for finite pulse duration and the influence of inhomogeneous broadening on the signal's temporal position and slope of the trailing edge are well known.

The 3rd-order polarization in the direction $2\vec{k}_2 - \vec{k}_1$ becomes

$$\vec{P}^{(3)}_{2\vec{k}_2-\vec{k}_1}(t) = -i\Theta(t-\tau)\Theta(\tau) \sum_{n,m,n',m'} \vec{\mu}^*_{nm} e^{-i(\epsilon^c_n-\epsilon^v_m-i\gamma^{cv}_{nm})(t-\tau)}(\vec{\mu}_{nm'}\cdot\vec{E}_{\vec{k}_2})$$
$$\times(\vec{\mu}_{n'm}\cdot\vec{E}_{\vec{k}_2})e^{i(\epsilon^c_{n'}-\epsilon^v_{m'}-i\gamma^{cv}_{n'm'})\tau}(\vec{\mu}_{n'm'}\cdot\vec{E}_{\vec{k}_1})^* \quad . \qquad (13)$$

Here pulse 1 acts first (at time 0) and creates a polarization which oscillates and starts to dephase for a time interval τ until pulse 2 arrives. The dipole matrix elements $\vec{\mu}_{mn} = \langle c, n|c\vec{r}|v, m\rangle$ contain the desired polarization selection rules. To account for the spectral width of the laser pulses, the $\vec{\mu}_{nm}$ are interpreted as effective matrix elements which incorporate the spectral weight of the related transition. $\epsilon^{c(v)}_{n(m)}$ are the conduction (valence) band electronic energies and the γ's are phenomenologically introduced dephasing rates. For the pair of uncoupled 3-level systems Eq. (13) becomes

$$\vec{\mathrm{P}}^{(3)}_{2\vec{\mathrm{k}}_2-\vec{\mathrm{k}}_1}(\mathrm{t},\tau) = \frac{-i}{2}\Theta(t-\tau)\Theta(\tau)E_2^2E_1^*$$
$$\times\Bigg[\begin{pmatrix}\cos\Theta_{12}\\0\end{pmatrix}\{|\mu_h|^2e^{-i(\epsilon^c-\epsilon^v_h-i\gamma^{cv}_h)(t-\tau)} + |\mu_l|^2e^{-i(\epsilon^c-\epsilon^v_l-i\gamma^{cv}_l)(t-\tau)}\}$$
$$\times\ \{|\mu_h|^2\, e^{i(\epsilon^c-\epsilon^v_h+i\gamma^{cv}_h)\tau} + |\mu_l|^2\, e^{i(\epsilon^c-\epsilon^v_l+i\gamma^{cv}_l)\tau}\}$$
$$+\begin{pmatrix}0\\-\sin\Theta_{12}\end{pmatrix}\{|\mu_h|^2\, e^{-i(\epsilon^c-\epsilon^v_h-i\gamma^{cv}_h)(t-\tau)} - |\mu_l|^2e^{-i(\epsilon^c-\epsilon^v_l-i\gamma^{cv}_l)(t-\tau)}\}$$
$$\times\ \{|\mu_h|^2e^{i(\epsilon^c-\epsilon^v_h+i\gamma^{cv}_h)\tau} - |\mu_l|^2e^{i(\epsilon^c-\epsilon^v_l+i\gamma^{cv}_l)\tau}\}\Bigg] \qquad (14)$$

The subscript l(h) indicates light(heavy) hole states and c(v) the conduction (valence) band states. μ_l and μ_h are effective matrix elements including the spectral weight of the laser pulses at lh and hh transitions. Equation (14) predicts the following polarization-dependent features of the DFWM signal:

1. For excitation of either the hh or lh transition alone, the orientation Θ_{sig} of the resultant 3rd-order nonlinear polarization vector $\begin{pmatrix}cos\Theta_{12}\\-sin\Theta_{12}\end{pmatrix}$ obeys the relation $\Theta_{sig} = -\Theta_{12}$ as expected for two independent two-level systems [6] with opposite circular polarization rules.

2. The time-integrated DFWM intensity is independent of Θ_{12} for excitation of either hh or lh excitons alone. However, it decreases at zero delay with increasing Θ_{12} if both transitions are excited. The magnitude of the decrease is determined by the relation between the sum and difference of the absolute values squared of the effective matrix elements from the hh and lh exciton transitions.

3. Quantum beats between lh and hh polarizations at the difference frequency $\epsilon^v_h - \epsilon^v_l$ are obtained if both transitions are excited. However, the phase of the quantum beats relative to $\tau = 0$ changes from 0° at $\Theta_{12} = 0°[|\mu_h|^2$ and $|\mu_e|^2$ terms add in Eq. (14)] to 180° at $\Theta_{12} = 90°$ ($|\mu_n|^2$ and $|\mu_l|^2$ terms subtract). This is a consequence of a shift in the lh and hh transition phases induced in the $\Theta_{12} = 90°$ configuration, since for each electron state, transitions from the two hole states originate from light of opposite circular polarization.

All these predictions are confirmed by experimental observations. Experiments performed for varying excitation densities reveal, however, substantial changes of the

DFWM signal features depending on the polarization configuration, if the laser beam intensity is reduced (see chap.7). Furthermore, several groups have reported polarization-dependent dephasing rates [10] and have observed a transition from a photon echo-like to a free-polarization decay-like signal in time-resolved DFWM if Θ_{12} is switched from 0 to 90° [11]. These results cannot be explained in the frame of the independent 3-level model and thus require a refined theoretical model taking into account coherent coupling between individual exciton states. In the next two paragraphs we will discuss extensions of the model considering the influence of exciton-exciton interaction and disorder-induced coupling.

5.1 Exciton-Exciton Coupling

The following discussion will be confined to the optical properties of the hh exciton transitions in order to simplify the mathematical expressions. All the conclusions obtained for the modifications of the polarization dependence of the DFWM caused by exciton-exciton coupling are extendable to 3-level systems as well.

The optical properties of the hh exciton in a GaAs QW are usually described by two independent two-level systems representing the different spin states, and which can be excited by σ^+ and σ^- light, respectively. Each of the noninteracting two-level systems is quantum mechanically described in a 2D one-particle Hilbert-space $H_+^{(1)}$ and $H_-^{(1)}$. An equivalent description exists in a 2-particle Hilbert-space $H^{(2)}$. Since the two excitons are distinguishable owing to their different spins, $H^{(2)}$ is given in this case by the direct product between the one-particle Hilbert spaces

$$H^{(2)} = H_+^{(1)} \otimes H_-^{(1)} . \tag{15}$$

The two independent 2-level systems for the hh excitons are equivalent to a 4-level system in $H^{(2)}$: the groundstate $|G>$ (no exciton), two single exciton states $|E^+>$ and $|E^->$ (either the σ^+ or the σ^- exciton is excited) and a two-exciton state $|E2>= (|E^+>|E^->)$ (simultaneously a σ^+ and a σ^- exciton exist). Figure 6 illustrates this transformation. Complete equivalence between the two independent 2-level systems and the 4-level system implies that the energy of $|E2>$ is exactly twice the energy of $|E^\pm>$ (no binding energy), the matrix element ν for the transitions from $|E^\pm>$ to $|E2>$ are equal to the matrix elements μ for transitions from $|G>$ to $|E^\pm>$. The same identity must be valid for the dephasing rates γ_μ and γ_ν from the two-exciton to the single exciton states and from the single exciton states to the ground state, respectively.

Experiments performed with linearly polarized excitation pulses are better discussed in terms of the 4-level scheme depicted in the right part of Fig. 6 which can be shown to be equivalent to the two uncoupled two-level systems and is obtained from the scheme in the middle part of Fig. 6 by a simple base transformation. The new 4-level system includes the groundstate, the two states $|E^x>= 2^{-1/2}(|E^+> + |E^->)$ and $|E^y>= 2^{-1/2}(|E^+> - |E^->)$ corresponding to the single exciton states excited by linear polarized light along the x- and y-axis, respectively, and the two-exciton state $|E2>= (|E^+>|E^->)$.

Phenomenological introduction of exciton-exciton interaction in the 4-level system leads to breaking of the symmetry between the upper and lower transitions. This may be accomplished by assuming either different matrix elements μ and ν for the lower and upper transitions, and/or different dephasing rates for the 2-exciton/exciton and

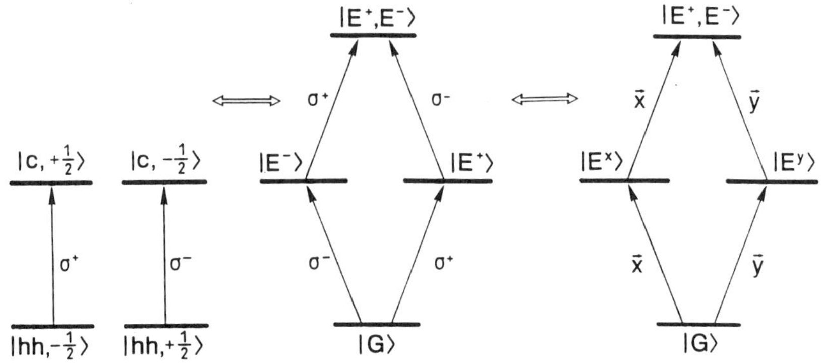

$$|G> = |hh,+\tfrac{1}{2}> |hh,-\tfrac{1}{2}>$$
$$|E^+> = |hh,+\tfrac{1}{2}> |c\ ,+\tfrac{1}{2}>$$
$$|E^-> = |c\ ,-\tfrac{1}{2}> |hh,-\tfrac{1}{2}>$$
$$|E^+,E^-> = |c\ ,-\tfrac{1}{2}> |c\ ,+\tfrac{1}{2}>$$

$$|E^x> = 2^{-1/2}(|E^+> + |E^->)$$
$$|E^y> = 2^{-1/2}(|E^+> - |E^->)$$

Fig. 6 Equivalence between two non-interacting 2-level systems (left) and non-renormalized 4-level systems with circular (middle) and linear (right) selection rules.

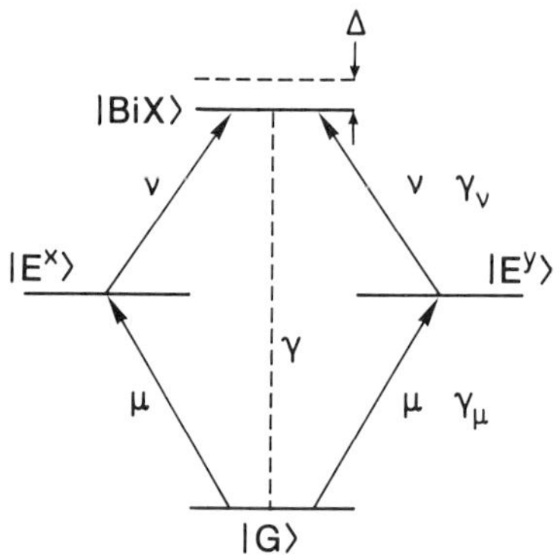

Fig. 7 Renormalized four-level system consisting of a ground state |G >, two exciton states $|E^x>$ and $|E^y>$ and a biexciton state |Bi X > with binding energy Δ.

the single-exciton/ground state transitions, and/or a finite binding energy Δ, i.e. the formation of a biexciton (see Fig. 7).

The influence of exciton/exciton interaction on the nonlinear optical response of 2D GaAs has been perceived only very recently. Kim *et al.* [39] and Weiss *et al.* [40] inferred from TR-DFWM that treatment of 2D excitons in semiconductors as noninteracting two-level systems is completely inadequate, since Coulomb-mediated many-body interactions provide the dominating contribution to the nonlinear response and determine the coherent emission in FWM experiments.

Biexcitonic effects play an important role in GaAs QW's in contrast to the bulk material because of the increased binding energy (~ 1meV [41]) of the excitonic molecule in two dimensions. In high quality samples with a free-exciton linewidth < 0.5 meV, PL spectra show an emission line red-shifted by about 1 meV with respect to the lowest hh exciton line and growing in amplitude almost quadratically with excitation intensity, which has been attributed to the radiative decay of the biexciton into an exciton and a photon [42,43]. The spectral overlap with the competing luminescence of bound excitons or excitons localized at interface islands may still call this assignment in question, however.

Coherent exciton/exciton polarization interaction has first been identified by Wegener *et al.* [44] and by Leo *et al.* [45] as the source of the DFWM-signal observed in samples with small inhomogeneous broadening at comparatively low excitation intensities for negative delays significantly larger than the laser pulse width. This effect is expected to appear for all renormalized 2-exciton systems. It will become dominant in case of biexciton formation where the signal can be interpreted as a resonant 2-photon absorption from beam 2 to the biexciton state [46]. The subsequent mixing of the polarization at the biexciton frequency with the field of pulse 1 generates the coherent emission in the direction $2\vec{k}_2 - \vec{k}_1$ superimposed to the normal transient grating contribution. During the past year, several groups have reported on the appearance of pronounced beating phenomena in DFWM signals obtained from coherently prepared exciton/biexciton systems [47-49]. In this case, the signal is a coherent superposition of the polarization created at the ground-state single-exciton state, exciton-biexciton and ground state/biexciton transitions. Although the oscillation have been called QB's, this classification seems still to be doubtful. Yaffe *et al.* [50] have observed a pronounced double peak structure in spectrally resolved nondegenerate FWM with CP pulses and attributed the second peak which is not detectable for the parallel polarized (PP) configuration to the biexciton transition. Drastic changes of the signal shape occur also in TR pump/probe experiments if the polarization of the beam is altered with respect to the pump and are most likely also a consequence of biexcitonic effects [51].

The 3rd-order polarization taking into account the additional contribution caused by exciton/exciton interaction is calculated by solving the optical Bloch equations. For δ-shaped laser pulses, the time-integrated signal intensity is given by the analytical expression [52]:

$$I(\tau) \sim \Theta(-\tau)exp(+2\tau\gamma) \times 2\mu^4\nu^4\left[\gamma_\mu^{-1} + \gamma_\nu^{-1} - 4\frac{\gamma_\mu+\gamma_\nu}{(\gamma_\mu+\gamma_\nu)^2+\Delta^2}\right] + \tag{16}$$
$$\Theta(+\tau)exp(-2\tau\gamma_\mu) \times \left[2\mu^4\nu^4\gamma_\nu^{-1} + cos^2\Theta_{12}\left(8\mu^8\gamma_\mu^{-1} - 16\mu^6\nu^2\frac{\gamma_\mu+\gamma_\nu}{(\gamma_\mu+\gamma_\nu)^2+\Delta^2}\right)\right]$$

Here the groundstate-exciton transitions are characterized by the parameters μ and γ_μ and the exciton-biexciton transitions by ν and γ_ν. Δ is the biexciton binding energy. Owing to the Heavyside step function Θ, the signals for negative and positive delay are described by the first and second term in Eq. (16), respectively, which decay with different time constants γ_μ and γ, which is the dephasing rate between the two-exciton state and the ground state. γ is taken as a free parameter in our calculation although it must be related to γ_μ and γ_ν in a microscopic theory.

Appearance of a signal for negative delay ($\tau < 0$) decaying with the time constant γ is an unambiguous proof for the presence of exciton-exciton interaction, since the first term in Eq. (16) vanishes for $\mu = \nu, \gamma_\mu = \gamma_\nu$ and $\Delta = 0$. The signal for positive delay displays the dependence of I(τ) on the angle Θ_{12} between the polarization vectors of the incident pulses. This feature also has to be interpreted as a result of exciton-exciton interaction since the $\cos^2\Theta_{12}$ term disappears for $\nu = \mu, \gamma_\mu = \gamma_\nu$ and $\Delta = 0$.

It should be noticed that the theory [Eq. (16)] predicts no modulation of the signal, i.e., neither quantum beats nor polarization interference, although two optical transitions with slightly different oscillation frequencies are simultaneously excited in the sample. This result evidently seems to contradict experimental data published by different groups [47-49]. The origin of this discrepancy is not yet understood.

The intensity and polarization angle of the signal at $\tau = 0^+$ are:

$$I(\tau = 0^+) \propto \Gamma_1 + cos^2(\Theta_{12})(2\Gamma_2 - 2\Gamma_3) \tag{17}$$

$$\Theta_{sig} = -\frac{1}{2} \arctan\left(\frac{sin(2\Theta_{12})(\Gamma_1 - \Gamma_3)}{\Gamma_3 - \Gamma_2 + cos(2\Theta_{12})(\Gamma_3 - \Gamma_2 - \Gamma_1)}\right) \tag{18}$$

with

$$\Gamma_1 = \frac{\nu^4}{\gamma_\nu}; \quad \Gamma_2 = 2\frac{\mu^4}{\gamma_\mu}; \quad \Gamma_3 = 4\mu^2\nu^2 \frac{\gamma_\mu + \gamma_\nu}{(\gamma_\mu + \gamma_\nu)^2 + \Delta^2}. \tag{19}$$

Figure 8 depicts plots of $I(\tau = 0^+)$ and Θ_{sig} versus Θ_{12} for the parameter set $\mu = 1, \nu = 0.8, \gamma_\mu^{-1} = 6$ps, $\gamma_\nu^{-1} = 4$ps, $\Delta = 0.3$meV. These curves reveal the following features

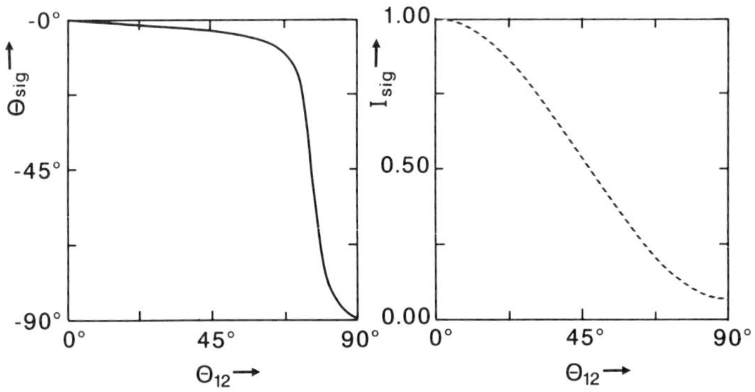

Fig. 8 Calculated peak intensity (right) and polarization (left) of the time-integrated DFWM signal vs. the polarization angle Θ_{12} of the exciting fields for $\mu = 1$, $\nu = 0.8$, $\gamma_\mu^{-1} = 6$ ps, $\gamma_\nu^{-1} = 4$ ps, $\Delta = 0.3$ meV.

caused by exciton/exciton interaction:

1. The ratio of the signal intensities at $\tau = 0^+$ observed for PP and CP excitation beams,

$$I_{\|}(\tau = 0^+)/I_{\perp}(\tau = 0^+) = \frac{\Gamma_1 + 2(\Gamma_2 - \Gamma_3)}{\Gamma_1} \tag{20}$$

can reach values >10 at low and moderate excitation densities where exciton/exciton coupling is important. As the smaller signal in the cross-polarized configuration is a consequence of exciton/exciton interaction we expect $I_\perp \approx I_\|$ at very low densities where this coupling becomes negligible because of the low density of populated single-exciton states. For very high excitation where exciton/exciton scattering dominates the exciton dephasing rate γ_μ approaches γ_ν and $\gamma_\mu + \gamma_\nu \gg \Delta$.

Since the differences between μ and ν are expected to be quite small (the presence of the first exciton will have only minor influence on the transition probability of any valence electron in the range within the large molecular radius), the signal intensity then becomes almost independent of Θ_{12} because $\Gamma_2 = \Gamma_3$.

2. The polarization angle of the signal Θ_{sig} is close to zero for $\Theta_{12} < 60°$, i.e. the signal is polarized almost parallel to the field of the second pulse. For $60° < \Theta_{12} < 80°$ the signal polarization becomes relatively undefined and for larger angles $\Theta_{sig} = -\Theta_{12}$.

3. Parallel to our own work, the same mechanism of exciton/exciton interaction has been applied by Saiki *et al.* [53] to explain unexpected polarization dependent spectral and temporal features of DFWM on the hh exciton transition in GaAs QW's. These authors could show that this model also predicts changes of the decay of the TI-signal with polarization angle. In the PP geometry a PE-like signal decaying with $1/4\ (\gamma_\mu)^{-1}$ is expected. In the CP geometry the response is also a PE but decaying now with $1/2(\gamma_\mu + \gamma_\nu)^{-1}$, which is much faster than $1/4\ (\gamma_\mu)^{-1}$ since the coherence time between the two-exciton state and the single exciton state is much shorter than the coherence time of the single exciton.

5.2. Disorder-Induced Coupling

Recently we have proposed an additional mechanism for the coherent coupling of the σ^+ and σ^- exciton transitions, namely mixing of either the conduction band or valence band levels by structural disorder in the sample with a spatially inhomogeneous coupling strength [9]. Compositional disorder present in the (Al,Ga)As barriers, interface roughness, dislocations, strain and impurities are microscopic physical mechanisms constituting local breaking of the symmetry, which is the fundamental presumption of the DFWM model of Schmitt-Rink, *et al.* [38]. We discuss here the case of coupling within the valence bands because the conduction band states are pure spin states, while the valence bands are partly mixed states and have a more complicated structure due to the QW confinement potential. It should be noted, however, that the coupling in principle may occur also between the conduction band levels without changing the predictions of the theoretical model. Interaction with the disorder lifts the degeneracy in the hole levels [54]. The situation is illustrated in Fig. 9. The allowed transitions for the diagonalized 2×4 level system with energies $\epsilon^c_{1/2}, \epsilon^c_{-1/2}, \epsilon^v_h + J, \epsilon^v_h - J, \epsilon^v_l + J, \epsilon^v_l - J$ are depicted by arrows. Owing to the spatially inhomogeneous nature of disorder, the coupling has to be characterized by a distribution of coupling strengths rather than by a fixed value of J. We assume a Gaussian distribution function $f(J)=\exp(-4\ln 2\ J^2/\Delta J^2)$ centered at an average coupling $J_0 = 0$. The width ΔJ will certainly depend sensitively on sample processing conditions and may vary considerably from sample to sample. A quantitative *ab initio* microscopic calculation of the coupling and a determination of ΔJ is still beyond the potential of the present theoretical model and must be the subject of future research. However, an upper bound for ΔJ can be estimated from the linewidth of the excitonic transition.

Although such mixing cannot remove the Kramer's degeneracy of one electron states, inclusion of the electron-hole exchange interaction can lift the degeneracy for excitonic states since Kramer's degeneracy is not required for two-particle states and splitting of the transitions is consistent with time-reversal symmetry.

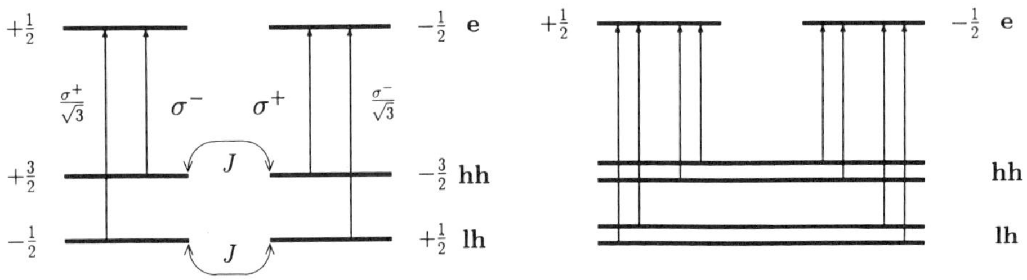

Fig. 9 Level scheme and selection rules for a GaAs QW with disorder induced coupling J (left) and diagonalized level scheme (right).

Solving the optical Bloch equations for the selection rules and eigenvalues of the coupled system we obtain for the 3rd-order nonlinear polarization

$$
\begin{aligned}
\vec{P}^{(3)}_{2\vec{k}_2-\vec{k}_1}(t,\tau) = & \frac{-i}{2}\,\Theta(t-\tau)\Theta(\tau)E_2^2E_1^* \\
& \times \Bigg[\begin{pmatrix}\cos\Theta_{12}\\ 0\end{pmatrix}\left\{|\mu_h|^2\, e^{-i(\epsilon^c-\epsilon^v_h-i\gamma^{cv}_h)(t-\tau)} + |\mu_l|^2\, e^{-i(\epsilon^c-\epsilon^v_l-i\gamma^{cv}_l)(t-\tau)}\right\} \\
& \times \left\{|\mu_h|^2\, e^{i(\epsilon^c-\epsilon^v_h+i\gamma^{cv}_h)\tau} + |\mu_l|^2 e^{i(\epsilon^c-\epsilon^v_l+i\gamma^{cv}_l)\tau}\right\} g(t-2\tau) \\
& + \begin{pmatrix}0\\ -\sin\Theta_{12}\end{pmatrix}\left\{|\mu_h|^2\, e^{i(\epsilon^c-\epsilon^v_h-i\gamma^{cv}_h)\tau(t-\tau)} - |\mu_l|^2 e^{i(\epsilon^c-\epsilon^v_l-i\gamma^{cv}_l)\tau(t-\tau)}\right\} \\
& \times \left\{|\mu_h|^2\, e^{i(\epsilon^c-\epsilon^v_h+i\gamma^{cv}_h)\tau} - |\mu_l|^2 e^{i(\epsilon^c-\epsilon^v_l+i\gamma^{cv}_l)\tau}\right\} g(t)\Bigg]
\end{aligned}
\tag{21}
$$

Here $g(t)$ is the Fourier transform of $f(J)$.

Figures 10 and 11 depict time-integrated and time-resolved response signals, respectively, calculated with the parameters $\gamma^{cv}_h = (6\text{ps})^{-1}$; $\gamma^{cv}_l = (4\text{ps})^{-1}$; $\epsilon^v_h - \epsilon^v_l = 6.28$ ps^{-1} and $\Delta J = 0.5$ ps^{-1}.

From Eq. (21) and Figs. 10 and 11 we derive the following features of the DFWM signal:

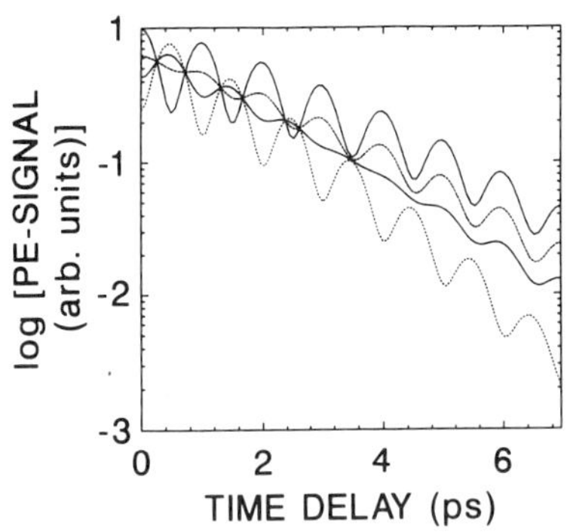

Fig. 10 Calculated time-integrated signal in the photon echo configuration for different orientations of the excitation fields. From above 0°, 45°, 60°, 90°.

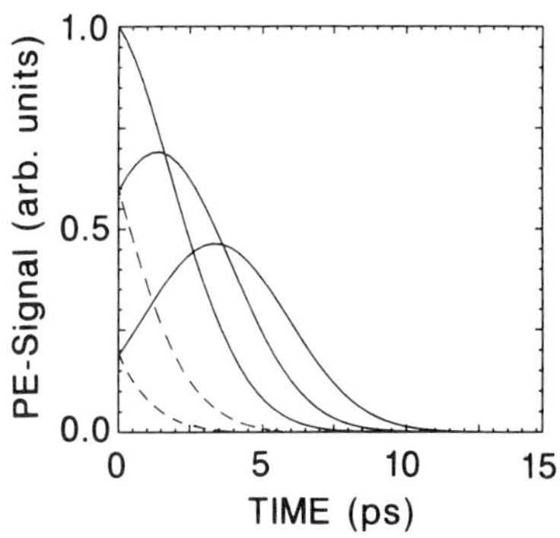

Fig. 11 Calculated time-resolved signal in photon echo configuration for delays of 0 ps, 2 ps, and 4 ps. Solid lines $\Theta_{12} = 0°$, dashed lines $\Theta_{12} = 90°$.

1. For parallel polarization of the excitation pulses ($\Theta_{12} = 0°$) the signal is given by the first term, which is centered at t=2τ as characteristic for a photon echo signal.

2. The decay of the time-integrated signal for $\Theta_{12} = 0°$ is determined by the phenomenologically introduced dephasing rates $\gamma^{cv}_{h,l}$.

3. For perpendicular polarization of the excitation pulses ($\Theta_{12} = 90°$) the signal corresponds to the second term. For determination of the temporal position, one must multiply the envelope profile centered at t=0 with the causality function $\Theta(t-\tau)$. The resulting signal has the typical characteristic of a photon echo but is emitted at $t \approx \tau$ like a free-polarization decay signal.

4. The signal decay rate observed in the case of perpendicular polarization of the exciting pulses for the time integrated signal is determined either by the phenomenologically introduced dephasing rates $\gamma^{cv}_{h,l}$ or by the width of the Fourier transform of the distribution function of f(J), depending on the relative magnitude of these quantities. The resulting signal decay rate is therefore always equal to or larger than the true dephasing rate measured in the parallel configuration.

5. The disorder-induced coupling of the excitons implies substantial changes of the QB phase between hh and lh excitons in dependence on Θ_{12} compared to the model of Ref. [38] at intermediate angles. The 180° change in QB phase between $\Theta_{12} = 0°$ and Θ_{12}=90° is in accordance with the previous model. However, for 45°, the beats do not cancel but are in-phase with those for $\Theta_{12} = 0°$, and moreover, the beat amplitude increases with τ (see Fig. 10). This result is explained by the different decays of the parallel and perpendicular contributions: at small delays, the in-phase and anti-phase beating amplitudes tend to cancel; for larger delays, the parallel contribution (in-phase beating) becomes dominant because of the faster decay of its perpendicular counterpart. For $\Theta > 45°$, the perpendicular anti-phase beat contribution dominates at small delays. The faster decay of this part leads to a remarkable shift from anti-phase to in-phase beating with increasing delay.

Hu *et al.* [50] have recently demonstrated that valence band mixing equivalent to coupling between s-and d-excitons yields a dephasing mechanism causing strong polarization and carrier density dependence of the DFWM signal. Whereas this model seems to predict the polarization and intensity of the signal in satisfying agreement with experimental observations, it fails to describe the drastic dependence of the signal emission time and decay constant on the polarization geometry.

6. EXPERIMENTAL SET-UP

The pulses for excitation of the transient gratings and probing of the subsequent decay, as well as for up-conversion of the DFWM signal in the time-resolved experiments were generated by a femtosecond Ti:sapphire laser system operating at a repetition rate of 76 MHz and providing transform-limited pulses as short as 150 fs. For many experiments the pulse duration was increased to about 1 ps by external spectral filtering of the output in order to suppress simultaneous excitation of the band to band transition, since the dephasing behavior of free excitons is strongly influenced by the presence of free electrons and holes even at modest excitation levels.

For all experiments, we used the 2-pulse, self-diffraction backward geometry. The polarization of the input pulses and of the time-integrated signal was adjusted by Glan-Thompson polarizers and $\lambda/2$ or $\lambda/4$ plates providing polarization definition of better than 200:1. The time-integrated signal was detected by a cooled GaAs photomultiplier or a Si-photodiode. Time-resolution of the diffracted signal for a certain delay τ was accomplished by up-conversion of the signal with a reference probe in a 2mm thick $LiIO_3$ crystal. The up-converted signal is monitored by a cooled GaAs-photomultiplier as a function of the time delay t of the reference pulse. It should be mentioned that these are the first time-resolved measurements of the weak signal diffracted from a SQW into the backward direction. The samples were maintained in a continuous flow helium cryostat at temperatures between 5-8 K.

7. EXPERIMENTAL RESULTS AND DISCUSSION

We now present results of a variety of experiments performed on various different samples in order to test the validity of the predictions made by the different theoretical coupling models.

7.1 Signal Decay for $\Theta_{12} = 0°$ and 90°

Substantial differences in excitonic dephasing rates measured in parallel and cross-polarized DFWM geometries have been reported previously [10,11,38] but the authors were unable to present a concise explanation of their observations. According to the disorder-induced-coupling model, it is expected that the difference between the two decay times will strongly vary from sample to sample due to the strong dependence of disorder on the growth conditions. This presumption is confirmed by the decay of the time-integrated DFWM signal curves depicted in Fig. 12. The left part, recorded on the hh-exciton transition of a high quality 20 nm single QW with a PL linewidth as narrow as 0.3 meV, reveals identical decay rates for the parallel and cross-polarized signal. The data shown in the right part were taken on the hh-exciton of a 12nm SQW with a remarkably inhomogeneously broadened linewidth of 2.1 meV. In this case the signal decay for the perpendicular polarization geometry is two times faster, although the true T_2 time reflected by the parallel configuration is even slightly longer than the value measured for the homogeneously broadened 20nm well. The ratio of the two signal decay times exhibits a pronounced density dependence since exciton/exciton scattering shortens $T_2^{\|}$ at considerably lower pulse intensities as compared to $T_2^{\perp}$. At $N=5 \times 10^9$ cm^{-2} its value approaches 1. The substantially different behavior of the two samples points to an extrinsic, rather than intrinsic origin for the change of the signal decay time constant with the polarization geometry. In order to gain further evidence for the correlation of these phenomena with sample quality, we have investigated the ratio $R = T_2^{\perp}/T_2^{\|}$ for a larger set of single QW's and one MQW grown in different labs.

The unavailability of a single quantity which directly measures sample quality, of course, makes such an investigation problematic, and the results can only be stated tentatively. We take the photoluminescence linewidth as an approximate measure of the disorder of the samples. Our findings are summarized in Fig. 13. We find that the samples with a PL linewidth less than 0.5 meV show equal dephasing times in the PP and CP configuration, whereas samples with larger linewidths (with the exception of the 30nm QW) exhibit remarkable differences in the two signal decay times, which amount to a factor of almost 2 in some cases. (Other groups have found differences as

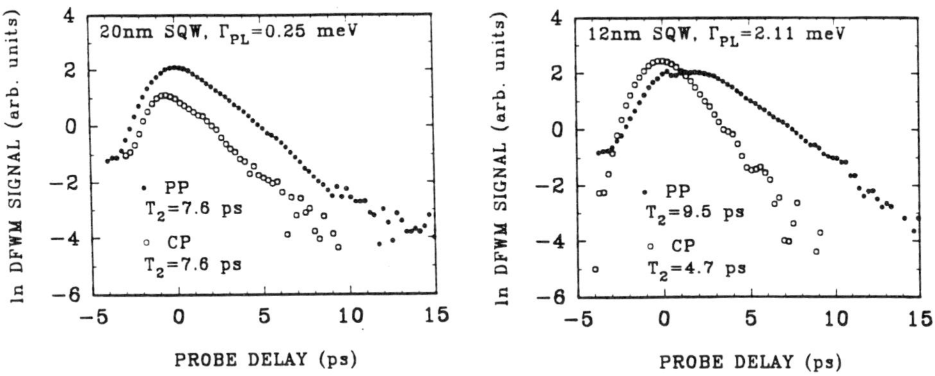

Fig. 12 Time-integrated DFWM signal for parallel and perpendicular polarization of the exciting pulses on a homogeneously broadened 20-nm-QW (left) and an inhomogeneously broadened 12-nm-Qw (right).

large as one order of magnitude [55].) The data of Fig. 13 show that the value of R is strongly sample-dependent. Considering all data points, the plot reveals no simple correlation of R with the thickness of the QW, but rather with the differing growth conditions of the various labs. In the set of single quantum wells (triangular points in Fig. 13) contained within the same sample, and therefore are likely to exhibit the same interfacial quality, we observe a clear well width dependence of R. Since in narrower wells the exciton is more sensitive to interfacial quality, this observation may point to the role of interface disorder in mixing of the exciton transitions. By and large, our findings tend to support disorder-induced mixing of the excitonic levels as a potential mechanism for generating the apparent polarization dependence of T_2 values.

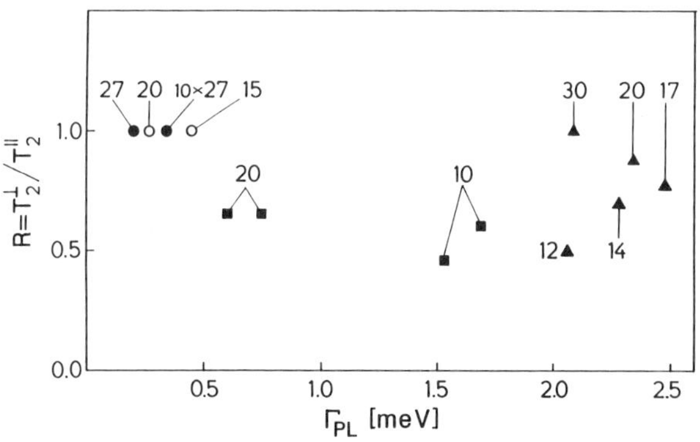

Fig. 13 Ratio of the experimental T_2 times measured for parallel and perpendicular polarization of the incident pulses versus photoluminescence linewidth for a variety of single QW samples and one multiple QW (black dot 10×27) grown in the laboratories of K. Ploog, MPI für Festkörperforschung, Stuttgart (black dots), H. Hey, Paul-Drude-Institut, Berlin (circles), K. Köhler, Fraunhofer Institut für Angewandte Festkörperphysik, Freiburg (squares), and Y. Horikoshi, NTT Basic Research Laboratories, Musashino-shi/Japan (triangles). The numbers give the quantum well thickness in nm.

7.2 Phase Shift of Quantum Beats

Figure 14 shows the time-integrated 2-pulse DFWM intensity in the $2\vec{k}_2 - \vec{k}_1$ direction for different Θ_{12} and no polarizer before the detector, measured with laser pulses of 1.3 ps duration (bandwidth 16 cm^{-1}) at a density of 1×10^9 excitons cm^{-2} on a 20 nm SQW with a PL linewidth of 0.74 meV. The center of the laser spectrum was chosen to simultaneously excite roughly equal quantities of lh and hh excitons. The beat period is 1.1 ps, and is in good agreement with the lh-hh splitting determined from cw-PL measurements.

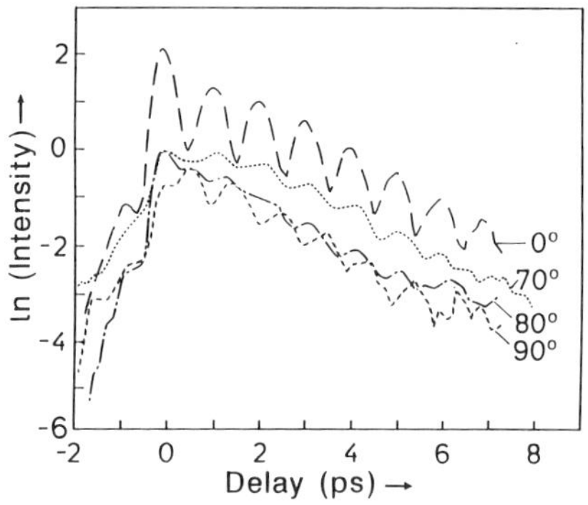

Fig. 14 Measured time-integrated signal in the photon echo configuration for different orientations of the excitation pulses. From above: 0°, 70°, 80°, 90°.

The dephasing time at $\Theta_{12} = 90°$ is only 70% of that measured at $\Theta_{12} = 0°$. Note first that the quantum beats at $\Theta_{12} = 90°$ are exactly 180° out of phase to those at $\Theta_{12} = 0°$. The crossover between in-phase and anti-phase beating is predicted by the uncoupled 3-level system model to occur at $\Theta_{12} = 45°$, where the beat amplitude should be zero. However, we find no such crossover. At $\Theta_{12} = 70°$, the beats remain in-phase with those at $\Theta_{12} = 0°$ and the relative beat amplitude increases with increasing delay time. At $\Theta_{12} = 80°$ we observe the appearance of anti-phase beats at small delay, followed by a region where the beat amplitude becomes very small, and finally, the reappearance of in-phase beating at longer delays. This behavior exhibits good agreement with the inhomogeneous coupling model, described in chap. 5.2. However, this theoretical model fails to explain the change of signal intensity with polarization configuration. In the experiments, the intensity $I^{\parallel}$ is four times higher than $I^{\perp}$, whereas the theory predicts the intensities to be the same. It is important that the critical angle for the change from anti-phase to in-phase beating is influenced by the relative intensities of the parallel and perpendicular signal. In our case the intensity enhancement of the parallel in comparison to the perpendicular signal obviously leads to a shift of the critical angle to larger values as compared to the angle calculated in the theoretical model ($\Theta_{12} = 60°$). Exactly this result is shown in Fig. 14. We therefore conclude that our experiments support the validity of this model.

7.3. Polarization Dependence of Time-Resolved DFWM Signals

Unexpected anomalies observed for the time-resolved emission of the hh-exciton DFWM signal in a MQW have been first reported by Cundiff *et al.* [11]. They found that the signal emission resembled a photon echo and free-polarization decay like behavior for parallel and perpendicular polarization, respectively. As this result contradicts the predictions of the model for noninteracting 2-level systems, the authors speculated that the prompt signal emission at $\Theta_{12} = 90°$ could be related to imperfections, but were unable to present a verification of their surmise by a concrete model. The disorder-induced coupling model presented in Chap.5.2 shows that the appearance of the signal at $t \sim \tau$ does not correspond to a true free polarization decay, but instead is caused by the inhomogeneity of the disorder coupling. In particular, this model predicts that this phenomenon should be tightly related to the occurence of different DFWM signal decay times for the parallel and cross-polarized geometry.

In order to establish this inherent relation between the two phenomena, we have performed time-resolved DFWM on a homogeneously broadened SQW for which the disorder-induced coupling between excitons seems to be quite small and on an inhomogeneously broadened MQW exhibiting substantial differences of $T_2^{||}$ and $T_2^{\perp}$.

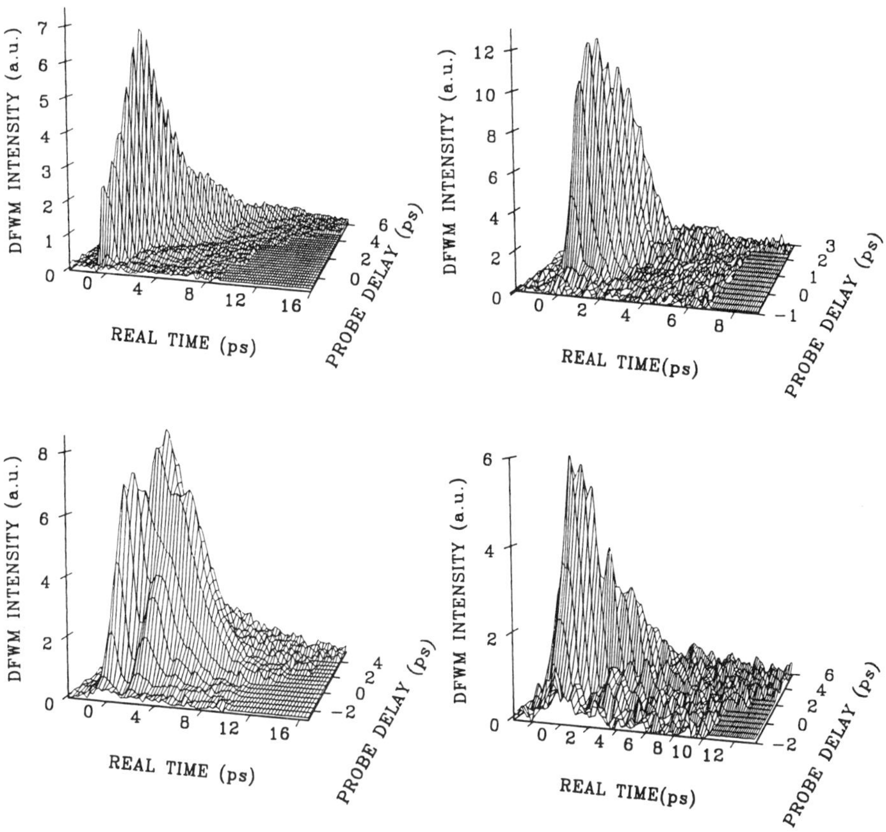

Fig. 15 Time-resolved DFWM-signals for parallel (left) and perpendicular polarization (right) of the exciting fields. Upper part: inhomogeneously broadened hh exciton transition of a MQW, lower part: homogeneously broadened SQW.

Figure 15 (upper part) presents a 3D plot of time-resolved DFWM signals for parallel and perpendicular polarization of the exciting fields on the hh-exciton transition of the MQW sample (25 periods of 9.3nm GaAs and 15nm $Al_{0.36}Ga_{0.64}As$, PL linewidth 1.8meV recorded via up-conversion for varying delays between $\vec{E}_1$ and $\vec{E}_2$. Time-integrated DFWM on this sample yielded signal decay times of 7.4ps (corresponding to a homogeneous linewidth of 0.1meV) and 3.7ps for $\vec{E}_2 \parallel \vec{E}_1$ and $\vec{E}_2 \perp \vec{E}_1$, respectively. These results confirm the presence of substantial inhomogeneous broadening and disorder-induced coupling in this sample. The time-resolved signals for the PP and CP configuration have both a typical PE signal shape with maxima appearing at $t = 1.85\tau$ and $t = 1.1\tau$, respectively. These observations agree fairly well with the predictions of the disorder-induced coupling model.

Next we describe results obtained from a 20nm SQW with an almost homogeneously broadened hh-exciton transition (Γ_{PL} = 0.3meV; $\Gamma_{hom} \approx$ 0.15meV). In this sample, disorder-induced coupling between excitonic states is expected to be negligibly small since no significant differences are detectable for $T_2^{\parallel}$ and $T_2^{\perp}$. The time-resolved DFWM curves presented in Fig. 15 (lower part) confirm this classification of the sample since the emitted signal reveal a free-polarization decay with the same time constant as observed for the time-integrated signal for both polarization geometries, supporting further the validity of the disorder-induced coupling model.

7.4. Polarization Dependence of Time-Integrated DFWM Signal Polarization and Intensity

Figure 16 shows Θ_{sig} versus Θ_{12} measured on a 27nm GaAs single QW (0.2 meV hh-exciton line width and no measureable shift between PL and PLE spectra) for two exciton densities. The laser pulse duration and bandwidth were 1.3ps and 16 cm^{-1}, respectively. At $N = 4\times10^9$ cm^{-2}, we find a well-polarized signal following $\Theta_{sig} = -\Theta_{12}$ predicted by the model of ref. [38] if the absolute values squared of the effective matrix elements from the hh- and lh-transitions $|\mu_h|^2$ and $|\mu_l|^2$ are different.

The behavior is completely different at the lower density of 4×10^8 cm^{-1}. For $\Theta_{12} < 60°$, the signal is polarized with a contrast ratio of only 10:1 and the dominant polarization component seems to approach the rule $\Theta_{sig} = 0$. For $60° \leq \Theta_{12} \leq 70°$ the signal is more or less unpolarized as indicated by the dashed line in Fig. 16. At large angles the $\Theta_{sig} = -\Theta_{12}$ rule is approached. Comparison of the experimental data of Fig. 16 with the theoretical curve in Fig. 8 reveals excellent agreement of the low intensity behavior with the predictions of the exciton/exciton coupling model. At higher intensities this exciton/exciton interaction is screened and consequently the polarization rule for the non-interacting 2-level systems in Ref. [38] provides the appropriate description.

Figure 17 displays the radiative intensity of the photon echo signal at zero delay as a function of Θ_{12} for two different densities. In both cases, we observe a reduction in signal intensity with increasing Θ_{12} as compared to $\Theta_{12} = 0°$. For the lowest excitation level ($4 \times 10^8 cm^{-2}$), the reduction amounts to roughly a factor of 30 between the intensities measured at $\Theta_{12} = 90°$ and 0°, and the shape of $I(\Theta_{12})/I(0°)$ shows excellent agreement with the $\cos^2\Theta_{12}$ dependence predicted by the exciton/exciton interaction model. For higher excitation intensities, a monotonic growth of the signal at $\Theta_{12} = 90°$ is found, as expected, due to the increasing importance of exciton screening effects. The ratio $I(90°)/I(0°)$ saturates at roughly 60% for densities $\geq 7\times10^9$ cm^{-2}. For larger densities, the decrease in signal with Θ_{12} seems to be significantly slower than the $\cos^2\Theta_{12}$-law. The origin of this discrepancy is not yet clear.

Finally, we would like to mention that Wang *et al.* [52] have independently reported on excitation density dependent changes of the ratio $I(\Theta_{12})/I(0°)$ very recently.

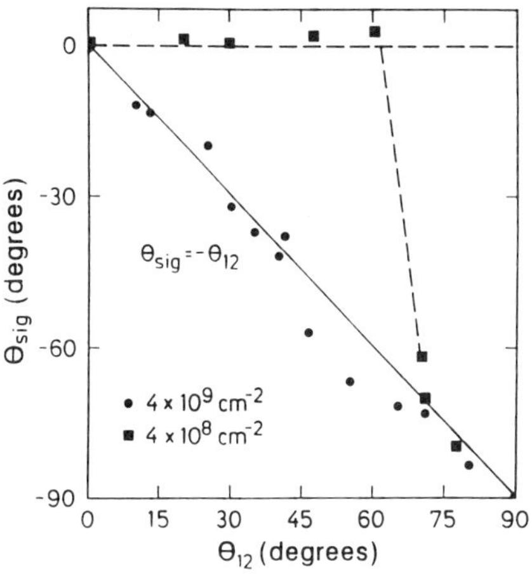

Fig. 16 The polarization of the DFWM signal θ_{sig} vs. θ_{12} from a 27 nm GaAs/AlGaAs QW for two different photoexcited exciton densities. The oblique dashed line indicates the region where the signal is almost unpolarized.

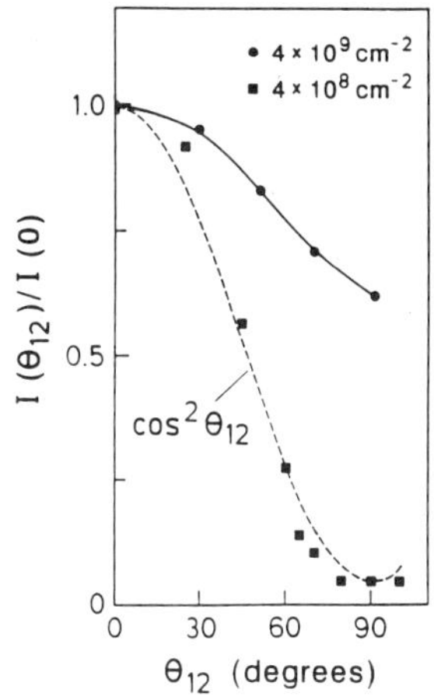

Fig. 17 Relative intensity, $I(\theta_{12})/I(0°)$, of the photon echo signal from a 27 nm GaAs/AlGaAs QW vs. θ_{12} at two exciton densities. The dashed line is a $\cos^2\theta_{12}$ fit with the $I(90°)$ contribution added.

They have shown that this behavior can be qualitatively explained by a changing dephasing rate caused by excitonic screening. This interpretation is in accordance with the explanation given above for the increase of the 90° signal with laser pulse intensity.

8. CONCLUSIONS

TI- and TR-DFWM with ps and fs optical pulses permit observation of the initial relaxation steps of photo-created electron/hole pairs in semiconductors. Employing these techniques, many important dephasing mechanisms for free-excitons in III-V compounds, such as exciton/exciton, exciton/electron(hole) and exciton/phonon scattering have been explored in recent years. Such investigations have shown remarkable differences in the dephasing efficiency of these processes for 2D and 3D excitonic systems. Extension to 1D (quantum wires) and 0D samples (quantum dots) is still hampered by the lack of suitable high quality samples. For quantum wires, a prolonged T_2 compared to QW's is expected owing to the reduced carrier scattering rate associated with the smaller density of final states available for scattered excitons. Preliminary experiments have shown [56] dephasing times for the 1D-exciton in GaAs quantum wires to be as short as those of corresponding QW's. This discrepancy is most likely due to the still relatively poor sample quality.

The influence of the polarization geometry on the intensity, polarization and the temporal features of the DFWM signal has gained attention only very recently. Systematic studies show remarkable variations of all these quantities with the polarization of the excitation pulses. These observations are unexplainable in the frame of non-interacting 2-level systems for the 2D exciton. We have demonstrated in this paper that a straight extension of this model, taking into account coupling between the 2-level systems by exciton/exciton interaction and/or sample disorder, leads to excellent agreement between theory and all experimental data presently available. Mutual exciton interaction proves to be a fundamental physical process in QW's, playing an important role in the regime of moderate and medium excitation.

Disorder-induced coupling, on the other hand, is critically dependent on sample quality. Perhaps polarization dependent DFWM can provide a much more specific characterization of sample disorder than the measurement of the PL linewidth.

Based on the detailed comparison of all experimental data with the theoretical models, we have to conclude that both exciton/exciton interaction and disorder-induced coupling have to be included into the theory in order to achieve agreement with all the phenomena observed in DFWM. The exciton/exciton interaction alone describes the change of the signal intensity and polarization with Θ_{12} and the dependence on excitation density fairly well. However, it additionally predicts a faster signal decay for the CP as compared to the PP geometry for inhomogeneously broadened transitions, in disagreement with our experimental results obtained on the 30nm SQW. Furthermore, this model fails to explain the time-resolved DFWM experiments since the signal emission is calculated to occur at $t = 2\tau$ for inhomogeneous broadening, independent of polarization geometry. In contrast, we measured a PE at $t = \tau$ for the CP configuration which is in agreement with previous publications [11]. These discrepancies are removed by the disorder-induced coupling model which, however, is unable to explain the alteration of signal intensity and polarization with Θ_{12}.

Acknowledgements

This article contains results obtained in cooperation with L. Schultheis, A. Honold and B.F. Feuerbacher. We are indebted to K. Ploog (Max-Planck-Institute für Festkörperforschung, Stuttgart and Paul-Drude Institut, Berlin), R. Hey, (P. Drude Institut), K. Köhler (Fraunhofer Institut für Angewandte Festkörperphysik, Freiburg) and Y. Horikoshi (NTT Basic Research Laboratories, S. Musashino-shi, Japan) for the supply of their samples. Stimulating and fruitful discussions with J. Shah, S. Cundiff, S. Schmitt-Rink, S.W. Koch, E.O. Göbel, and J. Feldmann are gratefully acknowledged. This work has been supported in part (D.B.) by the Deutsche Forschungsgemeinschaft.

REFERENCES

1. M.T. Asaki, C.-P. Huang, D. Garvey, J. Zhou, H.C. Kapteyn, and M.C. Murnane, "Generation of 11-fs pulses from a self-mode-locked Ti:sapphire laser", *Opt. Lett.* 18:977 (1993); P.F. Curley, Ch. Spielmann, T. Brabec, F. Krausz, E. Wintner, and A.J. Schmidt, "Operation of a femtosecond Ti:sapphire solitary laser in the vicinity of zero group-delay dispersion", *Opt. Lett.* 18:54 (1993); B. Proctor and F. Wise, "Generation of 13-fs pulses from a mode-locked $Ti:Al_2O_3$ laser with reduced third-order dispersion", *Appl. Phys. Lett.* 62:470 (1993).

2. R.L. Fork, B.I. Greene, and C.V. Shank, "Generation of optical pulses shorter than 0.1 psec by colliding pulse mode locking", *Appl. Phys. Lett.* 36:671 (1981); J.A. Valdmanis, R.L. Fork, and J.P. Gordon, "Generation of optical pulses as short as 27 femtoseconds directly from a laser balancing self-phase modulation, group-velocity dispersion, saturable absorption, and saturable gain", *Opt. Lett.* 10:131 (1985).

3. P.F. Curley and A.I. Ferguson, "Resonant frequency doubling of a self-starting coupled-cavity, mode-locked Ti:Al_2O_3 laser", *Opt. Lett.* 16:321 (1981).

4. Q. Fu, G. Mak, and H.M. van Driel, "High-power, 62-fs infrared optical parametric oscillator synchronously pumped by a 76-MHz Ti:Sapphire laser", *Opt. Lett.* 17:1006 (1992); R.J. Ellingson and C.L. Tang, "High-power, high-repetition-rate femtosecond pulses tunable in the visible", *Opt. Lett.* 18:438 (1993).

5. B.S. Wherrett, A.L. Smirl, and T.F. Boggess, "Theory of Degenerate Four-Wave-Mixing in Picosecond Excitation-Probe-Experiments", *IEEE J. Quant. Electr.* QE19:680 (1983); A.L. Smirl, T.F. Boggess, B.S. Wherrett, G.P. Perryman, and A. Miller, "Picosecond Transient Orientational and Concentration Gratings in Germanium", *IEEE J. Quant. Electr.* QE19:690 (1983); H.J. Eichler, P. Günther, and D.W. Pohl, "Laser-Induced Dynamical Grating", (Springer Verlag, Berlin-Heidelberg-New York, 1980).

6. I.D. Abella, N.A. Kurnit, and R.S. Hartmann, "Photon echos", *Phys. Rev.* 57:391 (1966); N.A. Kurnit, I.D. Abella, and S.R. Hartmann, "Observation of photon echo", *Phys. Rev. Lett.* 13:567 (1964).

7. V. Langer, H. Stolz, and W. von der Osten, "Observation of Quantum Beats in the Resonance Fluorescence of Free Excitons", *Phys. Rev. Lett.* 64:854 (1990); E.O. Göbel, K. Leo, T.C. Damen, J. Shah, S. Schmitt-Rink, W. Schäfer, J.F. Müller, and K. Köhler, "Quantum Beats of Excitons in Quantum Wells", *Phys. Rev. Lett.* 64:1801 (1990); B.F. Feuerbacher, J. Kuhl, R. Eccleston, and K. Ploog, "Quantum Beats between the Light and Heavy Hole in GaAs-AlGaAs Quantum Wells", *Sol. Stat. Comm.* 74:1279 (1990); K. Leo, T.C. Damen, J. Shah, E.O. Göbel, and K. Köhler, "Quantum Beats of Light and Heavy Hole Excitons in Quantum Wells", *Appl. Phys. Lett.* 57:19 (1990); K. Leo, T.C. Damen, J. Shah, and K. Köhler, "Quantum beats of free and bound excitons in GaAs/$Al_xGa_{1-x}As$ quantum wells", *Phys. Rev. B* 42:11359 (1990); K. Leo, J. Shah, T.C. Damen, A. Schulze, T. Meier, S. Schmitt-Rink, P. Thomas, E.O. Göbel, S.L. Chuang, M.S.C. Lao, W. Schäfer, K. Köhler, and P. Ganser, "Dissipative Dynamics of an Electronic Wave-packet in a Semiconductor Double Well Potential", *IEEE J. Quant. Electr.* 28:2498 (1992).

8. R. Eccleston, J. Kuhl, D. Bennhardt, and P. Thomas, "Intensity Dependent Four-Wave-Mixing Rules in Quantum Wells", *Sol. Stat. Comm.* 86:93 (1993).

9. D. Bennhardt, P. Thomas, R. Eccleston, E.J. Mayer, and J. Kuhl, "Polarization Dependence of Four-Wave-Mixing Signals in Quantum Wells", *Phys. Rev. B* 47:13485 (1993).

10. K. Leo, J. Shah, S. Schmitt-Rink, and K. Köhler, in: "Proceed. 7th Int. Symp. on Ultrafast Processes in Spectroscopy", Bayreuth, Germany, eds. A. Laubereau and A. Seilmeier (Inst. of Phys. Conf. Series 126, Bristol, Philadelphia, 1992) p. 411; H.H. Yaffe, Y. Prior, J.P. Harbison, and L.T. Florez, "Polarization and wavelength dependence of relaxation processes as measured by time delayed four-wave-mixing in quantum wells", in "Proceed. 2nd Conf. Quantum Electronics and Laser Science", 1991, Technical Digest Series, Vol. 11 (Optical Society of America, Washington, D.C., 1991), p. 196; O. Carmel, I. Bar-Joseph, IQEC, Vienna, post-deadline paper (1992).

11. S.T. Cundiff, H. Wang, and D.G. Steel, "Poalrization-dependent picosecond excitonic nonlinearities and the complexities of disorder", *Phys. Rev. B* 46:7248 (1992).

12. See e.g., "Optical Nonlinearities and Instabilities in Semiconductors", ed. by H. Haug (Academic Press, New York-London, 1988).

13. See e.g. R.G. Brewer, "Coherent optical transients", *Phys. Today* 5:50 (1977); T. Yajima and Y. Taira, "Spatial Optical Parametric Coupling of Picosecond Light Pulses and Transverse Relaxation Effect in Resonant Media", *J. Phys. Soc. Jpn.* 47:1620 (1977).

14. J. Feldmann, G. Peter, E.O. Göbel, P. Dawson, K. Moore, C. Foxon, and R.J. Elliot, "Linewidth Dependence of Radiation Exciton Lifetimes in Quantum Wells", *Phys. Rev. Lett.* 59:2337 (1987); R. Eccleston, B.F. Feuerbacher, J. Kuhl, W.W. Rühle, and K. Ploog, "Density-Dependent Exciton Radiative Lifetimes in GaAs Quantum Wells", *Phys. Rev. B* 45:11403 (1992); D.S. Citrin, "Radiative Lifetimes of Excitons in Quantum Wells: Localization and Phase-Coherence Effects", *Phys. Rev. B* 47:3832 (1993), D.S. Citrin "Homogeneous-Linewidth Effects on Radiative Lifetimes of Excitons in Quantum Wells", *Sol. Stat. Commun.* 84:281 (1992); V. Srinivas, J. Hrymiewicz, Y.-J. Chen, and C.E.C. Wood, "Intrinsic Linewidths and Radiative Lifetimes of Free-Excitons in GaAs Quantum Wells", *Phys. Rev. B* 46:10193 (1992).

15. A. Honold, L. Schultheis, J. Kuhl, and C.W. Tu, "Reflected degenerate four-wave-mixing on GaAs single quantum wells", *Appl. Phys. Lett.* 52:2105 (1988).

16. A.M. Weiner and E.P. Ippen, "Novel transient scattering technique for femtosecond dephasing measurements", *Opt. Lett.* 9:53 (1984).

17. See, e.g. J.C. AuJeung in "Optical Phase Conjugation", ed. R.A. Fisher (Academic, New York) 1983, p. 285, see also Ref. [28].

18. G. Noll, U. Siegner, S. Shevel, and E.O. Göbel, "Picosecond Stimulated Photon Echo Due to Intrinsic Excitations in Semiconductor Mixed Crystals", *Phys. Rev. Lett.* 64:792 (1990).

19. M.D. Webb, S.T. Cundiff, and D.G. Steel, "Observation of Time-Resolved Picosecond Stimulated Photon Echoes and Free Polarization Decay in GaAs/AlGaAs Multiple Quantum Wells", *Phys. Rev. Lett.* 66:934 (1991).

20. M. Koch, J. Feldmann, G. von Plessen, E.O. Göbel, P. Thomas, and K. Köhler, "Quantum Beats versus Polarization Interference: An Experimental Distinction", *Phys. Rev. Lett.* 69:3631 (1992).

21. M. Koch, J. Feldmann, E.O. Göbel, P. Thomas, and J. Shah, "Coherent Coupling of Exciton Transitions in different Quantum Well Islands", to be published.

22. For reviews see e.g., J. Kuhl, A. Honold, L. Schultheis, and C.W. Tu, "Optical Dephasing and Orientational Relaxation of Wannier-Excitons and Free Carriers in GaAs and GaAs/$Al_xGa_{1-x}As$ Quantum Wells", *Festkörperprobleme* 29:157 (1989); and S.T. Cundiff and D.G. Steel, "Coherent Transient Spectroscopy of Excitons in GaAs-AlGaAs Quantum Wells", *IEEE J. Quant. Electr.* 28:2423 (1992).

23. L. Schultheis, J. Kuhl, A. Honold, and C.W. Tu, "Picosecond Phase Coherence and Orientational Relaxation of Excitons in GaAs", *Phys.Rev. Lett.* 57:1793 (1986).

24. A. Honold, L. Schultheis, J. Kuhl, and C.W. Tu, "Collision braodening of two-dimensional excitons in a GaAs single quantum well", *Phys. Rev. B* 40:6442 (1989).

25. A. Honold, T. Saku, Y. Horikoshi, and K. Köhler, "Optical dephasing of light-hole excitons in GaAs single quantum wells", *Phys. Rev. B* 45:6010 (1992).

26. K. Leo, E.O. Göbel, T.C.Damen, J. Shah, S. Schmitt-Rink, W. Schäfer, J.F. Müller, K. Köhler, and P. Ganser, "Subpicosecond Four-Wave-Mixing in GaAs/$Al_xGa_{1-x}As$ Quantum Wells", *Phys. Rev. B* 44:5726 (1991).

27. L. Schultheis, M.D. Sturge, and J. Hegarty, "Photon echoes from two-dimensional excitons in GaAs-AlGaAs quantum wells", *Appl. Phys. Lett.* 47:995 (1985). M.D. Webb, S.T. Cundiff, and D.G. Steel, "Stimulated-picosecond-photon-echo studies of localized exciton relaxation and dephasing in GaAs/$Al_xGa_{1-x}As$ multiple quantum wells", *Phys. Rev. B* 43:12658 (1991).

28. V. Heuckeroth, D. Bennhardt, P. Thomas, and H. Vaupel, "Optical phase relaxation in disordered semiconductors due to electron-phonon coupling", Proceedings of the Vth Internat. Conf. on Hopping and Related Phenomena, Glasgow (1993), World Scientific Singapore.

29. M. Wegener, E.O. Göbel, G.Sucha, N. Sauer, T.Y. Chang, W. Schäfer, S. Schmitt-Rink, and D.S. Chemla, "Ultrafast dephasing of excitons at low temperatures in ternary InGaAs/InAlAs quantum wells", XVII Internat. Quantum Electronics Conf., IQEC 90, Anaheim (1990), Technical Digest,p.194.

30. U. Siegner, D. Weber, E.O. Göbel, D. Bennhardt, V. Heuckeroth, R. Saleh, S.D. Baranowskii, P. Thomas, H. Schwab, C. Klingshirn, J.M. Hvam, and V.G. Lyssenko, "Optical dephasing in semiconductor mixed crystals", *Phys. Rev. B* 46:4564 (1992).

31. L. Schultheis, A. Honold, J. Kuhl, K. Köhler, and C.W. Tu, "Optical Dephasing of Homogeneously Broadened Two-Dimensional Exciton Transitions in GaAs Quantum Wells", *Phys. Rev. B* 34:9027 (1986); L. Schultheis, A. Honold, J. Kuhl, K. Köhler, and C.W. Tu, "Phase Coherence and line broadening of free excitons in GaAs Quantum Wells", *Superlattices and Microstructures* 2:441 (1986).

32. T. Takagahara, "Localization and energy transfer of quasi-two-dimensional excitons in GaAs-AlAs quantum-well heterostructures", *Phys. Rev. B* 31:6552 (1985) and "Localization and homogeneous dephasing of quasi-two-dimensional excitons in quantum-well heterostructures", *Phys. Rev. B* 32:7013 (1985).

33. W.H. Knox, R.L. Fork, M.C. Downer, D.A.B. Miller, D.S. Chemla, C.V. Shank, A.C. Gossard, and W. Wiegmann, "Femtosecond Dynamics of Resonantly Excited Excitons in Room-Temperature GaAs Quantum Wells", *Phys. Rev. Lett.* 54:1306 (1985).

34. L. Schultheis, J. Kuhl, A. Honold, and C.W. Tu, "Ultrafast Phase Relaxation of Excitons via Exciton-Exciton and Exciton-Electron Collisions", *Phys. Rev. Lett.* 57:1635 (1986).

35. G. Tränkle, H. Leier, A. Forchel, H. Haug, C. Ell, and G. Weimann, "Dimensionality Dependence of the Band-Gap Renormalization in Two- and Three-Dimensional Electron-Hole Plasmas in GaAs", *Phys. Rev. Lett.* 58:419 (1987).

36. S. Schmitt-Rink, D.-S. Chemla, and D.A.B. Miller, "Theory of transient excitonic optical nonlinearities in semiconductor quantum-well structures", *Phys. Rev. B* 32:6601 (1985).

37. N. Peyghambarian, H.M. Gibbs, J.L. Jewell, A. Antonetti, A. Migus, D. Hulin, and A. Mysyrowicz, "Blue Shift of the Exciton Resonance due to Exciton-Exciton Interactions in a Multiple-Quantum-Well Structure", *Phys. Rev. Lett.* 53:2433 (1984); and D. Hulin, A. Mysyrowicz, A. Antonetti, A. Migus, W.T. Masselink, H. Morkoc, H.M. Gibbs, and N. Peyghambarian, "Well-size dependence of exciton blue shift in GaAs multiple-quantum-well structures", *Phys. Rev. B* 33:4389 (1986).

38. S. Schmitt-Rink, D. Bennhardt, V. Heuckeroth, P. Thomas, P. Haring, G. Maidorn, H. Bakker, K. Leo, D.-S. Kim, J. Shah, and K. Köhler, "Polarization dependence of heavy- and light-hole quantum beats", *Phys. Rev. B* 46:10460 (1992).

39. D.S. Kim, J. Shah, T.C. Damen, W. Schäfer, F. Jahnke, S. Schmitt-Rink, and K. Köhler, "Unusually Slow Temperal Evolution of Femtosecond Four-Wave-Mixing Signals in Intrinsic GaAs Quantum Wells: Direct Evidence for the Dominance of Interaction Effects", *Phys. Rev. Lett.* 69:2725 (1992).

40. S. Weiss, M.-A. Mycek, J.-Y. Bigot, S. Schmitt-Rink, and D.S. Chemla, "Collective Effects in Excitonic and Free Induction Decay: Do Semiconductors and Atoms Emit Coherent Light in Different Ways ?", *Phys. Rev. Lett.* 69:2685 (1992).

41. R.C. Miller, D.A. Kleinman, A.C. Gossard, and O. Munteanu, "Biexcitons in GaAs quantum wells", *Phys. Rev. B* 25:6545 (1982); D.A. Kleinman, "Binding energy of biexcitons and bound excitons in quantum wells", *Phys. Rev. B* 28:871 (1983); R. Cingolani, Y. Chen, and K. Ploog, "Biexciton formation in GaAs/$Al_xGa_{1-x}As$ multiple quantum wells: An optical investigation", *Phys. Rev. B* 38:13478 (1988).

42. R.T. Phillips, D.J. Lovering, G.J. Denton, and G.W. Smith, "Biexciton creation and recombination in a GaAs quantum well", *Phys. Rev. B* 45:4308 (1992).

43. D.R. Wake, J.C. Kim, and J.P. Wolfe, "Biexcitons in GaAs/AlGaAs Quantum Wells – their Approach to Chemical Equilibrium and Transport", Quantum Electronics Laser Science Conf., *QELS* 93, Baltimore, May 2-7, 1993, postdeadline paper.

44. M. Wegener, D.S. Chemla, S. Schmitt-Rink, and W. Schäfer, "Line shape of time-resolved four-wave mixing", *Phys. Rev. A* 42:5675 (1990).

45. K. Leo, M. Wegener, J. Shah, D.S. Chemla, E.O. Göbel, T.C. Damen, S. Schmitt-Rink, and W. Schäfer, "Effects of Coherent Polarization Interactions on Time-Resolved Degenerate Four-Wave Mixing", *Phys. Rev. Lett.* 65:1340 (1990).

46. C. Dörnfeld and J.M. Hvam, "Optical Nonlinearities and Phase Coherence in CdSe Studied by Transient Four-Wave Mixing", *IEEE J. Quant. Electr.* QE-25:904 (1989).

47. D.J. Lovering, R.T. Phillips, G.J. Denton, and G.W. Smith, "Resonant Generation of Biexcitons in a GaAs Quantum Well", *Phys. Rev. Lett.* 68:1880 (1992).

48. K.H. Pantke, D. Oberhauser, V.G. Lyssenko, J.M. Hvam, and G. Weimann, "Coherent generation and interference of excitons and biexcitons in GaAs/$Al_xGa_{1-x}As$ quantum wells", *Phys. Rev. B* 47:2413 (1993).

49. S. Bar-Ad and I. Bar-Joseph, "Absorption Quantum Beats of Magnetoexcitons in GaAs Heterostructures", *Phys. Rev. Lett.* 66:2491 (1991), S.Bar-Ad and I.Bar-Joseph, "Exciton Spin Dynamics in GaAs Heterostructures", *Phys. Rev. Lett.*

68:349 (1992); G. Finkelstein, S. Bar-Ad, O. Carmel, and I. Bar-Joseph, "Biexcitonic Effects in Transient Nonlinear Optical Experiments in Quantum Wells", *Phys. Rev. B* 47:12964 (1993).

50. H.H. Yaffe, Y. Prior, J.P. Harbison, and L.T. Florez, "Polarization dependence and selection rules of transient four-wave mixing in GaAs quantum-well excitons", *J. Opt. Soc. Am.* B10:578 (1993).

51. G. Smith, E.J. Mayer, J. Kuhl, D. Bennhardt, K. Bott, E. Heller, and P. Thomas, to be published.

52. K. Bott, O. Heller, D. Bennhardt, P. Thomas, E.J. Mayer, G. Smith, and J. Kuhl, to be published.

53. T. Saiki, M. Kuwata-Gonokami, T. Matsusue, and H. Sakaki, "Elementary Excitation Picture of Excitonic Resonant Nonlinearity in a GaAs Quantum Well", Quantum Electronics and Laser Science Conf., QELS-93, Baltimore, May 2-7 (1993); 1993 Technical Digest Series, Vol. 12.

54. G. Bastard, "Wave Mechanics Applied to Semiconductor Heterostructures" (Les Editions de Physique, Les Ulis, France 1988).

55. S. Cundiff, private communications.

56. E.J. Mayer, G. Smith, J. Kuhl, H. Lage, D. Heitmann, and K. Ploog, to be published.

57. H. Wang, K.B. Ferrio, D.G. Steel, "Polarization-dependent transient nonlinear optical response of heavy- and light-hole excitons in GaAs", Quantum Electronics and Laser Science Conference, QELS-93, Baltimore, May 2-7 (1993); 1993 Techn. Digest Series, Vol. 12.

58. Y.Z. Hu, R. Binder, S.W. Koch, "Valence-band mixing and photon echo in semiconductor quantum wells", Quantum Electronics and Laser Science Conf., *QELS-93*, Baltimore, May 2-7, (1993), 1993 Techn. Digest Series, Vol. 12; and "Photon echo and valence-band mixing in semiconductor quantum wells", *Phys. Rev. B* 47:15679 (1993).

COHERENT EXCITONIC AND FREE CARRIER DYNAMICS IN BULK GaAs AND HETEROSTRUCTURES

T. Kuhn,[1] E. Binder,[1] F. Rossi,[2] A. Lohner,[3] K. Rick,[3]
P. Leisching,[3] A. Leitenstorfer,[3] T. Elsaesser,[3,4] and W. Stolz[2]

[1]Institut für Theoretische Physik und Synergetik
Universität Stuttgart, 70550 Stuttgart, Germany
[2]Fachbereich Physik und Zentrum für Materialwissenschaften
Universität Marburg, 35037 Marburg, Germany
[3]Physik Department E11
Technische Universität München, 85748 Garching, Germany
[4]Max-Born-Institut für Nichtlineare Optik und
Kurzzeitspektroskopie, 12489 Berlin, Germany

INTRODUCTION

Coherent dynamics in atomic and molcular systems has been investigated for a long time. The first spin echo experiment[1] was performed in 1950 on protons in a water solution of Fe^{+++} ions. Pulses in the radio frequency range were generated by means of a gated oscillator with pulse widths between 20 μs and a few milliseconds. With these pulses dephasing times of the order of 10 ms have been measured. In the 1960s echo experiments were brought into the visible range.[2,3] A Q-switched ruby laser produced pulses of approximately 10 ns duration which were used to observe photon echoes from ruby. In this case the dephasing times were of the order of 100 ns. For the observation of such coherent dynamics the pulse width has to be shorter than the dephasing time. In semiconductors typical dephasing times are much shorter, they are in the range of a few picoseconds down to some femtoseconds. Therefore, experiments had to wait until the development of suitable lasers which were able to generate sub-picosecond pulses.

The investigation of nonequilibrium carriers generated by photoexcitation started with the analysis of the incoherent energy relaxation processes in the late 1960s using cw-lasers.[4] The measurement of the carrier temperature as a function of the laser intensity gave insight in the power loss from the carriers to the lattice, i.e. to carrier-phonon scattering rates. In the 1970s pulse sources became available[5] for the study of photoexcited carriers and this initiated time-resolved studies of the relaxation process. The technological progress led to shorter and shorter pulses down to values which have now reached 6 fs.[6] Many experiments have been performed[7] using different techniques: band-to-band luminescence[8–14] which monitors the product of electron and hole distribution functions, band to acceptor luminescence[15–18] which gives information on the carrier distribution function alone, and pump-and-probe measurements[14,19–24] where the differential transmission is proportional to the sum of electron and hole distribution functions.

The theoretical analysis of these relaxation phenomena is based on the semiclassical Boltzmann equation for the distribution functions of the carriers. Various methods have been applied for its solution: parametrized distribution functions,[25–29] an eigenfunction expansion,[30] a direct numerical integration,[31,32] and Monte Carlo simulations.[12,13,30,33–41] This latter technique has found the widest application for quantitative studies because it seems best suited to include on a microscopic level a variety of scattering mechanisms (carrier-phonon, carrier-carrier, carrier-plasmon, intervalley, etc.). Due to the combination of the experimental and theoretical investigations, a detailed understanding of the incoherent dynamics and a good knowledge of material parameters has been achieved.

However, the light field of a laser not only creates nonthermal carrier distributions, but induces also a coherent polarization. The investigation of the coherent dynamics in semiconductors started in the 1980s. Different aspects have been studied: the optical Stark effect,[42,43] the dephasing of free carriers[44] and excitons,[45–52] quantum beats related to various types of level splittings,[53–56] charge oscillations in double-quantum-well systems[57–59] and superlattices,[60,61] many-particle effects,[62–66] and the emission of coherent THz radiation.[59,61,67]

These experiments cannot be analyzed within the framework of the Boltzmann equation. The reason is that the coherence is related to a phase relation between electrons and holes, which is neglected in the semiclassical theory. The theoretical description requires a quantum-mechanical treatment where, in addition to the distribution functions of electrons and holes, also the interband polarization is taken into account as an independent kinetic variable. Several approaches have been used in the literature: Bogoliubov transformations,[68,69] nonequilibrium Green's functions,[70–82] band-edge equations based on the real space density matrix,[83–86] and the density matrix formalism in momentum space.[83,87–96] Nonequilibrium Green's functions and the density matrix formalism are strongly related: The density matrix coincides with the equal-time Green's function $G^{<}$ (in the Kadanoff-Baym notation[97]) or G^{-+} (in the Keldysh notation[98]). Here, we will adopt the density matrix method because it provides a more transparent approach to time-dependent processes. The variables are more directly related to physical observables than two-time Green's functions.

Due to the increased difficulty in obtaining a numerical solution when compared to the semiclassical case, in most cases the incoherent scattering processes are treated on a simplified level either using a relaxation time approximation or taking into account only one interaction mechanism. One of the aims of the present work is to generalize the Monte Carlo technique used in semiclassical transport theory to the case of the coherent dynamics and thus to make use of its advantages in treating complicated scattering mechanisms.[99,100]

This paper is divided into three sections: First we discuss the physical model describing the semiconductor interacting with a coherent light field and the relevant interaction mechanisms for the carriers. We derive the equations of motion and analyze the differences between the coherent and the semiclassical treatment of light-matter interaction. Then we generalize the model to describe four-wave-mixing experiments. We present experimental and theoretical results for the case of bulk GaAs excited close to the band gap. Finally we apply the theory to heterostructures. We calculate exciton states and the coherent dynamics in asymmetric double-quantum-wells.

COHERENT AND INCOHERENT DYNAMICS

In this section we derive the equations of motion describing the coherent and incoherent dynamics in a semiconductor under spatially homogeneous conditions.[87,89,95,96] We consider a bulk semiconductor with two parabolic bands, the conduction band and the heavy hole band. The system is described by a Hamiltonian which, as usual, can be decomposed into a part H_0 which is treated exactly, and a part H_1 describing interactions which is treated within some approximation scheme. In a semiclassical kinetic picture only the free carriers and phonons are included in H_0. All kinds of interactions, carrier-phonon, carrier-carrier, and in particular also the interaction of the carriers with the external light field, give contributions to H_1. They are treated as scattering processes between different states described by H_0 and the corresponding rates are obtained by using Fermi's Golden Rule. These approximations lead to the Boltzmann Transport Equation (BTE) for the carrier distribution functions.

Carrier-light interaction

For the investigation of the coherent dynamics introduced by a coherent laser pulse, the coupling between the carriers and the light field has to be included in H_0. In an electron-hole picture it is given by

$$\begin{aligned} H_0 &= \sum_{\mathbf{k}} \epsilon^e_{\mathbf{k}} c^\dagger_{\mathbf{k}} c_{\mathbf{k}} + \sum_{\mathbf{k}} \epsilon^h_{\mathbf{k}} d^\dagger_{\mathbf{k}} d_{\mathbf{k}} + \sum_{\mathbf{q}} \hbar\omega_{\mathbf{q}} b^\dagger_{\mathbf{q}} b_{\mathbf{q}} \\ &+ \sum_{\mathbf{k}} [\mathbf{M}_{\mathbf{k}} \cdot \mathbf{E}^{(+)}(t) c^\dagger_{\mathbf{k}} d^\dagger_{-\mathbf{k}} + \mathbf{M}^*_{\mathbf{k}} \cdot \mathbf{E}^{(-)}(t) d_{-\mathbf{k}} c_{\mathbf{k}}] \,. \end{aligned} \tag{1}$$

Here, $c^\dagger_{\mathbf{k}}, d^\dagger_{\mathbf{k}}, b^\dagger_{\mathbf{q}}$ ($c_{\mathbf{k}}, d_{\mathbf{k}}, b_{\mathbf{q}}$) denote creation (annihilation) operators of electrons, holes, and phonons, respectively, $\epsilon^e_{\mathbf{k}} = E_G + \hbar^2 k^2/2m_e$ and $\epsilon^h_{\mathbf{k}} = \hbar^2 k^2/2m_h$ the energies of electron and hole states, E_G the energy gap, m_e (m_h) the electron (hole) effective mass, and $\omega_{\mathbf{q}}$ the phonon dispersion relation. The external light field is treated on a semiclassical level and within the rotating-wave approximation. $\mathbf{M}_{\mathbf{k}}$ is the dipole matrix element and $\mathbf{E}^{(+)}$ ($\mathbf{E}^{(-)}$) the positive (negative) frequency part of the electric field of the light

$$\mathbf{E}^{(+)} = \mathbf{E}_0(t) e^{-i\omega_L t}, \qquad \mathbf{E}^{(-)} = \mathbf{E}^{(+)*} \,, \tag{2}$$

with central frequency ω_L and a pulse shape given by $\mathbf{E}_0(t)$.

The basic variables to describe the dynamics of the system are the distribution functions (intraband density matrices) of electrons and holes,

$$f^e_{\mathbf{k}} = \langle c^\dagger_{\mathbf{k}} c_{\mathbf{k}} \rangle \qquad \text{and} \qquad f^h_{\mathbf{k}} = \langle d^\dagger_{\mathbf{k}} d_{\mathbf{k}} \rangle \,, \tag{3}$$

and the interband polarization (interband density matrix),

$$p_{\mathbf{k}} = \langle d_{-\mathbf{k}} c_{\mathbf{k}} \rangle \qquad \text{and} \qquad p^*_{\mathbf{k}} = \langle c^\dagger_{\mathbf{k}} d^\dagger_{-\mathbf{k}} \rangle \,. \tag{4}$$

Using the Heisenberg equations of motion for electron and hole operators, we obtain a closed set of equations of motion for the distribution functions and the polarization

$$\frac{d}{dt} f^e_{\mathbf{k}} = \frac{d}{dt} f^h_{-\mathbf{k}} = g^0_{\mathbf{k}}(t) = \frac{1}{i\hbar} [\mathbf{M}_{\mathbf{k}} \cdot \mathbf{E}_0(t) e^{-i\omega_L t} p^*_{\mathbf{k}} - \mathbf{M}^*_{\mathbf{k}} \cdot \mathbf{E}^*_0(t) e^{i\omega_L t} p_{\mathbf{k}}] \,, \tag{5}$$

$$\frac{d}{dt} p_{\mathbf{k}} = \frac{1}{i\hbar} [(\epsilon^e_{\mathbf{k}} + \epsilon^h_{-\mathbf{k}}) p_{\mathbf{k}} + \mathbf{M}_{\mathbf{k}} \cdot \mathbf{E}_0(t) e^{-i\omega_L t} (1 - f^e_{\mathbf{k}} - f^h_{-\mathbf{k}})] \,. \tag{6}$$

The main difference between the coherent generation rate $g^0_{\mathbf{k}}$ in Eq. (5) and the results obtained from Fermi's Golden Rule is the fact that in the present approach the generation of carriers is a two-step process: The light field first creates a polarization and then the carriers are generated by the interaction of this polarization with the light field. This two-step process is the basis for four-wave-mixing experiments which will be discussed below.

Without interactions, Eqs. (5),(6) describe a set of independent two-level systems driven by an external field with a variable detuning between the transition energy and the light frequency. The dynamics is fully coherent. It is equivalent to a set of atoms in a strong light field, and the equations of motion are the well-known optical Bloch equations in the absence of damping. These equations are modified by the various interaction mechanisms. The common feature of all contributions in the interaction Hamiltonian is the fact that creation and annihilation operators appear at least in third order. Thus, in general we cannot obtain closed equations of motion for a finite set of expectation values as was the case for the distribution functions and the polarizations when considering only H_0. Instead we obtain a hierarchy of equations which has to be truncated within some approximation scheme.

These interaction mechanisms lead to additional contributions in the equations of motion which can be classified as: (i) a renormalization of the free-carrier energies $\epsilon_{\mathbf{k}}^{e,h}$ by a self energy $\hbar\Omega_{\mathbf{k}}^{e,h}$ according to

$$\mathcal{E}_{\mathbf{k}}^{e,h} = \epsilon_{\mathbf{k}}^{e,h} + \sum_{i,n} \hbar\Omega_{\mathbf{k}}^{e,h}\Big|^{(i,n)} , \tag{7}$$

(ii) a renormalization of the external field due to an internal field $\Delta_{\mathbf{k}}$ leading to an effective field $U_{\mathbf{k}}$ according to

$$U_{\mathbf{k}} = \mathbf{M}_{\mathbf{k}} \cdot \mathbf{E}_0(t)e^{-i\omega_L t} + \sum_{i,n} \Delta_{\mathbf{k}}\Big|^{(i,n)} \tag{8}$$

and (iii) incoherent contributions. The equations of motion then read

$$\frac{d}{dt}f_{\mathbf{k}}^{e} = g_{\mathbf{k}}(t) + \sum_{i,n} \frac{d}{dt}f_{\mathbf{k}}^{e}\Big|_{incoh}^{(i,n)} , \tag{9}$$

$$\frac{d}{dt}f_{\mathbf{k}}^{h} = g_{-\mathbf{k}}(t) + \sum_{i,n} \frac{d}{dt}f_{\mathbf{k}}^{h}\Big|_{incoh}^{(i,n)} , \tag{10}$$

$$\frac{d}{dt}p_{\mathbf{k}} = \frac{1}{i\hbar}(\mathcal{E}_{\mathbf{k}}^{e} + \mathcal{E}_{-\mathbf{k}}^{h})p_{\mathbf{k}} + \frac{1}{i\hbar}U_{\mathbf{k}}(t)(1 - f_{\mathbf{k}}^{e} - f_{-\mathbf{k}}^{h}) + \sum_{i,n} \frac{d}{dt}p_{\mathbf{k}}\Big|_{incoh}^{(i,n)} , \tag{11}$$

with the generation rate

$$g_{\mathbf{k}}(t) = \frac{1}{i\hbar}\left[U_{\mathbf{k}}p_{\mathbf{k}}^{*} - U_{\mathbf{k}}^{*}p_{\mathbf{k}}\right] . \tag{12}$$

Here, the superscript i refers to the various types of interactions while n refers to the respective order in perturbation theory. Neglecting memory effects[80,101–105] and terms involving higher powers in the polarization ("polarization scattering"),[106] as well as interference terms between different interactions,[107] the incoherent contributions for the distribution functions have the structure of Boltzmann scattering terms,

$$\frac{d}{dt}f_{\mathbf{k}}^{e,h}\Big|_{incoh} = \sum_{\mathbf{q}} \left[W_{\mathbf{k},\mathbf{k}+\mathbf{q}}^{e,h}f_{\mathbf{k}+\mathbf{q}}^{e,h}(1 - f_{\mathbf{k}}^{e,h}) - W_{\mathbf{k}+\mathbf{q},\mathbf{k}}^{e,h}f_{\mathbf{k}}^{e,h}(1 - f_{\mathbf{k}+\mathbf{q}}^{e,h})\right] , \tag{13}$$

where $W_{\mathbf{k}+\mathbf{q},\mathbf{k}}$ denotes a transition rate from $\mathbf{k}$ to $\mathbf{k}+\mathbf{q}$. The polarization decays due to a dephasing rate $\Gamma_{\mathbf{k}}$ according to

$$\frac{d}{dt}p_{\mathbf{k}}\Big|_{incoh} = -\Gamma_{\mathbf{k}}p_{\mathbf{k}} = -\frac{1}{2}\sum_{\nu=e,h}\left[\Gamma_{\mathbf{k}}^{\nu(in)} + \Gamma_{\mathbf{k}}^{\nu(out)}\right]p_{\mathbf{k}} \tag{14}$$

with $\Gamma_{\mathbf{k}}^{\nu(in)} = \sum_{\mathbf{q}} W_{\mathbf{k},\mathbf{k}+\mathbf{q}}^{\nu}f_{\mathbf{k}+\mathbf{q}}^{\nu}$ and $\Gamma_{\mathbf{k}}^{e,h(out)} = \sum_{\mathbf{q}} W_{\mathbf{k}+\mathbf{q},\mathbf{k}}^{\nu}(1 - f_{\mathbf{k}+\mathbf{q}}^{\nu})$. In the following the corresponding first- and second-order contributions entering Eqs. (9)-(11) will be derived for the case of carrier-phonon and carrier-carrier scattering.

Carrier–phonon interaction

The carriers are coupled to the dynamics of the lattice. The Hamiltonian describing carrier-phonon interaction is given by

$$H_1^{cp} = \sum_{\mathbf{q},\mathbf{k}} [\gamma_\mathbf{q}^e c_{\mathbf{k}+\mathbf{q}}^\dagger b_\mathbf{q} c_\mathbf{k} + \gamma_\mathbf{q}^{e*} c_\mathbf{k}^\dagger b_\mathbf{q}^\dagger c_{\mathbf{k}+\mathbf{q}} + \gamma_\mathbf{q}^h d_{\mathbf{k}+\mathbf{q}}^\dagger b_\mathbf{q} d_\mathbf{k} + \gamma_\mathbf{q}^{h*} d_\mathbf{k}^\dagger b_\mathbf{q}^\dagger d_{\mathbf{k}+\mathbf{q}}] , \tag{15}$$

The structure of this Hamiltonian is independent of the specific type of phonons (acoustic or optical) and of the interaction mechanism (polar or deformation potential). The former is specified by the dispersion relation $\omega_\mathbf{q}$, the latter by the interaction matrix element $\gamma_\mathbf{q}^{e,h}$. In the numerical analysis discussed in the following sections, we take into account the interaction with optical phonons via Fröhlich interaction and with acoustic phonons via deformation-potential coupling. Because the theoretical derivation in this section depends only on the structure of the Hamiltonian, it is the same for both types.

Using the Heisenberg equations of motion, we obtain

$$\frac{d}{dt} f_\mathbf{k}^e \Big|^{cp} = \frac{1}{i\hbar} \sum_\mathbf{q} \Big\{ \gamma_\mathbf{q}^e [\langle c_\mathbf{k}^\dagger b_\mathbf{q} c_{\mathbf{k}-\mathbf{q}} \rangle - \langle c_{\mathbf{k}+\mathbf{q}}^\dagger b_\mathbf{q} c_\mathbf{k} \rangle] + \gamma_\mathbf{q}^{e*} [\langle c_\mathbf{k}^\dagger b_\mathbf{q}^\dagger c_{\mathbf{k}+\mathbf{q}} \rangle - \langle c_{\mathbf{k}-\mathbf{q}}^\dagger b_\mathbf{q}^\dagger c_\mathbf{k} \rangle] \Big\} \tag{16}$$

and analogous equations for $f_\mathbf{k}^h$ and $p_\mathbf{k}$. Thus, the system of equations of motion does not remain closed, higher expectation values (phonon-assisted density-matrices) appear. The first-order contributions are now obtained by factorizing these expectation values on the right hand side of Eq. (16). They always contain a single phonon creation or annihilation operator and thus describe coherent phonons.[108,109] However, in a homogeneous system only $\mathbf{q} = 0$ coherent phonons would be possible, and the corresponding terms in Eq. (16) cancel. Thus, the first order terms vanish. To obtain the second order contribution, we first have to derive the equation of motion for the terms on the r.h.s. of Eq. (16). Neglecting the coupling to the light field (which would correspond to coupled phonon-photon processes and include the light-induced energy shifts), the result for the first term is given by

$$\begin{aligned} \frac{d}{dt} \langle c_\mathbf{k}^\dagger b_\mathbf{q} c_{\mathbf{k}-\mathbf{q}} \rangle &= \frac{1}{i\hbar} (-\epsilon_\mathbf{k}^e + \epsilon_{\mathbf{k}-\mathbf{q}}^e + \hbar\omega_\mathbf{q}) \langle c_\mathbf{k}^\dagger b_\mathbf{q} c_{\mathbf{k}-\mathbf{q}} \rangle \\ &+ \frac{1}{i\hbar} \sum_{\mathbf{q}'} \Big\{ \gamma_\mathbf{q}^e [\langle c_\mathbf{k}^\dagger b_\mathbf{q} b_{\mathbf{q}'} c_{\mathbf{k}-\mathbf{q}-\mathbf{q}'} \rangle - \langle c_{\mathbf{k}+\mathbf{q}'}^\dagger b_{\mathbf{q}'} b_\mathbf{q} c_{\mathbf{k}-\mathbf{q}} \rangle] \\ &+ \gamma_\mathbf{q}^{e*} [\langle c_\mathbf{k}^\dagger b_\mathbf{q} b_{\mathbf{q}'}^\dagger c_{\mathbf{k}-\mathbf{q}+\mathbf{q}'} \rangle - \langle c_{\mathbf{k}-\mathbf{q}'}^\dagger b_{\mathbf{q}'}^\dagger b_\mathbf{q} c_{\mathbf{k}-\mathbf{q}} \rangle] \Big\} \\ &+ \frac{1}{i\hbar} \sum_{\mathbf{k}'} [\gamma_\mathbf{q}^{e*} \langle c_\mathbf{k}^\dagger c_{\mathbf{k}'}^\dagger c_{\mathbf{k}'+\mathbf{q}} c_{\mathbf{k}-\mathbf{q}} \rangle + \gamma_\mathbf{q}^{h*} \langle c_\mathbf{k}^\dagger d_{\mathbf{k}'}^\dagger d_{\mathbf{k}'+\mathbf{q}} c_{\mathbf{k}-\mathbf{q}} \rangle] . \end{aligned} \tag{17}$$

The second-order contribution is now obtained by factorizing the expectation values of four operators (carrier-phonon density-matrices) into distribution functions and polarizations, e.g.

$$\langle c_\mathbf{k}^\dagger b_\mathbf{q} b_{\mathbf{q}'}^\dagger c_{\mathbf{k}-\mathbf{q}+\mathbf{q}'} \rangle = f_\mathbf{k}^e N_\mathbf{q} \delta_{\mathbf{q}\mathbf{q}'} \tag{18}$$

where $N_\mathbf{q}$ denotes the phonon distribution function. Then we have a closed set of equations of motion, however not only for the distribution functions and polarizations, but also for phonon-assisted density-matrices. It contains memory effects which may lead to energy-nonconserving transitions on short time-scales.[101–107] By assuming that the distribution functions are sufficiently slowly varying, a Markov approximation can be performed leading to the standard Boltzmann transition rates for phonon emission and absorption,

$$W_{\mathbf{k},\mathbf{k}+\mathbf{q}}^{e,h} = \frac{2\pi}{\hbar} \sum_\pm |\gamma_\mathbf{q}^{e,h}|^2 \delta(\epsilon_\mathbf{k}^{e,h} - \epsilon_{\mathbf{k}+\mathbf{q}}^{e,h} \pm \hbar\omega_\mathbf{q})(N_\mathbf{q} + \frac{1}{2} \pm \frac{1}{2}) , \tag{19}$$

and to carrier-phonon self-energies according to

$$\hbar\Omega_{\mathbf{k}}^{e,h}\Big|^{(cp,2)} = \sum_{\mathbf{q},\pm} |\gamma_{\mathbf{q}}^{e,h}|^2 \frac{\mathcal{P}}{\epsilon_{\mathbf{k}}^{e,h} - \epsilon_{\mathbf{k+q}}^{e,h} \pm \hbar\omega_{\mathbf{q}}} [(N_{\mathbf{q}} + \frac{1}{2} \pm \frac{1}{2}) f_{\mathbf{k+q}}^{e,h} + (N_{\mathbf{q}} + \frac{1}{2} \mp \frac{1}{2})(1 - f_{\mathbf{k+q}}^{e,h})] \,, \tag{20}$$

where $\mathcal{P}$ denotes the principal value. They describe a time-dependent polaron shift.

Carrier–carrier interaction

The carriers which are created by the laser field are charged particles and thus interact via the Coulomb potential $V_{\mathbf{q}}$. In a two-band model the interaction Hamiltonian is given by

$$H_1^{cc} = \sum_{\mathbf{k},\mathbf{k}',\mathbf{q}} V_{\mathbf{q}} \left[\frac{1}{2} c_{\mathbf{k+q}}^{\dagger} c_{\mathbf{k}'-\mathbf{q}}^{\dagger} c_{\mathbf{k}'} c_{\mathbf{k}} + \frac{1}{2} d_{\mathbf{k+q}}^{\dagger} d_{\mathbf{k}'-\mathbf{q}}^{\dagger} d_{\mathbf{k}'} d_{\mathbf{k}} - c_{\mathbf{k+q}}^{\dagger} d_{\mathbf{k}'-\mathbf{q}}^{\dagger} d_{\mathbf{k}'} c_{\mathbf{k}} \right] \tag{21}$$

The three parts refer to electron-electron, hole-hole, and electron-hole interaction, respectively. We have neglected terms which do not conserve the number of each carrier type, i.e. impact ionization and Auger recombination, as well as the interband exchange interaction[74]. Using the Heisenberg equations of motion, the resulting carrier-carrier contributions to the equations of motion for the distribution functions are given by

$$\frac{d}{dt} f_{\mathbf{k}}^{e}\Big|^{cc} = \frac{1}{i\hbar} \sum_{\mathbf{k}',\mathbf{q}} V_{\mathbf{q}} [-\langle c_{\mathbf{k-q}}^{\dagger} c_{\mathbf{k}'+\mathbf{q}}^{\dagger} c_{\mathbf{k}'} c_{\mathbf{k}} \rangle + \langle c_{\mathbf{k}}^{\dagger} c_{\mathbf{k}'+\mathbf{q}}^{\dagger} c_{\mathbf{k}'} c_{\mathbf{k+q}} \rangle \tag{22}$$

$$+ \langle c_{\mathbf{k+q}}^{\dagger} d_{\mathbf{k}'-\mathbf{q}}^{\dagger} d_{\mathbf{k}'} c_{\mathbf{k}} \rangle - \langle c_{\mathbf{k}}^{\dagger} d_{\mathbf{k}'-\mathbf{q}}^{\dagger} d_{\mathbf{k}'} c_{\mathbf{k-q}} \rangle] \,, \tag{23}$$

and analogous equations for $f_{\mathbf{k}}^{h}$ and $p_{\mathbf{k}}$. In contrast to the case of carrier-phonon interaction, here we have first order contributions obtained by factorizing the expectation values into distribution functions and polarizations. This constitutes the Hartree-Fock approximation[74,87,89] and the results can be expressed in terms of the self energies

$$\hbar\Omega_{\mathbf{k}}^{e,h}\Big|^{(cc,1)} = -\sum_{\mathbf{k}'} V_{\mathbf{k}-\mathbf{k}'} f_{\mathbf{k}'}^{e,h} \tag{24}$$

due to electron-electron and hole-hole interaction, and the internal field

$$\Delta_{\mathbf{k}}\Big|^{(cc,1)} = -\sum_{\mathbf{k}'} V_{\mathbf{k}-\mathbf{k}'} p_{\mathbf{k}'} \tag{25}$$

due to electron-hole interaction.

For the second-order contributions we now have to look for the deviations of the two-particle density-matrices from their factorized form. These deviations describe two-particle correlations. Thus, we have to set up equations of motion for quantities like

$$\langle c_{\mathbf{k}}^{\dagger} c_{\mathbf{k}'+\mathbf{q}}^{\dagger} c_{\mathbf{k}'} c_{\mathbf{k+q}} \rangle + f_{\mathbf{k}}^{e} f_{\mathbf{k+q}}^{e} \delta_{\mathbf{k},\mathbf{k}'} \quad \text{and} \quad \langle c_{\mathbf{k+q}}^{\dagger} d_{\mathbf{k}'-\mathbf{q}}^{\dagger} d_{\mathbf{k}'} c_{\mathbf{k}} \rangle - p_{\mathbf{k+q}}^{*} p_{\mathbf{k}} \delta_{\mathbf{k},-\mathbf{k}'} \tag{26}$$

By performing the same series of approximations as discussed for the case of carrier-phonon interaction, we obtain the second-order carrier-carrier scattering rates[89]

$$W_{\mathbf{k+q},\mathbf{k}}^{e,h(cc)} = \frac{2\pi}{\hbar} |V_{\mathbf{q}}|^2 \sum_{\nu=e,h} \sum_{\mathbf{k}'} f_{\mathbf{k}'}^{\nu} (1 - f_{\mathbf{k}'-\mathbf{q}}^{\nu}) \delta(\epsilon_{\mathbf{k}}^{e,h} + \epsilon_{\mathbf{k}'}^{\nu} - \epsilon_{\mathbf{k}'-\mathbf{q}}^{\nu} - \epsilon_{\mathbf{k+q}}^{e,h}) \,, \tag{27}$$

and a second-order contribution to the carrier-carrier self-energy which, however, we will not write explicitly.

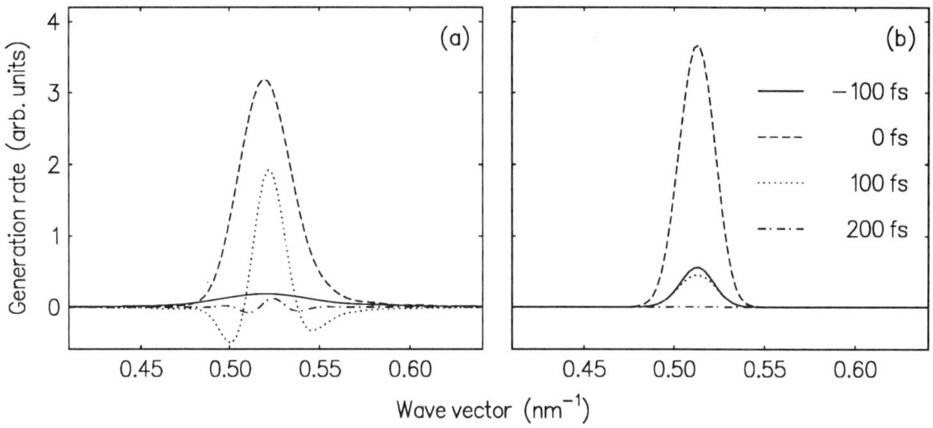

Figure 1. Generation rate as a function of the wave vector (a) for the full model and (b) for the semiclassical generation model at different times during the pulse for the case of excitation with a 100 fs pulse at 1.7 eV.

The presence of free carriers leads to a screening of the bare Coulomb potential. For a discussion of the screening problem the reader is referred to the literature.[32,76,110,111] Here we use quasistatic screening within the single plasmon pole approximation, where the screened Coulomb potential is given by[74,110,112]

$$V_{\mathbf{q}}^{s} = \frac{4\pi e^2}{\varepsilon_s \kappa^2 V} \frac{1+\alpha x^2}{1+x^2+\alpha x^4} \tag{28}$$

with

$$x = \frac{q}{\kappa} \quad \text{and} \quad \alpha = C\frac{\varepsilon_s \hbar^2 \kappa^4}{64\pi m_r n e^2}\,. \tag{29}$$

Here, m_r is the reduced effective mass, n the density, ε_s the background dielectric constant, V a normalization volume, and the RPA screening wave-vector κ is calculated self-consistently according to

$$\kappa^2 = \frac{4\pi e^2}{\epsilon_s V} \sum_{\nu=e,h} \sum_{\mathbf{k}} - \left(\frac{\partial \epsilon_{\mathbf{k}}^{\nu}}{\partial \mathbf{k}}\right)^{-1} \left(\frac{\partial f_{\mathbf{k}}^{\nu}}{\partial \mathbf{k}}\right)\,. \tag{30}$$

The constant C is usually treated as a fitting parameter, here taken to be 4.[112] When using the screened Coulomb potential, the Hartree Fock self-energy (Eq. (24)) must be supplemented by the Coulomb-hole term[74] according to

$$\hbar\Omega_{\mathbf{k}}^{e,h}\Big|^{(cc,1)} = -\sum_{\mathbf{k}'} V_{\mathbf{k}-\mathbf{k}'}^{s} f_{\mathbf{k}'}^{e,h} + \frac{1}{2}\sum_{\mathbf{k}'} \left[V_{\mathbf{k}'}^{s} - V_{\mathbf{k}'}\right] \tag{31}$$

Numerical method

For the numerical solution of the semiconductor Bloch equations we have adopted a combined technique: The equation for the polarization has been solved by direct numerical integration while the equation for the distribution function has been solved by means of a Monte Carlo simulation. The reason for this division is that, at least within the present approximations, the equation for the polarization is simpler since its incoherent contribution involves in general one integration less then the equation for the distribution functions. On

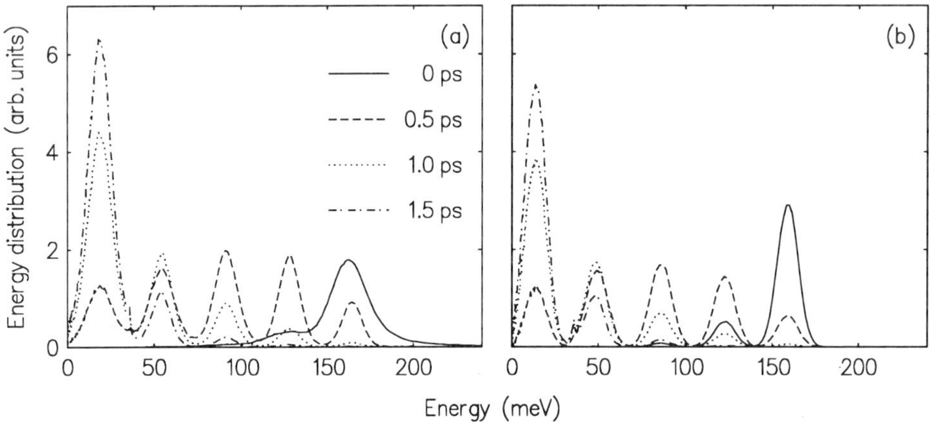

Figure 2. Energy distributions of electrons at different times during and after the pulse (a) for the full model and (b) for the semiclassical generation model.

the other hand, for the treatment of scattering processes in a Boltzmann equation the Monte Carlo has turned out to be the most efficient technique. Details of the method have been published elsewhere.[96] It should be noted that when taking into account higher powers in the polarization or cross terms involving different interactions, the equation for the polarization becomes more complicated.[89,107] Attempts to use a Monte Carlo technique also for the polarization are currently in progress.[113]

Results

We now want to investigate the role of the coherence in the carrier dynamics by comparing results obtained within the semiconductor Bloch equations with those obtained from a Boltzmann equation. In order to concentrate on the main effects, here we have neglected carrier-carrier scattering which, however, will be taken into account in the next section concerning four-wave-mixing experiments. Furthermore, we have neglected the carrier-phonon contribution to the self-energy since, in general, the Hartree-Fock term is the dominant part.

Let us first focus on the generation process for the case of an excitation by a 100 fs pulse with an energy of 1.7 eV. In Fig. 1(a) the generation rate (Eq. (12) multiplied by the density-of-states) is plotted as a function of the wave-vector at different times during the pulse. Figure 1(b) shows the corresponding rate in the semiclassical limit. In the latter case we have a Gaussian broadening in energy related to the Gaussian time dependence by the energy-time uncertainty. This broadening is independent of time. At 100 fs the rate is slightly smaller than at -100 fs due to phase-space filling. In the coherent case we observe a strong time-dependence in the shape of the curves. At short times the generation rate is very broad, becoming narrower with increasing time, and exhibiting even negative parts at later time. Also in this case the broadening is a consequence of energy-time uncertainty. However, the time which has to be taken for the estimation of the energy uncertainty is only the time from the beginning of the pulse up to the observation time. Thus, at short times carriers are generated in a much wider range of energies. As long as there are no processes which destroy the coherence, the carriers generated by energy-nonconserving transitions can recombine at later times due to the interaction with the light field leading to negative parts in the generation rate.

Figure 2 shows the corresponding energy distribution of electrons (including the density

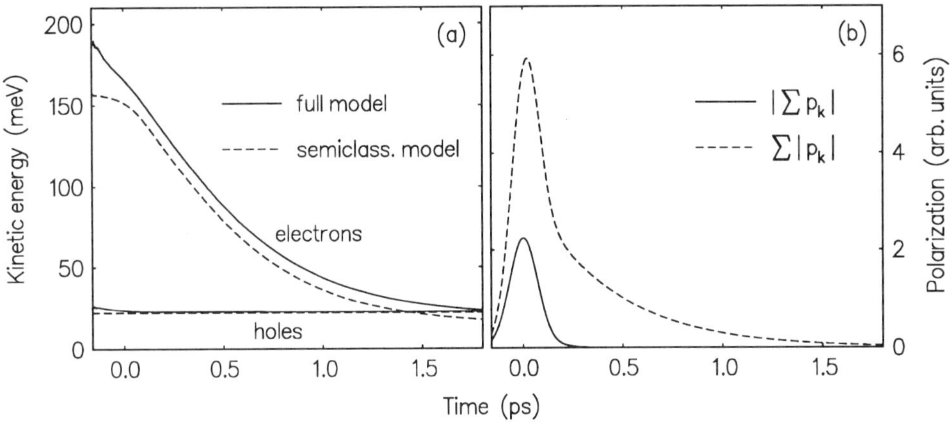

Figure 3. (a) Kinetic energies and (b) polarizations as functions of time.

of states) at different times for (a) the full and (b) the semiclassical generation model. In the semiclassical case the emission of optical phonons leads to replicas of the generated distribution shifted downwards by multiples of the phonon energy. In the Bloch case we again see the large broadening at short times. The interaction with phonons now leads to a loss of coherence so that some of the carriers generated out of resonance cannot recombine anymore. Therefore the distribution remains broadened and does not reach zero between the replicas as in the semiclassical case.

The relaxation process is most clearly visible by looking at the mean kinetic energies of electrons and holes. In Fig. 3(a) they are shown as a function of time for both cases. In particular during the pulse in the Bloch case the energies are remarkably higher than in the semiclassical case. This is due to the fact that, even if the broadening is symmetric in energy, the increasing density-of-states leads to an increased generation of carriers with larger energies. The same reason leads also to an increased recombination of these carriers, thus the difference between both cases decreases towards the end of the pulse. However, the electrons keep a slightly larger energy. This has mainly two origins: First, the asymmetric generation leads to an asymmetry in the phonon replicas; thus the energy of each replica is increased. Second, in the full model we have taken into account the Hartree-Fock self-energies which lead to a shrinkage of the gap during the generation process and effectively shift the excess energy of the laser to larger values.

Figure 3(b) shows the coherent (macroscopic) polarization $|\sum_{\mathbf{k}} p_{\mathbf{k}}|$ and the incoherently summed polarization $\sum_{\mathbf{k}} |p_{\mathbf{k}}|$ as functions of time. The coherent polarization decays rapidly due to inhomogeneous broadening in k-space. The incoherently summed polarization is a measure of the degree of coherence still present in the system. It decays due to incoherent scattering processes. Since we have taken into account only carrier-phonon scattering, the dephasing time is of the same order as the energy relaxation time.

Up to now we have used an excitation far from the gap. In this region the excitonic effects slightly modify the quantitative details of the dynamics due to the Coulomb enhancement, however there are no qualitative changes. This is different in regions close the gap. In Fig. 4(a) we have plotted the carrier density generated by a 1 ps pulse, where only the excess energy of the laser has been varied. Neglecting Hartree-Fock terms (dashed line) we find the square root behavior of the density-of-states, close to the gap slightly washed out due to the finite pulse duration. Switching on the Hartree-Fock terms, we now find a strong absorption at an excess energy of -4.5 meV due to the 1s-exciton and a small maximum at -1.1 meV

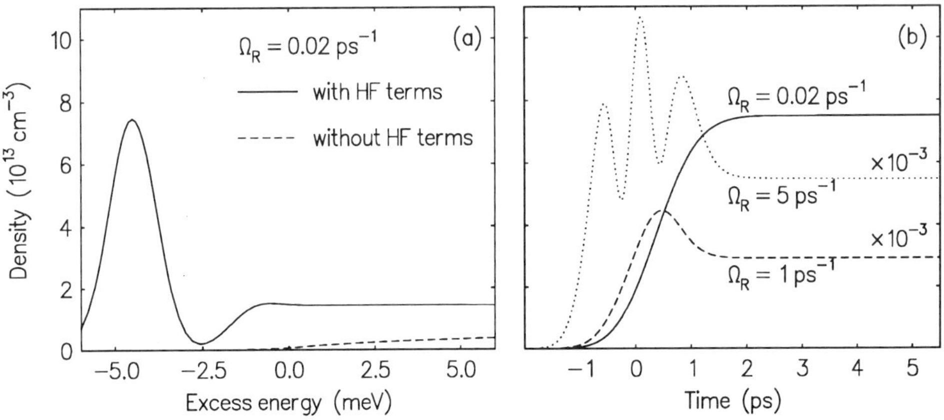

Figure 4. Carrier density generated by a 1 ps laser pulse (a) as a function of the excess energy of the laser with and without taking into account Hartree-Fock terms and (b) as a function of time for different laser intensities. (Note that for $\Omega_R = 1ps^{-1}$ and $\Omega_R = 5ps^{-1}$ the densities have been reduced by a factor of 1000.)

corresponding to the 2s-exciton. Above the gap, as is well known, the Coulomb enhancement leads to a density which is nearly independent of the excess energy.

For the case of excitation in resonance with the exciton, the carrier density as a function of time is plotted in Fig. 4(b). For a low laser intensity the density increases monotonically. If the Rabi frequency $\Omega_R = 2\mathbf{M_k} \cdot \mathbf{E}_{max}/\hbar$ becomes of the order of the inverse pulse duration we see the onset of Rabi oscillations. However, in contrast to the case of a two-level atom, here the density does not reach zero anymore. This fact is due many-body effects which move the band gap towards the exciton frequency. Thus we always generate also free carriers and the exciton does not behave exactly like a two-level system.

FOUR-WAVE-MIXING IN BULK SEMICONDUCTORS

Up to now we have discussed the role of the coherence introduced by a laser pulse on the carrier dynamics. We have seen that the light field leads to a coherent polarization in the system. This coherent polarization, on the other hand, emits again coherent light which, however, has the same direction as the incident laser pulse. Thus, in general it cannot be directly detected. For the detection of a coherently emitted light field there are two possibilities: its direction can be made different from the incident laser pulse or a largely different frequency range can be used. The former case is established using two pulses in a four-wave-mixing geometry. This is the subject of the present section. An example for the latter case will be discussed in the following section.

Theory

Four-wave-mixing experiments can be performed either in a three-beam or in a two-beam geometry. Here we will concentrate on the latter, the two-pulse self-diffraction case. The geometrical setup for the measurement of time-integrated, time-resolved, and spectrally resolved four-wave-mixing is shown schematically in Fig. 5. Two laser pulses with a variable delay time T_{12} are focused on the sample in the directions $\mathbf{k}_1$ and $\mathbf{k}_2$. If there is a temporal

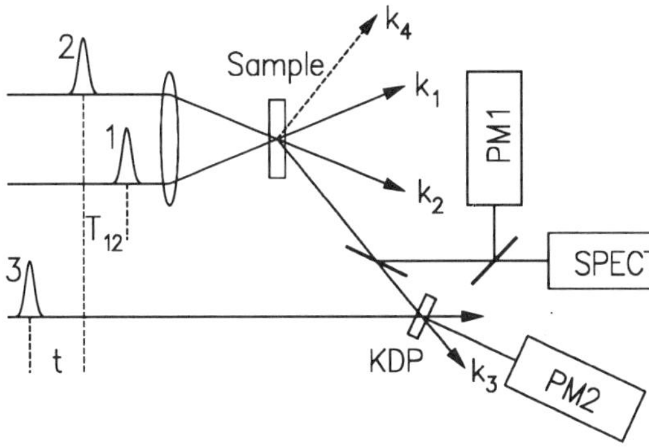

Figure 5. Schematic plot of the FWM setup. Pulses 1 and 2 create the transient grating, the diffracted signal in direction $\mathbf{k}_3$ is analyzed time-integrated (with photomultiplier PM1), time-resolved (with PM2 after upconversion with pulse 3 in the nonlinear KDP crystal), or spectrally resolved (with monochromator SPECT).

overlap between the pulses, they create an interference pattern on the sample which spatially modulates the generation rate and thus creates a transient grating inside the sample. This grating, in turn, partly diffracts the incoming pulses into the directions $\mathbf{k}_3 = 2\mathbf{k}_2 - \mathbf{k}_1$ and $\mathbf{k}_4 = 2\mathbf{k}_1 - \mathbf{k}_2$. However, as we have seen in the previous section, the generation process in fact is related to the interaction of the light with the polarization. Thus, a transient grating is generated also if there is no temporal overlap between the pulses, as long as the second pulse still sees a polarization created by the first pulse. Four-wave-mixing experiments are therefore an ideal tool to study the decay of the polarization.

To describe the physics, the semiconductor Bloch equations have to be generalized to include also the directions of the laser pulses. They lead to an inhomogeneous excitation of the sample.[112,114,115] However, due to the short time-scales involved (typically below or of the order of $1\,ps$) and due to the length-scales of the transient grating ($|\mathbf{k}_1 - \mathbf{k}_2|^{-1}$ is typically larger than or of the order of $1\,\mu m$), the effects of spatial transport can be neglected. Then, the basic quantities for the dynamics, which are the distribution functions of electrons and holes as well as the interband polarization, can be parametrized by a spatial coordinate $\mathbf{R}$. With $\mathbf{k}_1 = \mathbf{Q} + \mathbf{q}$ and $\mathbf{k}_2 = \mathbf{Q} - \mathbf{q}$, the external field is given by

$$\mathbf{E}^{(+)}(t) = \left[\mathbf{E}_0^{(1)}(t)e^{i\mathbf{qR}} + \mathbf{E}_0^{(-1)}(t)e^{-i\mathbf{qR}}\right] e^{i(\mathbf{QR}-\omega_L t)}, \tag{32}$$

where $\mathbf{E}_0^{(1)}$ and $\mathbf{E}_0^{(-1)}$ denote the temporal shapes of the two laser pulses, again taken to be Gaussian. The semiconductor Bloch equations can be decomposed by using a Fourier expansion.[112] Inserting the distribution functions and the polarization according to

$$f_{\mathbf{k}}^{e,h} = \sum_{n=-\infty}^{\infty} f_{\mathbf{k}}^{e,h\,(n)} e^{in\mathbf{qR}} \qquad \text{and} \qquad p_{\mathbf{k}} = e^{i\mathbf{QR}} \sum_{n=-\infty}^{\infty} p_{\mathbf{k}}^{(n)} e^{in\mathbf{qR}} \tag{33}$$

into the semiconductor Bloch equations, a coupled set of equations for the Fourier components is obtained where only the even components of the distribution functions and the odd components of the polarization remain. Restricting ourselves to the lowest order necessary to describe a four-wave-mixing experiment, we take $f_{\mathbf{k}}^{e,h\,(0)}$ describing the homogeneous part of the carrier distribution, $f_{\mathbf{k}}^{e,h\,(2)} = f_{\mathbf{k}}^{e,h\,(-2)*}$ describing the transient grating, $p_{\mathbf{k}}^{(1)}$ and $p_{\mathbf{k}}^{(-1)}$ describing the polarizations in the directions of the two pulses, and $p_{\mathbf{k}}^{(3)}$ and $p_{\mathbf{k}}^{(-3)}$ describing

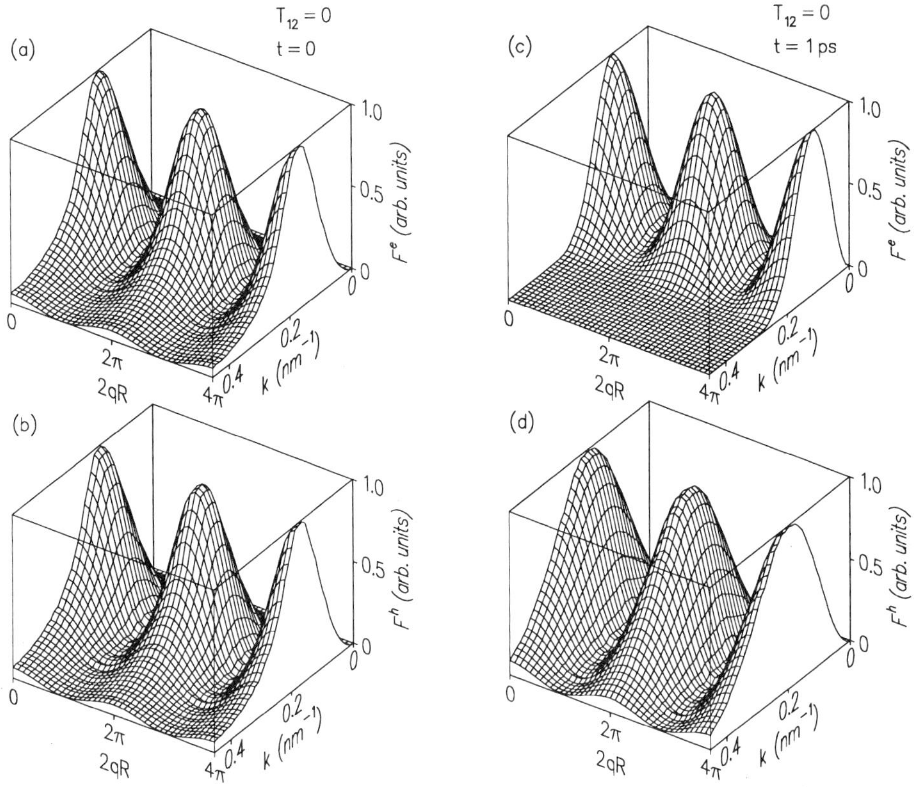

Figure 6. Distribution functions of electrons and holes as functions of k and the position R for delay time zero at time t = 0 and t = 1 ps for the case of excitation with two 100 fs pulses at 1.523 eV.

the diffracted polarizations. By expanding also the effective field according to

$$U_{\mathbf{k}} = e^{i\mathbf{QR}} \sum_{n=-\infty}^{\infty} U_{\mathbf{k}}^{(n)} e^{in\mathbf{qR}}, \tag{34}$$

the equations of motion for the Fourier components of the distribution functions are given by

$$\begin{aligned} \frac{d}{dt} f_{\mathbf{k}}^{e\,(0)} &= \frac{1}{i\hbar}\left[U_{\mathbf{k}}^{(1)} p_{\mathbf{k}}^{(1)*} + U_{\mathbf{k}}^{(3)} p_{\mathbf{k}}^{(3)*} + U_{\mathbf{k}}^{(-1)} p_{\mathbf{k}}^{(-1)*} + U_{\mathbf{k}}^{(-3)} p_{\mathbf{k}}^{(-3)*} - c.c.\right] \\ &+ \sum_{\mathbf{k}'}\left[W_{\mathbf{kk}'} f_{\mathbf{k}'}^{e\,(0)}(1 - f_{\mathbf{k}}^{e\,(0)}) - W_{\mathbf{k}'\mathbf{k}} f_{\mathbf{k}}^{e\,(0)}(1 - f_{\mathbf{k}'}^{e\,(0)})\right], \end{aligned} \tag{35}$$

$$\begin{aligned} \frac{d}{dt} f_{\mathbf{k}}^{e\,(2)} &= \frac{1}{i\hbar}\left[U_{\mathbf{k}}^{(3)} p_{\mathbf{k}}^{(1)*} + U_{\mathbf{k}}^{(1)} p_{\mathbf{k}}^{(-1)*} + U_{\mathbf{k}}^{(-1)} p_{\mathbf{k}}^{(-3)*} U_{\mathbf{k}}^{(1)*} p_{\mathbf{k}}^{(3)} - U_{\mathbf{k}}^{(-1)*} p_{\mathbf{k}}^{(1)} - U_{\mathbf{k}}^{(-3)*} p_{\mathbf{k}}^{(-1)}\right] \\ &+ \sum_{\mathbf{k}'}\left[W_{\mathbf{kk}'} f_{\mathbf{k}'}^{e\,(2)}(1 - f_{\mathbf{k}}^{e\,(0)}) - W_{\mathbf{k}'\mathbf{k}} f_{\mathbf{k}}^{e\,(2)}(1 - f_{\mathbf{k}'}^{e\,(0)})\right], \end{aligned} \tag{36}$$

$f_{\mathbf{k}}^{e\,(-2)} = f_{\mathbf{k}}^{e\,(2)*}$ and analogous equations for $f_{\mathbf{k}}^{h}$. The equations of motion for the Fourier components of the polarizations read

$$\begin{aligned} \frac{d}{dt} p_{\mathbf{k}}^{(1)} &= \frac{1}{i\hbar}\left[\left(\mathcal{E}_{\mathbf{k}}^{e} + \mathcal{E}_{-\mathbf{k}}^{h}\right) p_{\mathbf{k}}^{(1)} + U_{\mathbf{k}}^{(1)}\left(1 - f_{\mathbf{k}}^{e\,(0)} - f_{-\mathbf{k}}^{h\,(0)}\right)\right. \\ &- \left. U_{\mathbf{k}}^{(3)}\left(f_{\mathbf{k}}^{e\,(-2)} + f_{-\mathbf{k}}^{h\,(-2)}\right) - U_{\mathbf{k}}^{(-1)}\left(f_{\mathbf{k}}^{e\,(2)} + f_{-\mathbf{k}}^{h\,(2)}\right)\right] - \Gamma_{\mathbf{k}} p_{\mathbf{k}}^{(1)}, \end{aligned} \tag{37}$$

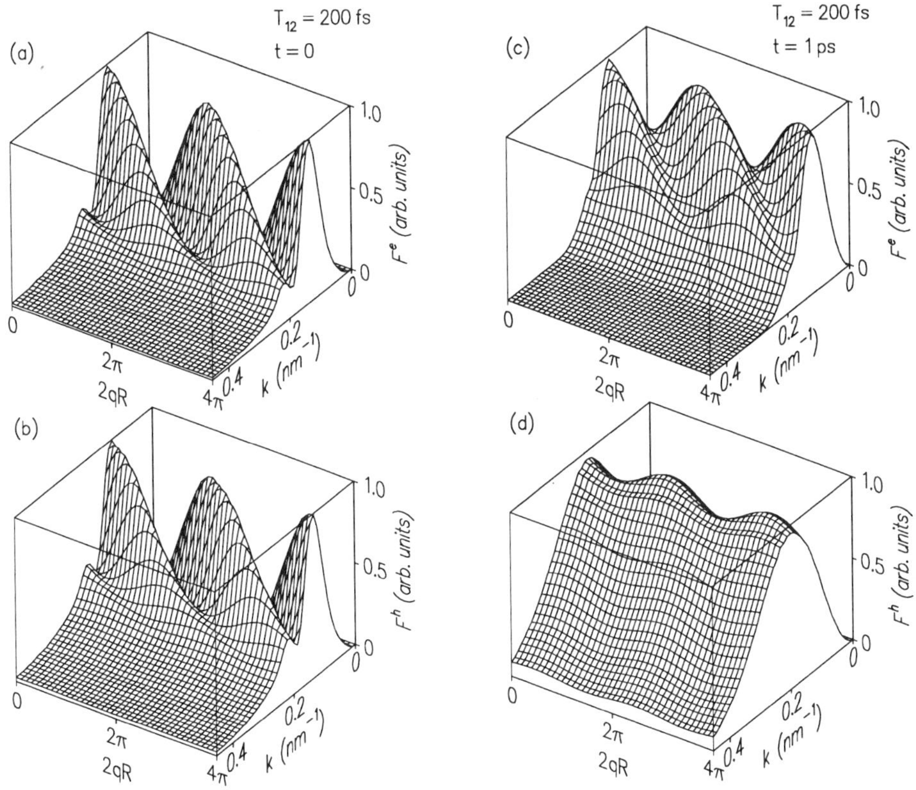

Figure 7. Same as Fig. 6 but for a delay time of 200 fs.

$$
\begin{aligned}
\frac{d}{dt} p_{\mathbf{k}}^{(3)} &= \frac{1}{i\hbar} \left[\left(\mathcal{E}_{\mathbf{k}}^{e} + \mathcal{E}_{-\mathbf{k}}^{h} \right) p_{\mathbf{k}}^{(3)} + U_{\mathbf{k}}^{(3)} \left(1 - f_{\mathbf{k}}^{e\,(0)} - f_{-\mathbf{k}}^{h\,(0)} \right) \right. \\
&- \left. U_{\mathbf{k}}^{(1)} \left(f_{\mathbf{k}}^{e\,(2)} + f_{-\mathbf{k}}^{h\,(2)} \right) \right] - \Gamma_{\mathbf{k}} p_{\mathbf{k}}^{(3)} , \qquad (38)
\end{aligned}
$$

and the equations for $p_{\mathbf{k}}^{(-1)}$ and $p_{\mathbf{k}}^{(-3)}$ are obtained from Eqs. (37) and (38) by changing signs in the superscripts denoting the Fourier components. In the present study we have neglected an additional coupling of the Fourier components due to space dependencies in the self energy, screening, and scattering rate which might become important in the case of a strong spatial modulation. The respective quantities are calculated only with the homogeneous parts $f_{\mathbf{k}}^{e,h\,(0)}$ of the distribution functions. The following scattering mechanisms have been included: carrier-carrier scattering, scattering with optical phonons via Fröhlich interaction and with acoustic phonons via deformation potential coupling. Since in the low-density limit and in a region close to the band gap the imaginary part of the self energy strongly overestimates the dephasing due to the neglect of excitonic effects, we have modelled the dephasing contribution due to carrier-carrier scattering by a constant rate Γ_{cc} here taken to be 0.5 ps in agreement with experimental results described below.

For a numerical solution of Eqs. (35) - (38) the previously described technique has to be generalized. Within the present approximations the scattering part of the equations of motion for the distribution functions (Eqs. (35),(36)) are exactly the same. However, while the homogeneous component $f_{\mathbf{k}}^{(0)}$ is always positive, the transient grating contributions $f_{\mathbf{k}}^{(\pm 2)}$ are complex quantities. Since the scattering rates $W_{\mathbf{k}\mathbf{k}'}$ are positive and Eq. (36) is linear

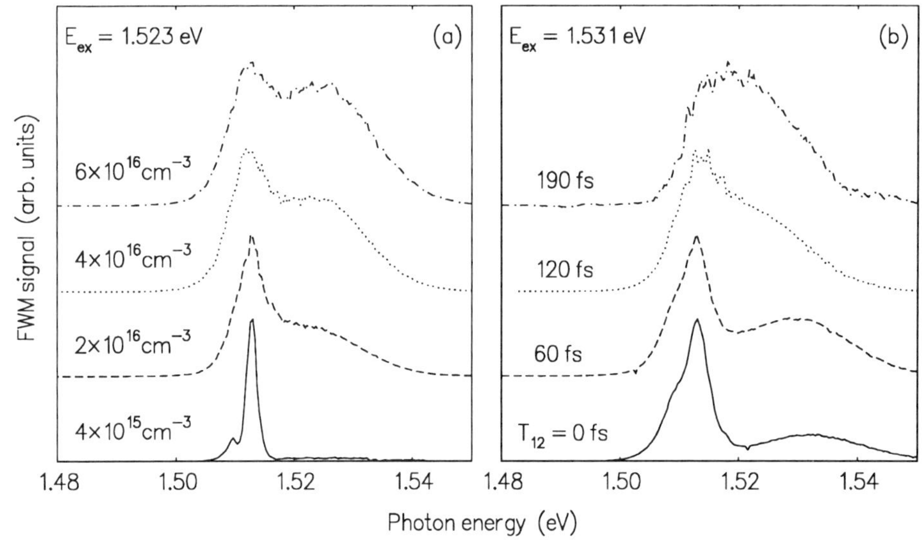

Figure 8. Measured spectrally resolved four-wave-mixing signals (a) at delay time zero for different carrier densities and (b) at different delay times for excitation with two 100 fs pulses.

in $f_{\mathbf{k}}^{(2)}$, the scattering does not mix real and imaginary parts. Furthermore, both real and imaginary parts can be split into two parts, a positive one and a negative one, which can be treated separately due to the linearity. Thus, we obtain effectively four Boltzmann-like equations for the four parts of $f_{\mathbf{k}}^{(2)}$ which can be solved by a usual Monte Carlo technique.

Transient gratings

The diffraction of the signal into the direction $\mathbf{k_3}$ is based on the creation and dynamics of a transient grating described by $f_{\mathbf{k}}^{e,h(\pm 2)}$ Therefore we will start with an analysis of the distribution functions of electrons and holes as a function of spatial coordinate and wave vector for the case of a 100 fs pulse centered 4 meV above the gap. In Fig. 6 the functions

$$F_{\mathbf{k}}^{e,h}(\mathbf{qR}) = k^2 \left[f_{\mathbf{k}}^{e,h(0)} + 2\mathrm{Re} f_{\mathbf{k}}^{e,h(2)} \cos 2\mathbf{qR} - 2\mathrm{Im} f_{\mathbf{k}}^{e,h(2)} \sin 2\mathbf{qR} \right] \tag{39}$$

are shown for the case of zero delay time at the time of the pulse maximum ($t = 0$) and after 1 ps. As can be expected, since the laser beams form an interference pattern, the carriers are generated according to a cosine-function in real space. Electrons and holes are generated with equal distribution functions, the subsequent relaxation process leads to a cooling of electrons both due to electron-phonon and electron-hole scattering and a heating of the holes due to electron-hole scattering. Even if the pulse is centered only slightly above the gap, there is a significant emission of optical phonons by the electrons since, in particular at short times, the generation rate extends to quite large energies (see Fig. 1).

Figure 7 shows the same distribution functions for a delay time of 200 fs. Now there is virtually no temporal overlap between the two pulses. The transient grating is now formed by the interaction of the second pulse with the polarization created by the first one. But due to the inhomogeneous broadening, different k-contributions have different phases. Therefore, the grating has now a different phase-shift for each k-value. The carrier density, on the other hand, is nearly independent of the position. The relaxation leads again to a cooling of the electrons and a heating of the holes. In the present case, however, since scattering processes

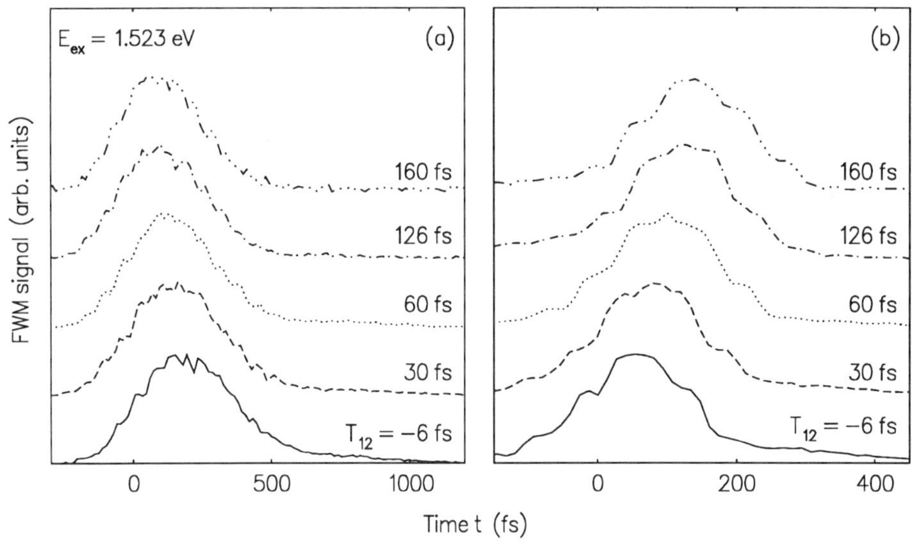

Figure 9. Measured time-resolved four-wave-mixing signals after filtering (a) at the exciton line and (b) at the free-carrier peak at different delay times.

tend to establish a local quasi-equilibrium distribution, also the spatial inhomogeneity is strongly reduced during the relaxation process.

Experiments

Let us now turn to the experimental results. The coherent dynamics in bulk GaAs close to the band gap has been investigated by temporally and spectrally resolved four-wave-mixing experiments. The setup of the experiments is shown in Fig. 5. The sample consists of a $0.5\ \mu m$ thick GaAs layer clad by AlGaAs barriers. Two transform-limited 100 fs pulses created by a mode-locked Ti:sapphire laser generate the transient grating. The diffracted signal is either spectrally dispersed with the help of a monochromator (spectral width 2 meV) or convoluted with a third laser pulse via sum-frequency mixing in a nonlinear KDP crystal.

Figure 8(a) shows the spectrally resolved signal for different values of the generated carrier density at delay time zero. The laser frequency is 1.523 eV, i.e. 4 meV above the gap. At low densities, however, the signal is located completely at the exciton. With increasing density an increasing free-carrier contribution centered at the laser frequency appears. The excitonic component is present up to the highest measured densities. This strong excitonic enhancement is due (i) to the large ratio between excitonic and interband oscillator strength[110,116] and (ii) to the slow dephasing of the excitonic polarization which persists much longer than the free-carrier signal. It should be noted that the equilibrium Mott density, i.e. the density where the exciton ground state merges into the continuum, at 10 K is $4 \times 10^{15}\ cm^{-3}$. The presence of an excitonic contribution above this density is due to the strong nonequilibrium in the carrier system.[66] Nonthermalized carriers screen the exciton much less efficiently than thermalized carriers.

In Fig. 8(b) the time-resolved signal for the case of excitation at 1.531 eV is plotted for different delay times. With increasing delay time the free-carrier contribution is increasing and finally completely dominates the spectrum.

The two components in the spectrally resolved signal show a different temporal behavior. For an isolated observation of the two components, the excitonic or the free-carrier signal has

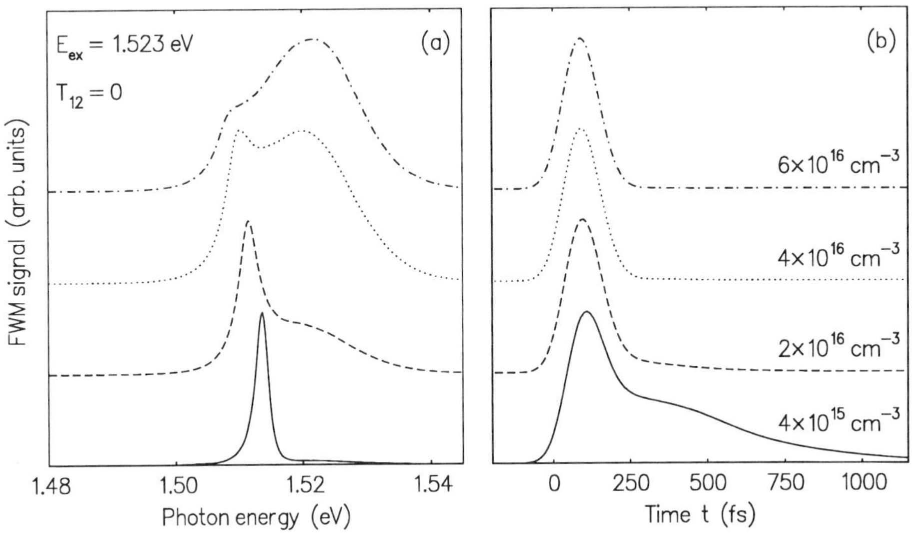

Figure 10. Calculated four-wave-mixing signals at delay time zero, (a) spectrally resolved and (b) time-resolved, for different carrier densities.

been selected by interference filters and subsequently convoluted with pulse 3. The excitonic signals for different delay times are shown in Fig. 9(a). They are nearly independent of the delay time, the maxima exhibit a slight shift towards earlier times. The free-carrier signal shown in Fig. 9(b), on the other hand, exhibits a clear photon echo behavior. The signal is decaying much faster in time due to the inhomogeneous broadening.

Many-body effects

The experimental results concerning the temporal as well as the spectral shape of the signal indicate that many-body effects play an important role.[63,65,66,90,112] We now want to see whether they can be understood in the framework of the semiconductor Bloch equations. Figure 10 shows the spectrally and temporally resolved signals for the same densities as in the experiment. The spectral curves are in good agreement with the experiments, there are only two deviations: first, the excitonic contribution disappears at a somewhat smaller density than observed in the experiment, and second, the exciton peak exhibits a slight red-shift with increasing density. Different source for these discrepancies are possible: First, at these short times the use of a quasistatic screening model is expected to overestimate the role of screening. However, a correct self-consistent treatment of the screened Coulomb potential would require the calculation of a dielectric function which depends explicitly on two times. Preliminary work in this direction already exists, but has not yet been included in a full analysis of the dynamics.[76,81] Second, the shape of the spectra, in particular the ratio between the excitonic and the free-carrier contribution depends on the dephasing time. Therefore, a better modeling of the microscopic dephasing processes is required which, however, is difficult in the transition region between excitonic and free-carrier behavior. The corresponding time-resolved signals (Fig. 10(b)) show two different contributions. A prompt one related to the free-carrier part in the spectrum and a delayed one related to the excitonic part which decreases with increasing density and is visible only for the lower densities.

In Fig. 11 the spectrally resolved signals are shown for different positive and negative delay times. With increasing negative delay time the excitonic part increases which indicates

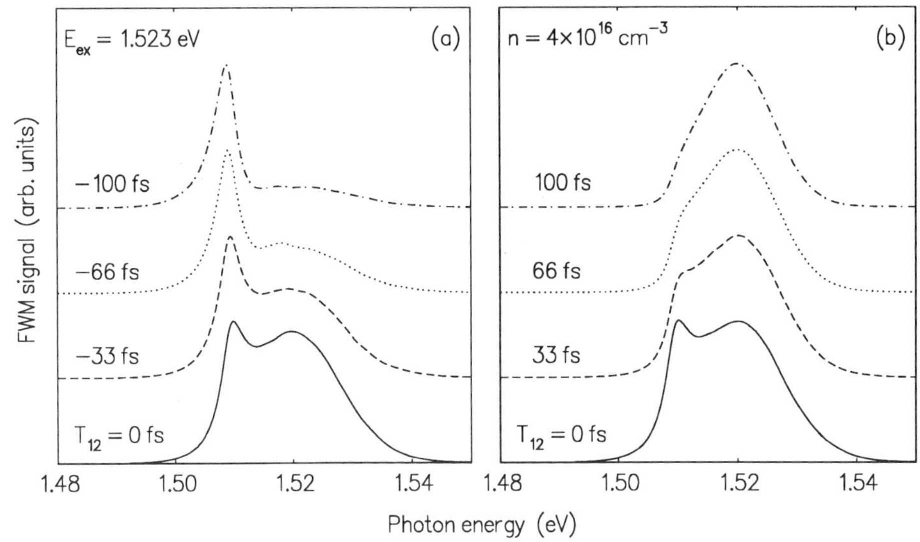

Figure 11. Calculated spectrally resolved four-wave-mixing signals at delay time zero for different (a) negative and (b) positive delay times.

that in this case the many-body effects dominate. In fact, for non-overlapping pulses only many-body effects can lead to a signal at negative delay times. For increasing positive delay times we find, as in the experiment, an increasing free-carrier contribution indicating that now it is the laser pulse which is diffracted by the transient grating.

Finally, in Fig. 12 temporally resolved signals are plotted which are obtained by filtering at the exciton peak (Fig. 12(a)) and at the free-carrier peak (Fig. 12(b)). However, since in the numerical results the exciton contribution is very small at a density of 4×10^{16} cm^{-3}, the value of the experiments in Fig. 9, here we have taken a density of 4×10^{15} cm^{-3}. As in the experiment we find an excitonic contribution with a position of the maximum nearly independent of the delay time. For short delay times also the numerical results exhibit a small shift towards earlier times, however it is less pronounced than in the experiments. The free-carrier contribution shows a clear photon-echo behavior due to the inhomogeneous broadening in k-space.

COHERENT DYNAMICS IN HETEROSTRUCTURES: THz-RADIATION

In the previous sections we have investigated the coherent dynamics in bulk semiconductors. However, during the last two decades many experiments have been performed using heterostructures.[46,48–51,56,58–61] While in a bulk semiconductor the dynamics is governed only by material parameters, heterostructures offer the possibility of tuning some properties by "band gap engineering". Of particular interest are single- or multiple-quantum well structures which, due to the controllable localization behavior of the carrier states, give rise to interesting new phenomena like resonances,[117,118] charge transfer,[117,119,120] coherent oscillations,[58,60] and the emission of electromagnetic radiation in the THz range.[59,61,67]

In this section we will concentrate on an excitation close to the band gap where the dynamics is mainly governed by excitonic effects. Furthermore we will limit ourselves to the case of low densities. We do not treat in detail scattering processes, in the simulations the dephasing is described by a phenomenological dephasing time. Thus we take into account

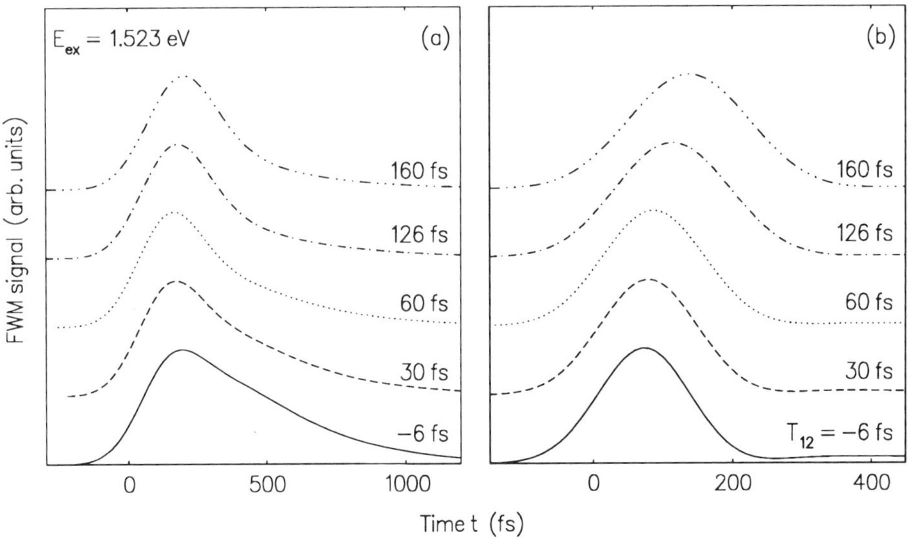

Figure 12. Calculated time-resolved four-wave-mixing signals after filtering (a) at the exciton line and (b) at the free-carrier peak at different delay times.

only the first-order (Hartree-Fock) terms of carrier-carrier interaction.

The system we want to investigate is an asymmetric double-quantum well structure as shown in Fig. 13. The interesting point is that the localization of the carriers can be controlled by an external electric field applied in the growth direction. Without applied field the lowest electron state is localized in the wide well, the second one in the narrow well. With increasing field the energy of the second one is lowered, it comes into resonance with the first one and, due to the tunnel interaction, at a field of 7 kV/cm both states become delocalized. Increasing the field further, the lowest state now becomes localized in the narrow well. The highest hole state remains always in the wide well, the second and third ones exhibit a similar crossing as in the electron case.

Theory

In a quantum well heterostructure the motion of carriers in one direction (the growth- or z-direction) is quantized, while in the other two dimensions the motion is free. The carrier states are then characterized by a discrete subband index i or j and a continuous two-dimensional wave-vector k. (In the following i will refer to electron subbands and j to hole subbands.) The Hamiltonian for the free carriers interacting with a light field in the frequency range of the gap energy is then given by

$$\begin{aligned} H_0 &= \sum_{i_1}\sum_{\mathbf{k}} \epsilon^e_{i_1,\mathbf{k}} c^\dagger_{i_1,\mathbf{k}} c_{i_1,\mathbf{k}} + \sum_{j_1}\sum_{\mathbf{k}} \epsilon^h_{j_1,\mathbf{k}} d^\dagger_{j_1,\mathbf{k}} d_{j_1,\mathbf{k}} \\ &+ \sum_{i_1,j_1}\sum_{\mathbf{k}} \left[\mathbf{M}^{eh}_{i_1 j_1,\mathbf{k}} \cdot \mathbf{E}^{(+)}(t) c^\dagger_{i_1,\mathbf{k}} d^\dagger_{j_1,-\mathbf{k}} + \mathbf{M}^{eh*}_{i_1 j_1,\mathbf{k}} \cdot \mathbf{E}^{(-)}(t) d_{j_1,-\mathbf{k}} c_{i_1,\mathbf{k}}\right] , \end{aligned} \tag{40}$$

where all symbols have the same meaning as in Eq. (1). They only contain additional subband indices. The basic variables for the description of the carrier dynamics are given by the distribution functions of electrons and holes,

$$f^e_{i_1,\mathbf{k}} = \left\langle c^\dagger_{i_1,\mathbf{k}} c_{i_1,\mathbf{k}} \right\rangle \qquad \text{and} \qquad f^h_{j_1,\mathbf{k}} = \left\langle d^\dagger_{j_1,\mathbf{k}} d_{j_1,\mathbf{k}} \right\rangle , \tag{41}$$

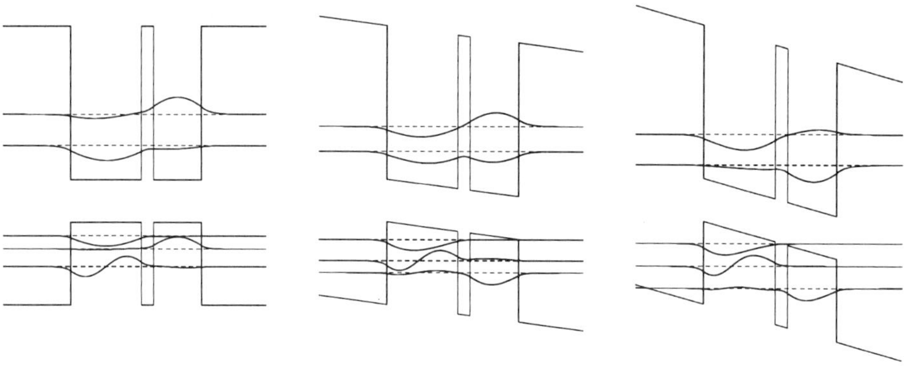

Figure 13. Envelope wave functions of the asymmetric double-quantum well for three values of the applied electric field. The well widths are 15 nm and 10 nm, the barrier width is 2.5 nm.

and the interband polarizations,

$$p^{he}_{j_1 i_1,\mathbf{k}} = \left\langle d_{j_1,-\mathbf{k}} c_{i_1,\mathbf{k}} \right\rangle \qquad \text{and} \qquad p^{he*}_{j_1 i_1,\mathbf{k}} = \left\langle c^{\dagger}_{i_1,\mathbf{k}} d^{\dagger}_{j_1,-\mathbf{k}} \right\rangle . \tag{42}$$

Furthermore, due to the lack of translational invariance in the z-direction, in addition we have intraband (intersubband) polarizations

$$p^{ee}_{i_1 i_2,\mathbf{k}} = \left\langle c^{\dagger}_{i_1,\mathbf{k}} c_{i_2,\mathbf{k}} \right\rangle \qquad \text{and} \qquad p^{hh}_{j_1 j_2,\mathbf{k}} = \left\langle d^{\dagger}_{j_1,\mathbf{k}} d_{j_2,\mathbf{k}} \right\rangle . \tag{43}$$

We will come back to these intraband polarizations later, first let us assume that they are small and can be neglected in the dynamics of distribution functions and interband polarizations. The equations of motion, determined by H_0, are then given by

$$\frac{d}{dt} f^{e}_{i_1,\mathbf{k}} = \sum_{j_1} g^{0}_{i_1 j_1,\mathbf{k}} , \qquad \frac{d}{dt} f^{h}_{j_1,-\mathbf{k}} = \sum_{i_1} g^{0}_{i_1 j_1,\mathbf{k}} , \tag{44}$$

$$\frac{d}{dt} p^{he}_{j_1 i_1,\mathbf{k}} = \frac{1}{i\hbar}(\epsilon^{h}_{j_1,-\mathbf{k}} + \epsilon^{e}_{i_1,\mathbf{k}}) p^{he}_{j_1 i_1,\mathbf{k}} + \frac{1}{i\hbar}\mathbf{M}^{eh}_{i_1 j_1,\mathbf{k}} \cdot \mathbf{E}^{(+)}(t)[1 - f^{e}_{i_1,\mathbf{k}} - f^{h}_{j_1,-\mathbf{k}}] , \tag{45}$$

where we have introduced the pair generation rate

$$g^{0}_{i_1 j_1,\mathbf{k}} = \frac{1}{i\hbar}[\mathbf{M}^{eh}_{i_1 j_1,\mathbf{k}} \cdot \mathbf{E}^{(+)}(t) p^{he*}_{j_1 i_1,\mathbf{k}} - \mathbf{M}^{eh*}_{i_1 j_1,\mathbf{k}} \cdot \mathbf{E}^{(-)}(t) p^{he}_{j_1 i_1,\mathbf{k}}] . \tag{46}$$

In order to describe excitonic effects which completely determine the optical properties of a semiconductor in a region close to the gap, we have to include carrier-carrier interaction. The corresponding Hamiltonian is given by

$$\begin{aligned} H^{cc}_{i} &= \frac{1}{2} \sum_{i_1,i_2,i_3,i_4} \sum_{\mathbf{k},\mathbf{k}',\mathbf{q}} V_{i_1 i_2 i_3 i_4}(\mathbf{q}) c^{\dagger}_{i_1,\mathbf{k}+\mathbf{q}} c^{\dagger}_{i_2,\mathbf{k}'-\mathbf{q}} c_{i_3,\mathbf{k}'} c_{i_4,\mathbf{k}} \\ &+ \frac{1}{2} \sum_{j_1,j_2,j_3,j_4} \sum_{\mathbf{k},\mathbf{k}',\mathbf{q}} V_{j_4 j_3 j_2 j_1}(\mathbf{q}) d^{\dagger}_{j_1,\mathbf{k}+\mathbf{q}} d^{\dagger}_{j_2,\mathbf{k}'-\mathbf{q}} d_{j_3,\mathbf{k}'} d_{j_4,\mathbf{k}} \\ &- \sum_{i_1,i_2,j_1,j_2} \sum_{\mathbf{k},\mathbf{k}',\mathbf{q}} V_{i_1 j_2 j_1 i_2}(\mathbf{q}) c^{\dagger}_{i_1,\mathbf{k}+\mathbf{q}} d^{\dagger}_{j_1,\mathbf{k}'-\mathbf{q}} d_{j_2,\mathbf{k}'} c_{i_2,\mathbf{k}} \end{aligned} \tag{47}$$

with the Coulomb matrix element

$$V_{n_1n_2n_3n_4}(\mathbf{q}) = \frac{2\pi e^2}{\varepsilon_s A q}\int \phi^*_{n_1}(z)\phi^*_{n_2}(z')e^{-q|z-z'|}\phi_{n_3}(z')\phi_{n_4}(z)dz\,dz' \tag{48}$$

where $\phi_n(z)$ are the envelope wave functions in the z-direction and A is a normalization area.

The first-order contributions to the equations of motion due to carrier-carrier interaction are obtained in complete analogy to the bulk case. Electron-electron and hole-hole interactions give rise to the self energy matrices

$$\begin{aligned} \hbar\Omega^e_{i_1i_2,\mathbf{k}}\Big|^{(cc,1)} &= -\sum_{i_3}\sum_{\mathbf{k}'} V_{i_1i_3i_2i_3}(\mathbf{k}-\mathbf{k}')f^e_{i_3,\mathbf{k}'} \\ \hbar\Omega^h_{j_1j_2,\mathbf{k}}\Big|^{(cc,1)} &= -\sum_{j_3}\sum_{\mathbf{k}'} V_{j_2j_3j_1j_3}(\mathbf{k}-\mathbf{k}')f^h_{j_3,\mathbf{k}'} \end{aligned} \tag{49}$$

and electron-hole interaction leads to the internal field matrix

$$\Delta_{i_1j_1,\mathbf{k}}\Big|^{(cc,1)} = -\sum_{i_2,j_2}\sum_{\mathbf{k}'} V_{i_1j_2j_1i_2}(\mathbf{k}-\mathbf{k}')p^{he}_{j_2i_2,\mathbf{k}'}\,. \tag{50}$$

Therefore, we can introduce the renormalized energy matrices

$$\mathcal{E}^e_{i_1i_2,\mathbf{k}} = \epsilon^e_{i_1,\mathbf{k}}\delta_{i_1i_2} + \hbar\Omega^e_{i_1i_2,\mathbf{k}}\Big|^{(cc,1)}\,, \tag{51}$$

$$\mathcal{E}^h_{j_1j_2,\mathbf{k}} = \epsilon^h_{j_1,\mathbf{k}}\delta_{j_1j_2} + \hbar\Omega^h_{j_1j_2,\mathbf{k}}\Big|^{(cc,1)}\,, \tag{52}$$

and the effective field matrix

$$U_{i_1j_1,\mathbf{k}} = \mathbf{M}^{eh}_{i_1j_1,\mathbf{k}}\cdot\mathbf{E}^{(+)}(t) + \Delta_{i_1j_1,\mathbf{k}}\Big|^{(cc,1)}\,. \tag{53}$$

The equations of motion now read

$$\frac{d}{dt}f^e_{i_1,\mathbf{k}} = \sum_{j_1} g_{i_1j_1,\mathbf{k}}\,, \qquad \frac{d}{dt}f^h_{j_1,-\mathbf{k}} = \sum_{i_1} g_{i_1j_1,\mathbf{k}}\,, \tag{54}$$

$$\frac{d}{dt}p^{he}_{j_1i_1,\mathbf{k}} = \frac{1}{i\hbar}\sum_{i_2j_2}[\mathcal{E}^h_{j_1j_2,-\mathbf{k}}\delta_{i_1i_2} + \mathcal{E}^e_{i_1i_2,\mathbf{k}}\delta_{j_1j_2}]p^{he}_{j_2i_2,\mathbf{k}} + \frac{1}{i\hbar}U_{i_1j_1,\mathbf{k}}(1 - f^e_{i_1,\mathbf{k}} - f^h_{j_1,-\mathbf{k}})\,, \tag{55}$$

with the generation rate

$$g_{i_1j_1,\mathbf{k}} = \frac{1}{i\hbar}\left[U_{i_1j_1,\mathbf{k}}\,p^{he*}_{j_1i_1,\mathbf{k}} - U^*_{i_1j_i,\mathbf{k}}\,p^{he}_{j_1i_1,\mathbf{k}}\right]\,. \tag{56}$$

Exciton states

Before investigating the dynamics let us look at the excitonic eigenstates. Since the interband polarization is the electron-hole pair wave function in the $(i, j, \mathbf{k})$-space,[121] this means that we have to look for stationary solutions of Eq. (55). In the low-density limit, i.e. $f^e_{i_1,\mathbf{k}} \approx 0$ and $f^h_{j_1,-\mathbf{k}} \approx 0$, the self-energy matrices are negligible and, making the ansatz

$$p^{he}_{j_1i_1,\mathbf{k}} = \tilde{p}_{j_1i_1,\mathbf{k}}e^{-i\omega t}\,, \tag{57}$$

we obtain the eigenvalue equation for $\tilde{p}_{j_1i_1,\mathbf{k}}$ according to

$$\hbar\omega\tilde{p}_{j_1i_1,\mathbf{k}} = (\epsilon^h_{j_1,-\mathbf{k}} + \epsilon^e_{i_1,\mathbf{k}})\tilde{p}_{j_1i_1,\mathbf{k}} - \sum_{\mathbf{k}'}\sum_{i_2,j_2} V_{i_1j_2j_1i_2}(\mathbf{k}-\mathbf{k}')\tilde{p}_{j_2i_2,\mathbf{k}'}\,. \tag{58}$$

This equation has been solved numerically on a discretized mesh in k-space for the asymmetric double-quantum well structure shown in Fig. 13. In Fig. 14 the lines show the transition energies between the free electron and hole subband edges. The solid and the dashed lines refer to the transition from the highest hole subband to the first and second electron subband, respectively. The anticrossing behavior with a resonance field of 7 kV/cm is clearly visible. The other lines refer to transitions involving higher hole subbands. Again we notice an anticrossing at a field of 6 kV/cm.

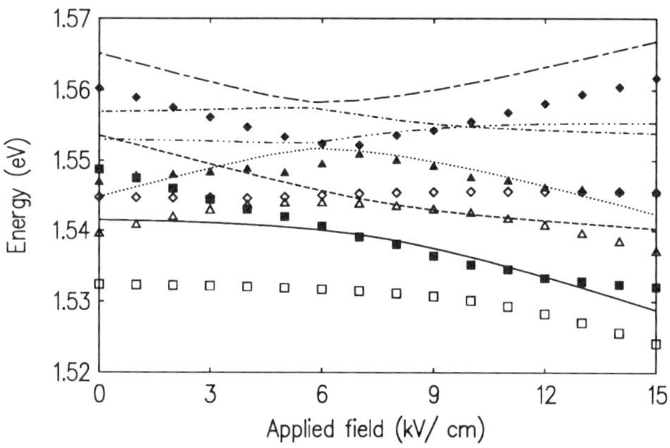

Figure 14. Transition energies between the electron and hole subband edges (lines) and corresponding 1s exciton energies (symbols) calculated for two electron and three hole subbands (see Fig. 13).

The symbols in Fig. 14 refer to the 1s-exciton states corresponding to the transitions*. Let us concentrate on the excitons corresponding to the highest hole subband (open and solid squares). They exhibit a similar behavior to that of the free-carrier energies (solid and dashed lines), however we notice that the resonance is now shifted to a field of 10.5 kV/cm. The reason can be understood by looking at the binding energies. At low field the open square shows a much larger binding energy than the solid square. This is due to the fact that in the former case electron and hole are localized in the same well (direct exciton), while in the latter case they are localized in different wells (indirect exciton) and thus have a reduced Coulomb interaction. On the other hand, at high fields the situation is reversed. This transition leads to the observed shift of the resonance.[59,117,118]

In Fig. 15 we report the six components of the exciton eigenfunctions $p^{he}_{ji,\mathbf{k}}$ as well as the projected pair density in the space of the electron and hole z-coordinate for the lowest exciton state at two values of the applied electric field. The pair density ρ is obtained from the real space exciton wave-function

$$\Phi^{exc}(z_e, z_h; \mathbf{r}) = \frac{1}{\sqrt{A}} \sum_{i_1, j_1} \sum_{\mathbf{k}} \tilde{p}^*_{j_1 i_1, \mathbf{k}} e^{i\mathbf{kr}} \phi_{i_1}(z_e) \phi_{j_1}(z_h) \, , \tag{59}$$

where z_e denotes the electron z–coordinate, z_h the hole z–coordinate and $\mathbf{r}$ the electron–hole

*It should be noted that for the higher states the classification is not unique since these states are resonant with band-to-band transitions. We have plotted those states which exhibit the strongest 1s-like contribution in the wave-function.

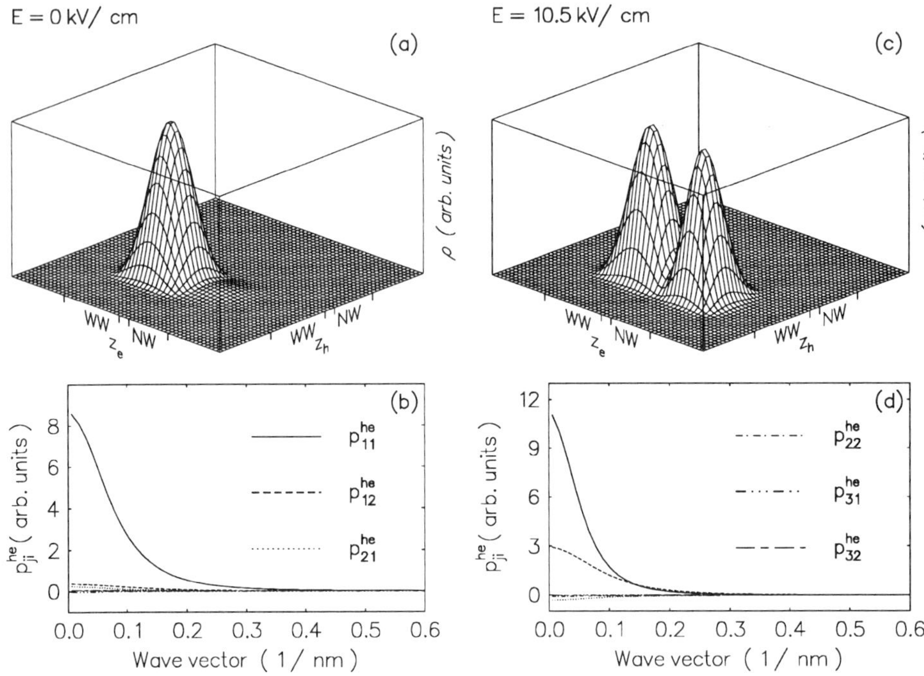

Figure 15. (a), (c) Electron-hole pair-density along the growth direction and (b), (d) exciton wave functions (interband polarizations) for two values of the electric field.

relative coordinate in the x–y–plane by projection onto the (z_e, z_h)-plane according to

$$\rho(z_e, z_h) = \int_A |\Phi^{exc}(z_e, z_h; r)|^2 d^2r = \sum_{\mathbf{k}} | \sum_{i_1, j_1} \tilde{p}^*_{j_1 i_1, \mathbf{k}} \phi_{i_1}(z_e) \phi_{j_1}(z_h)|^2 . \quad (60)$$

At zero field only one component of the wave-function is essentially non-zero; the exciton correspond to one electron and one hole subband. Both electron and hole are localized in the wide well. At the resonance field of 10.5 kV/cm, on the other hand, we find that not only p^{he}_{11} but also p^{he}_{12} is different from zero. This means that the exciton mixes both subbands. As a result the electron is completely delocalized, in contrast to the free carrier states, which are out of resonance at this electric field. The lowest free electron state is already mainly localized in the narrow well.

The excitonic states are eigenstates of the isolated system. In order to compare the results with experiments, we have to calculate observable quantities. From the exciton energies and eigenfunctions it is possible to calculate the absorption spectrum. Here, however, we will choose a different way which is more directly related to the experimental measurement of a photocurrent. We apply the same technique already discussed in the case of a bulk semiconductor: We calculate the density of carriers generated by a 1 ps laser pulse for different values of the photon energy of the laser. The photocurrent will be proportional to this density. The results, obtained by the numerical solution of Eqs. (54) and (55), are shown in Fig. 16 for the same fields as in Fig. 15. At zero field we observe two strong peaks corresponding to the p^{he}_{11} and the p^{he}_{22} exciton which are both direct excitons. At resonance both the p^{he}_{11} and the p^{he}_{12} exciton are delocalized and the oscillator strength is distributed almost equally among both. The highest peak is related to the p^{he}_{31} exciton which is now a direct exciton in the narrow well. The peak positions and heights are in good agreement with

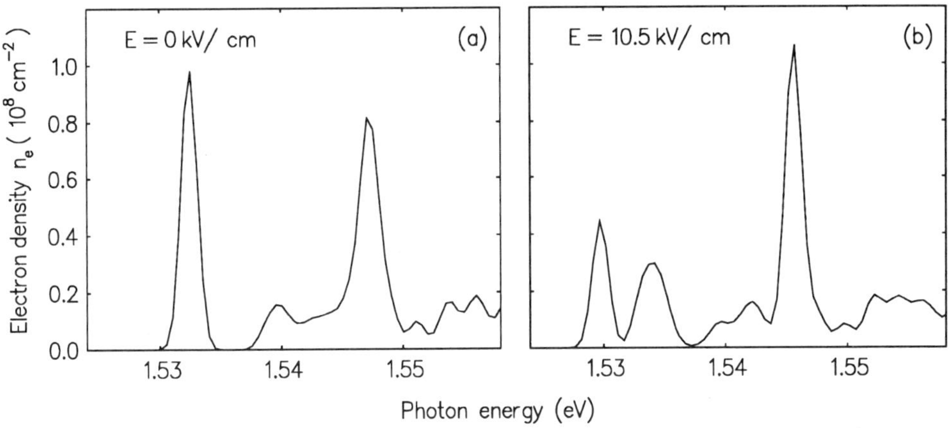

Figure 16. Carrier density generated by a 1 ps pulse as a function of the excess energy for two values of the electric field.

the measured photocurrent[59]. In the experiment there are additional peaks due to light-hole excitons which are not taken into account in the present model.

THz radiation

If the energetic width of a laser pulse is larger than the difference between two states, a coherent superposition of these states is generated. This is the basis of quantum beats which have been observed in a wide variety of systems.[53–56] Quantum beat experiments are typically performed in a four-wave-mixing geometry and thus can be analyzed theoretically in the same way as discussed for the bulk case.[122] Here we will concentrate on a different aspect, which is related to the spatial non-homogeneity of a system. If electron and hole states are localized in different regions in space, the creation of pairs leads to the build-up of a dipole moment. Due to this build-up coherent radiation is emitted, the optical rectification signal.[123] Furthermore, if a superposition of states is created, this dipole moment oscillates leading to the emission of radiation with a frequency determined by the energy splitting.[59,61,67]

Neglecting interband terms, which oscillate with a frequency of the order of the band gap and thus do not contribute to the emission we are interested in, the dipole moment of the system is given by

$$\mathbf{P} = \sum_{i_1,i_2}\sum_{\mathbf{k}} \mathbf{M}^{ee}_{i_1 i_2}\left\langle c^\dagger_{i_1,\mathbf{k}} c_{i_2,\mathbf{k}}\right\rangle - \sum_{j_1,j_2}\sum_{\mathbf{k}} \mathbf{M}^{hh}_{j_1 j_2}\left\langle d^\dagger_{j_2,\mathbf{k}} d_{j_1,\mathbf{k}}\right\rangle \tag{61}$$

with the intraband dipole matrix elements

$$\mathbf{M}^{ee}_{i_1 i_2} = e\int \phi^{e*}_{i_1}(z) z \phi^{e}_{i_2}(z) dz\, \mathbf{e}_z \quad \text{and} \quad \mathbf{M}^{hh}_{j_1 j_2} = e\int \phi^{h*}_{j_1}(z) z \phi^{h}_{j_2}(z) dz\, \mathbf{e}_z\,. \tag{62}$$

The dipole moment can be decomposed into a diagonal ($P^{(d)}_z$) and an off-diagonal ($P^{(od)}_z$) part according to.

$$P^{(d)}_z = \sum_{i_1}\sum_{\mathbf{k}} M^{ee}_{i_1 i_1} f^e_{i_1,\mathbf{k}} - \sum_{j_1}\sum_{\mathbf{k}} M^{hh}_{j_1 j_1} f^h_{j_1,\mathbf{k}}\,, \tag{63}$$

$$P^{(od)}_z = \sum_{i_1\neq i_2}\sum_{\mathbf{k}} M^{ee}_{i_1 i_2} p^{ee}_{i_1 i_2,\mathbf{k}} - \sum_{j_1\neq j_2}\sum_{\mathbf{k}} M^{hh}_{j_2 j_1} p^{hh}_{j_1 j_2,\mathbf{k}}\,. \tag{64}$$

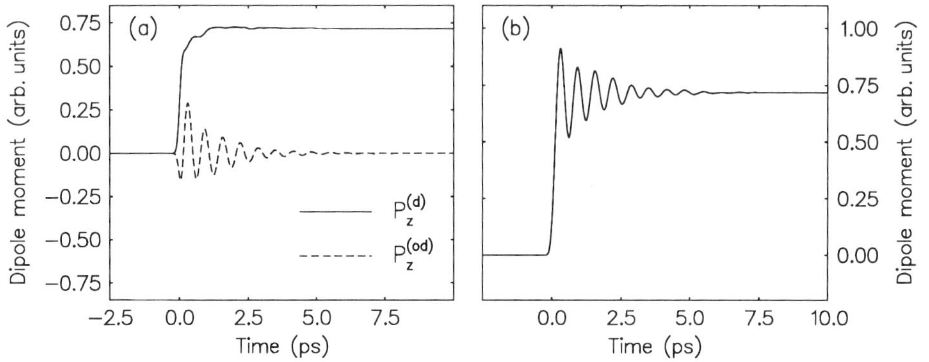

Figure 17. Dipole moment of the asymmetric double-quantum well system as a function of time for the case of excitation with a 150 fs pulse at 1.54 eV. (a) Diagonal and off-diagonal contributions separated, (b) total dipole moment.

The diagonal part gives rise to the rectification signal while the off-diagonal part leads to an oscillating emission.

Therefore, in order to calculate the dipole moment, we need, in addition to Eqs. (54) and (55), equations of motion for the intraband polarizations. Limiting ourselves again to the low-density cases, they are given by

$$\begin{aligned}\frac{d}{dt}p^{ee}_{i_1 i_2,\mathbf{k}} &= \frac{1}{i\hbar}\left(\epsilon^{e}_{i_2,\mathbf{k}} - \epsilon^{e}_{i_1,\mathbf{k}}\right) p^{ee}_{i_1 i_2,\mathbf{k}} \\ &+ \frac{1}{i\hbar}\sum_{j_1}\left[U_{i_2 j_1,\mathbf{k}}\, p^{he*}_{j_1 i_1,\mathbf{k}} - U^{*}_{i_1 j_1,\mathbf{k}}\, p^{he}_{j_1 i_2,\mathbf{k}}\right] , \qquad (65)\\ \frac{d}{dt}p^{hh}_{j_1 j_2 -\mathbf{k}} &= \frac{1}{i\hbar}\left(\epsilon^{h}_{j_2,-\mathbf{k}} - \epsilon^{h}_{j_1,-\mathbf{k}}\right) p^{hh}_{j_1 j_2,-\mathbf{k}} \\ &+ \frac{1}{i\hbar}\sum_{i_1}\left[U_{i_1 j_2,\mathbf{k}}\, p^{he*}_{j_1 i_1,\mathbf{k}} - U^{*}_{i_1 j_1,\mathbf{k}}\, p^{he}_{j_2 i_1,\mathbf{k}}\right] . \qquad (66)\end{aligned}$$

In Fig. 17 the diagonal and off-diagonal parts, as well as the total dipole moment are plotted as functions of time for the case of an applied electric field of 10.5 kV/cm and an excitation by a 160 fs pulse centered at 1.54 eV. The off-diagonal part exhibits oscillations and eventually decays due to dephasing processes which are described here by a phenomenological dephasing time T_2 = 3 ps. The diagonal part approaches a finite value due to the different localization of electron and hole wave functions.

Figure 18 shows the emitted signal as a function of time, and the corresponding spectrum. The temporal signal is given by the second derivative of the dipole moment. The dominant contribution to the spectrum comes from the oscillating part, the rectification part gives only a very small background. Before comparing quantitatively the calculated shapes with the experiment, the response function of the detector should be taken into account, which is not done in Fig. 18. However, it is interesting to notice that the calculated oscillation frequency of 1.56 THz corresponding to a splitting of 6.5 meV is in good agreement with the experiment.[59] It is remarkably larger than the splitting of the two lowest exciton states (4.5 meV) and demonstrates the important role played by higher excitonic states which are not taken into account in an analysis based on a three-level model.

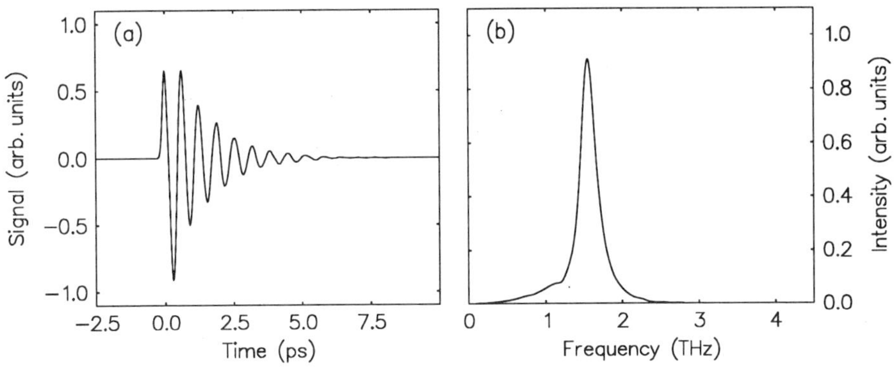

Figure 18. Emitted signal (a) as a function of time and (b) as a function of frequency obtained from the dipole moment shown in Fig. 17.

CONCLUSIONS

We have presented an analysis of the coherent dynamics in bulk semiconductors and heterostructures. The theory has been based on the semiconductor Bloch equations which have been solved numerically. In the first section we have discussed the difference between the coherent and the incoherent description of the generation process and its implications for the subsequent relaxation. Then, we have presented theoretical and experimental results concerning the four-wave-mixing in bulk GaAs. For excitations close to the band gap many-body effects are found to play an important role. They manifest themselves in the temporal as well as in the spectral shape of the diffracted signal. In particular it turned out that, due to nonequilibrium effects, the exciton dominates the spectrum up to a density much larger than the equilibrium Mott density.

Finally, we have investigated the coherent dynamics in asymmetric double-quantum well systems. For varying electric field excitons change their character from direct to indirect, the resonance field however is shifted with respect to the resonance field for free-carrier states. For the case where the laser pulse generates carriers in a superposition of states we have calculated the coherent emission of THz-radiation. The frequency is shifted with respect to the splitting of the lowest exciton states in good agreement with experimental findings.

ACKNOWLEDGMENTS

This work has been supported in part by the Deutsche Forschungsgemeinschaft (SFB 329). Valuable discussions with G. Mahler are gratefully acknowledged.

REFERENCES

1. E. L. Hahn, Spin echoes, Phys. Rev. **80**, 580 (1950).
2. N. A. Kurnit, I. D. Abella, and S. R. Hartmann, Observation of a photon echo, Phys. Rev. Lett. **13**, 567 (1964).
3. I. D. Abella, N. A. Kurnit, and S. R. Hartmann, Photon echoes, Phys. Rev. **141**, 391 (1966).
4. J. Shah and R. C. C. Leite, Radiative recombination from photoexcited hot carriers in GaAs, Phys. Rev. Lett. **22**, 1304 (1969).

5. C. V. Shank, R. L. Fork, R. F. Leheny, and J. Shah, Dynamics of photoexcited GaAs band-edge absorption with subpicosecond resolution, Phys. Rev. Lett. **42**, 112 (1979).
6. R. L. Fork, C. H. Brito Cruz, P. C. Becker, and C. V. Shank, Compression of optical pulses to six femtoseconds by using cubic phase compensation, Optics Lett. **12**, 483 (1987).
7. J. Shah, Photoexcited hot carriers: From cw to 6 fs in 20 years, Solid State Electron. **32**, 1051 (1989).
8. J. Shah, B. Deveaud, T. C. Damen, W. T. Tsang, A. C. Gossard, and P. Lugli, Determination of intervalley scattering rates in GaAs by subpicosecond luminescence spectroscopy, Phys. Rev. Lett. **59**, 2222 (1987).
9. H. J. Polland, W. W. Rühle, H. J. Queisser, and K. Ploog, Fröhlich interaction in two-dimensional GaAs/AlGaAs systems, Phys. Rev. B **36**, 7722 (1987).
10. K. Leo, W. W. Rühle, H. J. Queisser, and K. Ploog, Reduced dimensionality of hot-carrier relaxation in GaAs quantum wells, Phys. Rev. B **37**, 7121 (1988).
11. J. F. Ryan and M. Tatham, Picosecond optical studies of 2D electron - 2D phonon dynamics, Solid State Electron. **32**, 1429 (1989).
12. T. Elsaesser, J. Shah, L. Rota, and P. Lugli, Initial thermalization of photoexcited carriers in GaAs studied by femtosecond luminescence spectroscopy, Phys. Rev. Lett. **66**, 1757 (1991).
13. L. Rota, P. Lugli, T. Elsaesser, and J. Shah, Ultrafast thermalization of photoexcited carriers in polar semiconductors, Phys. Rev. B **47**, 4226 (1993).
14. H. Kurz, Femtosecond spectroscopy of hot carrier relaxation in bulk semiconductors, Semicond. Sci. Technol. **7**, B124 (1992).
15. R. G. Ulbrich, J. A. Kash, and J. C. Tsang, Hot-electron recombination at neutral acceptors in GaAs: A cw probe of femtosecond intervalley scattering, Phys. Rev. Lett. **62**, 949 (1989).
16. J. A. Kash, Carrier-carrier scattering in GaAs: Quantitative measurements from hot (e,A^0) luminescence, Phys. Rev. B **40**, 3455 (1989).
17. D. W. Snoke, W. W. Rühle, Y.-C. Lu, and E. Bauser, Nonthermalized distributions of electrons on picosecond time scale in GaAs, Phys. Rev. Lett. **68**, 990 (1992).
18. C. L. Peterson and S. A. Lyon, Observation of hot-electron energy loss through the emission of phonon-plasmon coupled modes in GaAs, Phys. Rev. Lett. **65**, 760 (1990).
19. J. L. Oudar, A. Migus, D. Hulin, G. Grillon, J. Etchepare, and A. Antonetti, Femtosecond orientational relaxation of photoexcited carriers in GaAs, Phys. Rev. Lett. **53**, 384 (1984).
20. H. Roskos, B. Rieck, A. Seilmeier, and W. Kaiser, Carrier cooling in nonpolar semiconductors studied with subpicosecond time-resolution, Solid State Electron. **32**, 1437 (1989).
21. C. W. W. Bradley, R. A. Taylor, and J. F. Ryan, Femtosecond electron and hole thermalization in AlGaAs, Solid State Electron. **32**, 1173 (1989).
22. T. Gong, P. M. Fauchet, J. F. Young, and P. J. Kelly, Femtosecond gain dynamics due to initial thermalization of hot carriers injected at 2 eV in GaAs, Phys. Rev. B **44**, 6542 (1991).
23. P. M. Fauchet and T. Gong, Femtosecond dynamics of hot-carriers in GaAs, in *Ultrafast Lasers Probe Phenomena in Semiconductors and Superconductors*, edited by R. Alfano, SPIE Proc. Vol. 1677 (SPIE, 1992) p. 25.
24. J.-P. Foing, D. Hulin, M. Joffre, M. K. Jackson, J.-L. Oudar, C. Tanguy, and M. Combescot, Absorption edge singularities in highly excited semiconductors, Phys. Rev. Lett. **68**, 110 (1992).
25. W. Pötz and P. Kocevar, Electronic power transfer in pulsed laser excitation of polar semiconductors, Phys. Rev. B **28**, 7040 (1983).
26. T. F. Zheng, W. Cai, P. Hu, and M. Lax, Simulation of ultrafast relaxation of photoexcited electrons via analytical distribution functions, Solid State Electron. **32**, 1089 (1989).
27. K. Leo and J. H. Collet, Influence of electron-hole scattering on the plasma thermalization in doped GaAs, Phys. Rev. B **44**, 5535 (1991).
28. A. A. Grinberg and S. Luryi, Nonstationary quasiperiodic energy distribution of an electron gas upon ultrafast thermal excitation, Phys. Rev. Lett. **65**, 1251 (1990).
29. A. A. Grinberg, S. Luryi, N. L. Schryer, R. K. Smith, C. Lee, U Ravaioli, and E. Sangiorgi, Adiabatic approach to the dynamics of nonequilibirium electron ensembles in semiconductors, Phys. Rev. B **44**, 10536 (1991).
30. K. El Sayed, T. Wicht, H. Haug, and L Bányai, Study of the Coulomb Boltzmann kinetics in a

quasi-twodimensional electron gas by eigenfunction expansions and Monte Carlo simulations, Z. Phys. B **86**, 345 (1992).
31. J. Collet and T. Amand, Athermal and thermal relaxation of high density electron-hole plasma in GaAs, Physica **134B**, 394 (1985).
32. R. Binder, D. Scott, A. E. Paul, M. Lindberg, K. Henneberger, and S. W. Koch, Carrier-carrier scattering and optical dephasing in highly excited semiconductors, Phys. Rev. B **45**, 1107 (1992).
33. M. A. Osman and D. K. Ferry, Monte Carlo investigation of the electron-hole-interaction effects on the ultrafast relaxation of hot photoexcited carriers in GaAs, Phys. Rev. B **36**, 6018 (1987).
34. P. Lugli, P. Bordone, S. Gualdi, P. Poli, and S. M. Goodnick, Hot phonons in quantum well systems, Solid-State Electron. **32**, 1881 (1989).
35. P. Lugli, P. Bordone, L. Reggiani, M. Rieger, P. Kocevar, and S. M. Goodnick, Monte Carlo studies of nonequilibrium phonon effects in polar semiconductors and quantum wells. I. Laser photoexcitation, Phys. Rev. B **39**, 7852 (1989).
36. C. J. Stanton, D. W. Bailey, and K. Hess, Monte Carlo modeling of femtosecond relaxation processes in AlGaAs/GaAs quantum wells, IEEE J. Quantum Electron. **QE-24**, 1614 (1988).
37. C. J. Stanton, D. W. Bailey, and K. Hess, Femtosecond pump, continuum probe nonlinear absorption in GaAs, Phys. Rev. Lett. **65**, 231 (1990).
38. D. W. Bailey, M. A. Artaki, C. J. Stanton, and K. Hess, Ensemble Monte Carlo simulations of femtosecond thermalization of low-energy photoexcited electrons in GaAs quantum wells, J. Appl. Phys. **62**, 4638 (1987).
39. D. K. Ferry, A. M. Kriman, H. Hida, and S. Yamaguchi, Collision retardation and its role in femtosecond-laser excitation of semiconductor plasmas, Phys. Rev. Lett. **67**, 633 (1991).
40. D. W. Bailey, C. J. Stanton, and K. Hess, Numerical studies of femtosecond carrier dynamics in GaAs, Phys. Rev. B **42**, 3423 (1990).
41. U. Hohenester, P. Supancic, P. Kocevar, X. Q. Zhou, W. Kütt, and H. Kurz, Subpicosecond thermaliztion and relaxation of highly photoexcited electrons and holes in intrinsic and p-type gaas and inp, Phys. Rev. B **47**, 13233 (1993).
42. A. Mysyrowicz, D. Hulin, A. Antonetti, A. Migus, W. T. Masselink, and H. Morkoç, "Dressed excitons" in a multiple-quantum-well structure: Evidence for an optical Stark effect with femtosecond response time, Phys. Rev. Lett. **56**, 2748 (1986).
43. W. H. Knox, D. S. Chemla, D. A. B. Miller, J. B. Stark, and S. Schmitt-Rink, Femtosecond ac Stark effect in semiconductor quantum wells: Extreme low- and high-intensity limits, Phys. Rev. Lett. **1989**, 1189 (1989).
44. P. C. Becker, H. L. Fragnito, C. H. Brito Cruz, R. L. Fork, J. E. Cunningham, J. E. Henry, and C. V. Shank, Femtosecond photon echoes from band-to-band transitions in GaAs, Phys. Rev. Lett. **61**, 1647 (1988).
45. Y. Masumoto, S. Shionoya, and T. Takagahara, Optical dephasing of excitonic polaritons in CuCl studied by time-resolved, nondegenerate four-wave mixing, Phys. Rev. Lett. **51**, 923 (1983).
46. L. Schultheis, M. D. Sturge, and J. Hegarty, Photon echoes from two-dimensional excitons in GaAs-AlGaAs quantum wells, Appl. Phys. Lett. **47**, 995 (1985).
47. L. Schultheis, J. Kuhl, A. Honold, and C. W. Tu, Picosecond phase coherence and orientational relaxation of excitons in GaAs, Phys. Rev. Lett. **57**, 1797 (1986).
48. A. Honold, L. Schultheis, J. Kuhl, and C. W. Tu, Reflected degenerate four-wave mixing on GaAs single quantum wells, Appl. Phys. Lett. **52**, 2105 (1988).
49. J. Kuhl, A. Honold, L. Schultheis, and C. W. Tu, Optical dephasing and orientational relaxation of Wannier excitons and free carriers in GaAs and GaAs/AlGaAs quantum wells, Adv. in Solid State Phys. **29**, 157 (1989).
50. E. O. Göbel, Ultrafast spectroscopy of semiconductors, Adv. in Solid State Phys. **30**, 269 (1990).
51. K. Leo, E. O. Göbel, T. C. Damen, J. Shah, S. Schmitt-Rink, W. Schäfer, J. F. Müller, K. Köhler, and P. Ganser, Subpicosecond four-wave mixing in GaAs/AlGaAs quantum wells, Phys. Rev. B **44**, 5726 (1991).
52. G. Noll, U. Siegner, S. G. Shevel, and E. O. Göbel, Picosecond stimulated photon echo due to

intrinsic excitations in semiconductor mixed crystals, Phys. Rev. Lett. **64**, 792 (1990).
53. D. Fröhlich, A. Kulik, B. Uebbing, A. Mysyrowicz, V. Langer, H. Stolz, and W. von der Osten, Coherent propagation and quantum beats of quadrupole polaritons in CuO, Phys. Rev. Lett. **67**, 2343 (1991).
54. E. O. Göbel, K. Leo, T. C. Damen, J. Shah, S. Schmitt-Rink, W. Schäfer, J. F. Müller, and K. Köhler, Quantum beats of excitons in quantum wells, Phys. Rev. Lett. **64**, 1801 (1990).
55. V. Langer, H. Stolz, and W. von der Osten, Observation of quantum beats in the resonance fluorescence of free excitons, Phys. Rev. Lett. **64**, 854 (1990).
56. K. Leo, T. C. Damen, J. shah, E. O. Göbel, and K. Köhler, Quantum beats of light hole and heavy hole excitons in quantum wells, Appl. Phys. Lett. **57**, 19 (1990).
57. K. Leo, J. Shah, E. O. Göbel, T. C. Damen, K. Köhler, and P. Ganser, Tunneling in semiconductor heterostructures studied by subpicosecond four-wave mixing, Appl. Phys. Lett. **56**, 2031 (1990).
58. K. Leo, Dynamics of wavepackets in GaAs/AlGaAs heterostructures, Adv. in Solid State Phys. **32**, 97 (1992).
59. H. G. Roskos, M. C. Nuss, J. Shah, K. Leo, D. A. B. Miller, A. M. Fox, S. Schmitt-Rink, and K. Köhler, Coherent submillimeter-wave emission from charge oscillations in a double-well potential, Phys. Rev. Lett. **68**, 2216 (1992).
60. J. Feldmann, Bloch oscillations in a semiconductor superlattice, Adv. in Solid State Phys. **32**, 81 (1992).
61. C. Waschke, H. G. Roskos, R. Schwedler, K. Leo, H. Kurz, and K. Köhler, Coherent submilimeter-wave emission from Bloch oscillations in a semiconductor superlattice, Phys. Rev. Lett. **70**, 3319 (1993).
62. K. Leo, M. Wegener, J. Shah, D. S. Chemla, E. O. Göbel, T. C. Damen, S. Schmitt-Rink, and W. Schäfer, Effects of coherent polarization interactions on time-resolved degenerate four-wave mixing, Phys. Rev. Lett. **65**, 1340 (1990).
63. S. Weiss, M.-A. Mycek, J.-Y. Bigot, S. Schmitt-Rink, and D. S. Chemla, Collective effects in excitonic free induction decay: Do semiconductors and atoms emit coherent light in different ways?, Phys. Rev. Lett. **69**, 2685 (1992).
64. D.-S. Kim, J. Shah, J. E. Cunningham, T. C. Damen, W. Schäfer, M. Hartmann, and S. Schmitt-Rink, Giant excitonic resonance in time-resolved four-wave mixing in quantum wells, Phys. Rev. Lett. **68**, 1006 (1992).
65. D.-S. Kim, J. Shah, T. C. Damen, W. Schäfer, F. Jahnke, S. Schmitt-Rink, and K. Köhler, Unusually slow temporal evolution of femtosecond four-wave-mixing signals in intrinsic GaAs quantum wells: Direct evidence for the dominance of interaction effects, Phys. Rev. Lett. **69**, 2725 (1992).
66. A. Lohner, K. Rick, P. Leisching, A. Leitenstorfer, T. Elsaesser, T. Kuhn, F. Rossi, and W. Stolz, Coherent optical polarization of bulk GaAs studied by femtosecond photon-echo spectroscopy, Phys. Rev. Lett. **71**, 77 (1993).
67. P. C. M. Planken, M. C. Nuss, I. Brener, K. W. Goossen, M. S. C. Luo, S. L. Chuang, and L. Pfeiffer, Terahertz emission in single quantum wells after coherent optical excitation of light hole and heavy hole excitons, Phys. Rev. Lett. **69**, 3800 (1992).
68. C. Comte and G. Mahler, Dynamic Stark effect in interacting electron-hole systems: Light enhanced excitons, Phys. Rev. B **34**, 7164 (1986).
69. C. Comte and G. Mahler, Excitonic reference state of a model semiconductor in the dynamic Stark regime, Phys. Rev. B **38**, 10517 (1988).
70. R. Zimmermann, K. Kilimann, W. D. Kraeft, D. Kremp, and G. Röpke, Dynamical screening and self-energy of excitons in the electron-hole plasma, phys. stat. sol (b) **90**, 175 (1978).
71. R. Zimmermann, On the dynamical stark effect of excitons: The low field limit, phys. stat. sol (b) **146**, 545 (1988).
72. S. Schmitt-Rink, C. Ell, and H. Haug, Many-body effects in the absorption, gain, and luminescence spectra of semiconductor quantum well structures, Phys. Rev. B **33**, 1183 (1986).
73. S. Schmitt-Rink, D. S. Chemla, and H. Haug, Nonequilibrium theory of the optical Stark effect and spectral hole burning in semiconductors, Phys. Rev. B **37**, 941 (1988).
74. H. Haug, Microscopic theory of the optical band edge nonlinearities, in *Optical Nonlinearities*

and Instabilities in Semiconductors, edited by H. Haug (Academic, San Diego, 1988) p. 53.
75. F. Bechstedt and S. Glutsch, Non-equilibrium screening and plasmons in a coherently pumped semiconductor, J. Phys.: Condens. Matter **3**, 7145 (1991).
76. M. Hartmann, H. Stolz, and R. Zimmermann, Kinetics of screening in optically excited semiconductors, phys. stat. sol (b) **159**, 35 (1990).
77. K. Henneberger, W. Schäfer, and F. Jahnke, Optical and transport nonlinearities in laser excited semiconductors, Physica Scripta **T35**, 129 (1991).
78. K. Henneberger and H. Haug, Nonlinear optics and transport in laser-excited semiconductors, Phys. Rev. B **38**, 9759 (1988).
79. A. V. Kuznetsov, Interaction of ultrashort light pulses with semiconductors: Effective Bloch equations with relaxation and memory effects, Phys. Rev. B **44**, 8721 (1991).
80. D. B. Tran Thoai and H. Haug, Coulomb quantum kinetics in pulse-excited semiconductors, Z. Phys. B **91**, 199 (1993).
81. H. Haug and C. Ell, Coulomb quantum kinetics in a dense electron gas, Phys. Rev. B **46**, 2126 (1992).
82. W. Schäfer, Theory of dense nonequilibrium exciton systems, in *Optical Nonlinearities and Instabilities in Semiconductors*, edited by H. Haug (Academic, San Diego, 1988) p. 133.
83. I. Balslev, R. Zimmermann, and A. Stahl, Two-band density-matrix approach to nonlinear optics of excitons, Phys. Rev. B **40**, 4095 (1989).
84. A. Stahl, RPA-dynamics of the electronic density matrix in a two-band semiconductor, Z. Phys. B **72**, 371 (1988).
85. A. Stahl, Coupled two-level systems and the dynamics of semiconductor electrons, phys. stat. sol (b) **159**, 327 (1990).
86. R. Zimmermann, Theory of dephasing in semiconductor optics, phys. stat. sol (b) **173**, 129 (1992).
87. S. Schmitt-Rink and D. S. Chemla, Collective excitations and the dynamical Stark effect in a coherently driven exciton system, Phys. Rev. Lett. **57**, 2752 (1986).
88. M. Lindberg and S. W. Koch, Theory of the optical Stark effect in semiconductors under ultrashort-pulse excitation, phys. stat. sol (b) **150**, 379 (1988).
89. M. Lindberg and S. W. Koch, Effective Bloch equations for semiconductors, Phys. Rev. B **38**, 3342 (1988).
90. W. Schäfer, F. Jahnke, and S. Schmitt-Rink, Many-particle effects on transient four-wave-mixing signals in semiconductors, Phys. Rev. B **47**, 1217 (1993).
91. E. Heiner, Evolution equations for highly excited direct semiconductors in the athermal stage, phys. stat. sol (b) **146**, 655 (1988).
92. E. Heiner and W. Kleinig, Solutions of optical Bloch equations for highly excited direct gap semiconductors in a time-dependent mean-field approach, Physica Scripta **46**, 88 (1992).
93. R. Zimmermann and M . Hartmann, Resonant and off-resonant light-matter interaction in semiconductors, phys. stat. sol (b) **150**, 365 (1988).
94. M. Wegener, D. S. Chemla, S. Schmitt-Rink, and W. Schäfer, Line shape of time-resolved four-wave mixing, Phys. Rev. A **42**, 5675 (1990).
95. T. Kuhn and F. Rossi, Analysis of coherent and incoherent phenomena in photoexcited semiconductors: A Monte Carlo approach, Phys. Rev. Lett. **69**, 977 (1992).
96. T. Kuhn and F. Rossi, Monte Carlo simulation of ultrafast processes in photoexcited semiconductors: Coherent and incoherent dynamics, Phys. Rev. B **46**, 7496 (1992).
97. L. P. Kadanoff and G. Baym, *Quantum Statistical Mechanics* (Benjamin, New York, 1962).
98. E. M. Lifshitz and L. P. Pitaevskii, *Physical Kinetics* (Pergamon, Oxford, 1981).
99. C. Jacoboni and L. Reggiani, The Monte Carlo method for the solution of charge transport in semiconductors with applications to covalent material, Rev. Mod. Phys. **55**, 645 (1983).
100. C. Jacoboni and P. Lugli, *The Monte Carlo Method for Semiconductor Device Simulations* (Springer, Wien, 1989).
101. R. Zimmermann, Transverse relaxation and polarizations specifics in the dynamical Stark effect, phys. stat. sol (b) **159**, 317 (1990).
102. R. Zimmermann, Carrier kinetics for ultrafast optical pulses, J. Lumin. **53**, 187 (1992).

103. R. Zimmermann and J. Wauer, Non-Markovian relaxation in semiconductors: An exactly soluble model, Proc. DPC (Boston, 1993), to be published in J. Lumin.
104. D. B. Tran Thoai and H. Haug, Band-edge quantum kinetics for coherent ultrashort-pulse spectroscopy in polar semiconductors, Phys. Rev. B **47**, 3574 (1993).
105. F. Rossi, T. Kuhn, J. Schilp, and E. Schöll, Analysis of the coupled coherent and incoherent dynamics in photoexcited semicondutors: A Monte Carlo approach, in *Proc. 21st ICPS, Beijing, China*, edited by P. Jiang and H. Zheng (World Scientific, Singapore, 1992) p. 165.
106. A. V. Kuznetsov, Coherent and non-Markovian effects in ultrafast relaxation of photoexcited hot carrier: A model study, Phys. Rev. B **44**, 13381 (1991).
107. J. Schilp, T. Kuhn, and G. Mahler, Energy relaxation and dephasing of photoexcited carriers: Memory effects and cross terms between different interactions, Proc. 8th HCIS (Oxford, 1993), to be published in Semicond. Sci. Technol.
108. G. C. Cho, W. Kütt, and H. Kurz, Subpicosecond time-resolved coherent-phonon oscillations in GaAs, Phys. Rev. Lett. **65**, 764 (1990).
109. W. Kütt, Coherent phonons in III-V-compounds, Adv. in Solid State Phys. **32**, 113 (1992).
110. H. Haug and S. Schmitt-Rink, Electron theory of the optical properties of laser-excited semiconductors, Prog. Quant. Electr. **9**, 3 (1984).
111. D. C. Scott, R. Binder, and S. W. Koch, Ultrafast dephasing through acoustic plasmon undamping in nonequilibrium electron-hole plasmas, Phys. Rev. Lett. **69**, 347 (1992).
112. M. Lindberg, R. Binder, and S. W. Koch, Theory of the semiconductor photon echo, Phys. Rev. A **45**, 1865 (1992).
113. F. Rossi, S. Haas, and T. Kuhn, Analysis of coherent and incoherent ultrafast dynamics in photoexcited semiconductors: A Monte Carlo approach, Proc. 8th HCIS (Oxford, 1993), to be published in Semicond. Sci. Technol.
114. T. Yajima and Y. Taira, Spatial optical parametric coupling of picosecond light pulses and transverse relaxation effect in resonant media, J. Phys. Soc. Jpn. **47**, 1620 (1979).
115. J. R. Kuklinski and S. Mukamel, Generalized semiconductor Bloch equations: Local fields and transient gratings, Phys. Rev. B **44**, 11253 (1991).
116. R. G. Ulbrich, Dense nonequilibrium excitations: Band edge absorption spectra of highly excited Gallium Arsenide, in *Optical Nonlinearities and Instabilities in Semiconductors*, edited by H. Haug (Academic, San Diego, 1988) p. 121.
117. G. Bastard, C. Delalande, R. Ferreira, and H. W. Liu, Assisted relaxation and vertical transport of electrons, holes and excitons in semiconductor heterostructures, J. Lumin. **44**, 247 (1989).
118. A. M. Fox, D. A. B. Miller, G. Livescu, J. E. Cunningham, and W. Y. Jan, Excitonic effects in coupled quantum wells, Phys. Rev. B **44**, 6231 (1991).
119. A. P. Heberle, W. W. Rühle, M. G. W. Alexander, and K. Köhler, Resonances in tunneling between quantum wells, Semcond. Sci. Technol. **7**, B421 (1992).
120. R. Ferreira, P. Rolland, Ph. Roussignol, C. Delalande, A. Vinattieri, L. Carraresi, M. Colocci, N. Roy, B. Sermage, J. F. Palmier, and B. Etienne, Time-resolved exciton transfer in GaAs/AlGaAs double-quantum-well structures, Phys. Rev. B **45**, 11782 (1992).
121. S.-L. Chuang, S. Schmitt-Rink, D. A. B. Miller, and D. S. Chemla, Exciton Green's-function approach to optical absorption in a quantum well with an applied electric field, Phys. Rev. B **43**, 1500 (1991).
122. Y. Z. Hu, R. Binder, and S. W. Koch, Photon echo and valence-band mixing in semiconductor quantum wells, Phys. Rev. B **47**, 15679 (1993).
123. S. L. Chuang, S. Schmitt-Rink, B. I. Greene, P. N. Saeta, and A. F. J. Levi, Optical rectification at semiconductor surfaces, Phys. Rev. Lett. **68**, 102 (1992).

MANY-BODY THEORY OF COHERENT OPTICAL EFFECTS IN SEMICONDUCTORS

R. Binder, Y.Z. Hu, A. Knorr, M. Lindberg, and S.W. Koch

Optical Sciences Center
and Physics Department
University of Arizona
Tucson, Arizona 85721

ABSTRACT

An overview of the physics and the microscopic theory of coherent light-matter interaction in semiconductors and semiconductor heterostructures is presented. After an introduction of the basic concepts, a series of ultrafast coherent phenomena is reviewed, including the optical Stark effect, photon echoes, Rabi oscillations, quantum beats, multi-wave mixing, and pulse propagation effects.

INTRODUCTION

It is generally believed that most coherent optical effects known from two-level atoms [1] have analogies in semiconductors. Theoretical calculations, however, predict significant differences in several cases. For example, resonant excitation of the exciton resonance leads to oscillations in the electron-hole density (generalized Rabi oscillations), where the oscillation frequency is approximately twice that of the atomic system [2]. This doubling of the Rabi frequency is a direct consequence of the Coulomb interaction between the charge carriers, which results in an induced optical field. The induced field also modifies significantly two-pulse photon echo, four-wave mixing, free-induction decay, and pulse propagation effects in semiconductors [3] - [7].

In this review we present an overview of the theoretical analysis of coherent optical phenomena in semiconductors. After an outline of the general theoretical framework we first discuss the ultrafast optical response for excitation below the exciton resonance (optical Stark effect). We then study generalized Rabi oscillations for on-resonant excitation and discuss the temporal behavior of four-wave mixing photon echo signals for a variety of different excitation conditions. Finally, we analyze femtosecond pulse propagation effects for resonant and nonresonant excitation conditions.

For the calculations we use, as a first step, the semiconductor Bloch equations for the model of a two-band semiconductor [8] - [17]. These equations not only describe the dynamical behavior of the light-induced polarization in semiconductors analogous to the well-known atomic two-level dynamics [1] but also include the Coulomb-interaction and the excitation-induced many-body effects in the quasi-statically screened Hartree-Fock approximation.

The two-band approximation reduces the complexity of numerical calculations while it keeps most of the physical insight into many-body effects such as the renormalizations of the Rabi frequency (induced material polarization) and of the transition energy (band gap). However, since all multi-valence-band and valence-band mixing effects [18] - [23] are completely neglected such studies cannot be used to analyze the dependence of the quantum-well response on the polarizations of the exciting light fields [24] - [30]. We will therefore extend the two-band analysis to include the valence-band mixing and the quantum-confinement induced bandstructure modifications.

The investigation of the light-polarization-dependent FWM and photon echo signals not only allows for the investigation of purely coherent phenomena, but also furnishes a sensitive tool to determine ultrafast incoherent scattering and screening processes. Therefore, we include in the theory dephasing and scattering terms, which are formally beyond the Hartree-Fock approximation. We show that the dephasing due to pump excited carriers has unique signatures in polarization dependent FWM signals [31].

In addition to these pump-probe and light-scattering studies, we also discuss resonant and nonresonant pulse propagation in semiconductors and semiconductor amplifiers. Whereas lossless propagation and self-induced transparency (SIT) are well studied in two-level systems, true SIT has not been found in interacting many-body systems. Our studies for resonant pulse propagation show that even for idealized conditions only incomplete SIT can be expected in semiconductors, as a consequence of the incomplete Rabi oscillations [7, 4]. However, several aspects of SIT are preserved, such as very long propagation distances [32] of strong pulses, as well as pulse break-up. For short pulses propagating in inverted semiconductors, such as semiconductor amplifiers, our studies yield not only the expected pulse amplification, but for stronger pulses we predict loss- and gainless propagation or even net absorption [33].

THEORETICAL BASIS

In this section we briefly review the theoretical framework for the description of coherent phenomena in optically excited semiconductors. We present the semiconductor Bloch equations (SBE) for a system with a general valence-band structure and parabolic conduction bands, and assume throughout this article that the system is spatially homogenous. We first evaluate the SBE to analyze four-wave-mixing (FWM) phenomena. At the end of this section we also include the Maxwell equations in the form of a propagation equation for the optical-field amplitude in order to describe pulse propagation effects in semiconductors.

To theoretically study the optical excitation of a semiconductor with zincblende structure we use the Hamiltonian

$$H = H_e + H_h + H_{int} + H_{ext}. \tag{1}$$

Here, the electron contribution is

$$H_e = \sum_{\vec{k},s} \varepsilon_k^e \, a^\dagger_{\vec{k},s} a_{\vec{k},s} \, , \tag{2}$$

where $\vec{k}$ is the Bloch wavevector and the kinetic energy $\varepsilon_k^e = \hbar^2 k^2/2m_e$ includes the effective electron mass m_e. The hole Hamiltonian is [18, 19, 34]

$$H_h = \sum_{\vec{k},jj'} T^h_{jj'}(\vec{k}) \, b^\dagger_{\vec{k},j} b_{\vec{k},j'} \, , \tag{3}$$

where j denotes the z-component of the hole total angular momentum at $\vec{k} = 0$. We neglect the spin-orbit-split-off band. Using the basis

$$|j\rangle = \{|3/2\rangle \, , |-1/2\rangle \, , |-3/2\rangle \, , |1/2\rangle \, \} \, ,$$

the Luttinger Hamiltonian has the form

$$H_h = \gamma_1 \frac{\hbar^2 k^2}{2m_0} I \; + \; \frac{\hbar^2}{2m_0} \mathcal{T}_4 \, , \tag{4}$$

with the unit matrix I and

$$\mathcal{T}_4 = \begin{pmatrix} Q & R & 0 & -S \\ R^* & -Q & S & 0 \\ 0 & S^* & Q & R^* \\ -S^* & 0 & R & -Q \end{pmatrix} . \tag{5}$$

The matrix elements of $\mathcal{T}_4$ are functions of $\vec{k}$ and the Luttinger parameters γ_1, γ_2, and γ_3:

$$Q = \gamma_2 \left(k_x^2 + k_y^2 - 2k_z^2\right) , \tag{6}$$

$$S = 2\sqrt{3}\gamma_3 (k_x - i k_y) \, k_z \, , \tag{7}$$

$$R = -\sqrt{3} \left[\gamma_2 (k_x^2 - k_y^2) \; - \; 2i\gamma_3 k_x k_y \right] . \tag{8}$$

We use the axial approximation ($\gamma_2 = \gamma_3 = (\gamma_2 + \gamma_3)/2 \equiv \gamma_{23}$ in Eq. (8)), which ignores so-called warping effects. Then the matrix elements become

$$Q = \gamma_2 \left(k_\|^2 - 2k_z^2\right) , \tag{9}$$

$$S = 2\sqrt{3}\gamma_3 k_\| \, k_z \; e^{-i\phi} \, , \tag{10}$$

$$R = -\sqrt{3}\gamma_{23} \, k_\|^2 \; e^{-2i\phi} \, . \tag{11}$$

Within an envelope-function description for quantum wells (see, for example, [35, 34]), the matrix elements of the Hamiltonian with the subband wavefunction of electrons and holes have to be evaluated. Also, we have to replace the z-component of the wavevector by the corresponding derivative, $k_z \to -i\partial/\partial z$. In very thin quantum wells, it is only necessary to take the lowest subband into account. With regard to the matrix elements of $\mathcal{T}_4$, this means that the element S (Eq. (10)) can be neglected since its matrix elements (diagonal in the lowest subband) vanish. Therefore, the 4×4 matrix $\mathcal{T}_4$ breaks up into two 2×2 matrices,

$$\mathcal{T}_4 = \begin{pmatrix} \mathcal{T}_2 & 0 \\ 0 & \mathcal{T}_2^* \end{pmatrix} , \tag{12}$$

with

$$\mathcal{T}_2 = \begin{pmatrix} Q & R \\ R^* & -Q \end{pmatrix} . \tag{13}$$

The interaction due to the Coulomb potential V is described by the Hamiltonian

$$\begin{aligned} H_{int} &= \frac{1}{2}\sum_{\vec{q}\neq 0}\sum_{\vec{k}\vec{k}'} V(\vec{q}) \Bigg[\sum_{ss'} a^\dagger_{\vec{k}+\vec{q},s} a^\dagger_{\vec{k}'-\vec{q},s'} a_{\vec{k}',s'} a_{\vec{k},s} \\ &+ \sum_{jj'} b^\dagger_{\vec{k}+\vec{q},j} b^\dagger_{\vec{k}'-\vec{q},j'} b_{\vec{k}',j'} b_{\vec{k},j} - 2 \sum_{sj} a^\dagger_{\vec{k}+\vec{q},s} b^\dagger_{\vec{k}'-\vec{q},j} b_{\vec{k}',j} a_{\vec{k},s} \Bigg] . \end{aligned} \tag{14}$$

The Coulomb potential in bulk semiconductors is $V(q) = 4\pi e^2/q^2$ and in quantum wells $V(q) = F(q) 2\pi e^2/q$, with the form factor $F(q)$ including the wavefunction of the lowest subband. In very thin quantum wells one can approximate $F(q) \approx 1$. The electron charge includes already background screening effects of the unexcited material, $e^2 = e_0^2/\epsilon_b$, where e_0 is the free electron charge and ϵ_b is the background dielectric constant of the material.

The coupling to the external field $\vec{E}$ is described within the dipole approximation,

$$H_{ext} = -\sum_{\vec{k};sj} \vec{\mu}_{sj}(\vec{k}) \cdot \vec{E}(t)\, a^\dagger_{\vec{k},s} b^\dagger_{-\vec{k},j} + \text{h.c.} \quad . \tag{15}$$

Here, $\vec{\mu}_{sj}$ is the dipole matrix element evaluated using the electron wavefunction with wavevector $\vec{k}$ and spin s and the hole wavefunction with wavevector $\vec{k}$ and angular momentum j.

Beause of the small wavevector of the $\vec{E}$-field, we assume that we can treat the system as spatially homogeneous. This leads to a simple *parametric* spatial dependence of the microscopic quantities such as the optical polarization function $P_{sj}(\vec{k})$ and the electron and hole distribution functions $f^e_s(\vec{k})$ and $f^h_{jj'}(\vec{k})$ due to the spatial variation of $\vec{E}$. Within a Hartree-Fock decoupling scheme, the equations of motion for the polarization and distributions functions are [5]

$$\begin{aligned} i\hbar\frac{\partial}{\partial t} P_{sj}(\vec{k}) &= \left\{\varepsilon^e_{\vec{k}} - i\gamma_{sj}(\vec{k})\right\} P_{sj}(\vec{k}) + \sum_{j'} T^h_{jj'}(\vec{k}) P_{sj'}(\vec{k}) \\ &- \vec{\mu}_{sj}(\vec{k}) \cdot \vec{E}\left[1 - f^e_s(\vec{k})\right] + \sum_{j'} \vec{\mu}_{sj'}(\vec{k}) \cdot \vec{E} f^h_{jj'}(\vec{k}) \\ &- \sum_{\vec{k}'} V(\vec{k}' - \vec{k}) \left[f^e_s(\vec{k}') P_{sj}(\vec{k}) + \sum_{j'} f^h_{jj'}(\vec{k}') P_{sj'}(\vec{k}) \right] \\ &- \sum_{\vec{k}'} V(\vec{k} - \vec{k}') P_{sj}(\vec{k}') \\ &+ \sum_{\vec{k}'} V(\vec{k} - \vec{k}') \left[f^e_s(\vec{k}) P_{sj}(\vec{k}') + \sum_{j'} f^h_{jj'}(\vec{k}) P_{sj'}(\vec{k}') \right] , \end{aligned} \tag{16}$$

$$\hbar\frac{\partial}{\partial t} f^e_s(\vec{k}) = 2\,\mathrm{Im} \sum_j \left[\vec{\mu}_{sj}(\vec{k}) \cdot \vec{E} + \sum_{\vec{k}'} V(\vec{k} - \vec{k}') P_{sj}(\vec{k}') \right]^* P_{sj}(\vec{k}) , \tag{17}$$

and

$$
\begin{aligned}
i\hbar\frac{\partial}{\partial t}f^h_{jj'}(\vec{k}) &= \sum_s \left(\vec{\mu}_{sj'}(\vec{k})\cdot\vec{E}\right)^* P_{sj}(\vec{k}) - \vec{\mu}_{sj}(\vec{k})\cdot\vec{E}\,P^*_{sj'}(\vec{k}) \\
&+ \sum_{j''}\left[T^h_{jj''}(\vec{k})f^h_{j''j'}(\vec{k}) - f^h_{jj''}(\vec{k})T^h_{j''j'}(\vec{k})\right] \\
&+ \sum_{\vec{k}',s} V(\vec{k}-\vec{k}')\left[P^*_{sj'}(\vec{k}')P_{sj}(\vec{k}) - P^*_{sj'}(\vec{k})P_{sj}(\vec{k}')\right] \\
&- \sum_{\vec{k}',j''} V(\vec{k}-\vec{k}')\left[f^h_{jj''}(\vec{k}')f^h_{j''j'}(\vec{k}) - f^h_{jj''}(\vec{k})f^h_{j''j'}(\vec{k}')\right] . \qquad (18)
\end{aligned}
$$

Here, $T^h_{jj'}$ are the matrix elements of the Luttinger Hamiltonian (13). For shortness we left out the arguments which are common to all functions, i.e. $P_{sj}(\vec{k}) \equiv P_{sj}(\vec{k};\vec{R},t)$, $f^e_s(\vec{k}) \equiv f^e_s(\vec{k};\vec{R},t)$, and $f^h_{jj'}(\vec{k}) \equiv f^h_{jj'}(\vec{k};\vec{R},t)$, and $\vec{E} \equiv \vec{E}(\vec{R},t)$. The total polarization density is

$$
\vec{P}(\vec{R},t) = \frac{1}{L^3}\sum_{\vec{k};js}\vec{\mu}^*_{sj}(\vec{k})P_{sj}(\vec{k};\vec{R},t) , \qquad (19)
$$

where L^3 is the sample volume.

In Eq. (16) we have already included a dephasing rate

$$
\gamma_{sj}(\vec{k}) \equiv \gamma_{sj}(\vec{k};n) = \sum_{a=c_s,v_j}\gamma_a(\vec{k};n) , \qquad (20)
$$

which is formally a contribution to the equations beyond the Hartree-Fock approximation. It describes both the dephasing processes in the low-excitation limit and the additional dephasing due the excitation by the pump and/or probe beam. The dephasing processes in the low-excitation limit, which in a linear absorption spectrum give rise to the non-zero linewidth of the exciton resonances, comprise in principle carrier-phonon scattering and, if the system is pre-excited, i.e. inverted prior to the pump/probe experiment, carrier-carrier scattering in the initial electron-hole plasma. The investigation of many coherent effects allows for the simple approximation $\gamma_{sj}(\vec{k}) \equiv \gamma^0_{sj} = \hbar/T_2$, where T_2 is the dephasing or transverse relaxation time. In addition to this approximation, which we use in most sections of this paper, we discuss in the section on excitation induced dephasing (EID) the investigation of polarization dependent coherent phenomena, which are based on the pre-excited and pump-induced density dependence of γ_{sj}. Specifically, we expand the dephasing around the pre-excitation density n_0,

$$
\begin{aligned}
\gamma_{sj}(\vec{k};n) &\simeq \gamma_{sj}(\vec{k}=0;n) \qquad (21) \\
&\simeq \gamma^0_{sj} + \Delta n\ \gamma'_{sj}(\vec{k}=0;n=n_0) , \qquad (22)
\end{aligned}
$$

with $\gamma^0_{sj} = \gamma_{sj}(n=n_0)$ and

$$
\Delta n(t) = \sum_{\vec{k}}\left[\sum_s f^e_s(\vec{k},t) + \sum_j f^h_{jj}(\vec{k},t)\right] - \left[n^e_0 + n^h_0\right] . \qquad (23)
$$

In general, the computation of $\gamma_{sj}(\vec{k};n)$ is very difficult. If, however, the system is in a quasi-equilibrium, one can compute the dephasing rate within the Screened-Hartree-Fock approximation (see, e.g., Ref. [36])

$$
\gamma_a(\vec{k};n) = \sum_{\vec{k}'}\left[f^a(\vec{k}') + g_B(\varepsilon^a_{k'} - \varepsilon^a_k)\right]\ \mathrm{Im}W(\vec{k}-\vec{k}',\varepsilon^a_k - \varepsilon^a_{k'}) , \qquad (24)
$$

where W is the screened Coulomb potential, the distribution functions f^a are Fermi functions and g_B is the Bose function. In the case of valence-band couping, ε_k^h would be the dispersion according to the diagonalized Luttinger Hamiltonian.

In a pump-probe experiment with non-colinear beams, the $\vec{E}$-field takes the form

$$\vec{E}(\vec{R},t) \approx e^{i(\vec{K}_0\vec{R} - \omega_0 t)} \left[\frac{\vec{\mathcal{E}}_0}{2} + \frac{\vec{\mathcal{E}}_1}{2}\, e^{i(\vec{K}_1 - \vec{K}_0)\vec{R}}\right], \tag{25}$$

where $\vec{\mathcal{E}}_0$ and $\vec{\mathcal{E}}_1$ denote the amplitudes of the pump and probe pulses, respectively. Here we have kept only the contribution $\propto \exp(-i\omega t)$ (rotating wave approximation, RWA). If the center frequency of the probe pulse, ω_1, is not the same as ω_0, then the envelope $\vec{\mathcal{E}}_1$ in Eq. (25) carries the modulation term $exp(-i(\omega_1 - \omega_0)t)$. The spatial Fourier transforms of the polarization and the carrier distribution functions within the RWA are

$$P_{sj}(\vec{k};\vec{R},t) = e^{i(\vec{K}_0\vec{R} - \omega_0 t)} \sum_{m=-\infty}^{\infty} e^{im(\vec{K}_1 - \vec{K}_0)\cdot\vec{R}}\, \mathcal{P}_{sj}(\vec{k};m,t) \tag{26}$$

and

$$f^a(\vec{k};\vec{R},t) = \sum_{m=-\infty}^{\infty} e^{i\,m(\vec{K}_1 - \vec{K}_0)\cdot\vec{R}}\, f^a(\vec{k};m,t)\,, \tag{27}$$

respectively. The polarization propagating in the pump direction carries the index $m = 0$, the one in the probe direction $m = 1$, and the one in the four-wave-mixing or photon echo direction $m = -1$, respectively. The density grating, set up by the coherent superposition of the pump and probe fields, is denoted by the index $m = \pm 1$. The corresponding Fourier transform of Eqs. (16) - (18) is

$$\begin{aligned}
i\hbar\frac{\partial}{\partial t}\mathcal{P}_{sj}(\vec{k};m) &= \left\{\varepsilon_k^e - \hbar\omega_0 - i\left(\gamma_{sj}^0 - \gamma_{sj}'\left[n_0^e + n_0^h\right]\right)\right\}\mathcal{P}_{sj}(\vec{k};m) \\
&+ \sum_{j'} T_{jj'}^h(\vec{k})\mathcal{P}_{sj'}(\vec{k};m) \\
&+ \gamma_{sj}'\sum_{m'}\left[\sum_s f_s^e(\vec{k};m-m') + \sum_j f_{jj}^h(\vec{k};m-m')\right]\mathcal{P}_{sj}(\vec{k};m') \\
&- \vec{\mu}_{sj}(\vec{k})\cdot\frac{\vec{\mathcal{E}}_m}{2} - \sum_{\vec{k}'} V(\vec{k}-\vec{k}')\mathcal{P}_{sj}(\vec{k}',m) \\
&+ \sum_{m'}\left[\vec{\mu}_{sj}(\vec{k})\cdot\frac{\vec{\mathcal{E}}_{m'}}{2}\, f_s^e(\vec{k};m-m')\right. \\
&\quad \left. + \sum_{j'}\vec{\mu}_{sj'}(\vec{k})\cdot\frac{\vec{\mathcal{E}}_{m'}}{2} f_{j'j}^h(\vec{k};m-m')\right] \\
&- \sum_{\vec{k}',m'} V(\vec{k}'-\vec{k})\left[f_s^e(\vec{k}';m-m')\mathcal{P}_{sj}(\vec{k};m')\right. \\
&\quad \left. + \sum_{j'} f_{jj'}^h(\vec{k}';m-m')\mathcal{P}_{sj'}(\vec{k};m')\right] \\
&+ \sum_{\vec{k}',m'} V(\vec{k}-\vec{k}')\left[f_s^e(\vec{k};m-m')\mathcal{P}_{sj}(\vec{k}';m')\right. \\
&\quad \left. + \sum_{j'} f_{jj'}^h(\vec{k};m-m')\mathcal{P}_{sj'}(\vec{k}';m')\right],
\end{aligned} \tag{28}$$

$$
\begin{aligned}
i\hbar\frac{\partial}{\partial t}f_s^e(\vec{k};m) &= \sum_{m'}\left[\sum_j \left(\vec{\mu}_{sj}(\vec{k})\cdot\frac{\vec{\mathcal{E}}_{m'-m}}{2}\right)^* \mathcal{P}_{sj}(\vec{k};m')\right. \\
&\quad \left. -\vec{\mu}_{sj}(\vec{k})\cdot\frac{\vec{\mathcal{E}}_{m'+m}}{2}\mathcal{P}_{sj}^*(\vec{k};m')\right] \\
&+ \sum_{\vec{k}',j,m'} V(\vec{k}-\vec{k}')\left[\mathcal{P}_{sj}^*(\vec{k}',m'-m)\mathcal{P}_{sj}(\vec{k};m')\right. \\
&\quad \left. -\mathcal{P}_{sj}(\vec{k}';m'+m)\mathcal{P}_{sj}^*(\vec{k};m')\right] , \qquad (29)
\end{aligned}
$$

and

$$
\begin{aligned}
i\hbar\frac{\partial}{\partial t}f_{jj'}^h(\vec{k};m) &= \sum_{m'}\left[\sum_s \left(\vec{\mu}_{sj'}(\vec{k})\cdot\frac{\vec{\mathcal{E}}_{m'-m}}{2}\right)^* \mathcal{P}_{sj}(\vec{k};m')\right. \\
&\quad \left. -\vec{\mu}_{sj}(\vec{k})\cdot\frac{\vec{\mathcal{E}}_{m'+m}}{2}\mathcal{P}_{sj'}^*(\vec{k};m')\right] \\
&+ \sum_{j''}\left[T_{jj''}^h(\vec{k})f_{j''j'}^h(\vec{k};m) - f_{jj''}^h(\vec{k})T_{j''j'}^h(\vec{k};m)\right] \\
&+ \sum_{\vec{k}',s,m'} V(\vec{k}-\vec{k}')\left[\mathcal{P}_{sj'}^*(\vec{k}',m'-m)\mathcal{P}_{sj}(\vec{k};m')\right. \\
&\quad \left. -\mathcal{P}_{sj'}^*(\vec{k};m'+m)\mathcal{P}_{sj}(\vec{k}';m')\right] \\
&- \sum_{\vec{k}',j'',m'} V(\vec{k}-\vec{k}')\left[f_{jj''}^h(\vec{k}';m-m')f_{j''j'}^h(\vec{k};m')\right. \\
&\quad \left. f_{jj''}^h(\vec{k};m-m')f_{j''j'}^h(\vec{k}';m')\right] . \qquad (30)
\end{aligned}
$$

In a quantum well all functions of the in-plane momentum vector $\vec{k}$ remain unchanged if $\vec{k}$ is rotated by 2π. As a consequence, one can introduce the representation [23]

$$
\vec{\mathcal{P}}_{sj}(\vec{k};m,t) = \sum_{\ell=-\infty}^{\infty} \vec{\mathcal{P}}_{sj}(k,\ell;m,t)\, e^{-i\ell\phi} \qquad (31)
$$

for $\vec{\mathcal{P}}$ and corresponding expansions for all other functions. Using this representation in Eqs. (28) - (30), we obtain coupled equations for s-exciton polarization, $\vec{\mathcal{P}}_{sj}(k,\ell = 0;m,t)$, and d-exciton polarization, $(\vec{\mathcal{P}}_{sj}(k,\ell = 2;m,t)$. This is a consequence of the ϕ-dependence of the Luttinger matrix elements, (9) and (11), which allows only $\Delta\ell = 0$ and $\Delta\ell = 2$ coupling. The consequences of this coupling will be discussed below in the context of photon echo signals in quantum wells.

The generalized semiconductor Bloch equations, Eqs. (16) - (18), which include band-structure effects via the Luttinger Hamiltonian reduce to the familiar SBE [37, 17] if one neglects the coupling between spin states and between heavy- and light-hole bands, and if one neglects the excitation induced dephasing. In order to stress in this case the formal similarity of the semiconductor Bloch equations with atomic Bloch equations, we re-write Eqs. (16) - (18) for the $c_s - v_j$ transition, introducing the renormalized dipole energy

$$
\hbar\Omega_{sj}(\vec{k}) = \vec{\mu}_{sj}(\vec{k})\vec{E} + \sum_{\vec{k}'} V(\vec{k}-\vec{k}')P_{sj}(\vec{k}') \qquad (32)
$$

and the excitation induced selfenergy

$$
\Sigma_{\vec{k}} = -\sum_{\vec{k}'} V(\vec{k}'-\vec{k})\left[f_s^e(\vec{k}') + f_j^h(\vec{k}')\right] . \qquad (33)
$$

With $\varepsilon_k^{sj} = \varepsilon_k^{c_s} - \varepsilon_k^{v_j}$, we obtain

$$i\hbar\frac{\partial}{\partial t}P_{sj}(\vec{k}) = \left\{\varepsilon_k^{(sj)} + \Sigma_{\vec{k}} - i\gamma_{sj}^0\right\} P_{sj}(\vec{k}) + (1 - f_s^e(\vec{k}) - f_j^h(\vec{k}))\, \Omega_{sj}(\vec{k}), \tag{34}$$

$$\hbar\frac{\partial}{\partial t}f_s^e(\vec{k}) = 2Im\left[\Omega_{sj}^*(\vec{k})P_{sj}(\vec{k})\right] , \tag{35}$$

and

$$\hbar\frac{\partial}{\partial t}f_j^h(\vec{k}) = 2Im\left[\Omega_{sj}^*(\vec{k})P_{sj}(\vec{k})\right] . \tag{36}$$

To incorporate static screening into these equations, we merely have to replace the Coulomb potential $V(q)$ by a statically screened potential $W(q)$, and to add the Coulomb-hole contribution $\sum_{\vec{q}}(W(q) - V(q))$ to the selfenergy (33). To further include real carrier scattering, one can, for example, employ the relaxation rate approximation and add the term $\frac{\partial}{\partial t}f^a(\vec{k})\Big|_{scatt} = \gamma_a(f^a(\vec{k}) - f^a_{Fermi}(\vec{k}))$, $a = e, h$, to Eqs. (35) and (36), respectively. A quantitative investigation of the relaxation rate approximation can be found in Ref. [38].

Generally, extracting four-wave mixing signals from a spatial Fourier transform, as done above, requires an optically thin sample so that propagation effects like diffraction or re-absoprtion can be neglected. In the case of an optical semiconductor waveguide this is not possible and one therefore has to solve the propagation equation (Maxwell equation) self-consistently with the SBE. To do this, we employ the slowly varying envelope approximation, in which the propagation equation reads

$$\frac{\partial}{\partial \zeta}\vec{\mathcal{E}}(\zeta,\eta) = i\, 4\pi\, \frac{\omega_0}{cn_b}\, \frac{1}{2}\vec{\mathcal{P}}(\zeta,\eta) , \tag{37}$$

where a moving coordinate frame has been introduced ($\eta = t - z/v_g$ and $\zeta = z$). Here, $\vec{\mathcal{P}}$ is the envelope of the total polarization density $\vec{P}$ (Eq. (19)), v_g is the group velocity, and group-velocity dispersion effects due to off-resonant transitions have been neglected. Note, however, that group-velocity dispersion effects and all other nonlinearities due to the resonant transitions are fully included in such an analysis, since the optical polarization is treated numerically.

In the following sections we present numerical solutions of the semiconductor Bloch equations for a variety of coherent phenomena.

THE OPTICAL STARK EFFECT

We begin the discussion of results with the optical Stark effect. This effect in semiconductors is generally observed as a transient blue shift of the exciton resonance for nonresonant excitation with a femtosecond pulse [39, 40, 41, 42]. To theoretically study this phenomenon we numerically solve Eqs. (34) - (36) without screening ($W \equiv V$) and without relaxation [37]. Fig. 1 shows computed absorption spectra for various delay times t_d for the case of bulk GaAs excited by a Gaussian pump pulse with an amplitude FWHM of $120fs$ whose central frequency is detuned $10E_R$ below the bandgap energy E_g. The peak height of the pump pulse is such that $\vec{\mu}_{cv}\mathcal{E}_0/2 = 2\ E_R$ at $t = 0$. For the probe pulse we also assume a Gaussian shape with the same duration as the pump pulse, but centered energetically at the exciton resonance. The peak height of the probe

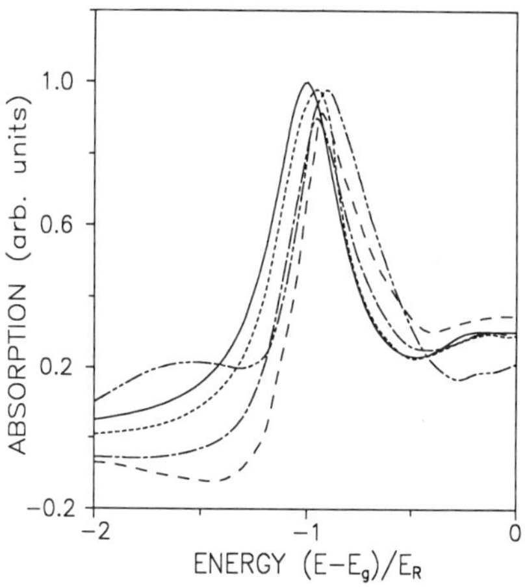

Figure 1. *Probe spectra for various delay times* t_d *in comparison to the linear exciton spectrum (solid line);* $t_d = -600$fs *(dash-double-dotted line),* $t_d = -60$ fs *(long dashed line),* $t_d = -0$ fs *(dash-dotted line),* $t_d = +60$ fs *(short-dashed line). The excitation pulse is centered* 10 E_R *below the bandgap energy* E_g *of the unexcited* GaAs *semiconductor, where* E_R *is the bulk exciton binding energy* (4.2 meV). *The pulse duration is* 120fs *(FWHM of the field amplitude). (From Ref. [37].)*

pulse is $\vec{\mu}_{cv} \cdot \vec{\mathcal{E}}_1/2 = 0.01E_b$ at $t = t_d$. For a convenient comparison of the results we normalize in this section the absorption spectra such that the 1s-exciton peak has a height of 1 before the excitation. The differential transmission spectra (DTS) are given, in the same units, as the difference between the absorption without excitation and the absorption with excitation at the delay time of interest. The DTS corresponding to Fig. 1 are shown in Fig. 2. Here we assumed a constant phenomenological dephasing

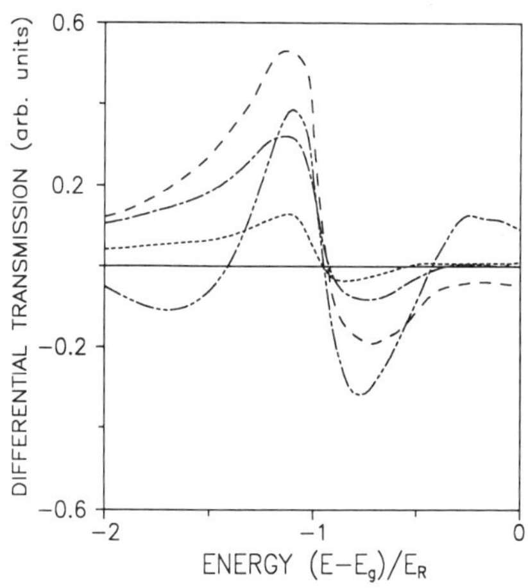

Figure 2. *Differential transmission spectra corrsponding to Fig. 1. (From Ref. [37].)*

time of 0.87ps, which corresponds to an exciton lineshape width before the excitation of $\gamma = 0.2E_R$. The fact that the pulse width is much shorter than the dephasing time turns out to be quite important for the spectra calculated. One consequence is that the exciton blue shift is much smaller than for stationary (cw-) excitation with the same peak intensity. Intuitively one can understand this by remembering that the probe spectrum is obtained as the Fourier transform of that polarization component which propagates in the probe direction. The time integration has to be performed until the polarization has decayed back to zero. Hence, the integration extends over a time much longer than the pump-pulse duration and the blue-shifting pump field is present only for a relatively short period of that time interval. Therefore, the resulting net blue shift for pulsed excitation is always smaller than *cw*-calculations [15] would predict. If we examine the situation in Fig. 1 a *cw*-calculation gives an exciton at $\hbar\omega_0 + [(\varepsilon_{1s} - \hbar\omega_0)^2 + 4(\vec{\mu}cv \cdot \vec{\mathcal{E}}_0/2)^2]^{1/2}$, which for $\vec{\mu}_{cv} \cdot \vec{\mathcal{E}}_0/2 = 2E_R$ yields a blue shift of $\simeq 0.85E_R$. The calculated shift for pulsed excitation, however, is only $0.1E_R$.

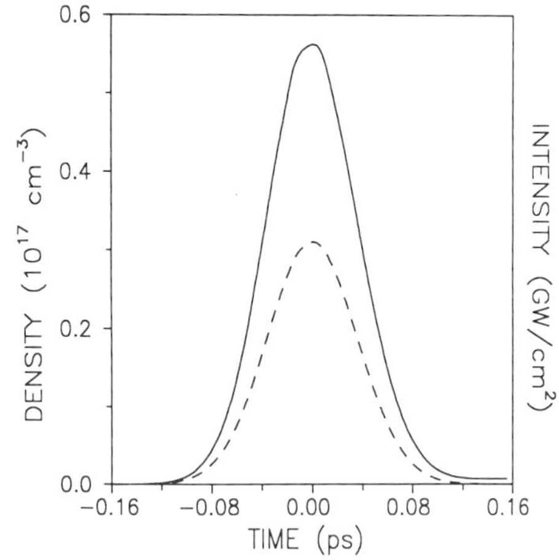

Figure 3. *Excited carrier density (solid line) and light-field intensity (dashed line) corresponding to Fig. 1. (From Ref. [37].)*

A consequence of the long dephasing time is that it allows for the coherent scattering of the pump-pulse into the probe direction as well as for an energy redistribution within the probe pulse, which in our case has the same duration as the pump. The DTS shows the effects of these processes as oscillations at negative delay times. As analyzed in detail in Ref. [8, 43, 13], the period of these oscillations is essentially given by t_d . For non-resonant excitation they are accompanied by the blue-shift due to the exciton Stark effect. For values of the time-delay where pump-probe overlap occurs, i.e., for $t_d \to 0$, the exciton resonance is increasingly bleached and less shifted than for larger negative time delays. These features are the consequence of excited carriers which give rise to significant phase-space filling effects. The temporal behavior of the excited density is displayed in Fig. 3. In addition to the bleaching for $t_d \to 0$ the low energy tail of the exciton exhibits a pronounced gain region as well. As we show in Ref. [37], one can explain the occurrence of this gain through a simple model calculation for the probe-spectrum of a two-level system without any many-body effects. In the limiting case of a stationary pulse the usual Mollow spectrum exhibits an absorption peak at the renormalized Rabi frequency. In the case where the pump field is suddenly switched off, e.g., at the half-width time Δt, the corresponding renormalization of the Rabi-frequency is changed and the probe spectrum exhibits oscillations with the period Δt. These oscillations manifest themselves in regions of increased or decreased transmission and are the basic reason for the gain region below the exciton resonance. Rough estimates show that the oscillations in energy space have the frequency $\Delta t/\hbar$ which corresponds in our case to a period of about $8E_R$. Unlike the gain region which occurs in the stationary Mollow spectrum of two-level systems, the gain discussed here is not caused by the scattering of pump photons into the probe direction. It vanishes with decreasing spectral width of the probe pulse because of the suppression of energy redistribution in a spectrally narrow pulse (see also Ref. [14]). Although these oscillations appear in the same general parameter regime as the coherent oscillations, they are caused by different physical mechanisms.

Another important aspect of the spectra shown in Fig. 1 is the negligible spectral overlap between the pump pulse and the exciton resonance. For sufficiently large detuning one observes the adiabatic following [44, 1] of the excited density with respect to the pulse, an example of which is shown in Fig. 3. As a result, the bleaching and shifting of the exciton vanishes at positive delay times (Fig. 1).

ON-RESONANT EXCITATION: GENERALIZED RABI OSCILLATIONS

Rabi oscillations of the excitation density occur in systems where the inverse dephasing time $\gamma/\hbar$ is smaller than the Rabi oscillation frequency, which, in atomic systems, is $\vec{\mu}_{cv} \cdot \vec{\mathcal{E}}_0/\hbar$. Under these conditions an optical pulse leaves the system in a state which is determined by the time-integral of the pulse (pulse area). A pulse area of 2π which resonantly excites an atomic two-level system causes exactly one Rabi flop. If the system excitation starts, for example, from the lower state before the pulse, it goes completely into the upper state during the pulse, and it completely returns to the ground state after the pulse. In our studies of Rabi flopping in semiconductors we want to clearly distinguish between coherent and incoherent effects. Therefore, we first investigate the limit $T_2 \gg \Delta t$, where Δt denotes the duration of the pulse. As an illustrative example we show in Fig. 4 the time dependence of the semiconductor density for a 2.2π-pulse, i.e. $(1/\hbar)\int dt \vec{\mu}_{cv}\vec{\mathcal{E}}(t) = 2.2\pi$, exciting the system at the exciton resonance or at slightly detuned frequencies. In contrast to an atomic system, the optical resonance transitions in a semiconductor are not simply given by the one-particle transitions since the exciton involves a large number of different $\vec{k}$-states. Each of these one-particle transitions has a slightly different detuning from the pump pulse. Different detunings imply different Rabi frequencies of the individual states, which may lead to pronounced interference effects, eventually even completely destructive interference and the absence of Rabi oscillations. However, as Fig. 4a shows, when excited at or above the exciton resonance, the semiconductor carrier density clearly exhibits temporal oscillations. Moreover, the number of Rabi flops is basically twice that expected for a two-level system. Furthermore, we note that a finite carrier density remains in the system after the pulse is gone. Fig. 4a also demonstrates that the amplitude of the Rabi oscillations is significantly reduced already for the relatively small detuning of -0.4 E_b, where $E_b = 2.85E_R$ ($\equiv -1.4E_b$) below the gap, long dashed curve in Fig. 4a. The oscillations occur only as small variations on an envelope which basically exhibits an adiabatic following character. The density remaining after the pulse is very small in comparison to the cases of resonant excitation.

To study the influence of damping and to further illustrate the role of the many-body Coulomb effects, we plot in Fig. 4b and c the results for resonant excitation with a 3π-pulse for different dephasings using the full semiconductor Bloch equations (Fig. 4b) and for the cases without exchange terms (solid line, Fig. 4c) and without all Coulomb terms (short dashed line, Fig. 4c). Furthermore, we show the system response with and without the exchange terms when the pump is tuned one binding energy below the exciton resonance (dash-dotted and long dashed curves in Fig. 4c). We see in Fig. 4b that the most pronounced effect of increasing dephasing is an increasing carrier density left in the system when the pump pulse is gone. This is fully expected since finite dephasing leads to a broadening of the exciton resonance and, hence, to increased spectral overlap between pump pulse and exciton absorption peak. With increasing overlap more and more real absorption occurs, so that the density dynamics becomes a

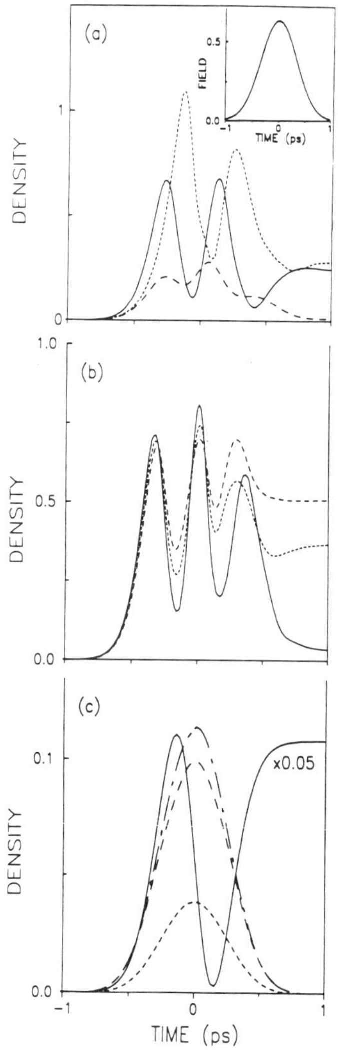

Figure 4. *Excited carrier density, $n_e = 2\sum_k f_k^e$, in units of a_B^{-2}, where a_B is the exciton Bohr radius for bulk GaAs, for a 50Å GaAs quantum well for various excitation frequencies: $\hbar\omega_0 = E_g - E_b$ (full line), $\hbar\omega_0 = E_g - 1.4E_b$ (long dashed line), $\hbar\omega_0 = E_g - 0.6E_b$ (short dashed line). The inset shows the pump field $\vec{\mu}_{cv} \cdot \vec{\mathcal{E}}_0/2$ in units of E_R as function of time for a pulse with an area of 2.2π.* (b) *Carrier density for resonant excitation with a 3π pulse without dephasing (solid line), with a dephasing time $\hbar/T_2 = 0.2E_R$ (long dashed line), and $\hbar/T_2 = 0.1E_R$ (short dashed line), respectively.* (c) *Solid line: computed carrier density without dephasing and exchange terms for otherwise the same conditions as the solid line in* (b). *Short dashed line: same conditions as solid line but $V \equiv 0$. Large detuning limit (1 E_b below the exciton resonance) with (dash-dotted line) and without (long dashed line) the exchange contributions. (From Ref. [2])*

superposition of the coherent and incoherent features, with the incoherent feature being an increase roughly proportional to the time integral over the pulse. More surprising is the difference between the results shown as solid lines in Fig. 4b and Fig. 4c. The solution without the exchange terms (Fig. 4c) exhibits 1.5 density oscillations, exactly as expected for the case of a two-level system under these conditions. However, as mentioned above, the solutions of the full semiconductor Bloch equations (Fig. 4b) show twice as many flops, demonstrating clearly that the field renormalization $\sum_{\vec{k}'} V(\vec{k}-\vec{k}')P(\vec{k}')$ is of the same magnitude as $\vec{\mu}_{cv} \cdot \vec{\mathcal{E}}/2$. Hence, the exchange terms are extremely important under resonant excitation conditions. The omission of these contributions even removes the physically required restriction for the distribution function $f^a(\vec{k})$ to vary only between 0 and 1. The short dashed line in Fig. 4c shows the completely unrealistic results when we omit all Coulomb terms in the Bloch equations.

The approximate doubling of the Rabi frequency should appear whenever $\vec{\mu}_{cv} \cdot \vec{\mathcal{E}}_0/2$ is of the same order of magnitude as the exciton binding energy. It has been found in Ref. [45] that in the limiting case $\vec{\mu}_{cv} \cdot \vec{\mathcal{E}}_0/2 \gg E_R$ this doubling does not occur.

PHOTON ECHO: TWO-BAND APPROXIMATION

We will now investigate two-pulse four-wave mixing (FWM) and photon echo effects. We start by considering the simple two-band structure, since coherent many-body effects have proven important already in this simple semiconductor model. The theoretical analysis of photon echoes in atomic gases is mainly based on a model which treats the individual atoms as independent oscillators. In a semiconductor the concept of independent oscillators is lost and free motion after an initial excitation is very complicated due to the nonlinearity of the system. Our calculations show that a very important factor in the understanding of photon echo effects in semiconductors is the excitation dependent band-gap shift. This renormalization leads to a resonance condition which changes during the excitation due to the changing carrier density. Even if initially the excitation takes place at the exciton resonance energy, it is possible that during the excitation the band states may end up in resonance. Thus the strength of the excitation plays an important role as well.

In Fig. 5 we show the time dependence of the signal for various delay times from $-400fs$ to $-150fs$ for the case when the center frequency of the excitation is at the $1s$-exciton resonance and the intensity of the probe pulse is 1% of the pump pulse intensity. The signal starts to rise when the second pulse hits the sample. In the case of independent two-level system an instantaneous signal indicates the dominance of a

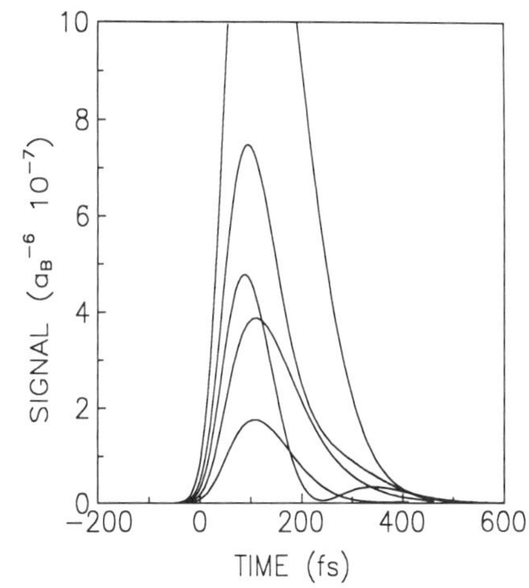

Fig. 5

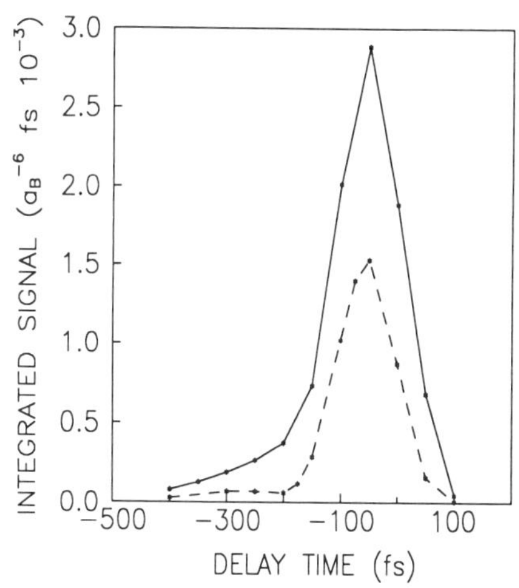

Fig. 6

Figure 5. *Time resolved signal in photon echo direction. The material parameters are chosen for* CdSe $(m_e = 0.125m_0, m_h = 0.431m_0, E_R = 16\text{meV}, \epsilon_b = 9)$, *and the dephasing rate is* $(T_2 = 200\text{ fs})$. *The center frequency of both pulses is at the exciton resonance* $(\hbar\omega_0 = \hbar\omega_1 = \varepsilon_{1s})$, *the pulse width is* 100 fs *(FWHM of the amplitude), the field strengths are* $\vec{\mu}_{cv} \cdot \vec{\mathcal{E}}_1/2 = 0.01E_R$, *and* $\vec{\mu}_{cv} \cdot \vec{\mathcal{E}}_0/2 = 0.1E_R$, *and the delay times* t_d *are* $-150, -175, -200, -300$ *and* -400 fs, *corresponding to decreasing signal maximum, respectively. Negative delay times indicate that pulse* $\vec{\mathcal{E}}_1$ *comes before* $\vec{\mathcal{E}}_2$. *(From Ref. [3].)*

Figure 6. *Time-integrated signal corresponding to Fig. 5 (solid line), and for* $\vec{\mu}_{cv} \cdot \vec{\mathcal{E}}_0/2 = 0.5\ E_R$ *instead of* $0.1\ E_R$ *(dashed line). (From Ref. [3].)*

single resonance, rather than a broadband continuum. In our case this resonance is the exciton, and the fact that the underlying one-particle states are continuously distributed is not reflected in the signal. The time at which the signal peaks is not related to the delay time. In a measurement situation this kind of signal could, however, easily be confused with a real photon echo if, by coincidence, the delay time t_d and the signal peak time are not too different or if only time-integrated measurements are performed. The decrease of the signal magnitude with increasing delay time is due to the dephasing taking place between the pulses. The corresponding time integrated signals as functions of delay time are shown in Fig. 6.

To see what happens when the continuum states are excited, we repeated our calculations for a condition in which the center pulse frequency is at the band edge. In Fig. 7 the signals with the same delay times and field strengths as in Fig. 5 are shown for delay times larger than the dephasing time. A clear echo feature can be

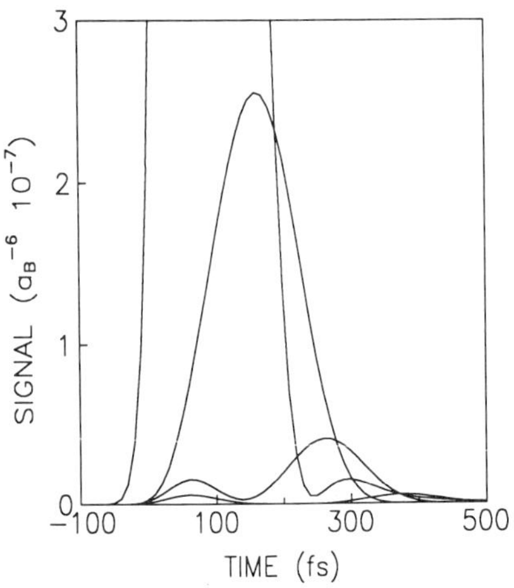

Fig. 7

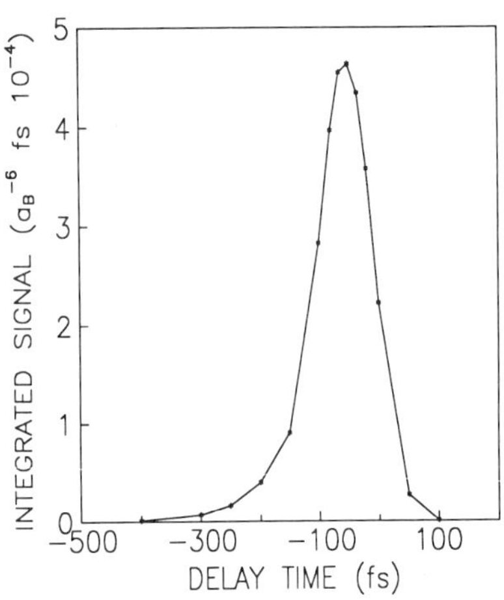

Fig. 8

Figure 7. *Same as Fig. 5, but for excitation at the band edge. The delay times are* -100, -200, -300, *and* -400 fs, *corresponding to decreasing signal maximum, respectively. (From Ref. [3].)*

Figure 8. *Time-integrated signal corresponding to Fig. 7 (From Ref. [3].)*

seen essentially one delay time after the second pulse, as expected also for the two-level systems. However, our signal shows an additional peak centered around $t = 0$ which is independent of the delay time. This component corresponds to the $1s$-exciton which is also excited due to the finite pulse bandwidth. Fig. 8 shows the integrated signal corresponding to Fig. 7 as a function of delay time. Fitting the curve formed by the maxima of the signals in Fig. 7 to the relation $|P|^2_{max} \propto \exp(-4t_d/T_2)$ we regain the used $T_2 \simeq 200fs$.

Even though the low excitation results presented so far exhibit some features which can be explained by considering the different exciton states as independent oscillators, our analysis shows that the interaction effects play a significant role in the dynamics. For example, if we neglect the many-body effects the strengths of the signals are drastically modified. Additionally, the band-gap renormalization can also be an important effect because it may completely modify the resonance conditions. To illustrate this feature,

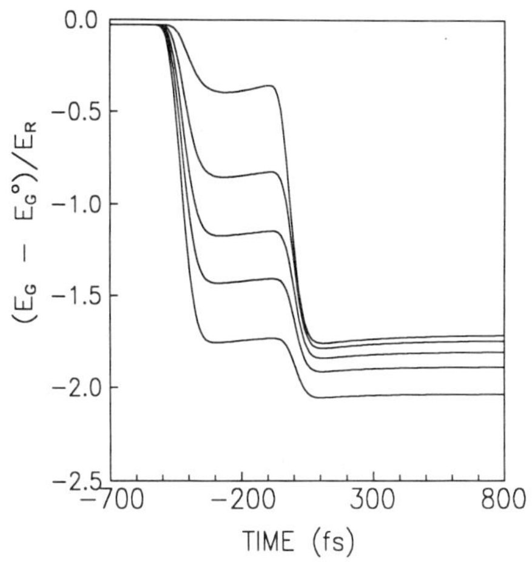

Fig. 9

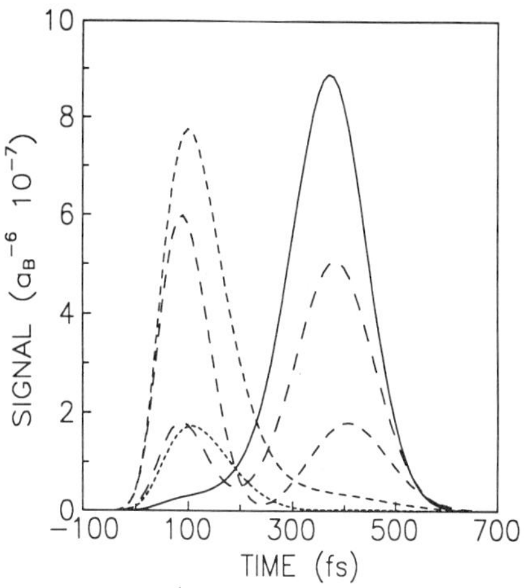

Fig. 10

Figure 9. *Renormalized band edge as a function of time.* E_g^0 *is the unrenormalized band gap. The delay time is* $t_d = -400$ fs *and the different curves are for* $\vec{\mu}_{cv} \cdot \vec{\mathcal{E}}_1/2 = 0.01E_R, 0.03E_R, 0.05E_R, 0.07E_R$, *and* $0.1E_r$, *from top to bottom, respectivley. All other parameters are the same as in Fig. 5 (From Ref. [3].)*

Figure 10. *Time-resolved signal in the photon echo direction for increasing strength of the first pulse* $\vec{\mathcal{E}}_1$. *Excitation occurs at the exciton resonance, the dephasing time is* $T_2 = 200$ fs, *the delay time is* $t_d = -400$ fs, *the pulse duration is for both pulses* 100 fs *(FWHM of the amplitude), and the peak amplitude of the second pulse is* $\vec{\mu}_{cv} \cdot \vec{\mathcal{E}}_0/2 = 0.1E_R$. *With increasing dash length, the curves are for* $\vec{\mu}_{cv} \cdot \vec{\mathcal{E}}_1/2 = 0.01E_R, 0.03E_R, 0.05E_R, 0.07E_R$, *and* $0.1E_R$ *(solid line). (From Ref. [3].)*

we plot in Fig. 9 the time dependence of the band edge corresponding to Fig. 5 (top curve). We see that the first pulse, which is very weak, leads to a shift of only about $0.3E_R$, whereas the second pulse causes a shift of the order of $2E_R$. In the case of such a large shift, a pulse which is originally centered at the 1s exciton resonance causes direct excitation of the continuum states after the bandgap has shifted by E_R. Nevertheless, this situation does not lead to a photon echo in Fig. 5, since the phase shifts necessary for the echo are produced in the time interval between the pulses, and during this time interval the band-edge shift is still too small to allow significant continuum excitation by the first pulse. However, for the same situation, only increasing the intensity of the first pulse, we shift and excite immediately part of the continuum so that the free propagation before the second pulse is sufficient to cause the needed phase shifts for the photon echo. To demonstrate this effect we keep the center frequency of the pulses at the 1s-exciton resonance, fix the delay time at $t_d = -400fs$, choose the strength of the second pulse as $\vec{\mu}_{cv} \cdot \vec{\mathcal{E}}_0/2 = 0.1E_R$ and vary the strength of the first pulse. In Fig. 10 we show the resulting signals as a function of time for various field strengths of the first pulse. As expected, for low intensities the signal has a single peak near zero time, whereas for large intensities the peak of the signal is at $400fs$, which is the correct photon echo position in this case. The curves for intermediate intensities show the transition from signals which are characteristic for "free induction decay" to those characteristic for a "photon echo". A double peaked structure is seen to appear between the limiting cases. The band edges plotted in Fig. 9 correspond to the field strengths of Fig. 10 and we can see that for increasing $\vec{\mu}_{cv} \cdot \vec{\mathcal{E}}_1$ the gap shift before the arrival of the second pulse increases and exceeds E_R. Consequently the band states are shifted in resonance with the field and the photon echo signal develops.

PHOTON ECHO: INCLUSION OF THE LUTTINGER HAMILTONIAN

To analyze polarization dependent features of photon echo experiments, we include the relevant aspects of the realistic valence-band structure in *III-V* compounds. For simplicity of discussion we adopt the conventional definition of heavy-hole ($j = \pm 3/2$) and light-hole ($j = \pm 1/2$) although the angular momentum is not a good quantum number for one-particle states with finite momentum. The matrix elements of $T_{jj'}(k, \ell - \ell')$ vanish unless $\ell - \ell' = 0$ and $j - j' = 0$, or $\ell - \ell' = \pm 2$ and $j - j' = \mp 2$ (see Eq. (13) and the ℓ-representation (31)). Therefore, an exciton with envelope angular momentum $l_z = \ell$ and valence-band angular momentum j is only coupled to an exciton with $\ell' = \ell \pm 2$ and $j' = j \mp 2$. For example, as shown in Fig. 11, an s-state heavy-hole exciton is coupled to a d-state light-hole exciton and an s-state lh-exciton is coupled to a d-state hh-exciton. (In Fig. 11 we have assigned the exciton orbital momentum ℓ to

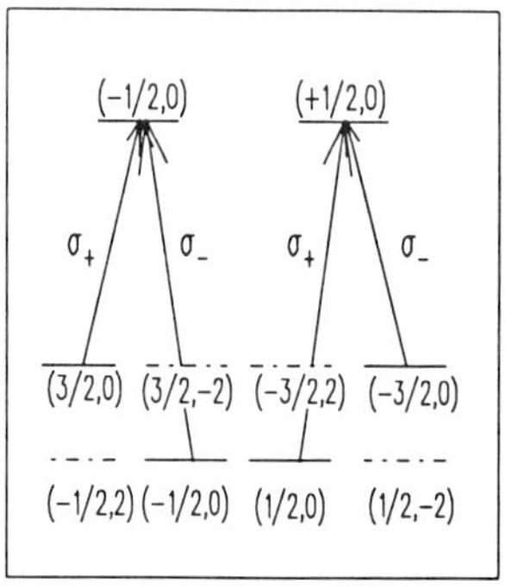

Figure 11. *Electron-hole-pair states resulting from the Luttinger Hamiltonian. The quantum numbers in brackets are the z-component, j, of the electron or hole angular momentum, respectively, and the envelope orbital momentum index ℓ. The arrows indicate the strong dipole-allowed transitions. (From Ref. [5].)*

the holes and the orbital momentum $\ell = 0$ to the electrons.) These electron-hole-pair states are two-fold degenerate. Hence, the complicated multiband structure reduces to a two-fold degenerate system, which can be labeled according to the electron spin in the respective conduction band. The $s = 1/2$ conduction band is optically coupled to the $s - hh$ band with $j = -3/2$ by a $\vec{\sigma}_-$ E-field, i.e., a circularly polarized field with polarization vector $\vec{\sigma}_- = (\vec{e}_x - i\vec{e}_y)/\sqrt{2}$. In such an excitation process the d-lh band is optically excited only because of the hh-lh coupling. Since the transition between the uncoupled d-lh band and the $s = 1/2$ conduction band is dipole forbidden, the d-lh component in this excitation process is considerably smaller than its hh counterpart. Hence, in comparison to the dominating s-hh contribution the d-lh contribution is negligible.

To analyze the field polarization effects, we numerically solve Eqs. (28) - (30) for $m = 0, \pm 1$, choosing the incident fields as circularly polarized ($\propto \vec{\sigma}_\pm$) and assuming that the propagation directions are nearly perpendicular to the plane of the quantum-well layers. Before we present our numerical results, we first discuss the situation of oppositely circularly polarized pump and probe pulses, choosing $\vec{\sigma}_+$ ($\vec{\sigma}_-$) polarized pump (probe). For this purpose, we analyze how these pulses interact with the subset of $s = -1/2$ conduction band and valence bands $j = 3/2$ and $j = -1/2$ (see Fig. 11). The possible echo signals are due to $P_{-1/2,3/2}(\vec{k}; -1)$ and $P_{-1/2,-1/2}(\vec{k}; -1)$,

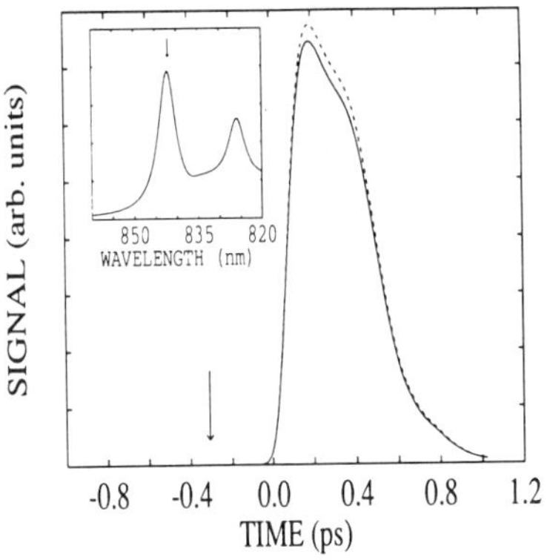

Fig. 12

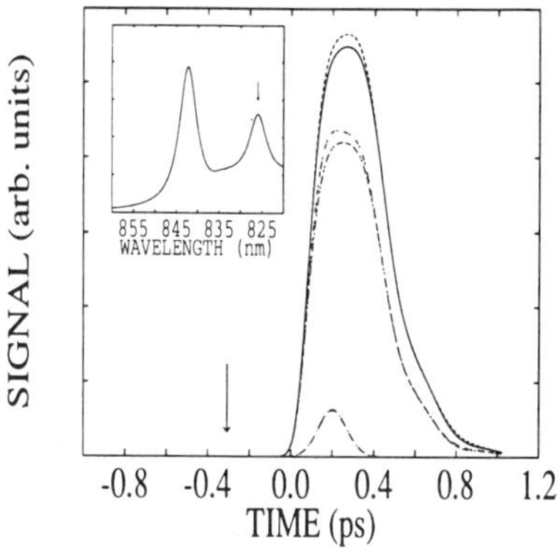

Fig. 13

Figure 12. *Computed four-wave-mixing signal in the direction* $2\vec{K}_0 - \vec{K}_1$ *for a* 100 Å *GaAs quantum well* $(\gamma_1 = 6.85, \gamma_2 = 2.1, \gamma_3 = 2.9, m_e = 0.067 m_0, \epsilon_b = 12.7)$. *The dephasing time is* $T_2 = 200$fs. *The light pulses have amplitudes* $\vec{\mu}_{-1/2;3/2} \cdot \vec{\mathcal{E}}_0/2 = 0.01\,\text{meV}$ *and* $\vec{\mu}_{-1/2;3/2} \cdot \vec{\mathcal{E}}_1/2 = 0.001\,\text{meV}$. *The delay time is* $t_d = 300$ fs *The central frequency (same for both pulses) is at the heavy-hole exciton resonance. The arrows indicate the center time of the pump* 0 fs *and probe* −300 fs, *respectively. The inset shows the corresponding linear absorption spectrum. The arrow in the inset marks the probe and pump center frequencies. The results shown as the dashed curve neglect the off-diagonal elements of the Luttinger Hamiltonian while the solid curve is computed using the full Luttinger Hamiltonian (Eq. (13)). (From Ref. [5].)*

Figure 13. *Same as Fig. 12 but both pump and probe pulses are tuned at the light-hole exciton resonance. The top two curves are the computed four-wave-mixing signals using the Luttinger Hamiltonian (solid line) and the simplified six-band model (dashed line). The two curves in middle are the light-hole contributions (the higher one is obtained from the six-band equations). The lowest curve centered around* 0.2 ps *(dash-dotted line) is the heavy-hole contribution which is almost identical in both calculations. (From Ref. [5].)*

where again the argument −1 in P indicates the FWM direction. We see from Eq. (28) that $P_{-1/2,3/2}(\vec{k};-1)$ stems from the scattering terms $\vec{\mu}_{-1/2;3/2} \cdot \vec{\mathcal{E}}_0\ f^h_{3/2,3/2}(\vec{k};-1)$ and $P_{-1/2;3/2}(\vec{k},-1) f^h_{3/2,3/2}(\vec{k};-1)$. However, the population pulsation $f^h_{3/2,3/2}(\vec{k};-1)$ has to be created by $(\vec{\mu}_{-1/2;3/2}\cdot\vec{\mathcal{E}}_0)\ (\vec{\mu}_{-1/2;3/2}\cdot\vec{\mathcal{E}}_1)^*$, which is zero since the pump and probe fields are oppositely polarized. The source terms in Eq. (28) for the signal $P_{-1/2,-1/2}(\vec{k};-1)$ are $\vec{\mu}_{-1/2;-1/2} \cdot \vec{\mathcal{E}}_0\ f^h_{3/2,-1/2}(\vec{k};-1)$, and $P_{-1/2;-1/2}(\vec{k},-1) f^h_{3/2,-1/2}(\vec{k};-1)$. Inspection of Eq. (30) shows, however, that $f^h_{3/2,-1/2}(\vec{k};-1) = 0$ in this case. It is straightforward to extend these arguments to other cases. Thus, within the Hartree-Fock approximation evaluated in a time-dependent $\chi^{(3)}$-analysis, the echo signals are exactly zero if pump and probe fields have opposite circular polarizations. Hence, our analysis indicates that for oppositely polarized pump and probe beams any experimentally observed photon echo signal has to be attributed to deviations of the system form a perfectly symmetric thin quantum well, excited in the low intensity limit. For example, random fluctuations of the quantum-well interfaces introduce a sample-dependent effective coupling between the *hh* *s*- and *d*-states. As a consequence of this disorder induced coupling

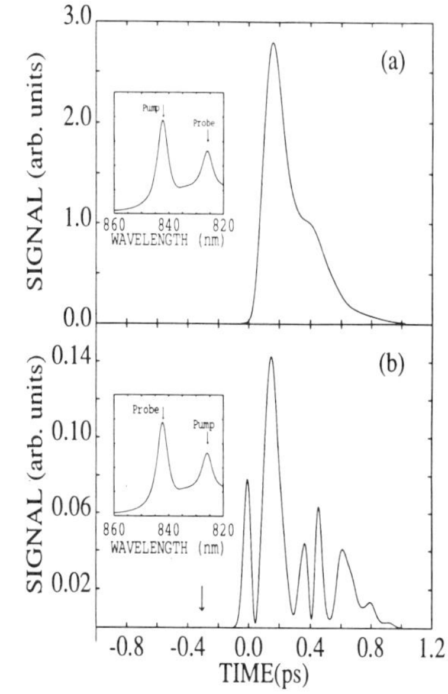

Figure 14: (a)*Same as Fig. 12, but the probe pulse center frequency is tuned into the light-hole exciton resonance.* (b) *Same as (a) but with exchanged pump and probe center frequencies. (From Ref. [5].)*

[29], four-wave-mixing signals become dipole allowed even if the pump and probe fields are oppositely circular polarized. Besides well-width fluctuations also other mechanism, such as long-range exchange interactions [28] as well as scattering by acoustic phonons [46] can cause deviations from the ideal results. To distinguish between the different effects it is essential to perform detailed time-resolved four-wave-mixing experiments in quantum wells under systematically varied sample and pump-probe conditions.

In addition to real spin transfer processes, there are other incoherent scattering processes, which do not need spin-flip processes and still would affect, for example, the $\vec{\sigma}_+$ exciton even if dipole transitions involve only the valence and conduction bands for the $\vec{\sigma}_-$ exciton. Before we discuss these scattering processes, we continue the study of the ideal coherent system for the case that the pump and probe fields have the same circular polarization. In this case the SBE for the six-band system (only s-state, no d-state excitons) reduce to the well-studied two-band Bloch equations. Previous investigations of exciton excitation in two-band bulk semiconductors have shown that the signal in the echo direction is dominated by the many-body exchange interactions [3]. Our numerical solution of the multiband SBE (with all coherent exchange terms included) closely follows the procedure outlined in Ref. [37]. Since we restrict ourselves to situations where the two subsystems shown in Fig. 11 are not coupled, we solve the equations separately for each subsystems and sum the contributions coherently to obtain the total signal. In most calculations we choose a quantum-well width of 100 Å with GaAs material parameters. To reduce the substantial numerical complexity we use perfect quantum confinement conditions and a purely two-dimensional Coulomb-potential. Furthermore, we assume Gaussian light pulses with a $100fs$ intensity full width at half maximum (FWHM). The maxima of the pulse amplitudes are $0.01meV$ for the pump and $0.001meV$ for the probe, and the dipole matrix elements of the hh transition are a factor $\sqrt{3}$ larger than those of the lh transitions. We use a delay time between the pulses of $t_d = -300fs$, i.e., the probe pulse comes $300fs$ earlier than the pump pulse which is centered at $t = 0fs$.

We evaluate Eqs. (28) - (30) for various pump and probe conditions with the same circular polarization. In Fig. 12 we show the computed signal in the FWM direction for pump and probe pulse frequencies centered at the *hh* exciton resonance. The FWM signal has the same polarization as that of the incident pulses. The solid curve in Fig. 12 shows the results of the effective mass approximation (i.e., hole states with angular momentum $\ell > 0$ are neglected). The dashed curve in Fig. 12 is obtained by numerically integrating the full multiband SBE, including the *s*- and *d*-exciton coupling. The inset is the computed linear absorption for the same material and pulse parameters. As discussed before, the *hh-lh* coupling quite generally is expected to be weak. Fig. 12 gives a quantitative estimate of this, clearly showing the validity of the effective-mass approximation in this situation. The signal in Fig. 12 is attributed to free-induction decay. We see from this calculation that the *lh* bands do not significantly contribute to the signal if the pump is in resonance with the 1*s*-*hh* exciton. The reasons are the large frequency detuning of the pulse with respect to the *lh* exciton and the $1/\sqrt{3}$ smaller matrix element of the *lh* transition.

In Fig. 13 we show the results for the case where the center frequencies of pump and probe pulses are at the 1*s*-*lh* exciton with otherwise unchanged parameters. The signal again exhibits free-induction-decay-like behavior. However, the contribution from the *hh* bands (centered around $0.2ps$) is almost one order of magnitude smaller than that from the *lh* band. In this situation the signal is mostly due to the *lh* exciton states, the *hh* continuum contribution is very weak even though it is resonantly excited.

Energetically nondegenerate four-wave-mixing results are shown in Fig. 14. Here the probe pulse is tuned to the 1*s*-*lh* exciton resonance and the pump is at the 1*s*-*hh* exciton. Similar to the results in Fig. 12, the *lh* bands are not excited in this case. The signal is still the free-induction decay from the *hh* exciton. In contrast to the rather simple signal shape shown in Fig. 14a, however, pronounced oscillations appear in the time-resolved signal if the *lh* exciton energy is chosen as the center frequency of the pump pulse and the probe is tuned into the hh exciton resonance (Fig. 14b). These quantum-beat oscillations are due to the interference between the 1*s*-*hh* exciton and its excited states. In Fig. 15 we see the results obtained when both pulses are tuned slightly above the *lh* bandgap. The *hh* continuum contribution (dashed line) yields a simple echo behavior, whereas the *lh* contribution contains oscillatory features. These oscillations are the quantum beats between the ground and excited states of the *lh* exciton. The two maxima from the *lh* signal are separated by approximately $220fs$ while the ideal quantum beat period is $200fs$. Similar quantum beat phenomena in the photon echo measurement of semiconductor quantum dots have already been predicted theoretically [6]. Our final four-wave-mixing example in Fig. 16 deals with quantum beats between *hh* and *lh* excitons. For this to occur, the pulse spectra must be broad enough to cover both exciton resonances. To simulate this situation in our calculations we leave all pulse parameters unchanged, but choose a slightly thicker quantum-well of 130 Å, where the *hh-lh* splitting is considerably smaller than in the previous examples. The inset in Fig. 16 shows that the pulse spectra engulf both exciton resonances in the linear absorption spectrum. The signal from the *lh* band is relatively simple while the *hh* band signal has some oscillating features which can be attributed to the quantum beats between the exciton ground state and excited states. The additional oscillations in the total signal are the quantum beats between the *hh* and *lh* states. The analytical value of this oscillation period, calculated from the energy difference between the *hh* and the *lh* resonance, is $276fs$ while the peak separations in Fig. 16 are $268fs$ and $291fs$ for the two oscillations on the decaying side of the signal.

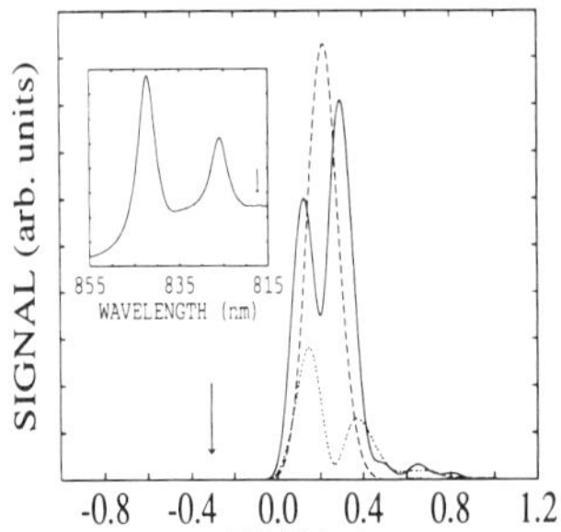

Fig. 15

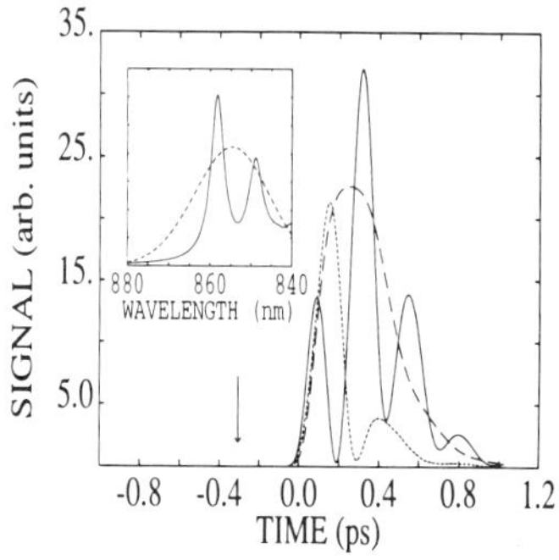

Fig. 16

Figure 15. *Same as Fig. 12, but both pump and probe have center frequencies at the band continuum* ($\hbar\omega_0 = \hbar\omega_1 = 1.518$ eV) *(From Ref. [5].)*

Figure 16. *Computed echo signal for* 130-Å GaAs *quantum well. The parameters are the same as Fig. 12, except pump and probe center frequencies are changed* ($\hbar\omega_0 = \hbar\omega_1 = 1.44$ eV). *The solid curve is the net echo signal, the short-dashed and long-dashed curve correspond to the heavy-hole and light-hole band contributions, respectively. The solid curve in the inset is the computed linear absorption of the same sample and the spectrum of probe and pump pulses is given in the dashed line. (From Ref. [5].)*

ULTRAFAST COUPLING OF SPIN STATES THROUGH EXCITATION INDUCED DEPHASING

In the previous section we saw that in a thin quantum well oppositely circularly polarized light pulses do not yield a four-wave mixing (FWM) signal, as long as there is no disorder or not enough time for real spin-flip processes of the holes to occur. This, however, does not imply that in general $\vec{\sigma}_+$-excitons and $\vec{\sigma}_-$-excitons are uncoupled. To investigate such a coupling without invoking spin-flip phenomena we drop in this section the earlier assumption that the pump pulse is so weak that the pump-induced excitation density does not contribute to additional excitation induced dephasing (EID) processes. For clarity, we neglect the light-hole bands, so that the relevant transitions are only the $\vec{\sigma}_+$-transition involving the $s = -1/2$ electron and the $\vec{\sigma}_-$-transition involving the $s = +1/2$ electron (Fig. 11). An effective coupling of the excitons corresponding to these two transition can be achieved in the following way. Assume, for example, the pump beam creates $\vec{\sigma}_+$-excitons. These excitons screen the Coulomb potential by means of intraband scattering. The screening of the Coulomb potential affects then also the $\vec{\sigma}_-$-excitons, because, quite generally, the screening of the Coulomb potential between any two carriers depends on the *total exciton density*, $n = \sum_{\sigma} n_{\sigma}$ (n_{σ} is the exciton density of $\vec{\sigma}$-excitons). The change of screening affects the binding energy of the exciton, the bandgap renormalization, and the dephasing rate. Our numerical studies have shown, that the change in the dephasing rate is the dominant effect that couples $\vec{\sigma}_+$ and $\vec{\sigma}_-$-excitons on a sub-*ps* timescale without spin-flip processes.

To investigate this effect in a FWM experiment, one has to use lineary polarized pump and probe pulses, because of the absence of a FWM signal for oppositely circularly polarized pulses [31]. Using linearly polarized beams, one always obtains a FWM signal, and the signal depends on the relative polarization of the beams. If the beams are parallel polarized, we denote the signal by $I_{||}$, and if the beam polarizations are perpendicular, we denote it $I_{\perp}$. Additional information can be obtained if the system is pre-excited with a certain exciton density.

We restrict ourselves in this section to a third-order expansion in the pump and probe field amplitudes of Eqs. (28) - (30). In this limit, the expansion of the dephasing up to linear order in the density (Eq. (22)) becomes exact and the γ'-term multiplying the density grating $\propto \sum_j f^h_{jj}(\vec{k}; m - m' = \pm 1)$ in Eq. (28) yields a $\vec{\sigma}_-$-$\vec{\sigma}_+$ coupling. An analytic perturbation analysis shows that the two gratings ($j = \pm 3/2$) are out of phase if the pulses are polarized perpendicularly. In that case the dephasing induced contributions to the signal $I_{\perp}$ are zero. For parallel polarized pulses, the gratings are in phase yielding a non-zero contribution to $I_{||}$. Consequently, the ratio $I_{||}/I_{\perp}$ is a sensitive function of γ'. In a pre-excited system γ', which is evaluated at the pre-excitation density n_0, may vary as a function of n_0 if the dephasing processes described by $\gamma(n)$ undergo saturation effects with increasing n_0. A typical variation of the dephasing rate (at $T = 20K$) is shown in the inset of Fig. 17. The system chosen in Fig. 17 is bulk

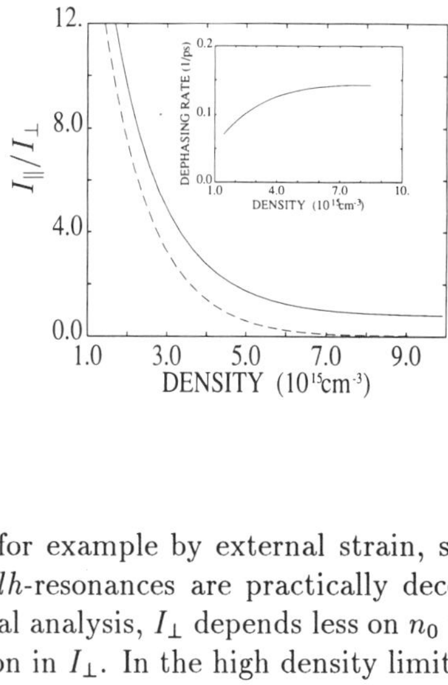

Figure 17. *Ratio of co- and cross-linearly polarized heavy-hole FWM signals in strained bulk GaAs as function of the pre-excited exciton density (solid line). The dashed line shows results (divided by 10) without exchange interactions.* Inset : *Density dependence of the underlying dephasing model. (From Ref. [31].)*

GaAs, where the *hh-lh* degeneracy is lifted, for example by external strain, so that, similar to thin quantum wells, the *hh-* and *lh*-resonances are practically decoupled. Fig. 17 shows that, similar to the perturbational analysis, $I_{\perp}$ depends less on n_0 than $I_{||}$ because there is essentially no EID-contribution in $I_{\perp}$. In the high density limit shown in Fig. 17 the EID contribution vanishes for both, $I_{\perp}$ and $I_{||}$, because $\gamma' \to 0$. Hence, the ratio $I_{||}/I_{\perp}$ approaches unity.

We also show in Fig. 17 results without exchange terms in the SBE. As discussed before (see also Ref. [3]), the exchange effects dominate the nonlinear optical signal for a two-band semiconductor. In our current model the contributions from both subsystems, $\vec{\sigma}_+$ and $\vec{\sigma}_-$, experience this effect. Only $I_{||}$, however, is additionally affected by EID since EID itself is influenced by the exchange effects, although only indirectly via the density grating. Removal of the exchange terms means that both, $I_{\perp}$ and $I_{||}$, are reduced.

Because $I_{\parallel}$ depends on the EID and the exchange effects, and the EID is not fully removed along with the exchange terms, and since $I_{\perp}$ is almost solely determined by the exchange terms the ratio $I_{\parallel}/I_{\perp}$ is enhanced without exchange terms. Investigation of EID therefore yields previously unaccessible information on ultrafast many-body effects in semiconductors with zincblende structure.

COHERENT PULSE PROPAGATION

To investigate the influence of the many-body effects on the coherent propagation of short pulses and to compare these results to the phenomenon of self-induced transparency (SIT) in atomic systems, we neglect in this section all relaxation processes and screening effects. Using the coupled MSBE (Maxwell Semiconductor Bloch Equations), Eqs. (37), (34), (35) and (36), we can derive the relation [7]

$$\frac{\partial}{\partial \zeta}\bar{S}(\zeta) \propto \sum_{\vec{k}} \left[f^e(\vec{k};\zeta,\eta=+\infty) - f^e(\vec{k};\zeta,\eta=-\infty) \right] \tag{38}$$

between the energy of the pulse and the density after the pulse-excitation. Here, we define the pulse energy density (energy/beam area) as

$$\bar{S}(\zeta) = \frac{cn_b}{8\pi} \int d\eta |E(\zeta,\eta)|^2 \,. \tag{39}$$

According to Eq. (39) the possibility for lossless propagation in an initially uninverted semiconductor is given for pulses which return the density to zero after the pulse. For pulse propagation in two-level systems this condition is satisfied if the pulse area is an integer multiple of 2π. These optical fields conserve their area while propagating and show pulse break-up for areas in excess of 2π. For example, a 4π-pulse breaks into two 2π-pulses, which travel without distortion through the medium after the break-up [47, 48, 1]. As discussed in Refs. [7] and [4], we could not find a pulse shape which returns the excitation density to zero in semiconductors. Hence, the simple 2π-solutions of the two-level system cannot be solutions of the Maxwell-SBE, because some carriers remain in the medium after the excitation. These remaining carriers lead to a gradual depletion of the pulse energy, indicating the impossibility of lossless propagation even under the idealistic conditions of vanishing carrier scattering and dephasing. To investigate the phenomenon of resonant pulse break-up in semiconductors we solve the coupled Maxwell-SBE for an input pulse of 4π. We show in Fig. 18 the temporal profile of the pulse intensity at different positions in the material. As can be seen from this figure, the propagating 4π-pulse eventually breaks into three quite well separated smaller pulses. Two of the three pulses have originally almost the same amplitude, whereas the third one is substantially smaller. During propagation, the amplitude of the first pulse increases at the expense of the two other ones. We note two main differences in comparison to the propagation of a 4π-pulse in a two-level system. First of all, a 4π-pulse in a semiconductor induces approximately four Rabi flops (effective 8π-pulse, compare Fig. 4). In a two-level system these four Rabi flops would lead to a breakup into four separated pulses, whereas we observe only three pulses in the semiconductor case. Additionally, small portions of the pulse energy are absorbed during the propagation. It was found by McCall and Hahn [47, 48] that the area of the optical field is the appropriate measure of the coherent interaction strength between light and matter. The famous area theorem shows that for the propagation of non-frequency-modulated pulses in two-level systems, pulse areas which are integer multiples of 2π

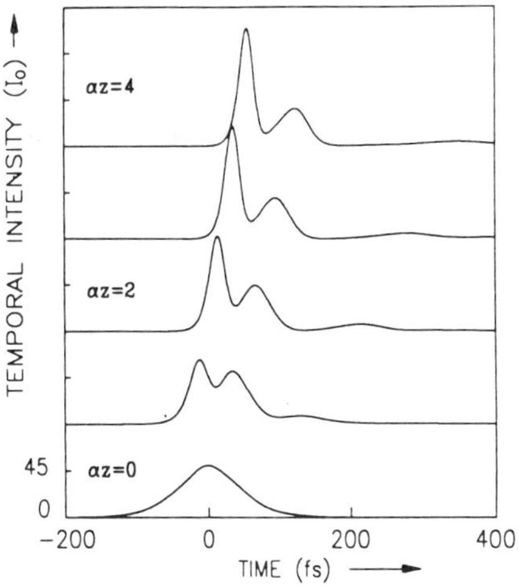

Fig. 18

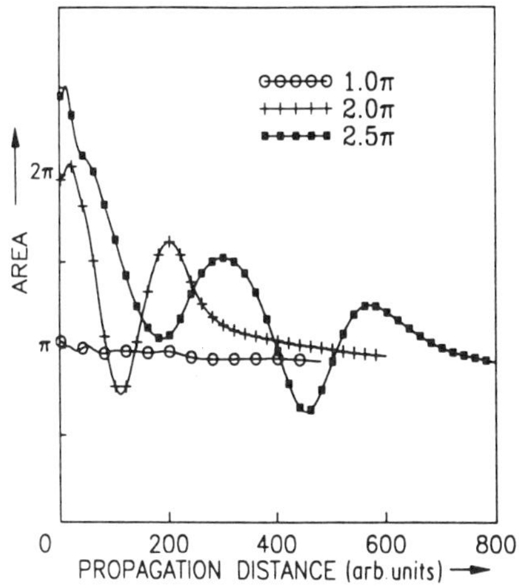

Fig. 19

Figure 18. *Temporal intensity profiles for an initially 4π-area and sech-shaped pulse at different positions in the semiconductor for the case of coherent propagation. Pulse break-up into three pulses is observed. This specific pulse would show total break-up into two 2π-pulses when traveling in a two-level system. (From Ref. [4].)*

Figure 19. *Effective area, defined in Eq. (40), of different real input pulses versus propagation distance. (From Ref. [4].)*

are stable steady state solutions. Since the refractive index for a homogeneously broadened two-level resonance is perfectly antisymmetric around the resonance energy, no net frequency modulation occurs for resonantly propagating pulses as long as pulse shape and excitation density are symmetric. The resulting temporal changes in the refractive index of the medium vanish. In semiconductors, on the other hand, the refractive index change is not symmetric and its zero crossing is blue shifted in comparison to the exciton resonance. Hence, all propagating pulses which resonantly excite the exciton accumulate a phase modulation. To take this into consideration, we introduce as the measure of the coherent interaction strength the modulus of the area A_c,

$$A_c(\zeta) = \sqrt{\left[\int d\eta \; \mathrm{Re}E(\zeta,\eta)\right]^2 + \left[\int d\eta \; \mathrm{Im}E(\zeta,\eta)\right]^2} \, . \tag{40}$$

For vanishing imaginary part, as well as for vanishing phase modulation of the pulse, the area defined in Eq. (40) reduces to the usual area definition. In Fig. 19 we plot the computed pulse area as function of the propagation distance for various input pulses. It can be seen that the area of a π-pulse is more or less conserved, whereas the area of larger pulses decreases to π in an oscillatory fashion. As a reminder, we mention that in a two-level-system area conservation occurs for all integer multiples of 2π-pulses, leading to several stable branches in a plot like Fig. 19 (see, for example, Ref. [1]). Obviously, the semiconductor scenario is substantially different. The conservation of a π-pulse instead of the 2π-pulse in atomic systems is not too surprising, since this is simply the consequence of the fact that a π-pulse in a semiconductor already generates one Rabi flop. However, even including pulses with an initial area of 4π, we did not find a higher branch in Fig. 19 with stable areas. We believe that this again is a consequence of the finite density remaining in the sample after each single Rabi flop. Due to the same reasoning, the π-branch in Fig. 19 is also only a quasi-stable solution. The pulse is initially successfully stabilizing its area but it cannot avoid the absorption losses which eventually lead to decay for longer propagation distances. Similar behavior has been found for pulse propagation under the simultaneous influence of two-level and Kerr nonlinearities [49].

PULSE PROPAGATION IN AMPLIFIERS

In the previous section we neglected relaxation, dephasing, and screening of the Coulomb potential, in order to allow comparison of coherent pulse propagation phenomena in semiconductors with two-level-systems. In reality, however, dephasing of the interband polarization and screening of the Coulomb potential are of importance for highly excited semiconductors. Especially in the case of optical amplifiers, which exhibit gain and therefore have a high excitation density prior to the arrival of the pulse, inclusion of dephasing and scattering terms in the SBE is important. Dephasing times for typical densities at room temperature are as short as $50fs$ ([38]). In this section, we study propagation of optical pulses in optical GaAs amplifiers by solving the Maxwell-SBE, Eqs. (34) - (36) We include quasi-static screening, a dephasing rate $\gamma = \hbar/60fs$ and the corresponding scattering terms within the relaxation rate approximation, with $\gamma_e \approx \gamma_h \approx \gamma$. Whereas most studies of amplifier nonlinearities concentrate on the center-frequency dependence of the pulse, we focus in the following on the intensity dependence. The center frequency will be the same throughout this section. The intensity I is chosen such that the *sech*-pulse with a FWHM of $150fs$ is fully engulfed in the gain region of the linear spectrum of the amplifier. The initial carrier density is $2.5 \times 10^{18} cm^{-3}$ at $T = 300K$. The sample length is taken as one linear gain length, $L_0 = g_0^{-1}$, where g_0 is the linear gain, so that the transmitted energy is amplified by a factor of e_1 for a small input pulse with the same frequency spectrum as the strong pulse.

Fig. 20 shows examples of our numerical results for various peak input intensities. The input (solid lines) and transmitted (dashed lines) pulse profiles for low $(I_0/4)$, intermediate $(I_0 = 1.8GW/cm^2)$, and high $(4I_0)$ peak input intensities are shown in Fig. 20a . The corresponding variation of the pulse energy, normalized to the input pulse energy, with scaled propagation distance $g_o\zeta$ is shown in Fig. 20b. For the low input intensity pulse (dashed line in Fig. 20b) amplification of the pulse energy is clearly seen. In contrast, for the intermediate intensity case (dotted line) the pulse energy propagates essentially unchanged, and for the high intensity case (dash-dotted line) the transmitted pulse suffers even a net absorption.

To analyze these phenomena, we recall that there is a continuum of electronic states in an inverted semiconductor. Prior to the pulse these states are partially populated by the electron-hole plasma according to Fermi-Dirac distributions. The one-particle states are only inverted in the spectral region between the renormalized bandgap up to the chemical potential, which is determined by the total carrier density. Hence, one always has the coexistence of *i*) a spectral gain region and *ii*) a region of optical absorption above the chemical potential. Our analysis shows that the results in Fig. 20 are caused by the interplay of the states in the gain and absorption regions. To appreciate this feature, one has to realize, that even when exciting into the gain region, the optical pulse interacts also with absorbing states. These states have a finite linewidth and therefore give rise to absorption of the pulse. For high-intensity pulses, when gain saturation becomes appreciable, this absorption can overcompensate the amplification since the

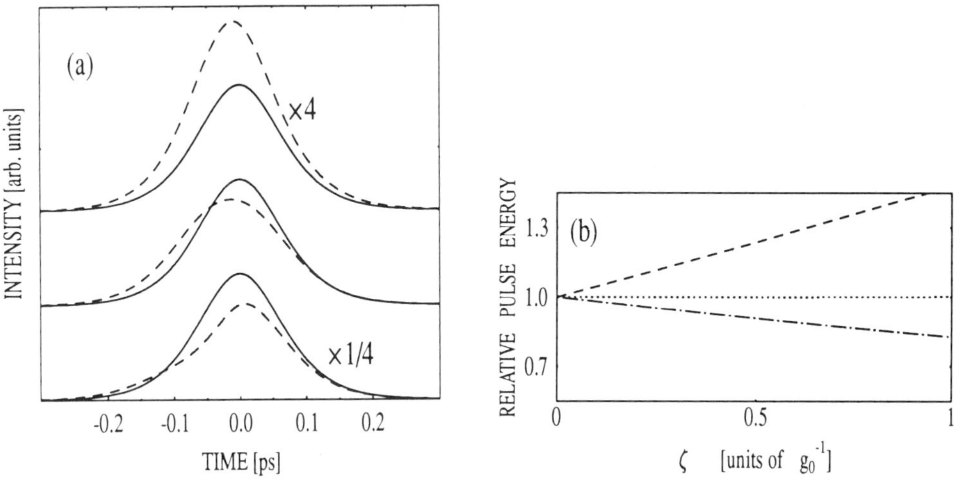

Figure 20. (a) *Input (solid) and output (dashed) pulse intensity profiles of propagation over one linear amplification length* g_0^{-1} *in a GaAs amplifier and three different peak input intensities* $I_0/4$ *(top),* $I_0 = 1.8\mathrm{GW/cm^2}$ *(middle), and* $I = I_0$ *(bottom) (for clarity the respective base lines for the different intensities are shifted).* (b) *The corresponding plots of pulse energy transmission (pulse energy divided by the input pulse energy) versus* $g_o\zeta$ *for* $I_0/4$ *(dash line),* $I_0 = 1.8\mathrm{GW/cm^2}$ *(dotted line), and* $I = 4I_0$ *(dot-dash line). An input pulse duration of* 150fs *FWHM was used in all calculations. (From Ref. [33].)*

weighting of the absorbing states is high (it increases $\propto k^2$ due to the semiconductor density of states).

The intermediate intensity ($I = I_0$) case shown in Fig. 20 represents a special situation. Here the amplifying states and the non-resonant absorbing states have equal weight and the density returns to its initial value after the pulse. This means that there is no net absorption or amplification, which is reflected by the fact that the transmitted pulse energy is almost identical to that of the input pulse. Under these conditions the pulse therefore undergoes lossless propagation.

SUMMARY

We presented a theoretical formulation and numerical analysis of coherent phenomena in semiconductors. Even though some general aspects resemble the observations of coherent effects in atomic systems, generally the semiconductor specific many-body interactions give rise to characteristic modifications and, in certain cases, qualitative changes. Whereas in the off-resonant semiconductor Stark effect the many-body modifications are generally small, these effects become larger for resonant excitation conditions. Phenomena like the semiconductor photon echo and FWM signals are even dominated by coherent many-body effects. The analysis of polarization dependent four-wave mixing signals allows investigation of ultrafast incoherent carrier scattering processes via excitation induced dephasing.

ACKNOWLEDGEMENTS

We thank D.G. Steel, H. Wang, E.M. Wright, and N. Peyghambarian and coworkers for collaboration and stimulating discussions. This work is supported by grants from the NSF, ARO/AFOSR, NEDO, and OCC, and through grants for CPU time at the Pittsburgh Supercomputer Center and the CCIT at the University of Arizona.

References

[1] L. Allen and J. H. Eberly, *Optical Resonance and Two-Level Atoms* (Dover Publications, Inc., New York, 1975).

[2] R. Binder, S. W. Koch, M. Lindberg, N. Peyghambarian, and W. Schäfer, Phys. Rev. Lett. **65**, 899 (1990).

[3] M. Lindberg, R. Binder, and S. W. Koch, Phys. Rev. A **45**, 1865 (1992).

[4] S. W. Koch, A. Knorr, R. Binder, and M. Lindberg, phys. stat. sol. (b) **173**, 177 (1992).

[5] Y. Z. Hu, R. Binder, and S. W. Koch, Phys. Rev. B **47**, 15679 (1993).

[6] S. W. Koch, Y. Z. Hu, and R. Binder, Physica B **189**, 176 (1993).

[7] A. Knorr, R. Binder, M. Lindberg, and S. W. Koch, Phys. Rev. A **46**, 7179 (1992).

[8] M. Lindberg and S. W. Koch, Phys. Rev. B **38**, 3342 (1988).

[9] W. Schäfer, K. H. Schuldt, and R. Binder, phys. stat. sol. b **150**, 407 (1988).

[10] S. Schmitt-Rink, D. S. Chemla, and H. Haug, Phys. Rev. B **37**, 941 (1988).

[11] R. Zimmermann and M. Hartmann, phys. stat. sol. b **150**, 365 (1988).

[12] W. Schäfer, in *Festkörperprobleme - Advances in Solid State Physics*, edited by U. Rössler (Vieweg, Braunschweig, 1988), Vol. 28, p. 63.

[13] S. W. Koch, N. Peyghambarian, and M. Lindberg, J. Phys. C **21**, 5229 (1988).

[14] I. Balslev, R. Zimmermann, and A. Stahl, Phys. Rev. B **40**, 4095 (1989).

[15] C. Ell, J. F. Müller, K. ElSayed, and H. Haug, Phys. Rev. Lett. **62**, 304 (1989).

[16] J. R. Kuklinski and S. Mukamel, Phys. Rev. B **44**, 11253 (1991).

[17] H. Haug and S. W. Koch, *Quantum Theory of the Optical and Electronic Properties of Semiconductors*, 2nd ed. (World Scientific, Singapore, 1993).

[18] J. M. Luttinger and M. Kohn, Phys. Rev. **97**, 869 (1955).

[19] J. M. Luttinger, Phys. Rev. **102**, 1030 (1956).

[20] A. Twardowski and C. Hermann, Phys. Rev. B **35**, 8144 (1987).

[21] B. Zhu, Phys. Rev. B **37**, 4689 (1988).

[22] M. Joffre, D. Hulin, A. Migus, and M. Combescot, Phys. Rev. Lett. **62**, 74 (1989).

[23] D. A. Broido and S. R. E. Yang, Phys. Rev. B **42**, 11051 (1990).

[24] *Optical Orientation*, edited by F. Meier and B. Zakharchenya (North-Holland, Amsterdam, 1984).

[25] T. C. Damen, L. Vina, J. E. Cunningham, J. Shah, and L. J. Sham, Phys. Rev. Lett. **67**, 3432 (1991).

[26] S. Bar-Ad and I. Bar-Joseph, Phys. Rev. Lett. **64**, 349 (1992).

[27] S. Schmitt-Rink, D. Bennhardt, V. Heuckeroth, P. Thomas , P. Haring, G. Maidorn, H. Bakker, K. Leo, D.-S. Kim, J. Shah, and K. Köhler, Phys. Rev. B **46**, 10460 (1992).

[28] M. Z. Maialle, E. A. de Andrada e Silva, and L. J. Sham, Phys. Rev. B **47**, 15776 (1993).

[29] D. Bennhardt, P. Thomas, R. Eccleston, E. J. Mayer, and J. Kuhl, Phys. Rev. B **47**, 13485 (1993).

[30] R. Eccleston, J. Kuhl, D. Bennhardt, and P. Thomas, Solid State Communications **86**, 93 (1993).

[31] H. Wang, K. Ferrio, D. Steel, Y. Hu, R. Binder, and S. Koch, Phys. Rev. Lett. **71**, 1261 (1993).

[32] W. Schäfer and K. Henneberger, Phys. Stat. Sol. B **159**, 59 (1990).

[33] A. Knorr, R. Binder, E. Wright, and S. Koch, Optics Letters **18**, 38 (1993).

[34] G. Bastard, *Wave Mechanics of Semiconductor Heterostructures* (Les Editions de Physique, Paris, 1989).

[35] M. Altarelli, in *Heterojunctions and Semiconductor Superlattices*, edited by G. Allen, G. Bastard, N. Boccara, M. Lannoo, and M. Voos (Springer-Verlag, Berlin, 1985), p. 12.

[36] H. Haug and S. Schmitt-Rink, Prog. Quantum Electron. **9**, 3 (1984).

[37] R. Binder, S. W. Koch, M. Lindberg, W. Schäfer, and F. Jahnke, Phys. Rev. B **43**, 6520 (1991).

[38] R. Binder, D. Scott, A. E. Paul, M. Lindberg, K. Henneberger, and S. W. Koch, Phys. Rev. B **45**, 1107 (1992).

[39] A. Mysyrowicz, D. Hulin, A. Antonetti, A. Migus, W. T. Masselink, and H. Morkoc, Phys. Rev. Lett. **56**, 2748 (1986).

[40] A. von Lehmen, D. S. Chemla, G. E. Zinker, and G. P. Heritage, Opt. Lett. **11**, 609 (1986).

[41] D. Fröhlich, R. Wille, W. Schlapp, and G. Weimann, Phys. Rev. Lett. **59**, 1748 (1987).

[42] N. Peyghambarian, S. W. Koch, M. Lindberg, B. Fluegel, and M. Joffre, Phys. Rev. Lett. **62**, 1185 (1989).

[43] M. Lindberg and S. W. Koch, J. Opt. Soc. Am. B **5**, 139 (1988).

[44] M. D. Crisp, Phys. Rev. A **8**, 2128 (1973).

[45] T. Östreich and A. Knorr, Phys. Rev. B (1993), accepted of publication.

[46] T. Uenoyama and L. J. Sham, Phys. Rev. B **42**, 7114 (1990).

[47] S. L. McCall and E. L. Hahn, Phys. Rev. Lett. **18**, 908 (1967).

[48] S. L. McCall and E. L. Hahn, Phys. Rev. **183**, 457 (1969).

[49] L. Matulic, G. E. Torres-Cisneros, and J. J. Sanchez-Mondragon, J. Opt. Soc. Am. B **8**, 1276 (1991).

SPIN-RELATED EFFECTS IN III-V SEMICONDUCTORS

R.T.Harley

Physics Department
Southampton University
Southampton SO9 5NH
UK

INTRODUCTION

For many purposes we can ignore the spin of electrons and holes in semiconductors. Thus an exciton in GaAs is often treated as a two level system although in reality the 1s orbital level of the electron and hole consists of several different states corresponding to different orientations of the spins of the particles. The degeneracy of the spin states can be removed by spin-orbit interaction, by low symmetry crystal fields, by exchange interactions and by applied magnetic fields. Clearly this fine structure needs to be considered in the context of coherent optical phenomena since spin reorientation may be significant for optical dephasing and splittings can give rise to quantum beating.

In this paper, in spite of its title, we shall be concerned mainly with GaAs/AlGaAs heterostructures and only in passing with bulk III-V semiconductors. We shall first review our knowledge of the fine structure of electrons, holes and excitons which has been obtained principally from optically detected magnetic resonance and from cw polarised optical measurements. These techniques are now being augmented by quantum beating and hole burning which promise to be very useful for short-lived optical states, for example in type I quantum wells. We shall finish with a review of spin relaxation time measurements at low temperatures. These have been obtained from cw polarised luminescence and from various time resolved optical techniques and are of considerable current interest.

EXCITON FINE STRUCTURE

In bulk GaAs with its cubic symmetry the conduction band at the Brillouin zone centre has s-orbital character with twofold degeneracy due to the electron's spin ($S=1/2$). The valence band has p-orbital character and therefore consists of six individual states when spin is included. The spin-orbit interaction separates these into the fourfold degenerate upper valence band ($J=3/2$) and the split-off valence band ($J=1/2$) some 340 meV lower in energy which is not considered further here. In axial rather than cubic symmetry as under uniaxial

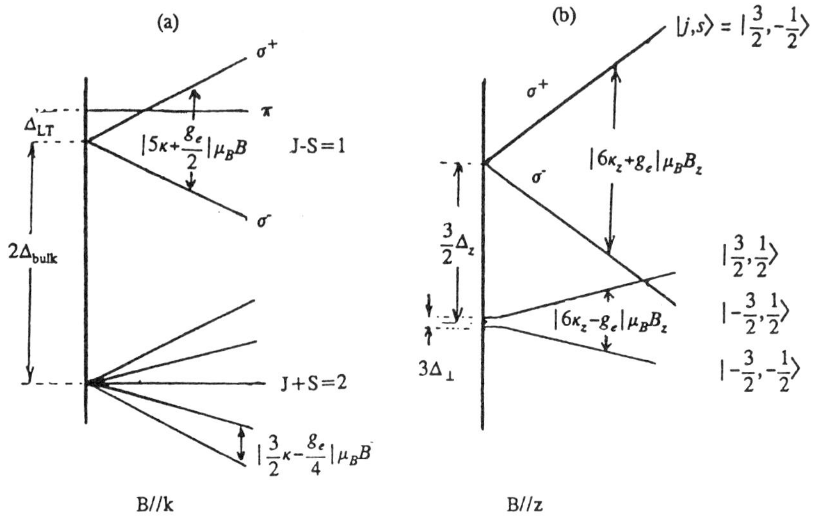

Figure 1. Fine structure of the 1s exciton level for (a) bulk zincblende semiconductor and (b) a heavy hole exciton in a quantum well grown on a (100) oriented substrate.

stress or in a planar heterostructure the valence band is further split into heavy hole ($j=\pm 3/2$) and light hole ($j=\pm 1/2$) states each twofold degenerate and separated by roughly 10 meV, depending on the details of the situation.

Application of a magnetic field to these states has two effects. First there is the quantisation of the orbital motion in the plane perpendicular to the field into Landau levels with energies separated by $\hbar\omega_c$ where $\omega_c = eB/m^*$ is the cyclotron frequency. Second, each Landau level shows a Zeeman splitting into spin components. Generally this spin splitting is much smaller than the inter-Landau level splittings and can therefore be described independently by assigning a g-factor to each Landau level. Without going into details, we recall that in k.p perturbation theory calculation of effective mass and of g-factor are closely related [1]. Thus at the zone centre the deviation of their values from the free electron values (ie the quantities ($g/2$ -1) and (m_0/m^* -1)) depend in comparable ways upon interband matrix elements of the momentum operator. Furthermore, nonparabolicity of the bands which is described by effective masses (and hence cyclotron frequencies) which vary with energy is also associated with an energy dependence of the g-values [2].

For an exciton, as is well known, the main effect of the Coulomb interaction is to bind the particles together giving a hydrogenic series of levels converging on the free particle energy (ie the band gap). Here we focus attention on the lowest energy level, ie the 1s orbital whose binding energy is typically a few meV. Ignoring the particle spins, a small applied magnetic field produces a quadratic shift of this state which merges for high fields, where the cyclotron splitting becomes comparable to binding energy, into the transition between n=1 Landau levels in conduction and valence bands [3]. Including spin the 1s exciton level consists of a set of states with different combinations of the particles' spins, as indicated in figure 1. In bulk GaAs the conduction band has $S=1/2$ and the valence band $J=3/2$ giving a total of eight approximately degenerate states, with small separations due to electron-hole exchange interaction [4].

The exchange interaction is a residual part of the Coulomb interaction and at first sight it seems odd to talk of exchange interaction between an electron and a hole since they would appear to be distinguishable particles. However the interaction is actually the sum of individual exchange interactions between the conduction electron and the electrons in filled valence band states. It can be divided into two parts, a short range (so-called analytical) part and a long range (nonanalytical) part which gives a splitting between longitudinal and transverse excitons [5]. The resulting level scheme for bulk zincblende semiconductor is shown in figure 1a consisting of a triplet with total spin (J-S) = 1 and a quintet (J+S) = 2 separated by the short range exchange energy. The optically allowed triplet is further split into longitudinal and transverse components by the long range exchange interaction [6].

In quantum wells the degeneracy of light and heavy hole valence bands is removed at k=0 by the axial quantum well potential resulting in well separated 1s heavy and light hole exciton levels. Each of these consists of four almost degenerate spin states which are split by exchange [7]. In a 2D system such as a quantum well the longitudinal - transverse splitting associated with the long range exchange interaction vanishes linearly with the exciton wavevector being essentially zero for $k \ll 1/L$ where L is the quantum well width [5]. Recent theoretical work has suggested that this k-dependent exchange interaction plays an important role in spin relaxation of excitons [8].

The Hamiltonian describing the fine structure of an exciton at k=0 in a quantum well can be written to a first approximation in the form [7]

$$H_{QW} = \sum_{i=x,y,z} [(g_{e,i}S_i - 2\kappa_i J_i)\mu_B B_i - \Delta_i J_i S_i] \tag{1}$$

with z corresponding to the growth axis of the structure. Here g_e is the electronic g-factor, κ is the parameter which describes the Zeeman energy of the hole and Δ is the short range exchange parameter. On the basis of k.p theory we expect that the electron g-factor will be almost isotropic [9]. In addition, in the limit of infinite well width we would expect Δ and κ to become isotropic and to extrapolate to the equivalent bulk values [6].

The light to heavy hole splitting is usually large enough that the two excitons can be treated separately, ignoring as a first approximation matrix elements of the above Hamiltonian between them. An effective spin Hamiltonian for the heavy hole exciton can then be written, with electron spin S=1/2 and effective heavy hole spin Σ= 1/2 used to represent the heavy hole states j=3/2 and j=-3/2. If we specialise to the case of a (100) oriented quantum well for which the ideal symmetry is D_{2d} then the most important terms become [7]

$$H_{hh} = g_e \mu_B \underline{S}.\underline{B} - g_h \mu_B \Sigma_z B_z - 3\Delta_z \Sigma_z S_z - \frac{3}{2}\Delta_\perp(\Sigma_x S_x - \Sigma_y S_y) \tag{2}$$

where $g_h = 6\kappa_z$. The transverse components of the heavy hole g-factor are zero because a transverse magnetic field does not have matrix elements between the two heavy hole states, j=+3/2 and j=-3/2 and the symmetry dictates $\Delta_x = -\Delta_y = \Delta_\perp$. The energy levels represented by (2) are shown in figure 1b. The z component of exchange separates the optically allowed and forbidden states and the transverse components produce a splitting of the forbidden states. Each of the parameters in (2) should be a function of the quantum well width, barrier heights etc. In particular the exchange should be strongly dependent on the degree of overlap between the electron and hole wavefunctions leading to strong enhancements in quantum wells.

Before reviewing the experimental results on the fine structure, we need to recall the origin of the optical selection rules for the excitonic transitions which are indicated in figure 1. Figure 2 is a schematic level diagram of the k=0 conduction and valence states in bulk and in a quantum well. The dipole-allowed transitions and their optical polarisations are shown and enable us to deduce the exciton selection rules [10]. As an example, consider the optical formation of a heavy hole exciton in a quantum well. One way to do this is with σ^+ polarisation promoting an *electron* from the -3/2 valence state to the -1/2 conduction band state. The hole created in this process has spin component j=3/2 and the exciton state is represented as |3/2,-1/2> in the |j,s> basis. A second possible way to create a heavy hole exciton uses σ^- polarised light and excites a |-3/2,1/2> exciton. The other two heavy hole

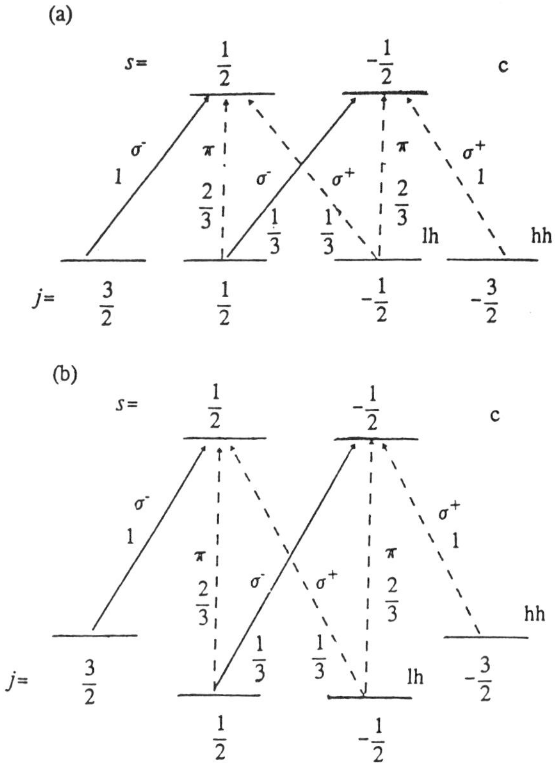

Figure 2. Schematic diagram of the conduction and valence band spin states at k=0 for (a) bulk material and (b) quantum well. The polarisations and relative strengths of the dipole allowed optical transitions are indicated.

exciton states, |3/2,1/2> and |-3/2,-1/2>, cannot be created optically. Similar arguments lead to the other selection rules indicated in figure 1.

MEASUREMENTS OF FINE STRUCTURE

The techniques for determining the g-factors and zero field splittings can be divided into three groups; magnetic resonance, conventional magnetooptical measurements and coherent optical techniques. Application of the last of these to semiconductor heterostructures is a new and promising area and will be the subject of other papers in this volume. We summarise here what has been learnt using the more conventional methods.

The classic measurements of electronic g-factors in bulk III-V materials were carried out by Weisbuch and Hermann using optically-detected electron spin resonance [1]. They found, for example that for GaAs g_e= -0.44 and that in AlGaAs the g-factor increases smoothly with concentration passing through zero for Al concentration 0.12. This behaviour is readily understood on the basis of k.p perturbation theory and the significance of the zero g-value is that at this point the band structure (or orbital) part of the g-factor just cancels the electron's intrinsic g-factor. Similar zero crossings occur in GaAs/AlGaAs quantum wells as a function of well width [11].

Electrically detected electron spin resonance has been used to obtain the electron g-factor in n-type modulation doped heterostructures at low temperatures [12]. In this situation, although the spin splittings observed in magnetotransport measurements are strongly enhanced by electron-electron interactions, the electron spin resonance gives the bare unenhanced electron g-factor for electrons at the Fermi level. The measurements demonstrated that the g-factor for GaAs increases (towards zero) for states above the bottom of the conduction band in a manner consistent with the nonparabolicity.

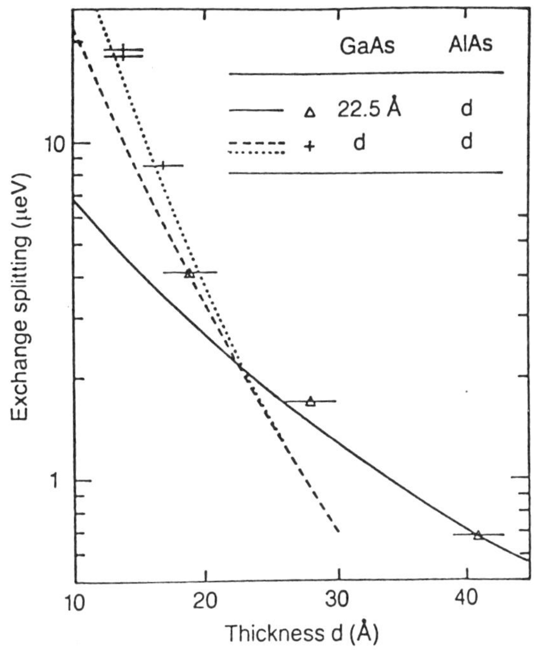

Figure 3. Exchange splitting observed in type II GaAs/AlAs quantum wells for samples with different GaAs and AlAs thicknesses. The calculation of the exchange enhancement due to quantum confinement is normalised to the data at d= 42 Å [7,14].

Optically detected magnetic resonance measurements have not been made generally on intrinsic (as opposed to defect bound) excitons in III-V bulk material or heterostructures because the lifetime of intrinsic excitons is too short. An exception is the case of type II GaAs/AlAs quantum wells [7,13] where the exciton lifetime is enhanced due to the spatial separation of electron and hole. The results of these measurements are indicated in figures 3 and 4. The observed signals correspond to microwave induced hole or electron spin flips within the exciton rather than to the exciton Zeeman splitting since the latter requires simultaneous spin flips of both the electron and the hole. The g-factor of the electron, that is at the X-point in the AlAs barrier, is found to be approximately 1.9 and the hole g-factor at the Γ-point in the GaAs well varies as indicated in figure 4 by triangles. We shall discuss the hole and electron g-values in GaAs wells when we consider type I quantum wells below. Figure 3 shows the short range electron-hole exchange splitting as a function of well and barrier widths together with a calculation of the expected variation based on the effective mass approximation for the wavefunction overlap [14]. As can be seen the interaction

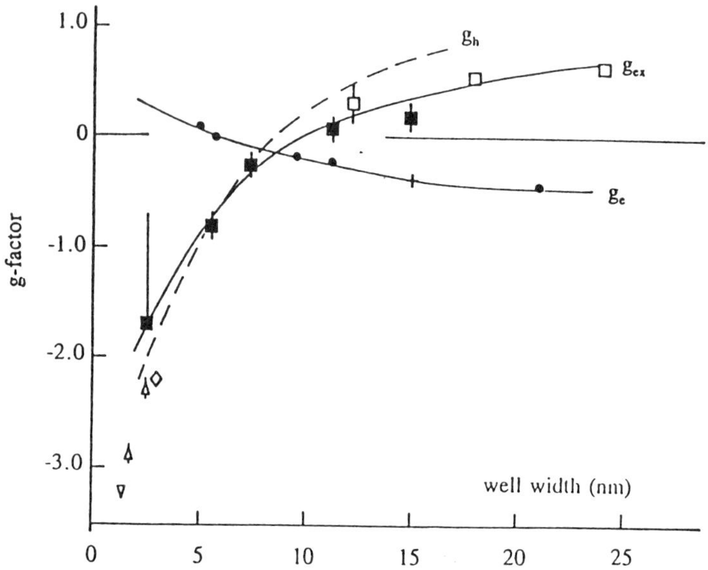

Figure 4. Measured g-factors for electrons (g_e, solid circles [11] and cross [12]) and for excitons (g_{ex}, solid squares [11] and open squares [16]) in type I GaAs/AlGaAs quantum wells. The solid curves are guides to the eye and the dashed curve represents the hole g-factor calculated from $g_h = g_{ex} - g_e$. The triangles are direct measurements of the hole g-factor in GaAs/AlAs type II quantum wells [7,13] and agree well with the variation of g_h from the type I measurements.

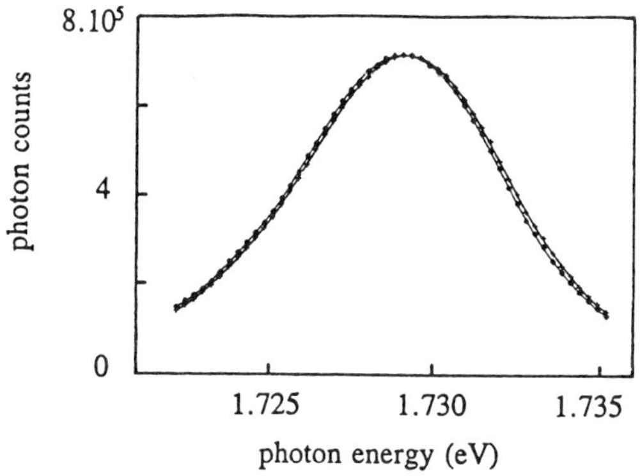

Figure 5. Measured σ^+ and σ^- Zeeman components of the heavy hole exciton luminescence at a field of 3.22 T and temperature 1.8K in a 2.57 nm well width GaAs/AlGaAs quantum well sample. Points are the measurements and the curves are the result of a polynomial fitting procedure which gives the splitting as 139$\pm$4 microvolts [11].

increases dramatically as the well and barrier widths are reduced and hence as the degree of overlap increases.

Values of the fine structure parameters for many bulk materials have been obtained from observation of exciton Zeeman spectra in absorption, reflectivity and emission over the past 20 years [6,15]. However this type of measurement is subject to many experimental difficulties and the results must be treated with caution. The most reliable data for GaAs appear to be those of Ekhardt et al [6] from careful analysis of magnetoreflectivity spectra.

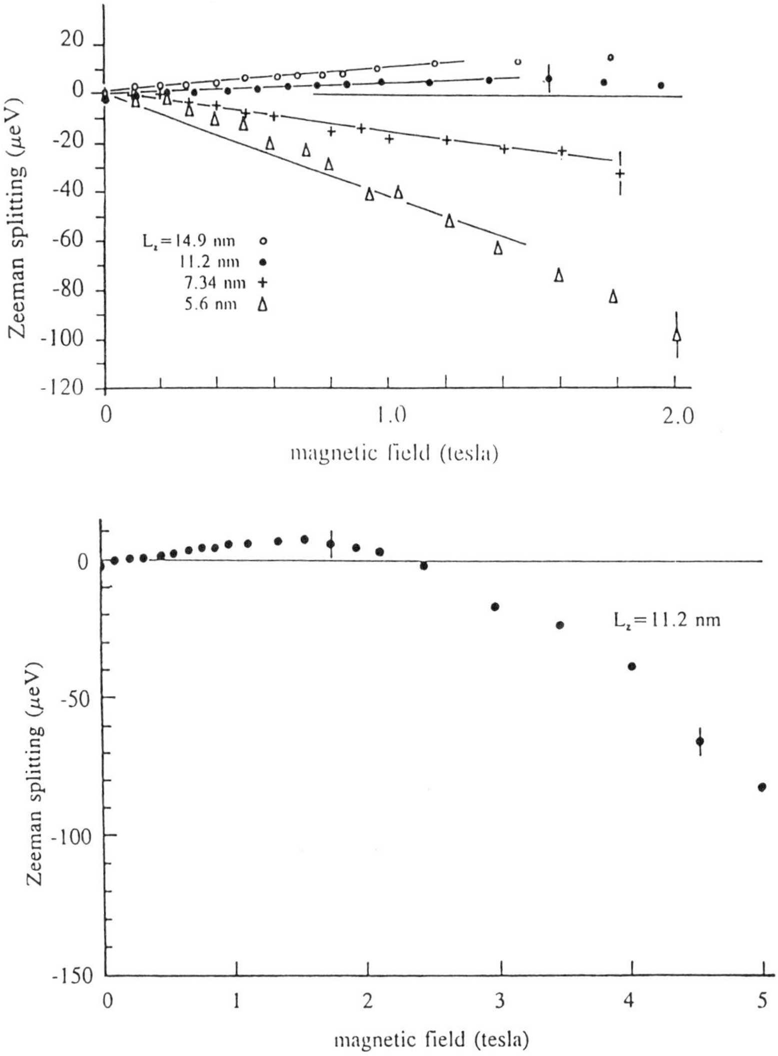

Figure 6. Measured heavy hole exciton Zeeman splittings for a range of GaAs/AlGaAs quantum well widths. The linear portion below 2 T gives the exciton g-factor. Notice the nonlinear behaviour at high fields particularly in wide wells [11].

Using $g_e = -0.46$ they were able to determine $\kappa(\text{bulk}) = 1.2$, a short range exchange parameter $\Delta(\text{bulk}) = 10 \pm 5$ microvolts and a longitudinal transverse splitting of the triplet (figure 1a) $\Delta_{LT} = 80 \pm 20$ microvolts. As we shall see the exchange interaction is consistent with extrapolation of measurements in quantum wells but this is not the case for the value of κ.

Zeeman spectroscopy of type I GaAs/AlGaAs quantum wells in emission at liquid helium temperature has been used to obtain the values of exciton g-factor g_{ex} shown in figure 4 [11,16]. Similar measurements have been made in InGaAs/GaAs quantum wells [17]. The method is based upon determination of the difference in the line position in σ^+ and σ^- polarisations for fields applied along the growth axis which is given by $g_{ex}\mu_B B_z$ with $g_{ex} = g_e + g_h$ (see figure 1). An example of the measurements appears in figure 5 and the Zeeman splittings for a series of well widths in figure 6. The difference is small compared to the inhomogeneous line width and careful analysis is needed to obtain reliable values for the Zeeman splitting. The sign of the splitting is readily obtained from these measurements. Notice that after an initial region of quasi-linear variation with field, which is used to obtain the g-values, the Zeeman splitting becomes markedly nonlinear and particularly for wider quantum wells may actually pass through zero and change sign. The possibility of such non-monotonic behaviour should be kept in mind when interpreting results of coherent optical experiments which may not be sensitive to the sign of the splitting. The nonlinearity is a result of field induced hybridisation of the n=1 heavy hole exciton with other higher lying exciton states of compatible symmetry [16] and is therefore a general phenomenon.

Various magnetooptical polarisation measurements have been used to investigate type I GaAs/AlGaAs quantum wells [11,18]. Figure 4 (solid circles) contains the electronic g-factors obtained as a function of well width from measurements of the Hanle effect in degenerate p-type wells. This involves cw measurement of the circular polarisation of photoluminescence from the sample at 1.8 K as a function of magnetic field applied transverse to the exciting light beam which is circularly polarised, incident along the growth axis and tuned to resonance with the band edge absorption. The technique gives the sign of the g-factor which was demonstrated to pass through zero for a well width near 5 nm. This variation can be understood quite well in terms of the nonparabolicity of the conduction band of GaAs [2]. As the well is made narrower the increased quantum confinement of the electron forces it into a state of higher energy in the GaAs conduction band. The observed change of effective mass [2] and the change in g-factor are consistent with one another and close to the predictions of k.p perturbation theory applied to bulk GaAs. The quantum wells apparently do little more than hold the electron up in a state away from the bulk band edge.

The measurements of electron and exciton g-factors can be combined to give the values of hole g-factor also plotted in figure 4 (dashed curve). A check of the reliability of this is to compare the values with the hole g-factors measured by magnetic resonance in type II wells of the same width [7,13] (see figure 4). We have thus reached the conclusion that the electron, hole and exciton g-factors in GaAs quantum wells show strong dependence on the well width. We have a good understanding of the electron's behaviour but the variation of the hole (and hence the exciton) g-factor has, at present, no satisfactory explanation. One significant point is that in bulk GaAs the accepted value of κ is 1.2 [6] giving $g_h = 6\kappa = 7.2$. Clearly the data of figure 4 would not easily extrapolate to this value at infinite well width.

A second set of cw magnetopolarisation measurements, based on level crossings in the exciton energy levels (see figure 1), have been used to investigate short range exchange interactions in GaAs type I quantum wells [17,18]. In this case the circular polarisation of photoluminescence was measured as a function of magnetic field applied parallel to the exciting light and to the growth axis at 1.8 K. The excitation was tuned to an energy high up in the conduction band of the sample so that, in the process of thermalisation before recombination, n=1 1s heavy hole excitons would form equally in each of the four possible spin states. Neglecting the existence of exciton level crossings, the emitted circular polarisation would be expected to increase monotonically as a result of partial thermalisation of populations between the spin states. However the measurements show anomalies of the type indicated in figure 7, which are assumed to occur at the level crossings and to result from changes in the transition rate between spin states at the crossings. Similar level crossing signals are observed in type II GaAs quantum wells [7,13]. Using the g-factors obtained from other measurements (see figure 4) it is possible to obtain the exchange splitting from the crossing fields. The results for different well widths and for both AlGaAs and AlAs barriers

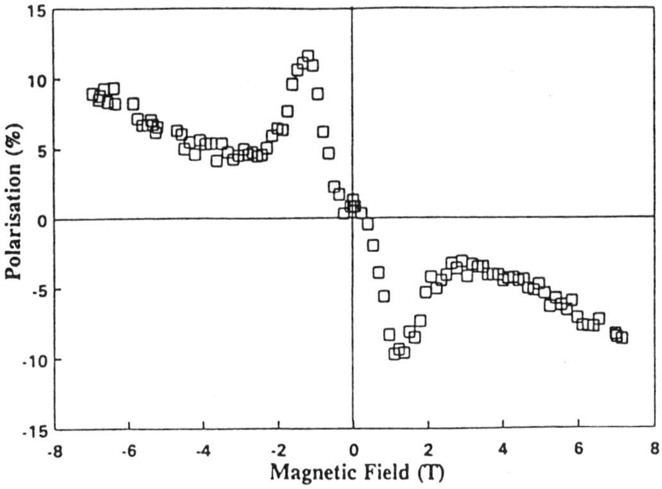

Figure 7. Measured degree of circular polarisation of the integrated heavy hole exciton luminescence in a 2.57 nm GaAs/AlGaAs quantum well sample at 1.8 K for excitation well above the band gap [17].

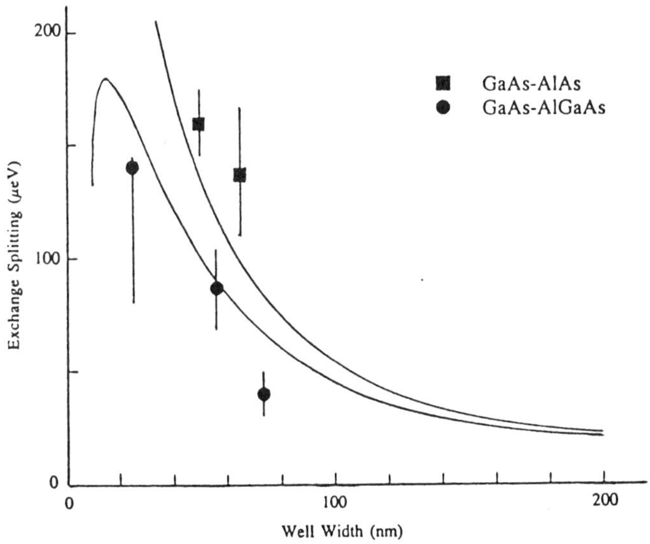

Figure 8. Experimental heavy hole exciton exchange splittings in type I GaAs quantum wells obtained from level crossing measurements [17]. The curves are calculations of the enhancement of electron-hole overlap due to quantum confinement and have been normalised to a value of exchange parameter $\Delta_z = 6$ microvolts at infinite width [19].

are shown in figure 8. Also shown in figure 8 is the result of calculation of the expected enhancements of the exchange interaction relative to that in bulk GaAs. The additional overlap provided by the AlAs barriers shows up clearly as an increase of exchange splitting at a given well width. The calculation is an effective mass approximation for the electron hole overlap in the exciton and has been normalised to a value of $\Delta_z = 6$ microvolts at infinite well width. This is in excellent agreement with the bulk value for the exchange interaction $\Delta(\text{bulk}) = 10 \pm 5$ microvolts [6].

Clearly the exchange interaction in these GaAs/AlGaAs quantum wells shows the expected enhancement due to quantum confinement. Figure 9 shows calculations by S.R.Andrews [19] of the enhancement of exchange interaction in a variety of other heterostructures including InGaAs/GaAs quantum wells and stepped barrier GaAs/AlGaAs wells [23]. It is apparent that the exchange enhancement varies markedly among these systems; being determined by electron - hole overlap it is reminiscent of the exciton binding energy but its variations are more extreme.

As described above, the ideal symmetry for a (001) oriented GaAs quantum well is D_{2d}, for which the optically allowed states of the 1s heavy hole exciton are degenerate in

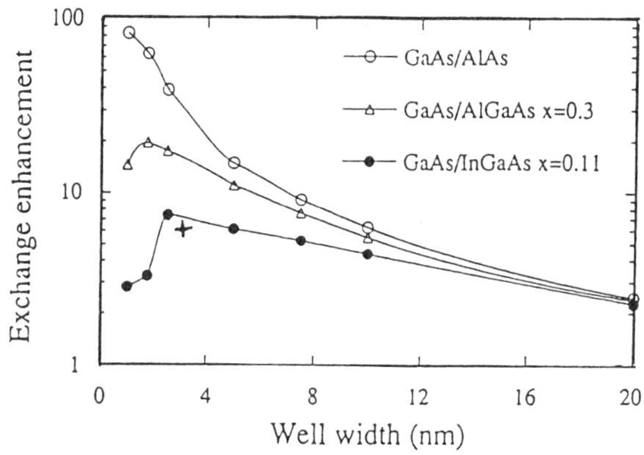

Figure 9. Calculated enhancement of heavy hole exciton exchange splittings for various heterostructures [19]. The cross indicates the enhancement for a 3 nm stepped barrier GaAs/AlGaAs well as used in reference 23.

zero magnetic field. This degeneracy arises from the fourfold (S_4) rotation/reflection symmetry about the (001) axis, and any distortion which removes this symmetry will raise the degeneracy. A splitting has been observed in type II [7] and in type I [17,18] GaAs quantum wells with well widths less than 8 nm using respectively magnetic resonance and magnetooptical measurements. In the type II systems the splitting is of order 0.8 microvolts and has been associated with characteristic growth features of the quantum wells. In the type I systems the splittings are of order 10 to 20 microvolts depending on well width and may similarly result from nonideal growth. It is of interest that interpretation of dephasing measurements in GaAs quantum wells described by Professor J. Kuhl at this workshop appears to require an inhomogeneous distribution of such splittings with a spread of order 60 microvolts.

SPIN RELAXATION

Study of figure 2 shows that when circularly polarised photons are absorbed by a semiconductor sample close to the band edge the electrons and holes created form populations which are at least partly spin-polarised. In a bulk sample electrons will be excited from both the heavy and light hole bands giving partial spin polarisation whereas in a quantum well only the heavy hole band will be involved giving complete spin polarisation. Following excitation these non-equilibrium spin populations will tend to relax and it is the processes involved which are referred to as spin relaxation. Generally the spin relaxation of the photoexcited carriers will be interrupted by their recombination and measurement of the degree of optical polarisation of the recombination radiation will give a direct measurement of the degree of spin polarisation.

The optical polarisation will be determined by the combined spin relaxation of the electrons and the holes and this will depend on the nature of the sample being studied. At low temperatures, in a pure sample with excitation resonant with the 1s exciton, the electrons and holes will always be combined in the exciton and their spin relaxation will thus be a combined excitonic process. In a p-type sample, where the population of non-photoexcited holes greatly exceeds the photoexcited populations, excitonic effects will be largely screened out and the hole spin population will be scarcely displaced from equilibrium. Therefore the optical polarisation of the recombination radiation should be determined completely by spin relaxation of the electrons. Similarly hole spin relaxation should determine the polarisation from an n-type sample. These conclusions have been questioned at various points in the literature, but it is now generally accepted that they hold for both bulk and quantum well samples at least for low excitation densities [21].

A comprehensive review of spin relaxation theory and measurements up to 1984, which includes most of the work on bulk semiconductors, and temperature dependence over a wide range is to be found in the book by Meier and Zakharchenya [10]. These experiments were carried out using cw optical techniques. More recently attention has focused on heterostructures and at the same time pulsed optical measurements have been introduced. The cw and pulsed measurements are to some extent complementary because of the different conditions under which they are made.

TECHNIQUES FOR SPIN RELAXATION MEASUREMENT

In a typical cw measurement, circularly polarised light is used to excite the sample at an energy resonant with the band gap or exciton transition. This creates a population of partially spin-polarised carriers or excitons. The degree of spin polarisation is determined by a dynamic equilibrium between spin relaxation and recombination processes and can be written

$$\langle S\rangle = S_0 \frac{\tau_s}{\tau_r + \tau_s} \tag{3}$$

where S_0 is the degree of polarisation at excitation determined by the selection rules as indicated in figure 2, τ_r is the recombination time (luminescence decay time) and τ_s is the spin relaxation time. Measurement of the degree of circular polarisation of the recombination radiation

$$P \equiv \frac{I(\sigma^+) - I(\sigma^-)}{I(\sigma^+) + I(\sigma^-)} \propto \langle S\rangle \tag{4}$$

gives directly the degree of spin polarisation of the photoexcited population and hence the relaxation time τ_s in units of the luminescence decay time. A separate determination of τ_r is then carried out in order to obtain τ_s. The main advantages of the cw techniques are that they are quick, simple and very sensitive allowing measurements to be made at very low excitation density (typically less than 4.10^8 cm^{-2} in a quantum well sample) and that they can be made with high spectral resolution. They have the disadvantage that in a case where the spin decay or the luminescence decay is not described by a single relaxation time, the value of τ_s will represent some overall average. It is becoming clear that the behaviour of excitons in quantum wells is in this category [8,24]. Furthermore examination of 3 and 4 will show that an accurate determination of spin relaxation time requires that it be of the same order of magnitude as the recombination time and this may restrict the usefulness of the method.

Pulsed techniques involve excitation of the sample with a pulse of circularly polarised light which generates a transient partially spin polarised population. The subsequent evolution of this population and its spin polarisation can then be monitored directly by time-resolved measurements of intensity and circular polarisation of luminescence [22], or via its effect on the absorption [23] or reflection of a delayed probe pulse of suitable polarisation. The clear advantage of this type of measurement is the resolution in real time of the spin decay which can reveal the exact time dependence independent of any assumptions. On the other hand the excitation density required by the measurements (usually in the range 10^9 to 10^{11} cm^{-2} in a quantum well sample [20,22-24]) is rather high and care is needed to avoid interaction effects. Furthermore the spectral resolution is limited by the pulse duration; decay times of interest turn out to be of order 10 ps or longer requiring pulses with spectral width of 1 meV or greater which is comparable to the full inhomogeneous width of exciton lines in typical quantum well samples. It has only recently become technically possible to time resolve the luminescence of resonantly excited excitons in GaAs quantum wells [24].

MECHANISMS OF SPIN RELAXATION AND ILLUSTRATIVE RESULTS

We shall consider in turn the possible mechanisms of spin relaxation of electrons, holes and excitons and point out the similarities and differences between bulk and quantum wells. It is not intended to give a comprehensive review of experimental data, rather we take illustrative examples.

For photoexcited *electrons* three mechanisms for spin relaxation have been identified. They are known by the names of their proposers as the Elliott-Yafet (EY), the D'yakonov-Perel (DP) and Bir-Aharonov-Pikus (BAP) mechanisms and are comprehensively reviewed by Pikus and Titkov in reference 10. It is expected that they will operate in quasi 2-D systems as well as in bulk material.

The EY mechanism relies on the fact that spin-orbit interaction mixes the states of opposite spin in the conduction band. Consequently processes which produce momentum scattering can also lead to spin relaxation. This leads to a spin relaxation rate proportional to the momentum scattering rate τ_p,

$$\tau_s^{-1} \sim \tau_p^{-1} \qquad (5)$$

however the process is quite weak since the conduction band is essentially of s-orbital character and the spin-orbit coupling is weak.

The basis of the DP mechanism is the small k-dependent splitting which develops between the two spin components of the conduction band for $k \neq 0$ as a consequence of the lack of a centre of symmetry in the zincblende structure. This splitting may be regarded as a fictitious internal magnetic field seen by the electron spin and which fluctuates as the electron undergoes momentum scattering processes. Generally the fluctuations are very rapid compared to the inverse of the splitting frequency so that a motional narrowing regime exists. The spin relaxation rate is therefore proportional to the momentum relaxation time,

$$\tau_s^{-1} \sim \tau_p \qquad (6)$$

The strength of both the EY and the DP mechanisms increases rapidly with the electron kinetic energy so that they tend to become important only at high temperatures. Analysis of the nature of the DP mechanism for quantum wells shows it to be modified due to the different k-dependence of the splitting from the bulk case [25].

The BAP mechanism comes into play when the photoexcited electrons are in contact with a high concentration hole gas so that appreciable electron-hole scattering occurs. The electron spins interact with those of the holes via the exchange interaction so that scattering is accompanied by spin relaxation. Clearly, except in special circumstances, in any experiment the photoexcited electrons will have to be in contact with a hole gas because it

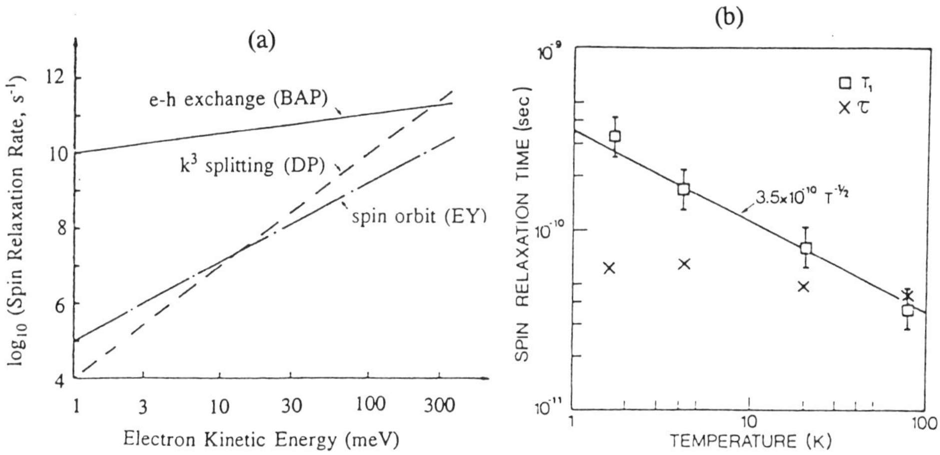

Figure 10. Electron spin relaxation in bulk p-type GaAs (4.10^{18} cm^{-3}). (a) Calculated rates for BAP, EY and DP mechanisms. (b) Experimental measurements of spin relaxation time (T_1) and of recombination time (τ) which demonstrate the dominance of the BAP mechanism in this case [26].

is necessary to use a p-type sample to avoid effects of hole spin polarisation. Furthermore the mechanism does not freeze out at low temperatures. Therefore this relaxation mechanism will usually be the dominant one for electron spins at low temperatures [22].

The actual values of spin relaxation time are rather sensitive to such factors as the sample doping. Fishman and Lampel [26] have investigated comparatively highly doped p-type GaAs (4.10^{18} cm^{-3}) and their results are illustrated in figure 10. Figure 10a shows calculations for the various mechanisms as a function of electron kinetic energy and hence temperature, and figure 10b shows some experimental measurements which display temperature dependence and order of magnitude expected for the BAP mechanism. Rather longer spin relaxation times at low temperatures (10^{-9} to 10^{-10} s) were obtained by Clark et al [27] for 10^{16} cm^{-3} p-type AlGaAs again dominated by the BAP mechanism.

Experiments in quantum wells also show the importance of the BAP mechanism for p-type samples at low temperatures. For example in a 6 nm thick 4.10^{11} cm^{-2} p-modulation doped GaAs well Damen et al [22] obtained $\tau_s = 150 \pm 25.10^{-12}$ s at 10 K. On the other hand specially designed heterostructures can produce spatial separation of the holes from the photoexcited electrons. Experiments along these lines in type II GaAs quantum wells [28] and in p-delta doped quantum wells [29] give values of electron spin relaxation time of order 2.10^{-8} s at low temperatures. Clearly the BAP mechanism has been suppressed and this much longer time is presumably a limit set by one of the other mechanisms.

Turning to the mechanisms for *hole* spin relaxation, the dominant effect is the p-orbital character of the valence band which gives strong spin-orbit admixture. In bulk material the light and heavy hole bands are degenerate at k=0 and there is strong hybridisation of the spin states at finite k. Consequently the spin relaxation time for holes is essentially the same as their momentum relaxation time which is very short, typically of order 10^{-11} s or less. In quantum wells and also in bulk material under uniaxial strain the degeneracy of the valence bands is removed and at k=0 the j=± 3/2 and ±1/2 states become pure eigenstates. The spin relaxation of heavy holes at k=0 will then be determined by other mechanisms similar to those effective for electrons [21].

Figure 11 shows data at 2K for n-type GaAs under (100) stress [30]; the relaxation time changes from 4.10^{-12} s at zero stress to a high stress asymptote of about 10^{-10} s. At the highest value of strain the light to heavy hole splitting is approximately 10 meV and the spin relaxation mechanism in this regime is believed to be analogous to the DP mechanism with momentum relaxation by neutral donors.

In a bulk n-type sample there will always be a limit set by impurity scattering but in a modulation doped sample this need not be the case and we may expect much longer heavy hole spin relaxation times. Indeed times of order 10^{-9} s have been observed at 2K in a 7.5

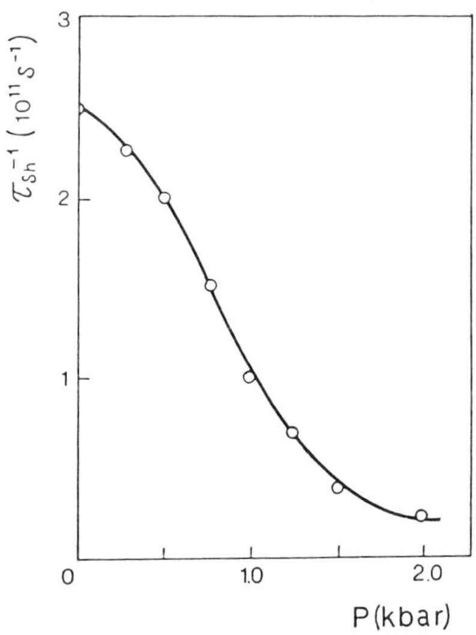

Figure 11. Spin relaxation rate of heavy holes in bulk n-type GaAs (10^{15} cm^{-3}) as a function of applied (100) stress. At the highest stress the heavy to light hole splitting is of order 10 meV [30].

nm thick modulation doped GaAs well with electron concentration varying from 4 to 12.10^{11} cm^{-2} [31]. In this example it is not clear what mechanism limits the hole spin relaxation.

In a quantum well away from k=0 the spin states become progressively admixed and the spin relaxation time is therefore expected to show a very strong dependence on the in-plane wavevector. Calculations of this effect by Ferreira and Bastard [32] are shown in figure 12. This strong k-dependence may be expected to produce a strong temperature dependence for the hole spin relaxation. It may also lead to rapid spin relaxation of the hole within an exciton since the hole's wavefunction within the exciton will be a superposition of a range of plane wave states with nonzero in-plane wavevector.

When considering the spin relaxation of *excitons* [8,24] it is necessary to distinguish the effects of different transitions within the exciton energy levels (see figure 1) and the way in which they will change the optical polarisation of recombination light. We shall concentrate on the quantum well case for which there is an increasing body of relevant experimental data; there is comparatively little for intrinsic excitons in bulk material. Referring to figure 1b, the intensity of the two polarisations (σ^+ and σ^-) will be determined by the populations of the two optically allowed exciton states. Hence relaxation from one of the optically allowed states to one of the non-allowed states will change only one polarisation component - say σ^+ - leaving the other component unaltered, whereas a transition between

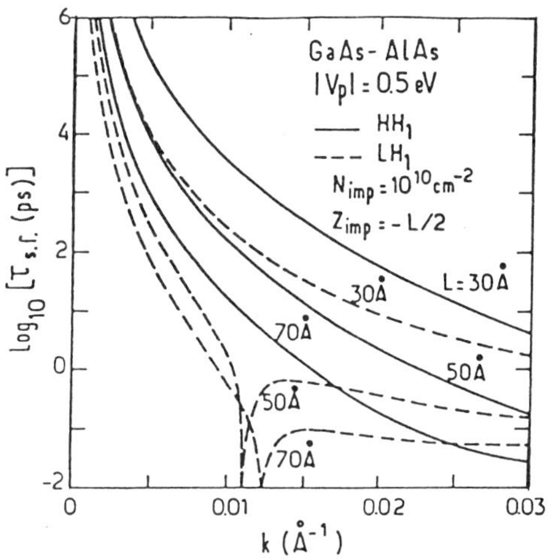

Figure 12. Calculated heavy and light hole spin relaxation times as a function of in-plane wavevector in GaAs/AlAs quantum wells and for a variety of well widths. It is assumed that the holes are scattered by a layer of impurities at the centre of the well [32].

the allowed states will increase one component and reduce the other equally. The former type of transition involves the flip of a hole or an electron alone and also requires emission or absorption of the exchange energy into the lattice. The latter type of transition requires simultaneous flips of both electron and hole spins, which might be called an exciton spin flip. At zero magnetic field this requires no energy interchange with the lattice and can therefore occur purely via matrix elements of the exchange interaction [8]. Each of these processes would change the net polarisation of the recombination radiation and would be expected to have a different characteristic time constant. Hence the decay may be quite complicated. It should be possible to separate the different processes by examining the intensities of the individual polarisation components which make up the net circular polarisation given by (4). In addition, as pointed out in references 8 and 24, the decay of luminescent intensity for resonantly excited excitons where only the optically allowed states are excited may have several characteristic times and indeed may be very different from that for nonresonant excitation where the four exciton states are fed equally.

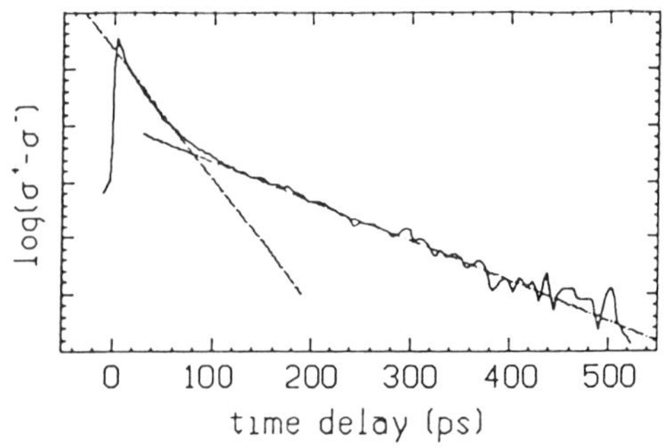

Figure 13. Decay of circular dichroism of heavy hole exciton absorption in a stepped barrier 3.0 nm GaAs/AlGaAs quantum well following excitation with a short σ^+ pulse at time zero. The decay shows a fast and a slow component which are believed to be characteristic of hole and exciton spin relaxation respectively [23,8].

At present we do not have a complete theory of these different times [8]. However as a first approximation it may be reasonable to assume that the electron's spin relaxation time within the exciton is quite long - of the same order as that of a free electron (10^{-8} s). For the reasons given above, we may expect the spin relaxation of the hole to be faster than that of a free hole and very recent calculations [8] suggest that the exchange driven exciton spin flips may be quite rapid also, the long range exchange component being the most important.

These expectations seem to be roughly in line with experiments. Figure 13 shows data obtained by Bar-Ad and Bar-Joseph [23] on a 3 nm GaAs stepped barrier quantum well which is an example of a two component decay. The fast component has been associated with hole spin relaxation (τ_h= 20 ps) whereas the slower component may be associated with exciton spin relaxation (τ_{ex}= 65 ps) [8]. Furthermore the spin relaxation of an exciton is generally faster than the relaxation of its constituent free particles measured in a similar but n- or p-type quantum well. A particularly clear demonstration of this is given in figure 14 [17] which shows the results of cw measurements in a single 5 nm, high purity GaAs quantum well grown in a structure which permitted the injection of either a hole or electron gas by means of electrical bias. At concentrations greater than about 4.10^{10} cm^{-2} of electrons or holes the relaxation is determined by the photoexcited minority carriers and times of order 10^{-9} s are observed. At lower concentrations excitons dominate the spin relaxation and a much shorter time $\sim 10^{-10}$ is obtained. (Even this short time may be an overestimate since the recombination time used was obtained from a non-resonant time-resolved measurement).

One important point to consider when comparing spin relaxation results is the temperature dependence. Detailed studies have not been reported for quantum well systems but as an indication figure 15 shows the temperature dependence of the two plateau regions of high concentration for the sample of figure 14 which represent the hole and electron spin relaxation times. They show differing temperature dependence but the important point is that each falls rapidly in the region below 15K, suggesting that for experimental purposes "low temperature" at least for this sample means within the liquid helium range.

CONCLUSIONS

The study of spin-dependent phenomena in semiconductors is at a very fascinating stage with interest focused particularly on heterostructures. Much has been learned about the g-factors and fine structure of electrons, holes and excitons in GaAs structures using conventional cw methods and it will be very interesting to see these methods applied to other

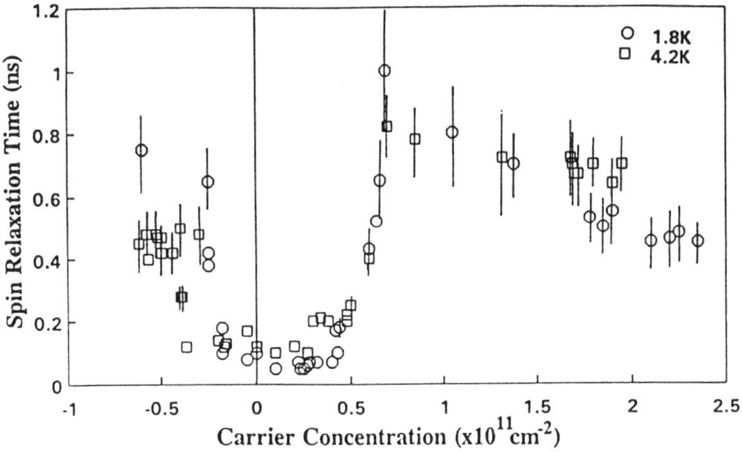

Figure 14. Measured spin relaxation times in a 5.75 nm thick GaAs/AlGaAs quantum well. To the right the sample is n-type giving heavy hole spin relaxation times, and to the left p-type giving electron spin relaxation times. For zero concentration the relaxation is excitonic [17,18].

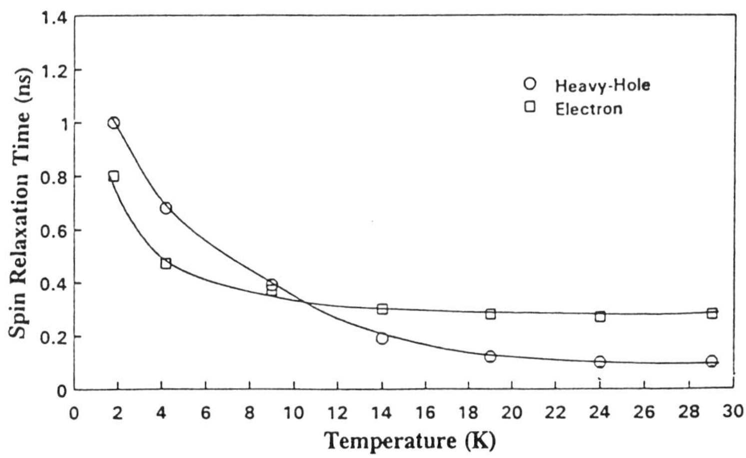

Figure 15. Temperature dependence of the two plateau regions of high carrier concentration for the same sample as figure 14 [17,18].

material systems. Coherent optical techniques capable of resolution within the inhomogeneous line profile are also now being applied to semiconductor heterostructures and promise to improve the precision of the data. Investigation of spin relaxation is a very active area with new results appearing at a considerable rate. The immediate challenge is to obtain a unified experimental picture from the pulsed and cw techniques. At present attention is focused on GaAs but in future other material systems will be investigated, as will effects of external perturbations such as magnetic field and variation of temperature.

ACKNOWLEDGEMENT

It is a pleasure to acknowledge the receipt of preprints of recent work from Professor Lou Sham and Dr Jagdeep Shah which have been useful in the preparation of this review.

REFERENCES

[1] See for example: C.Weisbuch and C.Hermann, Phys.Rev.B15 816 (1977); C.Hermann and C.Weisbuch, Phys.Rev. B15 823 (1977).

[2] E.D.Palik, G.S.Picus, S.Teitler and R.F.Wallis, Phys.Rev. 122 475 (1961); D.C.Rogers, J.Singleton, R.J.Nicholas, C.T.Foxon and K.Woodbridge, Phys Rev B34 4002 (1986); J.Singleton, R.J.Nicholas, D.C.Rogers and C.T.B.Foxon, Surf. Sci. 196 429 (1988).

[3] See for example: A.S.Plaut, J.Singleton, R.J.Nicholas, R.T.Harley, S.R.Andrews and C.T.B.Foxon, Phys.Rev. B38 1323 (1988).

[4] K.Cho, S.Suga, W.Dreybrodt and F.Willmann, Phys.Rev. B11 1512 (1975).

[5] L.C.Andreani, F.Bassani and A.Quattropani, Il Nuovo Cimento 10D 1473 (1988); L.C.Andreani and F.Bassani, Phys.Rev. B41 7536 (1990); U.Rossler, S.Jorda and D.Broido, Solid State Commun.73 209 (1990).

[6] W.Ekardt, K.Lösch and D.Bimberg, Phys.Rev. B20 3303 (1979).

[7] H.W.van Kesteren, E.C.Cosman,W.A.J.A.van der Poel and C.T.Foxon, Phys.Rev. B41 5283 (1990).

[8] M.Z.Maialle, E.A.de Andrada e Silva and L.J.Sham, Phys.Rev. (to be published).

[9] This point has been addressed by V.K.Kalevich and V.L.Korenev JETP Lett. 56 253 (1992) and by G.Hendorfer and J.Schneider, Semicond.Sci.Technol. 6 595 (1991).

[10] *Optical Orientation*, edited by F.Meier and B.P.Zakharchenya, (North Holland, Amsterdam, 1984).

[11] M.J.Snelling, G.P.Flinn, A.S.Plaut, R.T.Harley, A.C.Tropper, R.Eccleston and C.C.Phillips, Phys.Rev. B44 11345 (1991); M.J.Snelling, E.Blackwood, C.J.McDonagh, R.T.Harley and C.T.B.Foxon, Phys.Rev.B45 3922 (1992).

[12] M.Dobers, K. von Klitzing and G.Weimann, Phys.Rev. B38 5453 (1988).

[13] J.M.Trombetta, T.A.Kennedy, W.Tseng and D.Gammon, Phys.Rev. B43 2458 (1991).

[14] B.Rejaei Salmassi and G.E.W.Bauer, Phys.Rev. B39 1970 (1989).

[15] D.D.Sell, S.E.Stokowski, R.Dingle and J.V. Di Lorenzo, Phys.Rev.B7 4568 (1973); S.B.Nam, D.C.Reynolds, C.W.Litton, R.J.Almassy, T.C.Collins and C.W.Wolfe, Phys.Rev. B13 761 (1976); M.Suffizynski, J.Phys. C8 L400 (1975).

[16] W.Ossau, B.Jäkel, E.Bangert and G.Weimann, NATO Advanced Study Institutes Ser.B.Vol.183 (Plenum, New York, 1988) p.285; G.E.W.Bauer and T.Ando, Phys.Rev. B37 3130 (1988).

[17] E.Blackwood, PhD Thesis 1993 (Southampton University, Unpublished).

[18] M.J.Snelling, PhD Thesis 1991 (Southampton University, Unpublished).
[19] S.R.Andrews, Private Communication.
[20] As used in experiments by: S.Bar-Ad and I.Bar-Joseph, Phys.Rev.Lett. 66 249 (1991).
[21] This point is discussed in: T.Uenoyama and L.J.Sham, Phys.Rev.Lett. 64 3030 (1990).
[22] For example, T.C.Damen, L.Viña, J.E.Cunningham, J.Shah and L.J.Sham, Phys.Rev.Lett. 24 3432 (1991).
[23] S.Bar-Ad and I.Bar-Joseph, Phys.Rev.Lett. 68 349 (1992).
[24] A.Vinattieri, J.Shah, T.C.Damen, D.S.Kim, L.N.Pfeiffer and L.J.Sham (preprint).
[25] M.I.D'yakanov and V.Yu.Kacharovskii, Sov.Phys.Semicond.20 110 (1986); G.Bastard and R.Ferreira, Surface Science 267 335 (1992).
[26] G.Fishman and G.Lampel, Phys.Rev. B16 820 (1977).
[27] A.H.Clark, R.D.Burnham, D.J.Chadi and R.M.White, Phys.Rev. B12 5758 (1975).
[28] W.A.J.A.van der Poel, A.L.G.J.Severens, H.W.van Kesteren and C.T.Foxon, Superlattices and Microstructures, 5 115 (1989).
[29] J.Wagner, H.Schneider, D.Richards, A.Fischer and K.Ploog, Phys.Rev. B47 4786 (1993).
[30] A.N.Titkov, V.I.Safarov and G.Lampel, Proc.14th Int.Conf.of the Physics of Semiconductors, Edinburgh 1978 (Conference Series No.43, The Institute of Physics, Bristol) ed. B.L.H. Wilson p.1031; Also see reference 10 p.73.
[31] Ph.Rhoussingnol, P.Polland, R.Ferreira, C.Delalande, G.Bastard, A.Vinattieri,J.Martinez-Pastor, L.Carraresi, M.Colocci, J.F.Palmier and B.Etienne, Phys.Rev. B46 7292 (1992).
[32] R.Ferreira and G.Bastard, Phys.Rev. B43 9687 (1991).

QUANTUM BEAT SPECTROSCOPY OF EXCITONS IN SEMICONDUCTORS

W. von der Osten, V. Langer and H. Stolz

Fachbereich Physik
Universität-GH Paderborn
33095 Paderborn, Germany

1. INTRODUCTION

In this contribution, recent investigations are described in which quantum beat spectroscopy is employed to reveal and quantitatively characterize the coherence of exciton states in semiconductors and ionic crystals. Coherence in that case signifies the existence of a well-defined phase relationship between the time-dependent quantum mechanical wavefunctions. Following the creation of the exciton system in a coherent state by optical excitation with a short laser pulse, scattering will destroy the phase relation and, depending on the specific interactions, may lead to a rapid loss of coherence. Besides pure dephasing by elastic scattering, also energy relaxation by inelastic scattering destroys the coherence. In general, these processes advance simultaneously and finally are terminated by exciton recombination which takes the excited system back into the crystal ground state. Even though phase and energy relaxation in case of excitons are extremely fast proceeding on a pico- or subpicosecond time scale, the recent advances in ultrafast spectroscopy now allow to directly investigate the dynamics of these processes.

Conventional linear optical spectroscopic methods like absorption, luminescence or excitation spectroscopy are usually not sensitive to coherence. In the terminology of the density matrix of the system[1,2] they only enable to measure the diagonal elements of the density matrix which describe the time-dependent population of the states. To also gain information on the coherence, the nondiagonal elements have to be measured. This is possible by using a "phase sensitive" technique like quantum beat spectroscopy in which the coherence of states that is created by a short laser pulse manifests itself by a modulation of the resonance fluorescence intensity.

The "classical" experiment, in which coherence played an important role and from which a coherence time of an electronic state could be derived, has been performed by Hanle as early as 1924 [3]. He studied the depolarization of the strong 253.7 nm ($6^3P_1 \rightarrow 6^1S_0$) resonance fluorescence line of Hg-atoms excited with light linearly polarized perpendicular to a weak magnetic field. Detecting the signal along the field direction, he measured the degree of linear polarization for the emission from the $m_J = \pm 1$ Zeeman sublevels as function of field strength and, interpreting the measured dependence by the simple electron oscillator model, was able to deduce the coherence time. As the intensities were measured time-integrated, strictly speaking, from Hanle's experiment the coherence time multiplied by the Landé factor was obtained.

An important supposition for the coherence to be revealed in this experiment was the simultaneous excitation of the states by linearly polarized light that creates a coherent superposition with a well-defined phase. As excitation was accomplished by a cw light source (a Hg-lamp, by the way!), in order to create a coherent superposition of the two states the measurements were limited to the range where the Zeeman splitting is smaller than the homogeneous linewidth of the states i.e. $\Delta E(B) < \delta E_{hom}$ ("zero-field level crossing"; see e.g. ref. 4). This makes clear the advantage of quantum beating where simultaneous impulsive excitation of states with a spectrally broad pulse results in a coherent superposition even if the energy difference between the states is larger than the homogeneous width ($\Delta E(B) > \delta E_{hom}$). In addition, measuring the time dependence, quantum beat experiments provide the beating frequency as an additional piece of information permitting to determine the g-value and the coherence time independent of each other.

While in atomic and molecular physics quantum beat spectroscopy started with the advent of lasers to become soon a flourishing field and nowadays is well established (see e.g. ref. 5-7) for excitonic states in solids the potential of this method remained unrecognized until 3 years ago[8,9]. One reason was that coherence times of excitons, as compared to atoms, were expected to be considerably shorter and beyond the experimental possibilities. As will be demonstrated below, this is not at all the case. For some excitonic systems, coherence times as long as several hundreds of picoseconds are found easily accessible to time-resolving measurements. Actually, quantum beat spectroscopy became a very valuable tool in exciton physics[10-13]. One piece of information provided concerns very small splittings between energy states that can be measured with high precision and independent of inhomogeneous broadening of the transitions. From these measurements important exciton parameters like g-values or exchange energies can be derived. Above all, beat spectroscopy allows to develop a detailed and consistent picture of exciton relaxation. In particular, it enables to discriminate *Raman processes*, in which the coherence of the resonant intermediate state is retained, from *hot luminescence* occurring after phase relaxation. Measuring the time-dependent beating one can directly trace the temporal evolution of the exciton system from the initially prepared coherent state to the final incoherent ensemble and determine corresponding coherence times.

2. PRINCIPLE OF QUANTUM BEATING

The phenomenon of quantum beating is based on the concept of quantum mechanical superposition of states. Let us consider, for simplicity, an ensemble of three-level systems which in our case are of excitonic nature consisting of a ground state (0) and two excited states (1,2) with corresponding wavefunctions and energies Φ and E as shown in Fig. 1. Supposing allowed transitions $0 \rightarrow 1$ and $0 \rightarrow 2$, a coherent superposition of the excited states can be achieved by exciting the system with a short laser pulse (duration Δt_p) having a spectral width $\Delta E \approx \hbar/\Delta t_p$ larger than the splitting $|E_2 - E_1|$ of the excited states. This excitation results in a new quantum mechanical state. At time t = 0, it is given by the linear combination of the two eigenfunctions

$$|\Phi(t=0)\rangle = \alpha_1|\Phi_1(t=0)\rangle + \alpha_2|\Phi_2(t=0)\rangle, \tag{1}$$

where α_1, α_2 are probability amplitudes that represent the transition moments for absorption into each of the excited states. Considering its temporal evolution at some later time one may write

$$\begin{aligned}|\Phi(t)\rangle &= \alpha_1 \cdot \exp[-(i/\hbar)E_1 t - t/2\tau_1 - i\varphi_1(t)]\,|\Phi_1\rangle \\ &+ \alpha_2 \cdot \exp[-(i/\hbar)E_2 t - t/2\tau_2 - i\varphi_2(t)]\,|\Phi_2\rangle\,.\end{aligned} \tag{2}$$

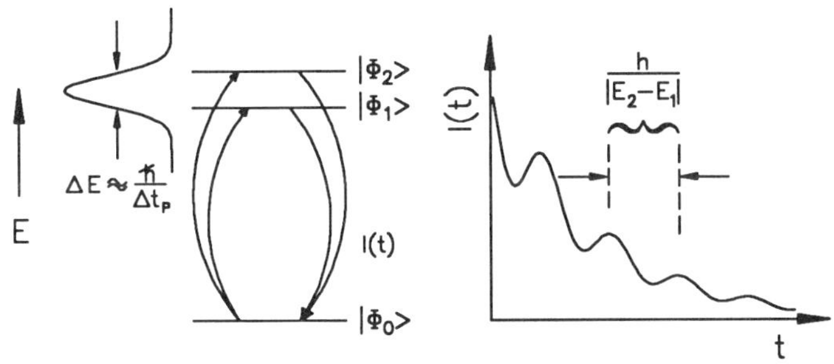

Figure 1. Three-level system consisting of ground and two excited states (left) to explain the occurrence of quantum beats in resonance fluorescence (right) induced by a short laser pulse. The pulse is assumed to be transform limited having spectral width and duration ΔE and Δt_p, respectively.

To more clearly reveal the dephasing, the time-dependent phase factors here are written explicitly taking α_1, α_2 to be real quantities and φ_1, φ_2 corresponding phases. In addition, a phenomenological damping term is introduced to account for the finite population lifetimes of the states for which we may approximately write $\tau_1 \simeq \tau_2 = T_1$ (T_1: energy relaxation time or population lifetime). Considering dipole-allowed transitions (dipole operator D), the emitted time-dependent intensity then is calculated from the dipole matrix element between ground and excited states according to

$$\begin{aligned}I(t) &\sim |\langle\Phi_0|D|\Phi(t)\rangle|^2 \\ &= (\alpha_1^2\,|M_{10}|^2 + \alpha_2^2\,|M_{20}|^2) \cdot \exp(-t/T_1) \\ &+ 2\alpha_1\alpha_2 \cdot \mathrm{Re}\,\{\, M_{10}\,M_{20} \cdot \exp[\,-(i/\hbar)|E_2 - E_1|t \\ &+ i(\varphi_1(t) - \varphi_2(t))]\} \cdot \exp(-t/T_1)\end{aligned} \tag{3}$$

with real matrix elements M_{ij} in obvious notation. Setting $|E_2 - E_1|/\hbar = \omega_{21}$, finally this expression can be cast into the form

$$I(t) \sim A \cdot \exp(-t/T_1) + B \cdot \exp(-2t/\tau_{coh}) \cdot \cos(\omega_{21}t) \quad (4)$$

where

$$1/\tau_{coh} = 1/2T_1 + 1/T_2' \quad (5)$$

and the intensity amplitudes A and B depend on the transition moments for excitation and deexcitation of the states.

According to eqs.(4) and (5), the intensity generated by coherent excitation is seen to decay exponentially with energy relaxation time T_1 and, because of the interference term, is modulated with the beating frequency ω_{21} (Fig. 1, right). It is this latter term which contains the information on the coherence of the system. It dies out with damping constant $2/\tau_{coh}$ (τ_{coh}: coherence time, often denoted optical dephasing time T_2) related to the homogeneous linewidth of the optical transition by

$$\delta E_{hom} = 2\hbar/\tau_{coh}\,. \quad (6)$$

To actually derive the expressions above and introduce the various times, besides calculating quantum mechanical expectation values for a single exciton system, statistical averaging over the ensemble is necessary. This may be done in a phenomenological way by making use of the density matrix formalism[1,7] which allows to treat the dynamics and coherence properties of the exciton system (for details see refs. 8, 14, 15). In particular, averaging the time-dependent phase term in eq.(3) results in the $1/T_2'$ contribution to the total decay rate of coherence $1/\tau_{coh}$ (T_2' : pure dephasing time). Since depopulation always results in loss of coherence, in the absence of pure dephasing eq.(5) predicts an upper limit for the coherence time $\tau_{coh} \leq 2T_1$.

Measuring the time-dependent intensity I(t) provides all quantities relevant for the exciton state. The beating frequency directly gives the energy difference between the excited states which, compared to spectral measurements, can be extremely small as only the homogeneous linewidth δE_{hom} puts a lower limit on it. In case of excitons, the splitting of the states is usually produced by some external perturbation like a magnetic field, the dependence of oscillation frequency upon field allowing to deduce magneto-optical parameters of the exciton states. Evidently quantum beats in resonance fluorescence reflect an intrinsic quantum mechanical interference as they are based on the coherence between wavefunctions of a single exciton. This is important to realize as it enables to measure, in the time domain, energy splittings independent of any inhomogeneous broadening which is usually present and obscures this splitting in the spectrum.

Concerning the dynamical behaviour of the exciton system, by measuring the intensity as a function of time all relevant relaxation times can be determined. According to eq.(4), τ_{coh} directly may be deduced from the coherent part of the signal while a zero-field measurement for which the interference term does not exist provides the energy relaxation time T_1 as well as, through eq.(5), the pure dephasing time T_2'.

Excitons are hydrogen-like excitations and in spite of the dependence of exciton energy upon wavevector, in good approximation, may be considered two-level systems with a multiple excited state due to degeneracy (see e.g. refs. 16, 17). This is because in the absorption process, direct or indirect, energy and wavevector are conserved so that resonant excitation with monochromatic light will create the exciton in a state with well-defined E and **K**. Compared to atoms, however, the relaxation processes of excitons are more complex. Besides radiative recombination, *inelastic scattering* by phonons into states with different energy and wavevector (E′, **K**′) may occur and also excitons can be effectively captured at defects or impurities. These processes all contribute to the energy relaxation rate $1/T_1$. In addition, *elastic scattering* of excitons results in pure dephasing characterized by the ("pure") phase relaxation rate $1/T_2'$. As long as high excitation densities and exciton-exciton collisions can be excluded, as in all cases discussed below, the elastic scattering essentially occurs at defects changing the wavevector of the exciton without affecting its energy (E, **K**′).

Before closing this section, two other remarks are appropriate:

1. The physical interpretation of quantum beating is completely analogous to that of Young's double slit experiment. In this experiment, an interference pattern will only become visible if the two slits are simultaneously opened and there is no way to discriminate through which of the two the photon has passed. In case of exciton emission from the coherent excited state, this would correspond to indistinguishable emission channels $1 \rightarrow 0$ and $2 \rightarrow 0$ into a common ground state in which case the total probability is given by squaring the added amplitudes for the single processes. Closing in Young's experiment one or the other slit, the pattern will disappear. For the exciton this represents the situation that only one of the transitions is allowed while the other is filtered out either by suitable polarization or spectral filtering. As the scattered channels now are distinguishable, beating will disappear. It is also clear from this brief note that quantum beats in fluorescence will only probe the coherence properties of excited states.

2. Quantum beats in resonance fluorescence as treated here must be distinguished from beating phenomena observed with coherent nonlinear techniques like four-wave mixing or absorption saturation (for references see e.g. contributions by J. Kuhl, G. von Plessen and J. Hvam to this volume and refs. 18-21). In contrast to fluorescence beats, in these pump-and-probe type experiments the third-order susceptibility of the medium is involved. This not only entails a more complex interpretation of data, it also may lead to altered relaxation times since much higher intensities are needed to produce the nonlinear signals. Even more important, in these experiments a macroscopic polarization is created. Therefore, *a priori* it is not evident if observed oscillations are due to a quantum interference or rather due to classical heterodyning of the polarization fields at the nonlinear detector. It was only recently that this distinction could be made e.g. by time-resolving the four-wave mixing signal[20,21].

3. TRANSFORM-LIMITED LIGHT SCATTERING OF EXCITONS

The method by which the fast exciton relaxation processes described below are investigated is *transient resonant light scattering*. In this linear optical method the exciton system is excited by a short light pulse, and the time dependence of the light scattering spectrum (or intensity $I(t,E_S)$) containing the desired information is recorded. To extract the full information from the scattered signal, transform-limited spectroscopy has to be employed which means that the analysis by the spectroscopic apparatus requires simultaneous optimum temporal and spectral resolutions[22,23]. More specifically, for interpreting the quantum beat results one needs to have a theory of resonant light scattering at hand that takes into account the details of excitation as well as the effect of the detection process.

Considering an exciton system interacting with a phonon bath, recently quantum mechanical perturbation theory was used to calculate the time-resolved scattering spectrum[15]. This calculation not only clarified the implications of the measuring process itself. By taking into account both elastic and inelastic relaxation processes of the exciton, it also could be demonstrated how the time dependence in connection with the polarization properties can be exploited to discriminate coherent Raman scattering from incoherent hot luminescence as necessary in interpreting the beat measurements. Referring the reader for details of the calculation and the complete results to previous publications (cf. refs. 11, 13, 24), we restrict ourselves to reproduce the expression for the coherent scattering contribution (or Raman component) which is given by

$$I^{CS}(t,\mathbf{e}_L,\mathbf{e}_S) = \left| \sum_m (\mathbf{M}_m \cdot \mathbf{e}_S)^* (\mathbf{M}_m \cdot \mathbf{e}_L) \cdot \psi_m(t) \right|^2 . \tag{7}$$

In this expression, $\psi_m(t)$ represents the complex time-dependent scattering amplitude given by

$$\psi_m(t) = \int_{-\infty}^{+\infty} dt_1 \int_{-\infty}^{+\infty} dt_2 \, F_L(t_2) \, F_S^*(t-t_1) \exp\left[i\omega_S t_1 - i\omega_L t_2 - i(E_m - (i/2)\,\delta E_{hom})(t_1 - t_2)/\hbar\right] \tag{8}$$

which via the envelope of the laser pulse ($F_L(t)$) and the transmittance function of the specific spectral filter ($F_S(t)$) used in the experiment contains the details of the excitation and detection system. The sum has to be performed over the resonantly excited exciton states (m) with energy E_m whereby the summation over the wavevector is neglected assuming direct excitons that interact with light only at $\mathbf{K} \simeq 0$. $\mathbf{M}_m$ are the transition matrix elements that account for the exciton-photon interaction and in case of Raman scattering considered later also comprises the exciton-phonon interaction. Finally, $\mathbf{e}_L$ and $\mathbf{e}_S$ are polarization unit vectors for the incident (laser) and scattered light.

Eq.(7) implies that the Raman component is polarized actually following the usual Raman selection rules. This is distinctly different for the hot luminescence component (intensity $I^{HL}(t,\mathbf{e}_L,\mathbf{e}_S)$) that occurs due to the elastic (dephasing) scattering processes of the exciton and is completely unpolarized (for details see ref. 13). Also contributions to I^{CS} from different states give rise to an interference term that corresponds to quantum beating as obvious from the order in which the summation and the squaring have to be performed. In case of I^{HL}, no such interference exists as all phase information is lost. Moreover, the inspection of $\psi_m(t)$ shows that the temporal evolution of $I^{CS}(t)$ is determined by the coherence time τ_{coh} (cf. eq.(6)) while the decay of the hot luminescence exhibits a different time dependence also involving the energy relaxation time T_1.

The actual form of $\psi_m(t)$ depends on the pulse shape and the type of spectral filter that is used for the signal analysis. To fit the experimentally observed time dependences below, we have used as a realistic model a single-sided exponential for the laser pulse amplitude and a Fabry-Perot resonator as spectral filter. For this case, analytical expressions can be obtained both for the coherent and incoherent parts of the scattering. These are not reproduced here, however, we mention that they show that excitation and detection can not be separated proving the necessity to include both in the description of the scattering process.

The expressions simplify greatly for the case of spectrally broad-band excitation and detection ("white light limit") which e.g. applies to the analysis of the quantum beat experiments in CdS described below[25]. Exploiting the optical selection rules in this case, it can be shown[15] that for linearly polarized excitation the Raman and hot luminescence contributions to the scattering process can be easily separated according to

$$\begin{aligned} I^{CS} &= I_{\parallel} - I_{\perp} \\ I^{HL} &= I_{\perp} \end{aligned} \tag{9}$$

with $I_{\parallel}$ ($I_{\perp}$) denoting the time-dependent scattered intensity polarized parallel (perpendicular) to that of excitation.

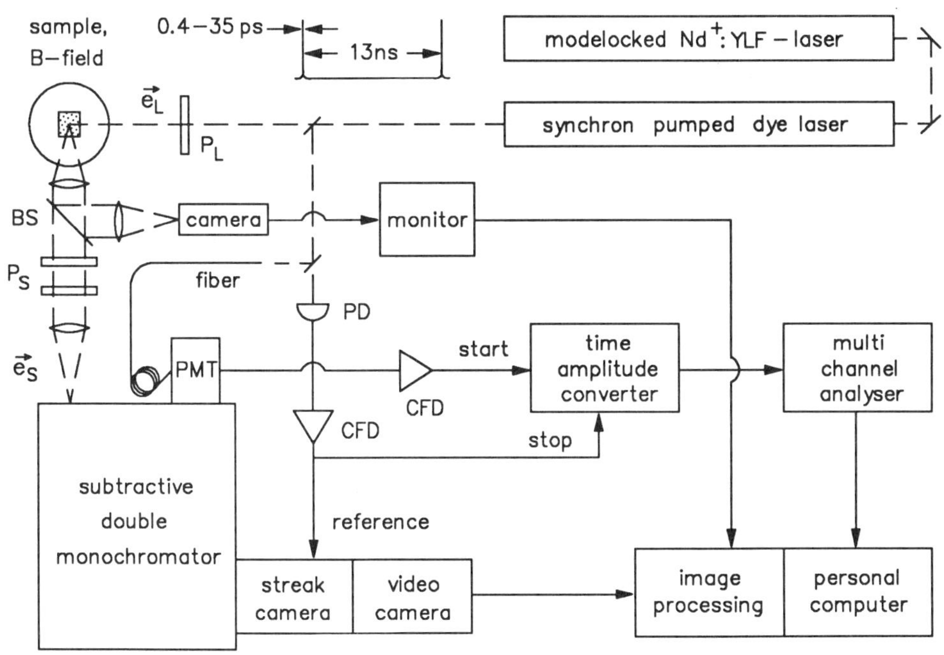

Figure 2. Experimental set-up for transform-limited picosecond spectroscopy. BS: beam splitter; CDF: constant fraction discriminator; $P_{L,S}$: polarizers; PD: photodiode; PMT: microchannel plate photomultiplier tube.

The experimental set-up to realize transform-limited time resolved spectroscopy for carrying out the quantum beat measurements is schematically illustrated in Fig. 2. The light pulses are generated by a tunable dye laser which is synchronously pumped by the frequency doubled or tripled radiation of a Nd^{3+}:YLF laser system according to the desired spectral range. Operating at a repetition rate of 76 MHz, the system provides pulses close to transform-limited between 0.4 and 35 ps tunable from 3.3 to 1.5 eV (375 to 820 nm). Most experiments described below were performed by means of pulses having a duration Δt_p = 5 ps and spectral width of about ΔE = 0.4 meV.

The key component of the apparatus to spectrally analyze the time dependent signals is a subtractively mounted double monochromator with 1 m focal length. Due to the mirror symmetric configuration of the two gratings and the light paths, it compensates for the light transit-time spread existing in each of the two single monochromators. By that, it avoids pulse broadening which in case of additive dispersion would add up to several hundreds of picoseconds. The time resolution of this set-up depends on the detector. With a fast microchannel-plate photomultiplier employing single photon counting, the time response to a 5 ps laser pulse is 20 ps, while 10 ps can be reached with a streak camera allowing to measure beating frequencies as high as 50 GHz.

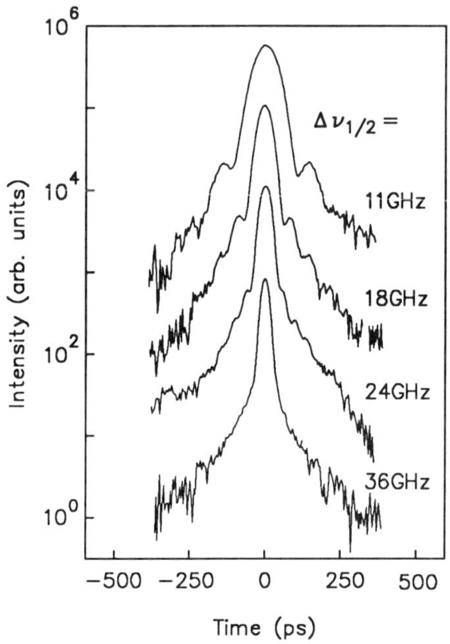

Figure 3. Time response of the set-up in Fig. 2 to 5 ps laser pulses for various band widths $\Delta\nu_{1/2}$ of the monochromator. The measurement was carried out at a laser wavelength 780 nm with the synchroscan streak camera as detector. Note the logarithmic intensity scale. (After ref. 24).

With this equipment, nearly perfect transform-limited performance down to this time range is achieved. This is demonstrated by Fig. 3 showing the time response to a 5 ps pulse directly falling onto the monochromator slit at various slit widths. As seen from the series of measurements, reduction of the bandwidth ($\Delta\nu_{½}$) gives rise to a strong broadening of the pulses in time ($\Delta t_{½}$) which is accompanied by the occurrence and shift of small side maxima. The quantitative analysis in terms of the frequency - time uncertainty product gives

$$\Delta\nu_{½} \cdot \Delta t_{½} = 0.886 \tag{10}$$

as expected for a rectangular slit. It nicely confirms that the time signals are nothing else but the Fourier transform of the spectral transmission function across the monochromator slit entirely consistent with the transform-limited characteristics of the experimental arrangement.

4. EXPERIMENTAL EXAMPLES

The first exciton at all to which the quantum beat method was applied and beating could be discovered was the free indirect ($\Gamma_6^+ \otimes L_{4,5}^-$) exciton in AgBr[8,9]. Due to the indirect character, for excitation at the exciton edge narrow scattering lines are resonantly enhanced involving two momentum-conserving TO(L) phonons. This process probes the intermediate exciton state and allowed to study its relaxation behaviour in great detail (see e.g. ref. 10). With a magnetic field applied, quantum beating could be revealed and coherent and incoherent contributions to the scattering separated. From analyzing the data thereby regarding the influence of the spectrometer and detector[15], the coherence time τ_{coh} = 500 ps came out surprisingly long probably due to the compact radius of the exciton (4 to 5 lattice constants). The energy relaxation time also was found to be T_1 = 500 ps implying T_2' = 1ns and equal contributions of elastic and inelastic scattering to the homogeneous linewidth (eqs.(5) and (6)). In addition, these measurements provided accurate numbers for the g-value and the electron-hole exchange energy for this exciton not obtained by any other measurement with this precision.

Another nice example are quantum beats in the exciton emission of type II GaAs/AlAs quantum wells and superlattices[26,27] for which pure dephasing times as long as 20 ns were found. The beating uncovered very small splittings of the exciton state in these systems the origin of which was later ascribed to exciton localization near the GaAs/AlAs interfaces and symmetry reduction[28,29].

In the following we focus on more recent experiments[12,14,25,30] that were performed in the semiconductors CdS and Cu_2O. In order to demonstrate the variety of information that can be obtained by quantum beat spectroscopy we have selected in Sections 4.1 and 4.2 a bound and a free exciton as examples. The bound excitons in CdS are largely atom-like systems but with degeneracies and splittings of energy states according to the valence and conduction band symmetries. Free excitons like those investigated in Cu_2O moreover possess kinetic energy adding another degree of freedom. Since the exciton in Cu_2O is direct, polariton effects are found to also play an important role[31,32]. These tremendously affect the beating and are discussed in Section 5.

4.1 Bound Excitons in CdS

Bound excitons seemed to be especially suitable systems for quantum beat spectroscopy because spatial localization was expected to largely reduce the number of relaxation channels and to result in coherence times accessible with the time resolution of our method. The investigated examples in CdS are complexes of the $\Gamma_9 \otimes \Gamma_7$ (A) exciton with a neutral and charged donor, respectively. At low temperature, they give rise to narrow line luminescence I_2 (D^0,X) and I_3 (D^+,X) on the low energy side of the free exciton (Fig. 4). The appearance of the spectra is much dependent upon type and content of defects in the samples and upon excitation conditions. The emission spectrum in Fig. 4 is produced by excitation in the I_3 absorption and besides several resonance fluorescence lines exhibits an LO phonon replica (phonon energy $\hbar\Omega$ = 37.8 meV) which was actually used to probe quantum beating.

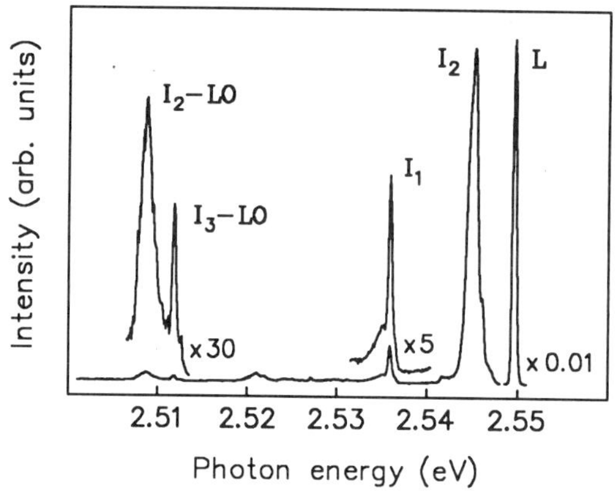

Figure 4. Bound exciton emission in CdS at T = 1.8K resonantly excited in the I_3 absorption at E_L = 2.5499 eV. L: Laser. The polarizations of the incident and scattered light $\mathbf{e}_L$, $\mathbf{e}_S$ are perpendicular to the crystallographic c-axis. Spectral bandwidth: 0.3 meV.

The electronic structure and the selection rules of these bound excitons are well established from conventional spectroscopy[33]. The inset of Fig. 5 displays the energy level scheme of the (D^+,X) complex consisting of a single ground state (symmetry Γ_1) and a fourfold excited state with symmetries Γ_5 (dipole-allowed singlet states Φ_1,Φ_2) and Γ_6 (dipole-forbidden triplet states Φ_3,Φ_4).

The effect of an external magnetic field upon these energy states as well as upon the matrix elements at low fields is governed by a linear Zeeman Hamiltonian

$$H_{Zeeman} = g_c \mu_B \mathbf{B} \cdot \mathbf{s} - g_v^{\parallel} \mu_B B_z \sigma_z . \tag{11}$$

Here s and σ_z denote the effective spin operator and spin z-component of the electron (s = 1/2) and the Γ_9 hole (σ = 1/2), respectively, the electron having an isotropic g-value g_c while the hole is completely anisotropic ($g_v^{\perp} = 0$) [33]. As shown in Fig. 5, with the magnetic field applied the excited states are split in two pairs. The magnitude of this splitting and the selection rules depend upon the angle θ between the field direction and the c-axis of the (hexagonal) crystal. While for $\theta \neq 0^o$ transitions from all four states are allowed (and may give rise to beating if coherently excited), selection rules for $\theta = 0^o$ predict transitions, left and right circularly polarized, respectively, only from the split Γ_5 sublevels (Φ_1,Φ_2). At low magnetic field, this splitting is given by

$$\Delta E_{12} = |g_c - g_v^{\|}|\, \mu_B B \,. \tag{12}$$

It is the beating between states Φ_1 and Φ_2 which were investigated in some detail[25] and will be discussed further.

The sample used in the actual experiments had cylindrical shape with the crystallographic c-axis oriented parallel to the cylinder axis[14]. All measurements were performed in Faraday geometry in nearly backscattering configuration with the fluorescence excited through a crystal surface perpendicular to the c-axis. The angle θ, variable by rotating the sample, defines a plane of reference to which the directions of light polarizations are referred.

Fig. 5 illustrates experimental data for B = 0 and B = 2T showing a clear beating structure that occurs superimposed to the fluorescence decay if the field is applied.

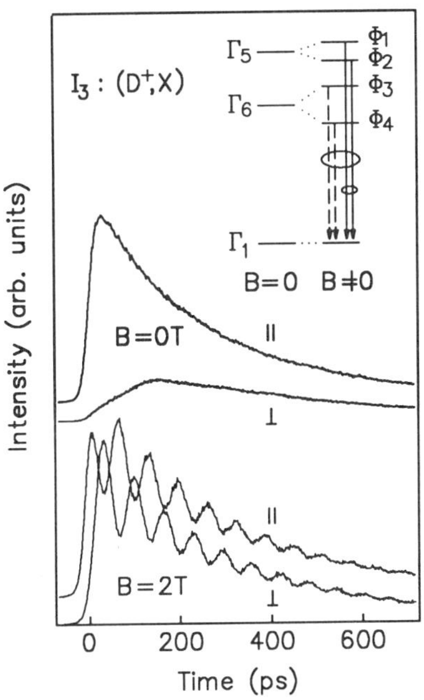

Figure 5. Time dependence and beating structure observed in the I_3-LO fluorescence in zero magnetic field and for B = 2T for $\theta = 0^o$. Excitation with light linearly polarized under 45° to the reference plane (see text). The scattered light is linearly polarized under 45° (∥) and −45° (⊥). The zero of time scale is derived from the system response to the laser pulse (not shown). Inset: I_3 energy levels with and without magnetic field. Transitions that can lead to beating are circled.

The width of the exciting laser pulse in these measurements would correspond to a system response of 20 ps (not shown) much shorter than the decay time of the signal. To avoid undesired elastic scattering due to the incident laser, the signal was detected in the 1LO replica rather than at the I_3 resonance itself*. To simultaneously excite the two sublevels Φ_1 and Φ_2, the exciting light has to be linearly polarized. The scattered light is analyzed with polarizations parallel ($I_{\|}$) and perpendicular ($I_{\perp}$) to this direction. According to the theory of light scattering outlined in Section 3, in zero magnetic field one expects the Raman component for detection with parallel polarization and the luminescence-like component perpendicular, the two showing different time behaviour. This is actually observed as seen from the upper pair of decay curves in Fig. 5, the small intensity $I_{\perp}$ indicating that pure dephasing by elastic scattering obviously is weak.

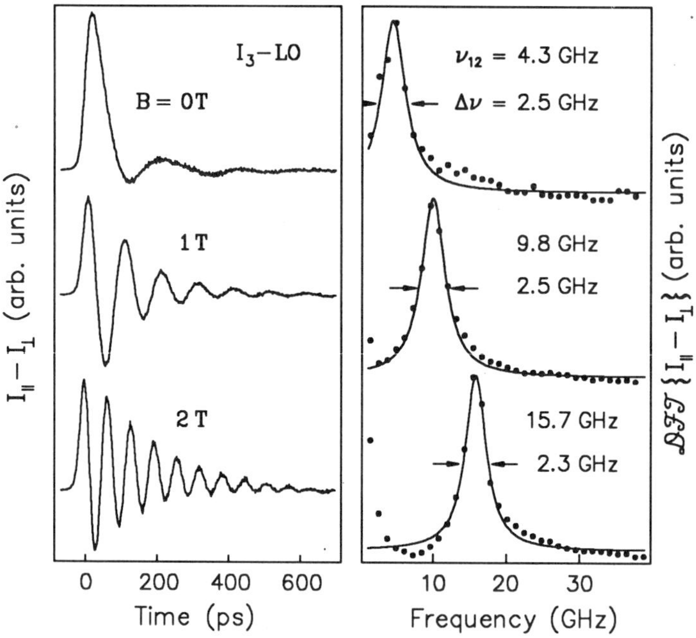

Figure 6. Left: Time dependence of the coherent contributions $I_{\|}-I_{\perp}$ to the I_3 - LO scattering intensity for different magnetic field strengths. Right: Corresponding spectra obtained by discrete Fourier transformation (full points) shown together with fits by a Lorentzian (full lines). Unlike in Fig. 5, the data are obtained with the incident light polarization parallel to the reference plane and corresponding polarizations ($\|$ and $\perp$) for the scattered light.

By applying the magnetic field, quantum beating develops in both parallel and perpendicular polarizations (lower curves). The relative amplitudes of the oscillation and the background directly reflect the ratio of coherent and incoherent scattering to

*From quantum beat measurements at the I_3 - 2LO replica, which gave completely identical results for the coherence time and energy splitting, we convinced ourselves that the phonon interaction is of no influence upon the exciton coherence.

the total luminescence. A characteristic property of the beating signals is the relative phase shift of π between oscillations in the two orthogonal polarizations. This is in complete accordance with the selection rules but also can be understood intuitively from the classical interpretation of the Hanle effect that will be given below (Section 4.2, Fig. 10).

The linear polarization in Fig. 5 was tacitly chosen in a particular direction (under 45° to the reference plane defined above) to avoid the occurrence of a low frequency beating that shows up already at B = 0T for all other directions of linear polarizations. This beating implies the presence of a zero-field splitting of the exciton state due to internal strain that actually causes the transition from the Γ_5 substates to the Γ_1 ground state to become linearly polarized. The data in Fig. 6 taken with the incident light polarization in the reference plane reveal this very clearly. On the lefthand side, we plotted the differences of intensities $I_{\parallel} - I_{\perp}$ which represent the coherent contribution to the light scattering (cf. Section 3). The righthand part shows the corresponding spectra obtained by discrete Fourier transformation. They allow us to directly deduce the magnitude of the splitting (from peak position ν_{12}) and the coherence time (from linewidth $\Delta\nu$). The strain splitting derived from the B = 0T spectrum is 4.3 GHz or 20 μeV. With the field applied, the magneto-quantum beats take over showing the expected increase in oscillation frequency and energy splitting with magnetic field strength.

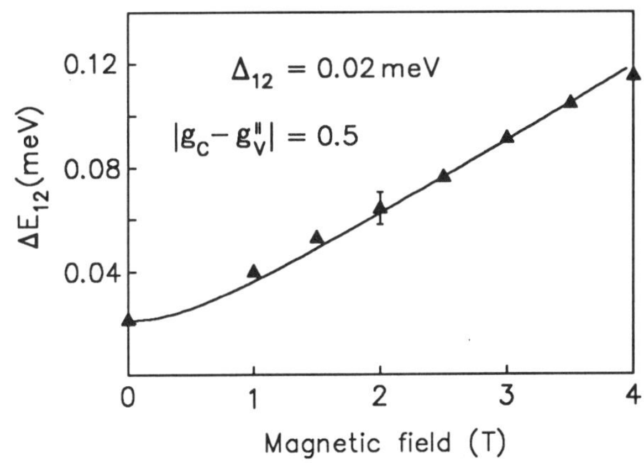

Figure 7. Energy splitting of states Φ_1 and Φ_2 of the I_3 bound exciton obtained from analyzing the measured quantum beat signals as function of magnetic field. Experimental data (triangles) are shown together with the fit (full line) as described in the text (from ref. 25).

The results of a complete analysis of the energy splitting as function of magnetic field are displayed in Fig. 7. To fit the experimental data, the Zeeman Hamiltonian in eq.(11) is used together with the effective Hamiltonian describing the strain dependent exchange splitting of the I_3 exciton states[25], the full line in Fig. 7 obtained by numerical diagonalization. Fitting the linear part with eq.(12), the zero-field splitting Δ_{12} = 0.02 meV and difference of g-values $|g_c - g_v^{\parallel}|$ = 0.5 are obtained (see figure). The splittings uncovered are extremely small. As the optical transition is inhomogeneously broadened, even the highest field (B = 4T) does not cause any splitting in the spectrum demonstrating the enormous potential of quantum beat spectroscopy.

The coherence time derived from the spectra is $\tau_{coh} = 2/\pi\Delta\nu \approx 280$ ps independent of the magnitude of the field (Fig. 6, right). Comparison with the energy relaxation time $T_1 \approx 320$ ps following from $(I_\parallel + I_\perp) \sim \exp(-t/T_1)$ gives $T_2' \approx 500$ ps (eq.(5)) showing that pure dephasing contributes nearly 60% to the total rate $1/\tau_{coh}$ or the homogeneous linewidth (eq.(6)).

We have also investigated the I_2 transition (Fig. 4) using appropriate geometry and polarizations, but were unable to detect beating[25]. Instead we found an extremely fast depolarization scattering between states of orthogonal polarization. The quantitative analysis in that case gave $T_2' \approx 20$ ps and $T_1 \approx 430$ ps indicating that elastic scattering (98%) is overwhelming and completely damps out the expected beats for $B \neq 0$. Similar results were achieved in our sample for an exciton-neutral acceptor complex (A^0,X) giving rise to line I_1 in Fig. 4.

The short dephasing times are surprising in view of our initial expectation that bound exciton states, due to their localization in the crystal lattice, would not be subject to strong elastic scattering. One rather might have thought that inelastic processes like radiative and nonradiative recombination would determine the homogeneous linewidth. Obviously for these bound excitons elastic interactions can exist even stronger than e.g. in the case of the free exciton in AgBr described above. These interactions might have to do with an energy transfer between the bound states but need further investigation.

4.2 The 1S Quadrupole Exciton in Cu_2O

For the investigations of excitons, Cu_2O is considered a prototype semiconductor as it exhibits a number of strongly marked exciton series in different parts of the spectrum (for a review see e.g. ref. 34). The 1S exciton treated in this paper corresponds to the lowest lying exciton state of the yellow series for which sharp line hydrogen-like absorption up to n=9 has been observed[35]. As a peculiar feature, this exciton leads to only weak absorption and emission in spite of the fact that it is due to a direct process at **K**=0. This is because the valence and conduction band states forming the exciton have the same parity (Γ_7^+ and Γ_6^+, respectively) and therefore direct transitions are electric-dipole forbidden. A group theoretical analysis predicts, besides an optically forbidden Γ_2^+ triplet state ("para-exciton"), a threefold degenerate singlet-triplet mixed state with Γ_5^+ symmetry ("ortho-exciton") which is only electric-quadrupole (EQ) allowed. It becomes electric-dipole (ED) allowed in higher order by participation of an odd-parity phonon, the strongest process involving a Γ_3^- phonon of energy $\hbar\Omega$ = 13.6 meV [36]. The spectrum in Fig. 8 displays some of these processes. Excitation here is undertaken slightly above the Γ_3^- phonon assisted exciton absorption edge, the spectrum showing the ortho-exciton resonance fluorescence and several 1- and 2-phonon scattering processes. In spite of numerous investigations of light scattering from the 1S ortho-exciton (for detailed discussions see e.g. ref. 37-39), the question if these processes are Raman- or hot luminescence-like could not be answered experimentally until quantum beat spectroscopy recently was employed[30].

Fig. 9 schematically illustrates the investigated 1- and 2-phonon scattering processes involving the EQ-allowed 1S state and the odd parity (Γ_3^-) optical phonon under different resonance conditions. 1-phonon scattering probing the exciton near **K** = 0 may be accomplished by excitation sharply in resonance with the EQ absorption at $E_L = E_Q$ (Fig. 9A: "incoming resonance") or in the onset of the indirect exciton absorption at $E_L = E_Q + \hbar\Omega$ (Fig. 9B: "outgoing resonance"). These processes each involve one EQ-allowed and one phonon-assisted ED transition in different order (EQ–ED(Γ_3^-) and ED(Γ_3^-)–EQ) and give rise to resonantly enhanced scattering at energies $E_Q - \hbar\Omega$ and E_Q, respectively[36].

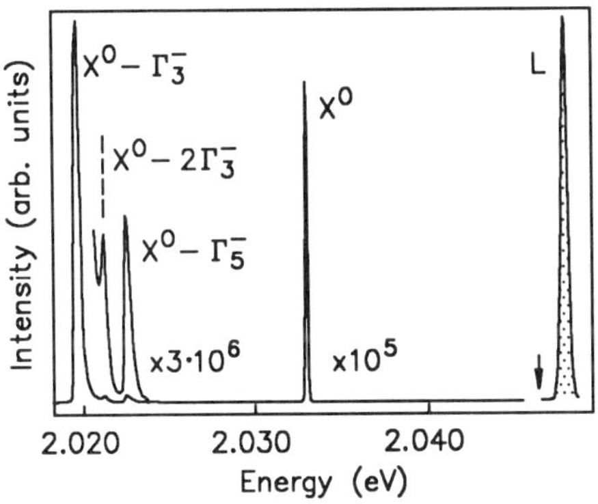

Figure 8. Light scattering spectrum of Cu_2O at T = 1.8K excited at E_L = 2.048 eV. The arrow marks the Γ_3^- phonon assisted absorption edge at 2.0465 eV. X^0: resonance fluorescence of the 1S ortho-exciton at energy E_Q = 2.0329 eV. The lines at lower energies correspond to resonantly enhanced 1- and 2-phonon scattering involving Γ_3^- and Γ_5^- phonons (compare Fig. 9); L: laser; $\mathbf{k}_L,\mathbf{k}_S \| [001]$; $\mathbf{e}_L,\mathbf{e}_S \| [100]$.

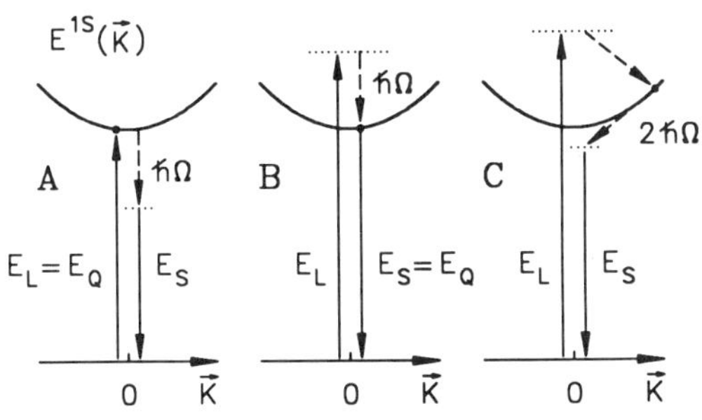

Figure 9. Schematic representation of the 1S ortho-exciton dispersion E(**K**) and possible scattering processes in Cu_2O. A, B: 1-phonon scattering with resonance for the incident ($E_L = E_Q$) and scattered ($E_S = E_Q$) photon, respectively. C: 2-phonon scattering. $\hbar\Omega$: energy of Γ_3^- phonon.

Quite differently, in 2-phonon scattering (Fig. 9C) both the absorption and the emission part is associated with a Γ_3^- phonon resulting in two indirect ED transitions (ED(Γ_3^-)–ED(Γ_3^-)). As known from Raman studies using cw excitation[37], this process becomes resonant at energies larger than the indirect absorption threshold at $E_Q + \hbar\Omega$. Considering that the Γ_3^- phonon is practically dispersionless, it is evident that, depending on excitation photon energy E_L, 2-phonon scattering in principle enables us to investigate the exciton state also at higher kinetic energies.

Even without magnetic field the quadrupole absorption is anisotropic and because of selection rules[40,41] the scattered light is strongly polarized. In our experiments the wavevector of the incident and scattered light ($\mathbf{k}_L,\mathbf{k}_S$) both were chosen along a [001] crystal direction. This configuration makes the selection rules particularly simple although, on the other hand, because of the forward scattering geometry experimentally one has to care for especially good straylight suppression in measuring the Raman spectra. In the actual experiments, the polarization of the incident laser light ($\mathbf{e}_L$) was chosen either along [100] or [110], that for the scattered light ($\mathbf{e}_S$) in each case parallel and perpendicular to these directions to permit the detection of the polarized and depolarized component (see Section 3).

In order to create quantum beating, the degeneracy of the exciton state was removed by applying a magnetic field $\mathbf{B}\|[001]$ (Faraday geometry). Its effect upon the exciton state may be described by an effective Hamiltonian

$$H_{Zeeman} = \frac{1}{2}(g_c + g_v)\,\mu_B\,\mathbf{F}\cdot\mathbf{B} \tag{13}$$

with $\mathbf{F}$ representing the total angular momentum operator of the coupled electron and hole spins, g_c and g_v the conduction and valence band g-values, and μ_B the Bohr magneton (see e.g. ref. 42). As displayed in the inset of Fig. 10, the ortho-exciton for $B \neq 0$ splits into substates $\Gamma_2^+(M=0)$, $\Gamma_3^+(M=1)$ and $\Gamma_4^+(M=-1)$. In Faraday geometry, only transitions from the $M=\pm1$ states into the Γ_1 ground state are optically allowed interacting with light of circular ($\sigma^\pm$) polarizations. These states show linear Zeeman splitting $\Delta E = |g_c+g_v|\mu_B B$ so that coherent excitation by a short linearly polarized laser pulse should give rise to quantum beats.

The scattered intensities for different polarizations are obtained from eq.(7) replacing $\mathbf{M}_m$ by the respective matrix elements $\mathbf{M}_m^{EQ}$ and $\mathbf{M}_m^{ED}$ of the EQ and ED transitions with the sum now running over the Γ_5^+ exciton substates (m=1,2,3). Calculating the matrix elements thereby considering coupling only to the Γ_3^- phonon and taking into account the mixing of the exciton states by the magnetic field, intensity expressions are obtained like those reproduced in Table 1 for the EQ–ED(Γ_3^-) process[14].

In Fig. 10 (left side), a corresponding experimental result is shown (incoming resonance; $\mathbf{e}_L\|[100]$). In accordance with the calculation ($I_\|=0$; $I_\perp=I_0$), for B=0T complete polarization occurs except for a small depolarized hot luminescence component that suggests pure dephasing to be extremely weak. With the magnetic field applied, the coherent nature of the scattering is revealed by the beating structure occurring for both polarizations superimposed to a small incoherent background intensity. Corresponding to the selection rules in Table 1, the oscillations are phase-shifted by π, their slowly decreasing amplitudes confirming that dephasing mainly comes from the depopulation of the exciton state. For very long times, the beats become entirely damped out giving essentially unpolarized scattering. This demonstrates the expected transformation from coherent Raman scattering into incoherent hot luminescence. The quantitative analysis of the damping and frequency of oscillations for this sample at B=0.25T gives τ_{coh} = 800 ps and ΔE = 26 μeV.

Completely identical beating signals and values are found in the case of outgoing resonance for comparable experimental conditions. This proves that in the excitation or deexcitation process the phonon is only needed to conserve energy and momentum, but is of no effect at all upon the coherence of the resonant intermediate state.

In Fig. 10 (right), measurements are presented to further test the calculated intensities by using different detection polarizations. Closer inspection and comparison with Table 1 gives excellent agreement with regard to the degree of modulation and the phase shift. We have arranged the beat signals so as to most clearly demonstrate the phase shift of $\pi/2$ between corresponding beat maxima when the analyzer is rotated in steps of 45° ($\pi/4$). In complete analogy with the classical interpretation of the Hanle effect (see e.g. ref. 4), this behaviour reflects the total exciton dipole moment which, in the presence of a magnetic field perpendicular to the dipole, rotates with a frequency $\Delta\omega/2$ ($\Delta\omega = |E_2 - E_1|/\hbar$: beating frequency*). Accordingly, the intensity varies with twice this frequency giving rise to a beating period $T=2\pi/\Delta\omega$ and the marked phase shift between the signals in Fig. 10 (right) when detected behind the linear analyzer in the indicated positions. It is to be pointed out that this behaviour is distinctly different from beating due to an energy splitting caused by a strain field as e.g. described for the I_3 transition in CdS [25] (Section 4.1). As in that case no dipole rotation is involved, rotating the (linear) analyzer does not result in a phase shift of the beat pattern. Instead, a "static" behaviour is observed merely consisting of a variation of the beating amplitude.

As to the selection rules, another interesting case is the disappearance of any modulation when the scattered light is analyzed with a circular polarizer (not shown). This result clearly shows that in the magnetic field the transitions involving the M = ±1 ortho-exciton states are σ-polarized (cf. also Table 1). Circularly polarized detection therefore selects only one of the two substates so that no beating occurs. This is in complete analogy to Young's double slit experiment discussed in Section 2 when one slit is closed and no interference pattern observed.

Two other results for the 1S ortho-exciton came from measuring the beating signals as function of magnetic field strength and of temperature[12,30]. Analyzing the beating frequency for fields up to B=1T, in excellent agreement with the prediction above a strictly linear dependence is found for the Zeeman splitting of the M = ±1 exciton substates[14] (not shown). Fitting these data, an accurate effective g-value $|g_c - g_v| = 1.78 \pm 0.05$ is derived slightly larger than that reported earlier from high-field measurements[42].

Measuring and evaluating the beating signal at different temperatures[14], the coherence time of τ_{coh} = 800 ps at 2.1K is found to rapidly shorten to 180 ps at 10K with no resolvable oscillations above that temperature. Except below 4K, $1/\tau_{coh}$ exhibits a linear dependence upon temperature suggesting energy relaxation by acoustic phonon scattering to be the dominant dephasing process[43]. A quantitative analysis in terms of acoustic phonon scattering rates entirely confirms this and in particular shows that ortho-to-para exciton scattering can be neglected**. Using the measured energy relaxation time T_1 (2.1K) = 910 ps, a pure dephasing time

---

*In case of atoms $\Delta\omega$ would correspond to twice the Lamor frequency.
**The Γ_2^+ para-exciton state is lower by 11.9 meV than the Γ_5^+ ortho-exciton[39].

Table 1. Calculated time dependent polarized 1-phonon scattering intensities in the case of incoming resonance (EQ–ED(Γ_3^-)). Incident light: $\mathbf{k}_L \parallel [001]$, $\mathbf{e}_L \parallel [100]$. Scattered light observed in Faraday geometry: $\mathbf{k}_S \parallel \mathbf{B} \parallel [001]$; $\mathbf{e}_S$ as shown. For clarity $T_2' = 0$, i.e. $\tau_{coh} = 2T_1$ is assumed (from ref. 14).

$\mathbf{e}_S$		$I(t, \mathbf{e}_L, \mathbf{e}_S)$ for $B \neq 0$
[100]	(0°)	$I_0\,(1 - \cos \omega t)\exp(-2t/\tau_{coh})$
[010]	(90°)	$I_0\,(1 + \cos \omega t)\exp(-2t/\tau_{coh})$
[110]	(45°)	$I_0\,(1 + 1/2 \cdot \sin \omega t)\exp(-2t/\tau_{coh})$
[1$\bar{1}$1]	(135°)	$I_0\,(1 - 1/2 \cdot \sin \omega t)\exp(-2t/\tau_{coh})$
[1i0]	(σ^+)	$I_0 \cdot \exp(-2t/\tau_{coh})$
[1$\bar{i}$0]	(σ^-)	$I_0 \cdot \exp(-2t/\tau_{coh})$

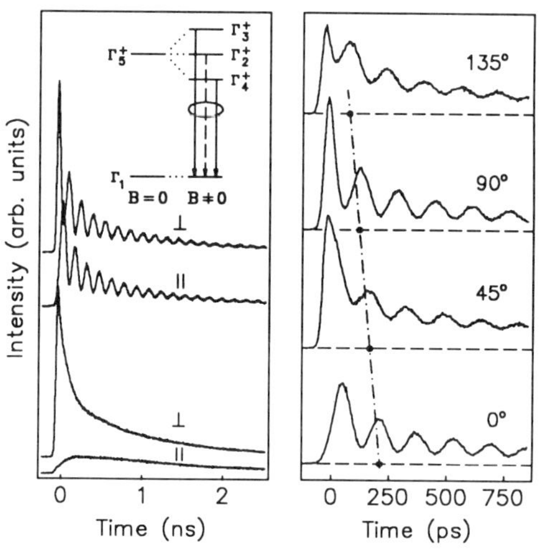

Figure 10. Left: Time dependence of the polarized $1\Gamma_3^-$ phonon scattering intensity for B=0 (below) and B=0.25T (above) in Cu_2O at 2.1K for resonant excitation at $E_L = E_Q$. Incident light: $\mathbf{k}_L \parallel [001]$; $\mathbf{e}_L \parallel [100]$. Scattered light: $\mathbf{k}_S \parallel \mathbf{B} \parallel [001]$; $\mathbf{e}_S \parallel [100]$ ($\parallel$) and [010] ($\perp$). Inset: energy levels of the 1S ortho-exciton with and without magnetic field. Right: Quantum beating observed for different polarizations of the scattered light to demonstrate the selection rules in Table 1. B=0.25T. Corresponding beat maxima are marked to make plain the phase shifts of subsequent signals.

$T_2' = 1430$ ps is obtained (eq.(5)) demonstrating that at these low temperatures inelastic and elastic scattering contribute with about equal strength to the homogeneous linewidth. This is like for the free exciton state in AgBr. In Cu_2O it is due to the fact that acoustic phonon scattering is largely frozen out, the lack of any other relaxation processes explaining the long coherence times found.

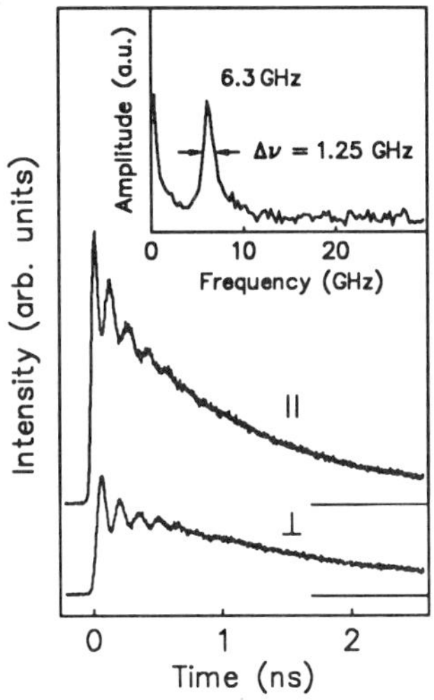

Figure 11. Quantum beating in the polarized $2\Gamma_3^-$ phonon scattering intensity for B=0.25T at T=1.8K. Excitation at E_L = 2.0467 eV slightly above the 1S ortho-exciton absorption. For other parameters see Fig. 10. Inset: Fourier spectrum calculated from $(I_{\parallel} - I_{\perp})$.

To uncover the exciton relaxation behaviour at higher energies, experiments were performed with the $2\Gamma_3^-$ phonon line that emerges when the excitation is tuned into the exciton band (see spectrum in Fig. 8). A typical result is presented in Fig. 11. Compared to 1-phonon scattering, much larger contributions from incoherent processes are found with only small oscillating structure sitting on top of it. We believe the background, to a large part, is due to emission from thermalized exciton states populated via the nonresonant $1\Gamma_3^-$ phonon-assisted absorption process which simultaneously takes place. This also would explain the deviations in the ratio of polarized background intensities found experimentally from the calculated one[14].

The beat signal in all measurements disappears for exciton kinetic energies $\epsilon = E_L - (E_Q + \hbar\Omega(\Gamma_3^-)) > 0.3$ meV due to the rapidly increasing importance of relaxation processes. This strong energy dependence of the coherence time becomes obvious from the data in Fig. 11 that were taken at $\epsilon \approx 0.2$ meV. The coherence time at this energy is already down to $\tau_{coh} = 510$ ps, while it is $\tau_{coh} = 820$ ps at $\epsilon \approx 0$. This latter value nicely agrees with τ_{coh} derived from 1-phonon scattering which probes the exciton state very near this energy. As expected, the Zeeman splittings obtained from 1- and 2-phonon scattering are in complete agreement as well.

Coming back to the initially raised question concerning the nature of the resonant 1- and 2-phonon processes, the experiments reveal that they are basically due to Raman scattering which, at later times, transforms into hot luminescence. We found the coherence times not only sample dependent but also to exhibit strong local variations across the sample. Besides intrinsic scattering, obviously strain and macroscopic irregularities are major factors that cause the exciton to lose coherence and inhibit quantum beats to develop requiring high quality samples to obtain reproducible results.

5. PROPAGATION BEATS FROM EXCITON-POLARITONS

Throughout the discussion of quantum beats in Cu_2O we used the exciton picture thereby neglecting the exciton-photon interaction which exists in case of direct exciton transitions. As a consequence of the coupling between the exciton and the light field, polaritons are being formed and are the appropriate eigenstates to describe the propagation of electromagnetic waves through the dielectric medium in the presence of excitonic resonances (see e.g. ref. 44). The most dramatic effect of the coupling consists in the splitting of the dispersion in two branches right at the crossover point of the original uncoupled photon-exciton dispersion curves. Due to the quadrupole character of the 1S ortho-exciton in Cu_2O, the interaction with the light field is weak (oscillator strength $f \approx 10^{-9}$). This leads to only a small splitting of the polariton branches ($\Delta E \approx 20\,\mu eV$) and a narrow region of strong dispersion in **K**-space (Fig. 12, left).

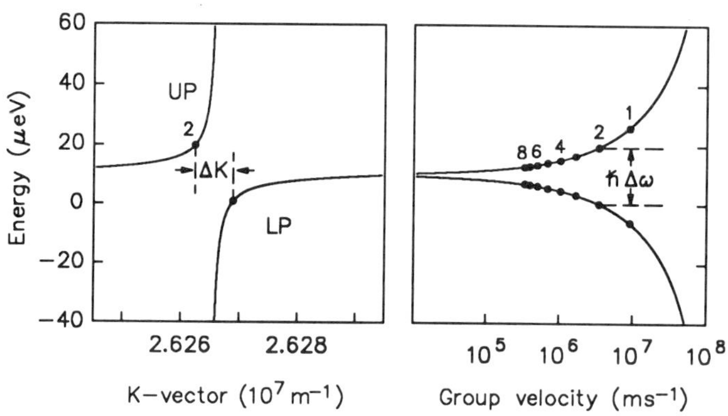

Figure 12. Calculated dispersion of the 1S exciton polariton in Cu_2O (left) and corresponding plot of energy vs. group velocity (right). Light propagation along $\mathbf{k}_L \| \mathbf{k}_S \| [110]$. The zero of energy is at $E_Q = 2.0329$ eV, the oscillator strength $f_{110} = 3.6 \cdot 10^{-9}$. The full circles correspond to the minima in the experimental beating signal in Fig. 13 and are accordingly numbered by n. UP, LP: upper and lower polariton branch (after ref. 31).

Regarding quantum beats, for such a system two novel aspects enter:

i) The exciton is *inherently* split in two branches which may be coherently excited by a short laser pulse of proper spectral width.

ii) Polaritons are propagating modes and, hence, propagation effects are expected to affect the usual beating pattern.

Both these expectations were nicely met by a series of measurements in which pulse propagation in resonance with the quadrupole polariton was investigated[31,32]. The experimental conditions are similar to those in the magneto-quantum beat measurements (Section 4.2) but differ in one important detail. Rather than detecting the 1- and 2-phonon Raman signals the intensity of the excitation pulse itself, having passed through the crystal, is recorded as function of time. It means that both excitation and detection has to be accomplished sharply in resonance with the quadrupole polariton, i.e. $E_L = E_Q = E_S$. Fig. 13 shows experimental data that were obtained using directions of wavevector and light polarization for which the polariton is optically allowed. Compared to the incident laser pulse (Δt_p = 30 ps), the transmitted pulse extends over much longer time with intensity coming out of the sample still after 2.5 ns. The striking feature is the resolved beating structure which is superimposed to this slow decay. Unlike for the magneto-quantum beats, their period now is time dependent increasing from about 100 ps at small times by roughly a factor of five within the first 2.5 ns. Also, these oscillations depend on sample thickness and direction of light propagation through the crystal.

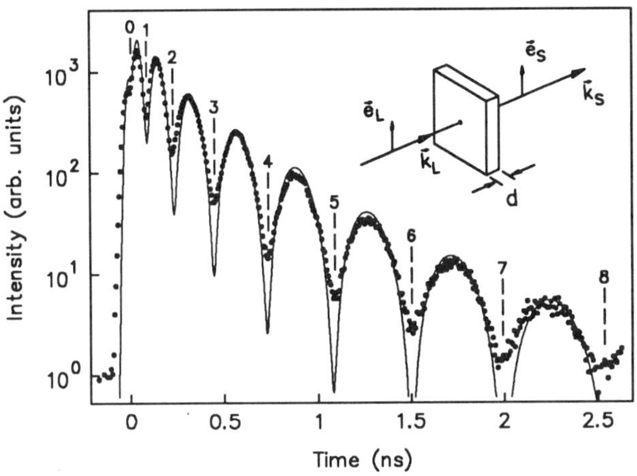

Figure 13. Propagation beats in Cu_2O for excitation with a 30 ps laser pulse (system response: 55 ps) tuned to the 1S exciton resonance. Sample thickness d = 0.91 mm; $\mathbf{k}_L,\mathbf{k}_S \| [110]$, $\mathbf{e}_L,\mathbf{e}_S \| [001]$; T = 1.8K. Experimental data (points) are shown together with a fit (full line) as described in the text. The indicated values of n denote the minima of intensity according to the case of destructive interference (eq.(16); after refs. 31 and 46).

To quantitatively analyze the beat structure, one has to start out from the dispersion relation of the quadrupole resonance that determines the propagation properties of the polariton. Near resonance it is described by[45]

$$K^2(\omega) = \frac{\omega^2}{c^2}\left[\epsilon_b + \frac{fc^2K^2}{\omega_0^2 + aK^2 - \omega^2 - i\omega\Gamma}\right] \tag{14}$$

with $\epsilon_b = 6.5$ denoting the background dielectric constant and $\hbar\omega_0 = E_Q = 2.0329$ eV the 1S exciton energy. The term $aK^2 = \omega_0\hbar K^2/M$ accounts for the spatial dispersion with an effective exciton mass $M = m_e + m_h = 2.7\, m_0$. Γ is a phenomenological damping rate (equal to $\delta E_{hom}/\hbar$) and f the oscillator strength which depends on the direction of wavevector **K**. To fit the experimental data in Fig. 13, the time dependent electric field at the rear of the sample (thickness d) can be calculated from the exciting pulse amplitude $E_0(\omega)$ at the entrance according to a Fourier integral[31]

$$E_S(t,d) \propto \int_{-\infty}^{+\infty} E_0(\omega) \sum_j a_j(K_j,\omega) \exp[i(K_j d - \omega t)]\, d\omega \tag{15}$$

where the sum accounts for the two polariton branches (j = UP,LP) excited with amplitudes $a_j(K_j,\omega)$. Using a pulse shape adapted to the experiment and taking into account the time resolution of the detector[46], from eq.(15) finally the intensity can be obtained giving an excellent fit to the measured data (full line in Fig. 13). The oscillator strength and the damping constant contained in the dispersion are taken as adjustable parameters and in the example shown were chosen $f_{110} = 3.7 \cdot 10^{-9}$ and $\hbar\Gamma = 0.9\ \mu$eV. Either consistent results were obtained for polariton propagation along [111] in which case an oscillator strength $f_{111} = 1.4 \cdot 10^{-9}$ is found close to the expected ratio $f_{110}/f_{111} = 3$ (ref. 47).

In an intuitive picture the gross characteristics of the time structure of the transmitted pulse can be visualized as due to interference of pairs of polariton wavepackets (pair n = 2 marked in Fig. 12, left) coherently created by the laser pulse, each at one of the polariton branches[31]. The polaritons in one pair have same group velocity v_g but differ in energy and wavevector by $\hbar\Delta\omega$ and ΔK. Travelling through the crystal (thickness d) to the rear end, they experience a relative phase shift

$$\Delta\Phi(d) = \Delta K \cdot d - \Delta\omega \cdot t. \tag{16}$$

Taking e.g. the minima of intensity in Fig. 13 (at times $t_n = d/v_g$) that would correspond to destructive interference, the condition $\Delta\Phi(d) = (2n+1)\pi$ (n = 0, 1, 2, ...) allows to translate the time varying beat period into a dependence of energy splitting vs. group velocity (solid dots in Fig. 12, right). We have denoted corresponding values by n, the excellent agreement with $E(v_g)$ calculated from the dispersion in eq.(14) (full line) confirming the validity of the picture. Recalling that the beat period is inversely related to the splitting of states (Section 2), it is quite obvious from this representation that polaritons with high group velocity arriving at early times give rise to a small beating period and vice versa, consistent with the experimental observation.

Especially this latter interpretation makes obvious that the polaritons in Cu_2O coherently propagate over macroscopic distances of the order of millimeter which is the average sample size. This is consistent with the long coherence times derived from the magneto-quantum beats. Not only that the samples are of excellent structural quality and scattering at defects is weak. Also, scattering from acoustic phonons which is the major cause of polariton dephasing is strongly reduced at low temperature ($T \leq 2K$) due to the small number of available final states.

Quite different from the case of resonant pulse transmission, beating in the 1- and 2-phonon scattering line, without magnetic field, does not occur in spite of the polariton splitting described above[13,14]. This is because polariton states with different wavevectors interact with different phonons not ending in the same final state and therefore not resulting in indistinguishable relaxation channels (Section 2). Still, the Raman process is affected by the polariton character of the quadrupole exciton in that its time decay depends on the spectral width of excitation (Fig. 14). Again, the reason for that is the strong $E(v_g)$ dependence near resonance (cf. Fig. 12). Accordingly, broad-band excitation will excite polaritons up to large group velocities that, via polariton-phonon scattering, will contribute to the Raman signal at early times and give rise to a fast and nonuniform decay as actually observed (Fig. 14 left, upper). In contrast, narrow- bandwidth excitation creates polaritons with low v_g from around the intersection of the photon and exciton dispersions which lead to the uniform (nearly exponential) slower decay displayed in Fig. 14 left (below).

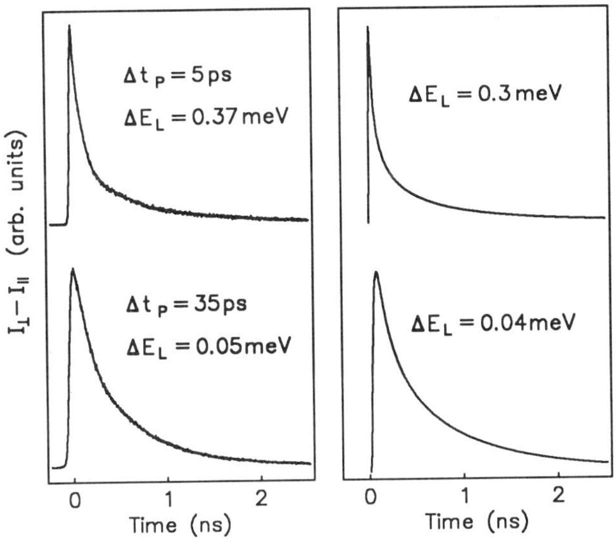

Figure 14. Time dependence of the coherent contribution $I_\perp - I_\parallel$ to the $1\Gamma_3^-$ phonon scattering line for different spectral widths ΔE_L of the exciting laser pulse (B=0T). Left: measured decay. Right: calculation taking into account the polariton character of the exciton as described in the text (refs. 13, 14).

Quantitatively this effect can be modelled by introducing a time-of-flight (tf) broadening[48] $\Gamma_{tf}(\mathbf{K}) = v_g(\mathbf{K})/d$ that must be added to the homogeneous linewidth giving the total polariton damping constant

$$\Gamma_{tot} = \Gamma + \Gamma_{tf}(\mathbf{K}) \,. \tag{17}$$

Rewriting eq.(7) to include the polariton effects, for the time dependent polarized $1\Gamma_3^-$ phonon Raman signal one obtains

$$I^{CS}(t,e_L,e_S) \propto \sum_{\mathbf{K}j} \sum_{\ell=1,2} \left| \sum_m (\mathbf{M}_{\ell,m}^{ED} \cdot \mathbf{e}_S)^* (\mathbf{M}_m^{EQ} \cdot \mathbf{e}_L)\, \psi_{m,\mathbf{K}j}(t) \right|^2 . \tag{18}$$

As obvious from the matrix elements already introduced before, this expression describes the incoming resonance appropriate to Fig. 14. The function $\psi_{m,Kj}(t)$ contains the effect of the measuring apparatus (see Section 3). For the present case, it follows from eq.(8) by replacing E_m and δE_{hom} by the polariton dispersion $E_m(K_j)$ with j = UP,LP and the total damping constant Γ_{tot} (eq.(17)), respectively. Corresponding to the spectral width of laser excitation, the contributions of the lower and upper polariton branches have to be taken into account as expressed by the outer (incoherent) summation, the sum over ℓ accounting for the twofold degeneracy of the Γ_3^- phonon.

The righthand side of Fig. 14 displays the scattered intensities numerically calculated from eq.(18)[15]. Very close to the actual experiment, the laser pulse was approximated by a single-sided exponential with the spectral widths ΔE_L indicated. As a spectral filter a Fabry-Perot was assumed having a bandwidth ΔE_S = 0.25 meV. Using the oscillator strength and the damping constant for the polariton deduced from the propagation beat experiments above, very good agreement is achieved between the calculation and the experimental finding. It demonstrates that polariton effects have to be included to quantitatively analyze light scattering from direct exciton states.

6. FUTURE POSSIBILITIES

The results presented in this contribution demonstrate that quantum beats in resonance fluorescence are a powerful means to investigate the dynamics of excitons and to develop a quantitative picture of phase and energy relaxation. Besides Raman also low energy Brillouin scattering processes may be investigated and used to reveal the coherence of the intermediate states. The applicability of the method essentially depends on the coherence times involved which to a large extent are a function of sample quality. As shown by the selected examples, both free and bound exciton states can be studied and offer a variety of possible systems.

Additional knowledge on the coherence of the excitations can come from pulse propagation experiments like those described for the exciton-polariton in Cu_2O. Contrary to usual polariton time-of-flight measurements (cf. e.g. ref 49,50) and other more recent pulse transmission studies[48,51,52], an essential requirement in this type of experiment is that the spectral width of the exciting light pulse is broader than the absorption of the attenuating medium. Depending upon the absorption strength, the pulse spectrum develops a dip, the interference of the residual spectral components leading to the beat structure after the pulse has traversed the medium. Evidently the occurrence of beating is independent of the nature of states so that propagation beats can also be expected in the presence of localized excitations. Actually, we were recently able to find in AgBr an interesting new beating structure in resonance fluorescence[53,54]. It is preliminarily interpreted as due to closely spaced states of small nonmetallic silver clusters formed in the material by light irradiation. The time structure in this case is complex. It may be connected with the inhomogeneous broadening of the states but requires further investigations.

The limitations of our method lie in the temporal and spectral resolution needed. They can be overcome, however, by employing correlation techniques replacing the direct detection of the time-dependent signals.

Acknowledgments

The authors like to thank the Deutsche Forschungsgemeinschaft for continuous support of this work. They especially appreciate the cooperation with Prof. D. Fröhlich, B. Uebbing and A. Kulik of the University of Dortmund in the propagation beat experiments and the supply of the high quality Cu_2O samples.

REFERENCES

1. K. Blum, *Density Matrix Theory and Applications* (Plenum Press, New York, 1981).
2. S. Stenholm, *Foundations of Laser Spectroscopy* (Wiley, New York, 1984).
3. W. Hanle, Z. Physik **30**, 93 (1924).; Ergeb. exakt. Naturwissenschaften **4**, 214 (1925).
4. W. Demtröder, *Laser Spectroscopy* (Springer, Berlin, 1991).
5. E.B. Alexandrov, Opt. Spectrosc. **17**, 521 (1964).
6. S. Haroche, in *High Resolution Spectroscopy*, edited by K. Shimoda, Topics in Appl. Phys. **13** (Springer, Heidelberg, 1976) p.253.
7. A. Corney, Phil. Trans. R. Soc. London A **307**, 573 (1982).
8. V. Langer, H. Stolz and W. von der Osten, Phys. Rev. Lett. **64**, 854 (1990).
9. V. Langer, H. Stolz and W. von der Osten, J. Luminesc. **45**, 406 (1990).
10. W. von der Osten in *Light Scattering in Solids VI*, edited by M. Cardona, Topics in Appl. Phys. **68** (Springer, Berlin, 1991) p.361.
11. H. Stolz in *Festkörperprobleme/Advances in Solid State Physics XXXI*, edited by U. Rössler (Vieweg, Braunschweig, 1991) p.219.
12. V. Langer, H. Stolz and W. von der Osten, J. Luminesc. **53**, 303 (1992).
13. H. Stolz, phys. stat. sol. (b) **173**, 99 (1992).
14. V. Langer, thesis, Universität-GH Paderborn 1992.
15. H. Stolz, Habilitationsschrift, University-GH Paderborn 1993 (to be publ. in Springer Tracts in Modern Physics).
16. R.S. Knox, *Theory of Excitons*, Solid State Physics Suppl. 5 (Academic Press, New York, 1963).
17. R.J. Elliott, in *Polarons and Excitons in Polar Semiconductors and Ionic Crystals*, edited by J.T. Devreese and F. Peeters, NATO ASI Series, Series B: Physics Vol. 108 (Plenum Press, New York, 1984) p.271.
18. J. Kuhl, in *Festkörperprobleme/Advances in Solid State Physics XXIX*, edited by U. Rössler (Vieweg, Braunschweig, 1989) p.157.
19. E.O. Göbel in *Festkörperprobleme/Advances in Solid State Physics XXX*, edited by U. Rössler (Vieweg, Braunschweig, 1990) p.269.
20. E.O. Göbel, M. Koch, J. Feldmann, G. von Plessen, T. Meier, A. Schulze, P. Thomas, S. Schmitt-Rink, K. Köhler and K. Ploog, phys. stat. sol. (b) **173**, 21 (1992).
21. M. Koch, D. Weber, J. Feldmann, E.O. Göbel, T. Meier, A. Schulze, P. Thomas and S. Schmitt-Rink, Phys. Rev. B **47**, 1532 (1993).
22. J.H. Eberly and K. Wódkiewicz, J. Opt. Soc. Am. **67**, 1252 (1977).
23. A. Freiberg and P. Saari, IEEE J. Quantum Electron. **19**, 622 (1983).
24. H. Stolz, V. Langer and W. von der Osten, J. Luminesc. **48&49**, 72 (1991).
25. H. Stolz, V. Langer, E. Schreiber, S. Permogorov and W. von der Osten, Phys. Rev. Lett. **67**, 679 (1991).
26. W. van der Poel, A. Severens and C.T. Foxon, Opt. Commun. **76**, 116 (1990).
27. P. Lavallard, C. Gourdon and R. Planel, J. Luminesc. **53**, 367 (1992).
28. S. Permogorov, A. Naumov, C. Gourdon and P. Lavallard, Solid State Commun. **74**, 1057 (1990).
29. I.L. Aleiner and E.L. Ivchenko, JETP Lett. **55**, 692 (1992).
30. V. Langer, H. Stolz, W. von der Osten, D. Fröhlich, A. Kulik and B. Uebbing, Europhys. Lett. **18**, 723 (1992).
31. D. Fröhlich, A. Kulik, B. Uebbing, A. Mysyrowicz, V. Langer, H. Stolz and W. von der Osten, Phys. Rev. Lett. **67**, 2343 (1991).
32. D. Fröhlich, A. Kulik, B. Uebbing, V. Langer, H. Stolz and W. von der Osten, phys. stat. sol. (b) **173**, 31 (1992).
33. D.G. Thomas and J.J. Hopfield, Phys. Rev. **128**, 2135 (1962).
34. V.T. Agekyan, phys. stat. sol. (a) **43**, 11 (1977).

35. K. Shindo, T. Goto and T. Anzai, J. Phys. Soc. Japan **36**, 753 (1974).
36. A.Z. Genack, H.Z. Cummins, M.A. Washington and A. Compaan, Phys. Rev. B **12**, 2478 (1975).
37. P.Y. Yu and Y.R. Shen, Phys. Rev. B **12**, 1377 (1975).
38. R.M. Martin and L.M. Falicov, in *Light Scattering in Solids*, edited by M. Cardona, Topics in Appl. Phys. **8** (Springer, Berlin, 1975) p.80.
39. J.S. Weiner and P.Y. Yu, Solid State Commun. **50**, 493 (1984).
40. J.L. Birman and R. Berenson, Phys. Rev. B **9**, 4512 (1974).
41. J.L. Birman, Phys. Rev. B **9**, 4518 (1974).
42. D. Fröhlich and R. Kenklies, phys. stat. sol. (a) **111**, 247 (1982).
43. Y. Toyozawa, Progr. Theor. Phys. **20**, 53 (1958).
44. F. Bassani and L.C. Andreoni, in *Proceedings of the International School of Physics "Enrico Fermi"*, Course 96, edited by U.M. Grassano and N. Terzi, Varenna, Italy (1985).
45. V.M. Agranovich and V.L. Ginzburg, in *Crystal Optics with Spatial Dispersion and Excitons*, edited by V.M. Agranovich and V.L. Ginzburg, Solid State Sciences **42** (Springer, Berlin, 1984).
46. A. Kulik, Diplomarbeit, University Dortmund 1992.
47. S. Nikitine, J.B. Grun and M. Certier, Phys. Kondens. Materie **1**, 214 (1963); P.D. Bloch, B. Meyer and C. Schwab, J. Phys. C: Sol. State Phys. **13**, 267 (1980).
48. J. Aaviksoo, J. Luminesc. **48&49**, 57 (1991).
49. R.G. Ulbrich and G.W. Fehrenbach, Phys. Rev. Lett. **43**, 963 (1979).
50. Y. Masumoto, Y. Unuma, T. Tanaka and S. Shionoya, J. Phys. Soc. Japan **47**, 1844 (1979).
51. J. Aaviksoo, J. Lippmaa and J. Kuhl, J. Opt. Soc. Am. **5**, 1631 (1988).
52. J. Aaviksoo, J. Kuhl and K. Ploog, Phys. Rev. A **44**, R5353 (1991).
53. E. Schreiber and W. von der Osten, J. Luminesc. **50**, 211 (1991).
54. M. Kiene, Diplomarbeit, University Paderborn 1992.

INVESTIGATION OF FREE-CARRIER SCATTERING IN SEMICONDUCTORS USING COHERENT SPECTROSCOPY

Jagdeep Shah and Dai-Sik Kim

AT&T Bell Laboratories
Holmdel, NJ 07733, USA

1. INTRODUCTION

Scattering of free carriers by phonons, free carriers and coupled plasmon-phonon modes is of fundamental interest in solid state physics, and also of interest in semiconductor devices such as bipolar transistors. Investigations of the rate of energy loss of high energy electrons incident on thin metallic films date back to 1930 [1, 2]. In more recent times scattering processes of free carriers, excitons and other excitations in semiconductors have been investigated by studying their relaxation dynamics with ultrafast lasers. There are a number of recent reviews of these ultrafast studies in semiconductors and their nanostructures [3, 4].

Most of these studies have concentrated on the energy relaxation processes for free carriers, which include emission of optical phonons via Fröhlich and deformation potential interactions, and interaction with other carriers and plasmons. It is also possible for carriers to undergo momentum scattering without losing or gaining significant energy. If the free carriers remain within the optically-coupled region following such a scattering process, then the usual pump-and-probe or luminescence experiments are insensitive to such scattering and hence cannot provide any information about it. While it is possible to devise more sophisticated ultrafast experiments which do access such scattering processes, the simplest approach to studying these processes might be through coherent spectroscopy. Since interband polarization is dephased by all electron or hole scattering processes, a measurement of dephasing of optically-created interband polarization by coherent nonlinear optical techniques (e. g. hole-burning and four-wave-mixing) provides a direct means of investigating free-carrier scattering processes.

We begin this review by a brief discussion of some basic concepts (Section 2) and some relevant early experiments on free-carrier energy loss (Section 3). In Section 4 we discuss coherent spectroscopy experiments on undoped semiconductors and quantum wells, and follow that (Sec. 5) with a discussion of coherent spectroscopy scattering experiments on modulation-doped quantum wells. We conclude this review with a brief summary and a discussion of outstanding issues.

Coherent Optical Interactions in Semiconductors
Edited by R.T. Phillips, Plenum Press, New York, 1994

2. BASIC CONCEPTS

Free-carrier scattering is extremely complex and depends not only on the type of carrier (electron or hole) but also on the energy and density of the photoexcited carriers, on the temperature of the crystal, and on the density and type of background carriers present in the system. For semiconductors like GaAs, electron-phonon scattering occurs via Fröhlich interaction for scattering within the Γ valley, and via deformation potential interaction for intervalley scattering by large wavevector phonons. Electron-electron and electron-hole scattering is generally very important and is treated in various approximations, including static screening, time-dependent static screening, dynamic screening. The molecular dynamics approach has also been applied to this problem. Electron-acoustic phonon scattering can be important for low energy electrons and for inter-subband scattering in quantum wells. Hole-phonon scattering is dominated by optical deformation potential scattering for large hole energies, and by acoustic deformation potential scattering for hole energies less than the optical phonon energy. Intervalley scattering is not important for holes, but inter-valence band transitions in the bulk and inter-subband transitions in quantum wells play an important role. Hole-electron and hole-hole scattering plays an extremely important role and can usually be treated in the static screening approximation if the hole kinetic energy is low. A good review of many of these processes is available in textbooks and review articles [5-11].

One aspect of carrier scattering processes that is often overlooked is that it can depend rather strongly on the initial distribution created by optical excitation. Thus, one may access different situations by varying the pulsewidth of the exciting laser. Similarly, cw or steady-state excitation presents a very different physical situation because it builds up a reservoir of carriers at the bottom of the conduction band and at the top of the valence band.

Carrier-carrier scattering is mediated by the Coulomb interaction for which small angle scattering resulting in little or no change in the energy dominates. The same is true for scattering by ionized impurities. These Coulomb-mediated processes may dominate the total scattering rate, and hence may play an important role in dephasing of the interband polarization even if they contribute little to the energy relaxation rate. Thus, coherent and incoherent transients often provide information about different physical phenomena. In particular, coherent spectroscopy may provide the best means of probing carrier-carrier scattering processes in high-quality semiconductors in which impurity-related scattering processes may be neglected.

3. INCOHERENT SPECTROSCOPY OF FREE-CARRIER SCATTERING

Ultrafast spectroscopy of semiconductors, until recently, concentrated on the incoherent response of the semiconductors to an ultrashort optical pulse. These studies included pump-and-probe transmission, reflection and Raman scattering studies, as well as time-resolved luminescence studies. These studies, as well as some cw spectroscopic studies, provided information primarily about the *energy relaxation* of photoexcited carriers. We provide a *brief* review of these studies to give an idea about the kind of information obtained from such studies. At the end of this section, we briefly discuss two experiments which provide information about the *momentum relaxation* of free carriers.

3.1. Electron Relaxation within the Γ Valley

In one of the first experiments to investigate this process, Auston et al. [12] measured the change in the reflectivity (ΔR) at 615 nm from bulk GaAs at room temperature excited with 1 ps UV pulses at 307.5 nm, as a function of time delay between these pulses. They found an initial decrease in the reflectivity followed by an increase, with ΔR saturating at a positive value. They interpreted the results in terms of cooling of thermalized, hot electrons at a rate of approximately 750 meV/electron-ps.

In a very different experiment, Kash et al. [13] investigated pump-and-probe Raman scattering from LO phonons in GaAs at low temperatures. They found that the risetime of the anti-Stokes Raman signal was approximately 2 ps. They interpreted the results in terms of sequential emission of 12 LO phonons (cascade model) and deduced an average emission time of a LO phonon of $\approx$ 165 fs. These experiments were performed at a density slightly less than $10^{17}\,cm^{-3}$. If electron-electron collisions lead to a hot, thermalized electron gas in times much shorter than 2 ps, then the results would be interpreted as the cooling of this gas at a rate of about 220 meV/electron-ps.

In contrast to these measurements, the results of Knox et al. [14] provide a direct measure of the carrier distribution function (DF). In a pump-and-probe transmission measurements with 100 fs pulses, an intrinsic GaAs quantum well sample (well width = 96 Å) at 300 K was excited about 20 meV above its lowest interband transition. The differential transmission spectrum (DTS) shows a peak due to bleaching of the interband absorption. This peak moves with the center of the exciting laser spectrum. Fig. 1 shows the DTS at various time delays for a fixed energy. From these data it was deduced that the non-thermal electron distribution becomes thermalized in about 200 fs for an excitation density of 2×10^{10} cm^{-2} / well.

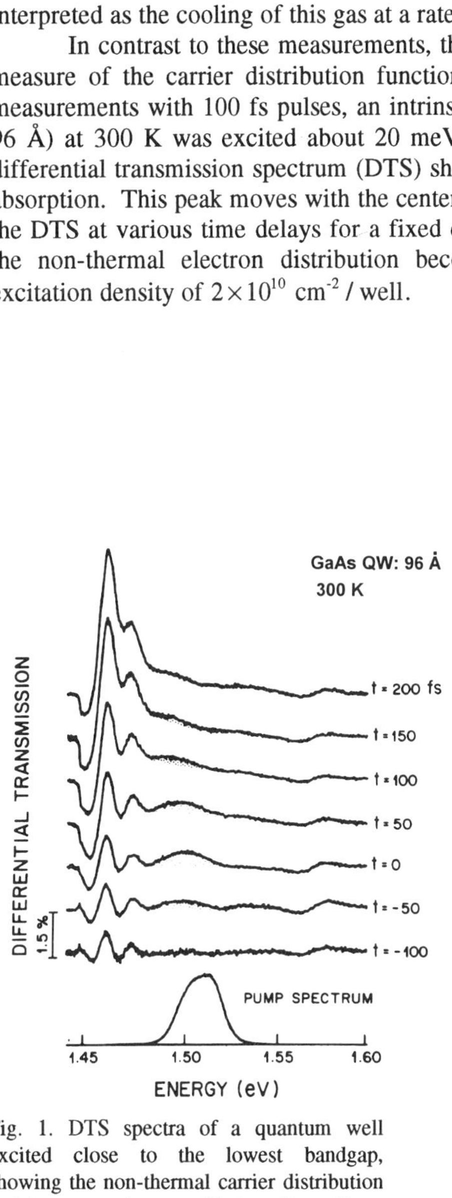

Fig. 1. DTS spectra of a quantum well excited close to the lowest bandgap, showing the non-thermal carrier distribution and its approach to equilibrium (from Knox et al. [14]).

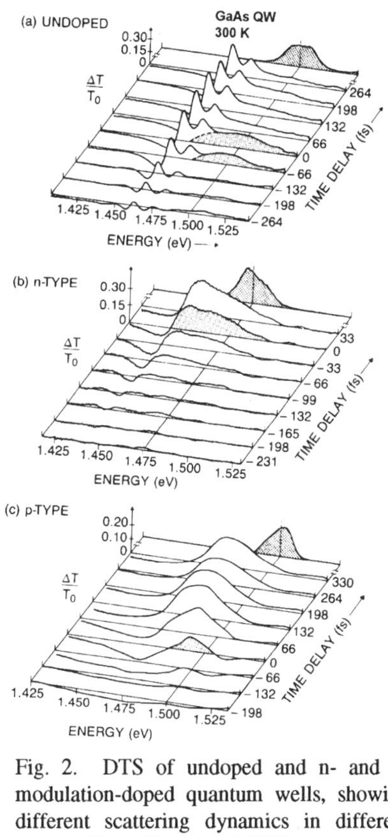

Fig. 2. DTS of undoped and n- and p-modulation-doped quantum wells, showing different scattering dynamics in different systems (from Knox et al. [15]).

Knox et al. [15] also investigated the effect of background electron and hole populations by studying n- and p-modulation-doped samples at 300 K. Their results (Fig. 2) show that the presence of a dense electron gas (3×10^{11} cm^{-2} / well) leads to an extraordinarily rapid thermalization (<10 fs) but the presence of an equal density of holes does not speed up thermalization. These interesting results are not yet fully understood. Two additional points should be noted about these results. First, at the high temperatures used in these experiments, the electrons and holes are nearly *non-degenerate*. Secondly, these and most other pump-and-probe transmission results are dominated by the *electron* distribution function and provide little information about hole distributions.

Ultrafast luminescence studies also provide information about electron and hole relaxation processes. In an investigation of bulk GaAs at 300 K, Elsaesser et al. [16] found that, for an excitation density of $\approx 2\times10^{17}$ cm^{-3}, the luminescence spectrum is extremely broad even at time delays less than 100 fs, as shown in Fig. 3. This was attributed to extremely rapid carrier-carrier collisions. Comparison with Monte Carlo simulations using various approximations showed that static and time-dependent static screening approximations could not explain the data adequately, but the molecular dynamics approach provided a better fit to the data. Recently there have been treatments of carrier-carrier scattering using full dynamical screening [17, 18]. However, screening for the non-equilibrium plasma remains one of the open questions in this field.

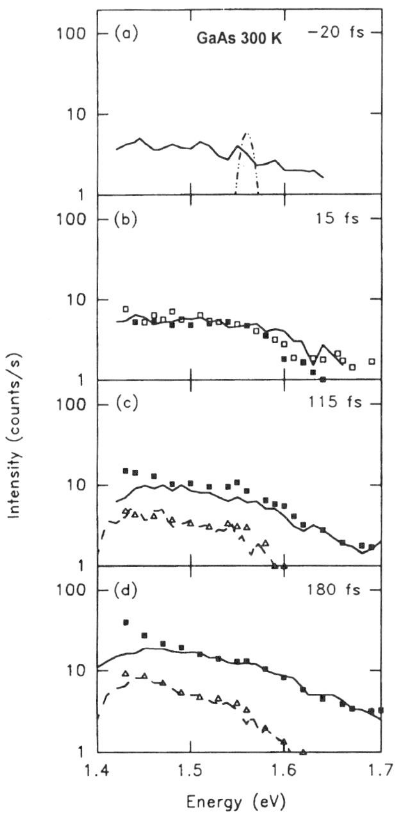

Fig. 3. Femtosecond luminescence spectra of GaAs at 300 K at various time delays, showing the broad, non-thermal electron distribution (from Elsaesser et al. [16]).

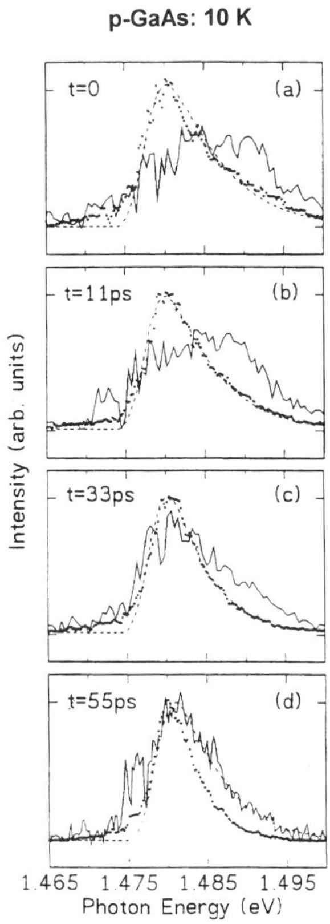

Fig. 4. Picosecond electron-acceptor luminescence spectra in p-GaAs, showing a non-thermal electron distribution lasting for several tens of picoseconds at very low excitation density (from Snoke et al. [19]).

To illustrate that carrier scattering in semiconductors is a very complex problem, we note the recent results of Snoke et al. [19] in which they measured time-resolved electron-to-acceptor luminescence from GaAs excited by a 5 ps pulse close to the bandgap. For the very low density used in these experiments ($\approx 10^{14}$ cm^{-3}), they found (Fig. 4) that the electron distribution remains non-thermal for several *tens of picoseconds,* in sharp contrast to the 100 fs times obtained by the experiments discussed above. However, from a theoretical analysis of carrier-carrier scattering, these authors showed that their results are consistent with the results of Elsaesser et al. [16] if the difference in density is considered.

Carrier-carrier scattering rates were also deduced by Kash et al. [20] from cw conduction band-to-acceptor luminescence [21-23] using phonon emission time as a clock. We only remark that, in terms of the initial distribution that is created, as well as in terms of the presence of carriers at the conduction-band minimum and valence-band maximum, the experimental conditions are very different between the cw and pulsed experiments. Therefore, caution must be exercised in comparing the results obtained by different techniques. We will not discuss these results but refer the interested reader to a recent review by Kash and Tsang [24].

3.2. Intervalley Scattering of Electrons

For excitation photon energy much larger than the bandgap energy, photoexcited electrons may be able to emit or absorb a large wavevector phonon via deformation potential interaction and transfer to one of the subsidiary conduction band valleys (L or X). The process of transfer to the L or the X valley is expected to be very fast, partly because of the large density of states in these valleys. This process is of importance in the physics of high field transport where transfer to the low-mobility subsidiary valley leads to a negative differential resistance. The dynamics of conductivity resulting from this transfer has been investigated by Nuss et al. [25]. The same transfer process also leads to the phenomenon of velocity overshoot which has been investigated by Fork et al. [26] and recently by Sha et al. [27] using pump-and-probe spectroscopy.

The process of transfer of photoexcited electrons to the subsidiary valleys and their subsequent slow return leads to an interesting slow rise of near-bandgap luminescence in GaAs at 300 K [28]. The return is slow because the density of final states in the Γ valley is small. From a comparison with Monte Carlo simulation of the experiment, the value for the intervalley deformation potential was deduced. This value is larger than the value deduced at helium temperatures from cw band-to-acceptor luminescence spectroscopy [23]. This difference can be largely accounted for by the increased transfer due to phonon absorption at higher temperatures. Femtosecond pump-and-probe transmission studies have also been used [29] to investigate the intervalley transfer process in GaAs.

3.3. Non-thermal holes

As pointed out above, ultrafast optical experiments are dominated by the electron distribution function and hence provide information primarily about the dynamics of electron relaxation. The dynamics of holes can be investigated by using n-doped samples [30, 31]. We briefly discuss here the results of Tomita et al. [31] who excited an n-modulation-doped quantum well sample at 10 K with excitation photon energy close to the bandgap and excitation density much smaller than the doping density. Under these conditions, the electron distribution at the bandedge is perturbed less than 10 %, and the luminescence intensity at the bandgap is governed by the hole population at the top of the valence band. The time evolution of the bandgap luminescence intensity shows a substantial increase after the end of the 100 fs excitation pulse. Tomita et al. [31] argued that this can only be explained by a substantial *non-thermal* hole population for the first 800 fs. By a comparison with Monte Carlo simulations it was shown that hole-electron interactions play a dominant role in the hole thermalization and that static screening provided a reasonable description of the process. As discussed above, a comparison with dynamic screening calculation would be interesting when such a calculation becomes available for non-thermal distributions.

3.4. Polarization-Rotation spectroscopy

The experiments discussed above provide information about energy relaxation rates. As discussed in Section 2, there are scattering processes that do not contribute to the energy relaxation of free carriers. These include small-angle carrier-carrier scattering, scattering of free carriers with impurities, and scattering with acoustic phonons of low energy. Oudar et al. [32] reported an elegant variation on the pump-and-probe transmission measurement to measure the momentum scattering of free carriers.

A linearly polarized pump pulse creates an anisotropic distribution of carriers in momentum space, and leads to a corresponding anisotropy in the absorption saturation of a weak probe pulse. A measurement of the polarization of the transmitted probe pulse as a function of the time delay between the pump and the probe pulses provides information about the dynamics of momentum relaxation of the free carriers. Using 100 fs pulses at 806 nm, GaAs at 77 K was excited to a density of 6×10^{17} cm^{-3}; free carriers were created with an excess energy less than an optical phonon energy. For these excitation conditions, Oudar et al. [32] found that the orientational relaxation time is 190 fs, most likely due to carrier-carrier collisions.

Portella et al. [33, 34] reported measurements on GaAs at 300 K using 9 fs pulses at 820 nm, creating free electrons with an excess energy of ~ 110 meV. They deduced momentum relaxation times varying from 60 fs to 25 fs as the excitation density was reduced from 8×10^{17} cm^{-3} to 8×10^{16} cm^{-3}. Since 9 fs pulses have a spectral width of about 200 meV, momentum relaxation by phonon emission and absorption is included in these measurements. This may be one reason why these values are shorter than those reported by Oudar et al. [35].

3.5. Momentum Scattering: Negative Absolute Mobility

Momentum scattering of free carriers in modulation-doped quantum wells was deduced by a luminescence imaging experiment by Höpfel et al. [36, 37]. A p-modulation-doped quantum well structure was weakly excited by picosecond pulses from a dye laser. An electric field parallel to the interfaces was applied to the quantum wells. As the minority electrons move under the influence of the applied electric field and other forces in the system, the position of the emitting region moves parallel to the interface planes. It was observed that the electrons actually move towards the *negative* electrode, thus exhibiting negative absolute mobility. This interesting phenomenon results from momentum scattering between the optically-excited minority electrons and the majority holes in the system. The large number of holes moving towards the negative electrodes *drag* the electrons with them, against the force exerted by the applied electric field. From an analysis of this unusual phenomenon, Höpfel et al. [36, 37] deduced *electron-hole* momentum scattering times of the order of 80 fs for minority electrons in p-modulation-doped quantum wells, and approximately 3 ps for minority holes in n-modulation-doped quantum wells. The large difference in the two scattering times results from the difference in masses, in the degeneracy and in the degree to which the two types of wells approach ideal two-dimensionality [37].

3.6. Summary

This completes a brief survey of measurements of carrier-carrier scattering in bulk semiconductors and their nanostructures. Most of these incoherent spectroscopy techniques determine the energy relaxation rates. In contrast, coherent spectroscopy techniques are sensitive to any scattering process that dephase the polarization created by the laser pulse, and hence provide a powerful means for determining some of the fastest scattering processes in semiconductors. We discuss some of these measurements in Sections 4 and 5.

4. COHERENT SPECTROSCOPY OF FREE CARRIERS IN UNDOPED SEMICONDUCTORS

The free-carrier continuum states in a semiconductor can be viewed as a collection of inhomogeneously-broadened two level systems in momentum space; i. e. the energy of the states is different for different k values. Many nonlinear optical techniques [38] have been developed for investigating inhomogeneously-broadened two-level systems. Two of these have been applied to a study of the dephasing of interband polarization in semiconductors. In a hole-burning experiment, one measures the transmission spectrum of a sample with a broadband optical probe following the excitation of the sample by a laser whose spectral bandwidth is narrower than the homogeneous linewidth of the process under investigation. The spectral width of the hole in the transmission spectrum provides information about the homogeneous linewidth and hence the dephasing rates. The second technique that has been applied is that of the photon echo [39]. It is well-known from the simple theory of a two-level systems [40] that if such a system is excited by two short pulses along k_1 and k_2, with pulse 2 along k_2 delayed by time τ_d with respect to pulse 1 along k_1, then the diffracted signal along 2 k_2 - k_1 comes at a time τ_d after pulse 2. Furthermore, the time-integrated diffracted signal decays with a time constant $T_2/4$ where T_2 is the dephasing time of the system. We discuss the results of these experiments on undoped semiconductors in this section.

4.1. Spectral Hole-Burning Experiments in GaAs

Oudar et al. [35] investigated the spectral dependence of absorption saturation around the excitation pulse wavelength in GaAs at 15 K. The sample was excited with a 0.5 ps pulse with a spectral width of 5 meV, centered about 20 meV above the bandgap. The change in the transmission spectrum of the sample resulting from the excitation of the sample by this pump pulse was measured with a 100 fs probe pulse and a multichannel detector. They found a broad spectral hole near the pump wavelength at short delay between the pulses. With increasing delay, the hole broadened further, and the transmission spectrum evolved into a spectrum corresponding to a thermalized hot distribution. They analyzed the data using a simple model which characterized the spectral width of the hole by $1/T_2$ and the relaxation rate of a non-thermal distribution towards a thermalized hot distribution by $1/T_1$, and deduced $T_2 \approx 300$ fs and $T_1 \approx 250$ fs.

This represents the first investigation of free-carrier dephasing phenomenon in semiconductors. The results were analyzed in terms of a simple model of dephasing and energy relaxation rates that seemed quite reasonable. However, we know a lot more about these processes nearly ten years later, so one must use caution in accepting the interpretation resulting from this model. For example, the pump pulsewidth is longer than the times deduced from the model, implying that considerable relaxation occurs during the pump pulse and a simple model might be inappropriate. Later measurements in quantum wells by Knox et al [14] with 100 fs pump and probe pulses led to shorter energy relaxation times at 300 K. Furthermore, the spectral hole was found to be shifted to lower energies even at very short times, an effect that has been recently discussed by Knox [41]. Finally, complicated effects around non-thermal Fermi edge singularities have been recently reported by Foing et al. [42].

Some of these problems may be sidestepped by using photon echo experiments which are more directly sensitive to dephasing rates of free carriers. We now discuss these experiments in undoped semiconductors.

4.2. Photon Echo Experiments in GaAs and GaAs Quantum Wells

The first experiments were performed by Becker et al. [43] using 6 fs amplified pulses [44] at 620 nm, the shortest pulses obtained to date. The FWHM of the spectrum of a 6 fs Gaussian pulse is about 300 meV and the center of the laser spectrum excited electron-hole pairs with an excess energy > 500 meV, conditions very different from those of Oudar et al. [32].

Fig. 5 shows the time-integrated photon echo signal obtained by Becker et al. [43] . The decay of the photon echo signal can be approximated by an exponential whose slope changes with excitation density. The decay constant varies from 10.7 fs to 3.5 fs, corresponding to T_2 varying from about 44 fs to 14 fs as the density increases from 2×10^{17} cm^{-3} to 7×10^{18} cm^{-3}.

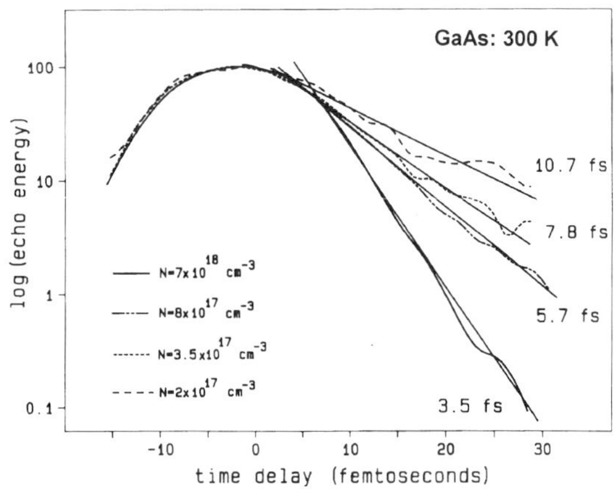

Fig. 5. Time-integrated photon echo signal from GaAs at 300 K as a function of time delay, for three different excitation densities (from Becker et al. [43]).

These experiments were repeated for 96 Å-thick GaAs quantum wells at 300 K by Bigot et al. [45]. The 8-10 fs near-IR pulses used for excitation had a spectral width of about 220 meV centered at about 1.56 eV. When the excitation density was varied from 1×10^{10} cm^{-2} per well to 1×10^{11} cm^{-2} per well, the exponential decay constants of the time-integrated photon echo signal varied from 50 fs to 16 fs. Fig. 6 and 7 show the variation of the decay constant with the density for bulk GaAs and GaAs quantum wells.

These measurements performed with the shortest available pulses show that the dephasing of the polarization excited in these experiments is extremely fast. The times deduced from these experiments are, in general, faster than the energy relaxation times measured by incoherent spectroscopy, consistent with the idea that momentum relaxation times can be much faster than energy relaxation times in these systems. However, this is just the beginning of our investigation of free-carrier scattering processes by coherent spectroscopy. It seems that there are many aspects of the problem that are not understood.

One question that needs to be answered is the role of the spectrum of the excitation pulse itself. The uncertainty principle requires that such short pulses have a very broad spectrum. Do the scattering processes themselves depend on the distribution of free carriers created by the photoexcitation pulse? There is clearly much more phase space available for scattering if the polarization is produced by a 10 fs pulse as compared to 10 ps pulse. I don't believe that we know the answers to such questions in detail.

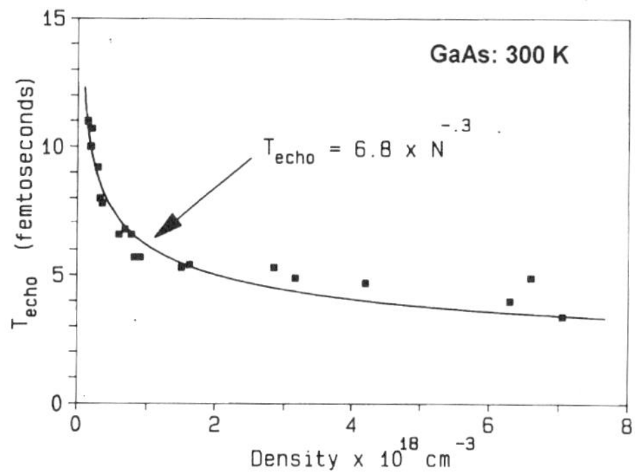

Fig. 6. Dependence of the decay time constant of the photon echo signal on the excitation density for bulk GaAs (from Becker et al. [43]) .

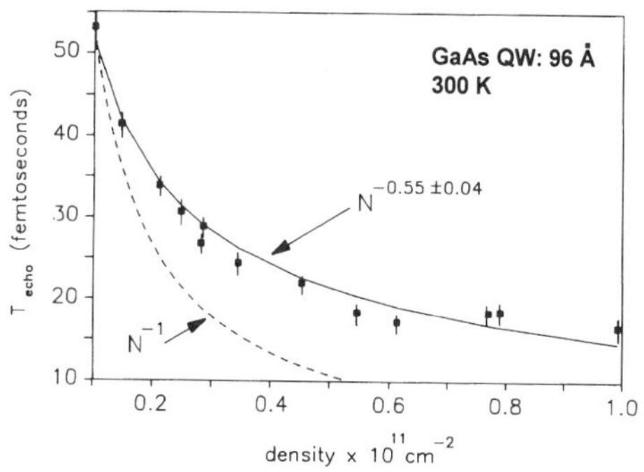

Fig. 7. Same as Fig. 6 but for GaAs quantum wells (from Bigot et al. [45]).

In analyzing these results obtained with such short pulses, one must also consider a possible breakdown of the usual Markovian approximation. This approximation basically assumes that the "bath" responds instantaneously and there are no memory effects. In the case of a molecular system in a liquid solution, the "bath" is the solvent. In the case electronic system in solids, the "bath" is the phonons with which the electronic system interacts. The Markovian assumption is a good assumption in solids for times longer than 100 fs, but must be examined carefully for shorter pulses. A theoretical discussion of memory effects in solids is presented by Hartmann and Schäfer [46]. An experimental investigation of non-Markovian behavior is presented by Bigot et al. [47]. I am not aware of any experimental investigation of this phenomenon in solids.

Another question that is of importance is the role of excitons in these experiments. Using 100 fs pulses, Kim et al. [48] investigated four-wave-mixing (FWM) signals from high-quality quantum wells at Helium temperatures using very low densities ($\leq 1\times10^9$ cm^{-2} per well). Thus the experimental conditions are very different from the experiments just described. Kim et al. [48] found a very large resonance in the FWM signal when the center of the incident laser spectrum coincided with the heavy-hole exciton transition energy. This is an effect that needs to be further investigated to see if it has any influence on our understanding of the results on free-carrier dephasing obtained with < 10 fs pulses whose spectrum inevitably encompasses some excitonic transitions. We discuss here two possible approaches to this problem proposed recently. We first discuss, in the next section, the use of *spectrally resolved* FWM experiments to determine what transitions are important. In Sec. 5, we discuss the use of modulation-doped quantum wells to destablilze the excitons so only free-carrier effects are important.

4.3. Spectrally-Resolved FWM Experiments

In a FWM or photon echo experiment, one generally measures time-integrated, spectrally-integrated diffracted signals. Measurement of the diffracted signal in a time-resolved and/or spectrally-resolved manner [49] provides additional information about the nonlinear response of the system.

Kim et al. [49] showed that the spectrum of the diffracted signal from a high quality 170 Å quantum well sample is dominated by excitons even when most of the spectrum of the exciting laser overlaps with the continuum states of the quantum wells (Fig. 8). This was presented as a further argument in favor of the strong role played by excitons in the nonlinear response of the system.

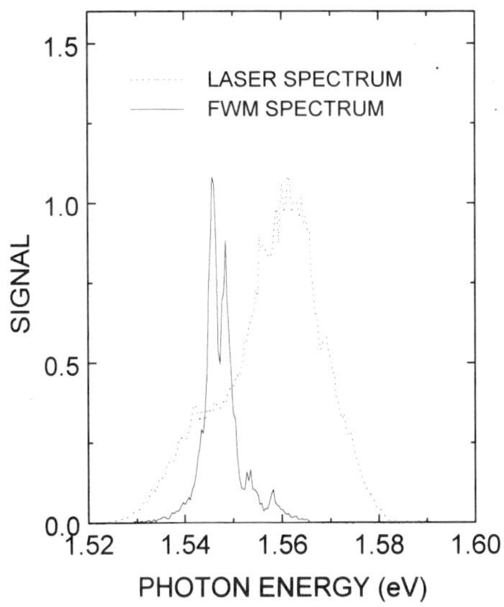

Fig. 8. Time-integrated FWM spectrum (solid curve) at zero delay from a 170 Å GaAs quantum well sample at 10 K obtained using an intentionally distorted laser spectrum (dashed curve) (from Kim et al. [49]).

In a more recent experiment, Rappen et al. [50] examined dephasing of excitons and free-carriers at the direct bandgap of Ge at 1.8 K using 100 fs near-bandgap excitation, and found that the FWM spectrum shows excitonic as well as free-carrier features. The analysis of the exciton dynamics yields a decay time of 100 fs, and T_2=200 fs if homogeneously broadening is assumed. They find that the maximum of the free-carrier feature occurs about 100 fs after the peak of the excitation pulse. This is similar to the slow rise in the excitonic signal reported by Kim et al. [51] and attributed to Coulomb interaction between the excitons. Rappen et al. [50] also attribute the slow rise to Coulomb interaction between the continuum states. They need a free-carrier dephasing time of about 800 fs in order to explain their results on the basis of semiconductor Bloch equations [52]. More recently, Lohner et al. [53] have investigated temporally and spectrally resolved excitonic and free-carrier polarizations in GaAs at low temperatures. They conclude that excitonic effects dominate at low densities ($\leq 10^{16}$ cm^{-3}) but that exciton screening and many-body effects play an increasingly important role at higher densities. They deduce a decay time limited by their time resolution of 70 fs and attribute it to carrier-carrier scattering. These experiments are valuable because they provide new information about the nonlinear response of intrinsic semiconductors. However, the nonlinear response is very complex and several different factors must be considered carefully, as shown by the following results.

Fig. 9 shows the time-resolved (but spectrally-integrated) FWM signals from the same 170 Å QW sample for three different τ_d (Kim et al. [51]). The peak near 1 ps is nearly independent of τ_d and results from the slow rise of the interaction-induced signals from excitons. We concentrate here on the additional peak near t=0 observed only near τ_d=0. While the FWM signal for $\tau_d > 0$ results from the diffraction from the *polarization grating*, this additional peak is associated with the diffraction of the laser by the *population grating* ([51]). This interpretation is supported by the lack of this peak in cross-polarization. Fig. 10 shows the intensity dependence of the time-resolved FWM signals ([54]) from the same sample. The peak at t = 0 resulting from the population grating increases strongly with increasing intensity for reasons that are not entirely clear at this time.

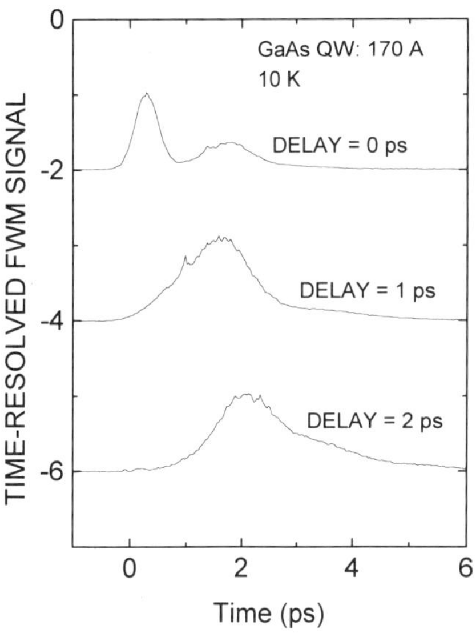

Fig. 9. Time-resolved FWM response of the 170 Å GaAs QW sample at three different τ_d (from Kim et al. [51]).

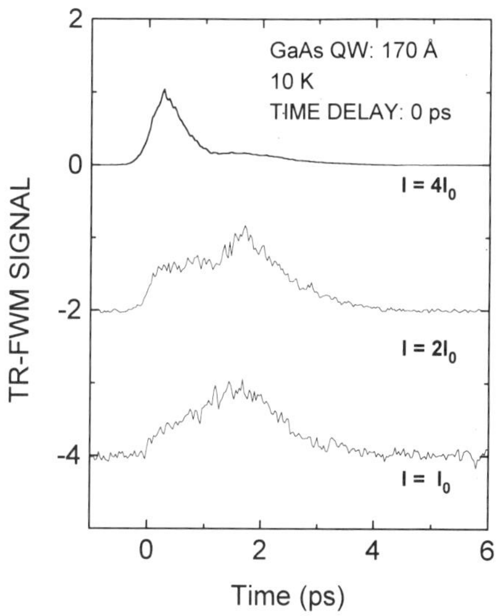

Fig. 10. Time-resolved FWM signal at different laser intensities (from Kim et al. [54].

Fig. 11 shows the *spectrum* of the time-integrated FWM signal at these time delays and compares them with the spectrum of the exciting laser. Although the excitonic feature dominates at all delays, the spectrum at zero delay has broad wings which resemble the laser spectrum. These broad wings disappear for $\tau_d > 0$. Note that the broad wings at

τ_d=0 peak at *lower energy than the exciton,* closer to the peak of the laser spectrum. Therefore, we interpret the results as the diffraction of the laser by the population grating generated at $\tau_d = 0$. In particular, the free-carrier continuum does not play a significant role in these results. If the laser peaked in the free-carrier continuum, we would expect the peak of the spectral wings to be at that energy. However, such an observation would still not imply that free-carrier effects dominate at early times. This points to the difficulty of identifying diffraction resulting from free carrier states if their dephasing times are shorter than the pulsewidth.

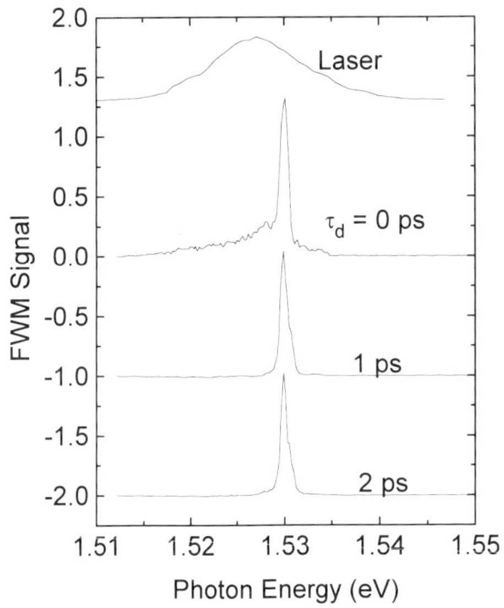

Fig. 11. Spectra for time-integrated FWM signals from a 170 Å intrinsic GaAs quantum well sample at 10 K (from Kim et al. [54]).

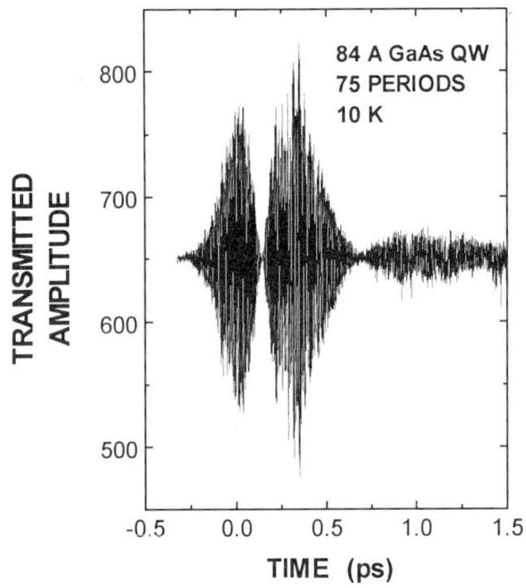

Fig. 12. The electric field amplitude of a pulse transmitted through an 84 Å undoped GaAs quantum well sample with 75 periods (total thickness of 6,300 Å) at 10 K. The center of the laser spectrum was at the HH exciton energy (from Kim et al. [55]).

Both the intensity dependence, and the FWM spectra at various delays illustrate the difficulty of separating excitonic and free-carrier contributions. Excitons dominate at low densities, but are screened at higher densities so that free-carrier effects become more important at higher densities. However, at higher excitation levels, the population grating effects dominate so that diffraction of the laser spectrum by the population grating becomes important. These effects are further complicated by the recent observation by Kim et al. ([55]) that even in moderately thick samples (~ 6,000 Å total thickness) typically used in nonlinear experiments, there is a severe distortion of the transmitted pulses on the time scale of the dephasing time. Fig. 12 shows interferometric measurements [55] of the electric field of a 100 fs pulse after transmission through a 84 Å undoped GaAs quantum well sample with 75 periods (total thickness of 6,300 Å) at 10 K. The transmitted pulse is severely distorted because of the re-radiation by the polarization created in the sample by the incident laser pulse. A close examination of the interferometric pattern shows that the re-radiation is 180 degrees out-of-phase as expected, and therefore leads to the nodes observed in Fig. 12.

These results have significant implications for the problem of distinguishing between excitonic and free-carrier effects. The electric field in the sample is the sum of the incident and induced fields. Since the induced fields persist for times longer than the incident field, it is possible to create *population grating* for τ_d longer than the pulsewidth. Since a population grating diffracts the laser spectrum, as seen above, this further complicates the distinction between excitons and free carriers.

4.4. Summary

There is a strong interest in the investigation of dephasing of interband polarization in undoped semiconductors resulting from carrier-carrier scattering. A number of investigations have been reported. However, the problem is very complex and dependence on many parameters, such as excitation density, excitation energy, and temperature, remains to be investigated in detail. There are also outstanding questions on the theoretical side, such as the nature of screening due to a non-thermal plasma, that need further investigation. This is likely to remain an interesting area of investigation for the coming years.

5. COHERENT SPECTROSCOPY OF FREE CARRIERS IN MODULATION-DOPED QUANTUM WELLS

One of the reasons for the complicated nonlinear optical behavior of undoped semiconductors is that excitons as well as free carrier continuum states play a role in the response of the system and it is difficult to distinguish between the two contributions. One way to get around this problem is by using doped semiconductors so that excitons are destabilized by the doping and free-carrier effects dominate. Modulation-doping in quantum wells provides a means of separating the charged impurities from the free carriers and leads to free-carriers with extremely high mobilities. This provides an ideal means of investigating free-carrier nonlinear response without any interference from excitons and impurities.

5.1. Time-Integrated FWM Signals

Kim et al. [49] found that the decay time of the time-integrated FWM signal from n-modulation-doped GaAs quantum wells depends strongly on the laser photon energy. The response is limited by the system time-resolution of ~ 100 fs when the laser excess energy above the Fermi energy is 12 meV, but it increases to ~ 1000 fs when the excess energy is zero (Fig. 13). Kim et al. [49] measured the decay rate as a function of density of each laser photon energy and extrapolated it to zero excitation density to obtain the zero-density decay rate as a function of photon energy. Fig. 14 shows a plot of the zero-density decay rate for two different samples, plotted as a function of δE normalized to the Fermi energy where δE is the difference between the laser photon energy and the chemical potential. The decay time of the time-integrated FWM signal varies from about 10 ps to .1 ps. If the system is treated as an inhomogeneously-broadened system, then the T_2 is as long as 40 ps.

The results of Kim et al. [49] represent the first measurements of the nonlinear response of a Fermi liquid in semiconductors. Knox et al. [15] earlier investigated the nonlinear response of modulation-doped quantum wells by pump-and-probe transmission spectroscopy at room temperature where the electron system is non-degenerate.

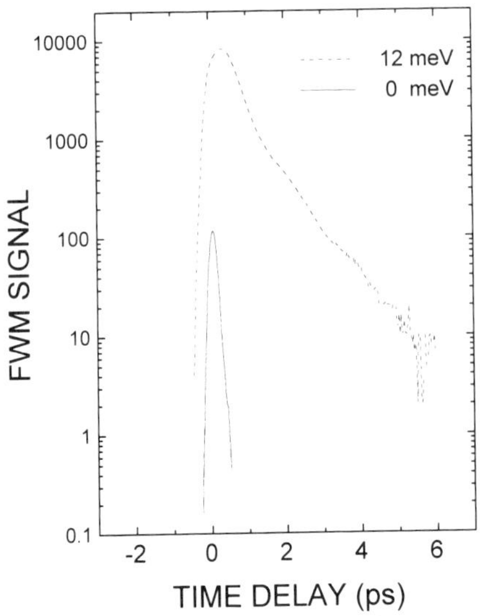

Fig. 13. Time-integrated FWM signals vs. time delay, for an n-modulation-doped GaAs quantum well structure for two different excitation photon energies (from Kim et al. [49]).

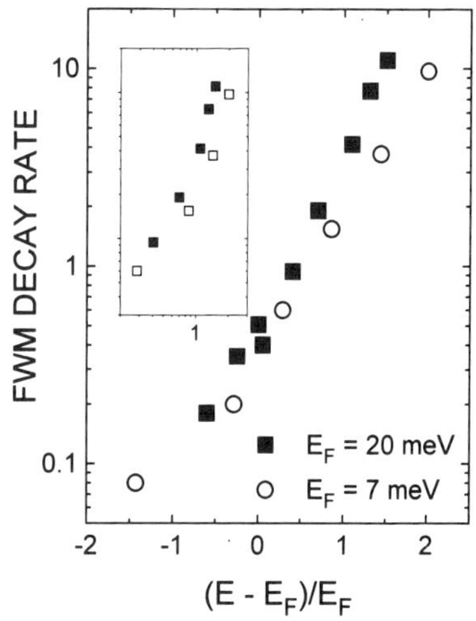

Fig. 14. Zero-density decay rate for different excess photon energy for two different samples. The inset shows a log-log plot for positive excess energies (from Kim et al. [49]).

Kim et al. [49] interpreted their results in terms of the Landau Fermi liquid theory. This theory has been successfully used to explain transport results in Fermi liquids. According to this theory, the fraction of the electrons in the Fermi liquid that can scatter an electron above the Fermi energy is limited by the Pauli exclusion principle and therefore depends on the energy of the electron. The closer the electron is to the Fermi energy, the smaller the fraction of the Fermi liquid that can scatter it. Kim et al. [49] argued that the dephasing of interband polarization is primarily due to the scattering of the photoexcited electron with the Fermi sea of electrons. Therefore, as the laser photon energy is increased, a larger fraction of the Fermi sea can scatter the photoexcited electrons, leading to shorter dephasing times for the interband polarization. The scattering rate for a 3D Fermi liquid is proportional to $\left[(E-E_F)/E_F\right]^2$ where E_F is the electron Fermi energy and E is the electron energy. In a 2D system, there is an additional logarithmic dependence [56] that makes a negligible contribution over a small energy range. A log-log plot of the scattering rate vs. excess energy (inset of Fig. 14) shows that this quadratic dependence is indeed observed.

The $(1/E_F)^2$ term in the above expression results from the increased screening of the electron-electron as the Fermi energy is increased. In a metal, the Fermi energy is fixed and there is no practical way to verify this dependence. In a modulation-doped quantum well, the Fermi energy can be varied by varying the doping density. For a given excess energy $\varepsilon=(E-E_F)$, a quantum well with a higher Fermi energy would have a smaller electron-electron scattering rate. Re-plotting the data in Fig. 14 as a function of ε we find that the dephasing rate for a given ε is indeed smaller for the sample with a larger density. In Fig. 14 the energy axis is normalized to the Fermi energy so that the two samples give the same scattering rate at a given normalized energy. Another advantage of the semiconductor system is that the interparticle separation parameter r_s is much smaller than for a typical metal so that some of the approximations made in the theory of metals are more valid for these semiconductors than some metals.

While these results provide a good qualitative understanding of the behavior of a 2D Fermi liquid, one must be cautious in deducing quantitative information from the results, as discussed earlier [49]. In particular, the measured decay rates of the FWM signals cannot be directly related to the scattering times of the system. The reason for this is that the experiment measures the behavior of the FWM signal averaged over a distribution of excess energies. In order to make quantitative comparisons, one must begin with (assumed or calculated) energy dependence of the scattering rate, and calculate the expected FWM signal for the actual distribution of polarizations created by the laser pulse. This is a difficult problem that has not yet been investigated.

Hawrylak et al. [57] have recently calculated the response of a 2D Fermi liquid and calculated various components of the dephasing rates attributable to electrons, heavy holes and light holes. They find that collective effects such as plasmon emission and interaction with coupled plasmon-phonon modes, and the complicated valence band structure make important contributions to the dephasing rates. Different approximations in the theory give dephasing rates differing by more than 50 %. The effects of spectral distribution discussed in the previous paragraphs must be considered before a meaningful comparison can be attempted.

5.2. Time-Resolved FWM Signals

As we discussed above, the FWM signal from an inhomogeneously-broadened system is in the form of a photon echo which occurs at time τ_d after the second pulse, where τ_d is the time delay between the two pulses. Fig. 15 shows time-resolved FWM signals from the heavily modulation-doped GaAs quantum well sample discussed in Sec. 5.1. Different traces correspond to different values of τ_d and time t is measured from the second pulse incident on the sample. We see that the peak of the FWM signal occurs close to $t=\tau_d$ for all values of τ_d. This is a strong argument in favor of treating the continuum states as inhomogeneously-broadened two-level systems in momentum space.

The sample with lower doping shows a prompt component in time-resolved signals, as shown in the inset of Fig. 15. This was explained in terms of residual excitonic effects in the low-density sample. The Fermi energy in this sample was 7 meV, of the order of the exciton binding energy. Therefore, the excitons are not completely destabilized. In contrast, the Fermi energy for the heavily doped sample is 20 meV, much larger than the exciton binding energy. Fig. 16 shows the behavior of the photon echo for the sample with smaller doping. The echo broadens considerably as the time delay τ_d is increased. The reason for this behavior is as follows: for small delays, all states contribute to the echo and the width of the echo pulse in time-domain is inversely proportional to the spectral width of the laser. With increasing time delays, only polarizations with long dephasing times contribute. The spectral width of the states contributing to the echo thus continuously decreases. Since the width in time-domain is inversely proportional to the spectral width, the photon echo broadens as seen in Fig. 16. This effect is less important for the heavily-doped sample because a larger range of states contributes at larger delays.

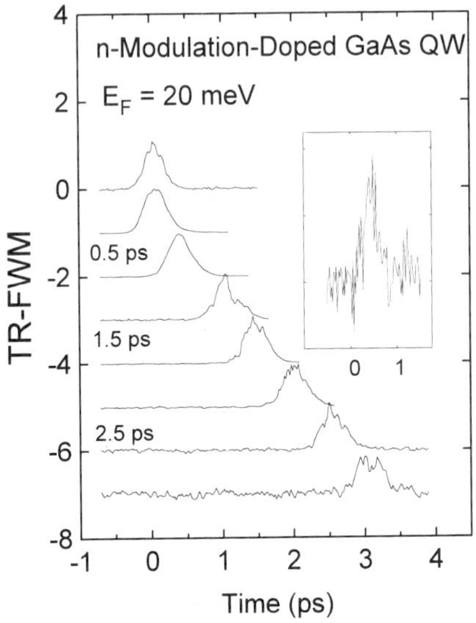

Fig. 15. Time-resolved FWM signals from an n-modulation sample for co-circularly polarized beams. The peak occurs at t= τ_d. The inset shows the signal for a sample with lower doping where a prompt signal is present (from Kim et al. [58]).

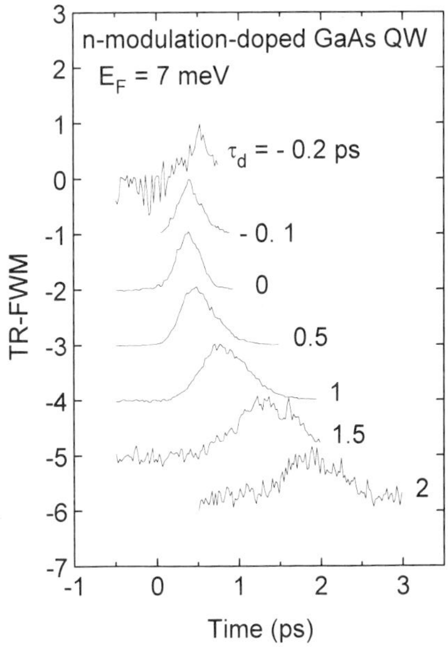

Fig. 16. Same as Fig. 15, but for the sample with lower doping concentration, showing the broadening of the photon echo signal (from Kim et al. [58]).

5.3. Summary

Modulation-doped quantum wells destabilize the excitons and allow investigation of free carrier scattering without interference of excitonic effects. The results provide a qualitative verification of the ideas of Landau Fermi liquid theory. A more quantitative comparison must await a more rigorous calculation of the FWM signals for a spectrally broad pulse. Time resolved measurements show clear photon echoes, thus verifying the idea of treating the continuum states as being inhomogeneously-broadened in momentum space. Varying the doping density provides a means of continuous transition from an exciton-dominated response to free-carrier-dominated response.

6. SUMMARY

Free-carrier scattering processes are very complex because they depend on many parameters such as free-carrier type, free-carrier density, free-carrier distribution function, temperature, and degeneracy. This complexity leads to seemingly different results in different experiments. Most of the incoherent spectroscopy experiments are affected only by the *energy loss rate* and provide a measure of this. The nature of screening plays a crucial role and a better understanding of screening, especially in the presence of non-equilibrium carrier distributions, would be very valuable. Coulomb-mediated free-carrier scattering rates are very large for small-angle scattering which does not affect energy loss rates but affect the momentum distribution of carriers. Coherent spectroscopy experiments provide a means of determining momentum relaxation rates in semiconductors. For intrinsic semiconductors, excitonic effects play an important role in ultrashort pulse experiments even when the center of the laser spectrum is in the continuum states of the semiconductor. One way to obtain information about free-carrier scattering without interference from exciton effects is to study modulation-doped samples which destabilize excitons. Such studies show long dephasing times of free carriers close to the Fermi edge, consistent with the Landau Fermi liquid theory. Studies on modulation-doped quantum wells have the potential of providing quantitative information about the nature of 2D Fermi liquids and thus distinguish between various theoretical approaches. Investigation of coherent properties of interband polarization induced by ultrafast lasers will continue to be an active area of research, using more sophisticated techniques to isolate and understand the complex dynamical processes.

Acknowledgments

It is a pleasure to thank S. Schmitt-Rink, W. Schaefer, J. E. Cunningham, T. C. Damen, M. Hartmann, F. Jahnke, K. Koehler, D. A. B. Miller, and L. Pfeiffer for collaboration on different aspects of the work presented here.

REFERENCES

[1] G. Rudberg, Proc. Roy. Soc. (London) **A27,** 111 (1930).
[2] D. Pines, "Elementary Excitations in Solids", W. A. Benjamin (1964).
[3] "Hot Carriers in Semiconductor Nanostructures: Physics and Applications", edited by Jagdeep Shah, Academic Press, Boston (1992).
[4] "Spectroscopy of Nonequilibrium Electrons and Phonons", edited by C. V. Shank and B. P. Zakhachrenya, Elsevier Science Publishers B. V., Amsterdam (1992).
[5] E. M. Conwell, in "Solid State Physics, Supplement 9", edited by F. Seitz, D. Turnbull and H. Ehrenreich, Academic Press, New York (1967).
[6] B. K. Ridley, "Quantum processes in semiconductors", Clarendon Press, Oxford (1982).
[7] B. R. Nag, "Theory of electrical transport in semiconductors", Pergammon Press (1972).
[8] B. R. Nag, "Electron Transport in Compound Semiconductors", Springer-Verlag, Heidelberg (1980).
[9] K. Seeger, "Semiconductor Physics: An Introduction", Springer-Verlag, Berlin (1982).
[10] Jagdeep Shah, and R. F. Leheny, in "Semiconductors probed by ultrafast laser spectroscopy", edited by R. R. Alfano, Academic Press, New York45-75 (1984).
[11] B. K. Ridley, in "Hot Carriers in Semiconductor Nanostructures: Physics and Applications", edited by Jagdeep Shah, Academic Press, Boston (1992).
[12] D. H. Auston, S. McAfee, C. V. Shank, E. P. Ippen and O. Teschke, Solid State Electronics **21,** 147-150 (1978).
[13] J. A. Kash, J. C. Tsang and J. M. Hvam, Phys. Rev. Lett. **54,** 2151-2154 (1985).
[14] W. H. Knox, C. Hirliman, D. A. B. Miller, J. Shah, D. S. Chemla and C. V. Shank, Phys. Rev. Lett. **56,** 1191-1193 (1986).
[15] W. H. Knox, D. S. Chemla, G. Livescu, J. E. Cunningham and J. E. Henry, Phys. Rev. Lett. **61,** 1290-1293 (1988).
[16] Thomas Elsaesser, Jagdeep Shah, Lucio Rota and Paolo Lugli, Phys. Rev. Letters **66,** 1757-1760 (1991).
[17] J. F. Young, P. L. Kelly and N. L. Henry, Semicond. Sci. Techno. **B7,** 148 (1992).
[18] R. Jalabert, and S. Das Sarma, Phys. Rev. **B41,** 3651 (1990).
[19] D. W. Snoke, W. W. Ruehle, Y. -C. Liu and E. Bauser, Phys. Rev. Lett. **68,** 990-993 (1992).
[20] J. A. Kash, Phys. Rev. **B 40,** 3455 (1989).
[21] B. P. Zakhachrenya, D. N. Mirlin, V. I. Perel and I. I. Reshina, Sov. Phys. Usp. **25,** 143 (1982).

[22] S. A. Lyon, J. Lumin. **35,** 121-154 (1986).
[23] R. G. Ulbrich, J. F. Kash and J. C. Tsang, pHYS. rEV. lETT. **62**, 949 (1989).
[24] J. A. Kash, and J. C. Tsang, in "Spectroscopy of Nonequilibrium Electrons and Phonons", edited by C. V. Shank and B. P. Zakhachrenya, Elsevier Scientific, Amsterdam (1992).
[25] M. Nuss, D. H. Auston, and F. Capasso, Phys. Rev. Lett. **58,** 2355-2358 (1987).
[26] C. V. Shank, R. L. Fork, B. I. Greene, F. K. Reinhart and R. A. Logan, Appl. Phys. Letters **38,** 104 (1981).
[27] W. Sha, T. B. Norris, W. J. Schaff, and K. E. Meyer, Phys. Rev. Lett. **67,** 2553-2556 (1991).
[28] Jagdeep Shah, Benoit Deveaud, T. C. Damen, W. T. Tsang, A. C. Gossard and P. Lugli, Phys. Rev. Letters **59,** 2222-2225 (1987).
[29] P. C. Becker, H. L. Fragnito, C. H. Brito-Cruz, J. Shah, R. L. Fork, J. E. Cunningham, J. E. Henry and C. V. Shank, Appl. Phys. Lett. **53,** 2089-2090 (1988).
[30] X. Q. Zhou, K. Leo and H. Kurz, Phys. Rev. **B 45,** 3886-3889 (1992).
[31] Akihisa Tomita, Jagdeep Shah, J. E. Cunningham, Stephen M. Goodnick, P. Lugli and Shun L. Chuang, Phys. Rev. **B,** (Aug. 15, 1993).
[32] J. L. Oudar, A. Migus, D. Hulin, G. Grillon, J. Etchepare and A. Antonetti, Phys. Rev. Lett. **53,** 384-387 (1984).
[33] M. T. Portella, J. -Y. Bigot, R. W. Schoenlein, C. V. Shank and J. E. Cunningham, in "Ultrafast Phenomena VII, Proc. 7th Int'l Conf.,", edited by C. B. Harris, E. P. Ippen, G. A. Mourou and A. H. Zewail, Springer-Verlag285-287 (1990).
[34] M. T. Portella, J. -Y. bigot, R. W. Schoenlein, J. E. Cunningham, and C. V. Shank, Appl. Phys. Lett. **60,** 2123-2125 (1992).
[35] J. L. Oudar, D. Hulin, A. Migus, A. Antonetti and F. Alexandre, Phys. Rev. Lett. **55,** 2074-2077 (1985).
[36] R. A. Höpfel, Jagdeep Shah, P. A. Wolff and A. C. Gossard, Phys. Rev. Lett. **56,** 2736-2739 (1986).
[37] R. A. Höpfel, Jagdeep Shah, P. A. Wolff and A. C. Gossard, in "Proc. of 18th Int'l Conf. on Physics of Semiconductors", edited by O. Engstrom, World Scientific, Singapore1319-1322 (1987).
[38] Y. R. Shen, John Wiley and Sons, New York (1984).
[39] N. A. Kurnit, I. D. Abella and S. R. Hartmann, Phys. Rev. Lett. **13,** 567 (1992).
[40] T. Yajima, and Taira, Y., J. Phys. Soc. Jpn. **47,** 1620 (1979).
[41] W. H. Knox, in "Hot Carriers in Semiconductor Nanostructures: Physics and Applications", edited by Jagdeep Shah, Academic Press, Boston (1992).
[42] J. -P. Foing, D. Hulin, M. Joffre, M. K. Jackson, J. -L. Oudar, C. Tanguay and M. Combescot, Phys. Rev. Lett. **68,** 110-113 (1992).
[43] P. C. Becker, H. L Fragnito, C. H. Brito-Cruz, R. L. Fork, J. E. Cunningham, J. E. Henry and C. V. Shank, Phys. Rev. Lett. **61,** 1647-1650 (1988).
[44] R. L. Fork, C. H. Brito-cruz, P. C. Becker, and C. V. Shank, Opt. Lett. **12,** 483-484 (1986).
[45] J. -Y. Bigot, M. T. Portella, J. W. Schoenlein, and C. V. Shank, Phys. Rev. Lett. **67,** 636-639 (1991).
[46] M. Hartmann, and W. Schaefer, Phys. Stat. Sol. **(b) 173,** 165-176 (1992).
[47] J. -Y. Bigot, M. T. Portella, R. W. Schoenlein, C. J Bardeen, A. Migun and C. V. Shank, Phys. Rev. Lett. **66,** 1138-1141 (1991).
[48] D. -S. Kim, Jagdeep Shah, J. E. Cunningham, T. C. Damen, W. Schafer, M. Hartmann and S. Schmitt-Rink, Phys. Rev. Lett. **68,** 1006-1009 (1992).
[49] D. -S. Kim, Jagdeep Shah, J. E. Cunningham, T. C. Damen, S. Schmitt-Rink, and Schafer, Phys. Stat. Sol. **(b) 173,** 11-20 (1992).
[50] T. Rappen, U. Peter, M. Wegener and W. Schaefer, QELS'93 Technical Digest 23 (1993).
[51] Dai-Sik Kim, Jagdeep Shah, C. Damen, W. Schaefer, F. Jahnke, S. Schmitt-rink, and K. Koehler, Phys. Rev. Lett. **69,** 2725-2728 (1992).
[52] H. Haug, and S. Koch, "Quantum Theory of the Optical and Electronic Properties of Semiconductors", World Scientific, Singapore (1990).

[53] A. Lohner, K. Rick, P. Leisching, A. Leitenstorfer, T. Elsaesser, T. Kuhn, F. Rossi and W. Stolz, to be published.

[54] D. -S. Kim, Jagdeep Shah, T. C. Damen, and W. Schaefer, unpublished (1993).

[55] D. -S. Kim, Jagdeep Shah, D. A. B. Miller, T. C. Damen, and W. Schaefer, submitted to Phys. Rev. Lett. (1993).

[56] P. Hawrylak, Phys. Rev. Lett. **59,** 485 (1987).

[57] Pawel Hawrylak, Jeff F. Young and Peter Brockmann, to be published (1993).

[58] D. -S. Kim, Jagdeep Shah, J. E. Cunningham, T. C. Damen, Stefan Schmitt-rink and Wilfried Schaefer, Phys. Rev. Lett. **28,** 2838-2841 (1992).

NONLINEAR COHERENT OPTICAL EFFECTS IN SEMICONDUCTORS

Duncan G. Steel[1], Hailin Wang[1], Min Jiang[1], Kyle Ferrio[1], and Steven Cundiff[2]

[1]Harrison M. Randall Laboratory of Physics
The University of Michigan, Ann Arbor, MI 48109
[2]Fachbereich Physik, Philipps-Universität
Renthof 5, W-3550 Marburg, Germany

INTRODUCTION

The resonant illumination by coherent optical radiation of an electronic excitation creates a quantum mechanical coherent superposition of the initial and final states of the transition. The nonlinear optical properties of this state can be used to provide a sensitive measurement of many fundamental parameters of the material including transport, elastic and inelastic scattering, energy level splittings between nearly degenerate states, energy relaxation, as well as associated information such as Landé g-factors, the degree of state mixing and symmetry.

In general nonlinear laser spectroscopy is a very established and well understood means to probe these phenomena. The basis for the nonlinear optical response in a crystal, however, can be considerably more complex than for simple isolated and noninteracting atoms. In this paper, we will examine the nature of the coherent nonlinear optical response of Wannier excitons in semiconductors at low exciton density. These excitons are strong resonances which appear in the absorption spectrum just below the bandedge in direct bandgap semiconductors and dominate the linear and nonlinear optical response in that spectral region. In high quality material, free from apparent disorder, we see that the nonlinear response of three dimensional excitons is dominated by strong spin-independent exciton-exciton scattering which also leads to loss of the optical coherence and a modification of the expected polarization coupling. In quasi-two-dimensional systems, the nonlinear response shows the strong effect of disorder. Spectral hole burning based on four wave mixing (FWM) reveals strong phonon coupling and the accompanying energy migration. The high spectral resolution afforded by nonlinear spectroscopy, however, also enables the detection of Zeeman splitting in small magnetic fields as well as the measurement of various relaxation rates.

Interest in the nonlinear optical response of semiconductors increased considerably several years ago with the demonstration of a strong room temperature nonlinear susceptibility with picosecond to nanosecond response times in GaAs/AlGaAs multiple quantum wells due to potential importance to device applications. This work was supplemented by more fundamental experimental and theoretical studies, however, as it became apparent that the fundamental physical description was incomplete. Here, we present a number of experimental results and simple physical pictures which will provide new insight into the fundamental nonlinear response as well as demonstrating the complexity that results in the presence of disorder. In addition, the results will show the power of coherent nonlinear laser spectroscopy in these systems for determination of

Coherent Optical Interactions in Semiconductors
Edited by R.T. Phillips, Plenum Press, New York, 1994

fundamental material parameters such as spectral diffusion rates associated with disorder and the excitonic g-factor.

THE COHERENT NONLINEAR OPTICAL RESPONSE OF THREE DIMENSIONAL EXCITONS IN BULK GaAs

There have been numerous studies discussed in the recent literature describing the fundamental nature of the nonlinear response of excitons [1, 2]. These investigations have emphasized quantum well structures and the results have been explained in the context of exchange and screening that arise from the intrinsic manybody nature of the problem. These models have tended to be scalar pictures of the electromagnetic coupling. Interestingly, in coherent nonlinear optical studies, there has been strong evidence that this picture is inadequate on short time scales when the vector nature of the interaction is taken into account.

For linear optical interactions the optical polarization properties and dipole selection rules have been well established in these systems, especially in the case of GaAs. The valence and conduction bands in GaAs at the zone center are characterized by angular momentum $J = 3/2$ and $J = 1/2$, respectively. The appropriate energy level structure is shown in Fig. 1. In this case, the simple dipole selection rules result in $\sigma_+(\sigma_-)$ polarized light coupling m to m+1 (m-1) transitions and linearly polarized light exciting m to m transitions. While this picture remains valid for nonlinear optical interactions, we show below that the simple application of this scheme to excitonic nonlinearities based on the classical model for optical nonlinearities using noninteracting two level systems leads to conflict with the experiments. The results show that the optical nonlinearity results from dynamical *spin-independent* coupling between different excitons resulting in line broadening and an apparent coupling between the different magnetic substates [3].

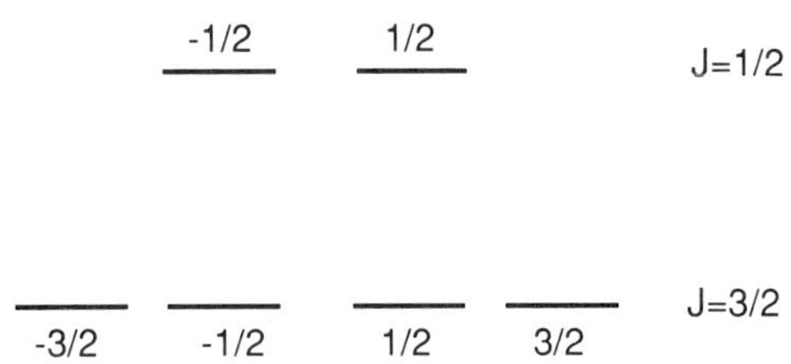

Figure 1. Energy level structure for bulk GaAs crystals at zone center.

The experiments for these studies were carried out at 6 K in a very high quality homogeneously broadened GaAs sample grown by molecular beam epitaxy with a 200 nm GaAs layer sandwiched between $Al_{0.3}Ga_{0.7}As$ layers (necessary to provide support). Strain was applied to lift the degeneracy between the heavy (hh) and light (lh) hole excitons [4, 5]. Figure 2 shows the energy level diagram appropriate for this system, where the red shift is larger for the lh. In this picture, the z-axis is also along the growth direction, normal to the surface. The corresponding linear absorption spectrum is shown in the inset of Fig. 3. The absorption width for the lh exciton is 0.2 meV.

The effect of the optical nonlinearity on the imaginary part (i.e., the absorptive component) of the susceptibility is most easily seen in a simple differential transmission (DT) measurement. Measurements were performed on the hh resonance to facilitate comparison with quantum well measurements and to avoid complications with light hole interactions. The laser pulses for the DT measurement had an auto correlation width of 1.5 ps and a spectral width of 4 meV, centered 1 meV below the hh resonance. The probe beam was spectrally resolved with 0.1 meV resolution by an optical multichannel analyzer.

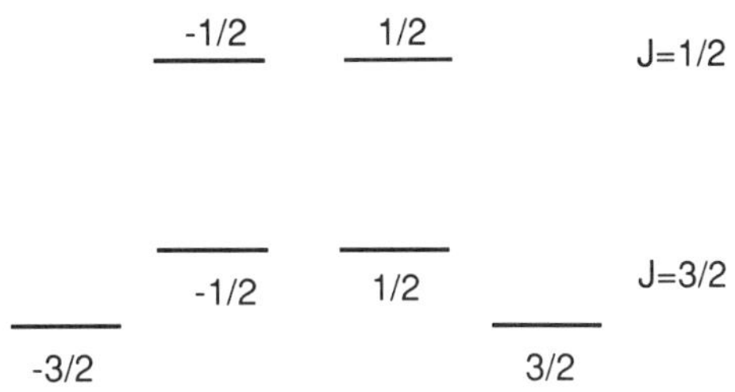

Figure 2. Energy level structure for GaAs at k=0 in the presence of uniaxial strain. The z-axis is along the growth direction

Figure 3 shows a typical DT spectrum of the hh exciton where the pump beam generates an exciton density of 3×10^{15} /cm^3 and is 3 ps advanced with respect to the probe to avoid coherent effects. The DT spectrum of the hh is nearly symmetric and shows only a negligible shift in the exciton energy (less than 0.1 meV). The change in the oscillator strength is determined by the spectral integral of the hh DT response. The spectrum in Fig. 3 has a net area less than 5% of the positive area, showing that the dominant nonlinear optical response comes from changes in the line width (i.e., dephasing rate).

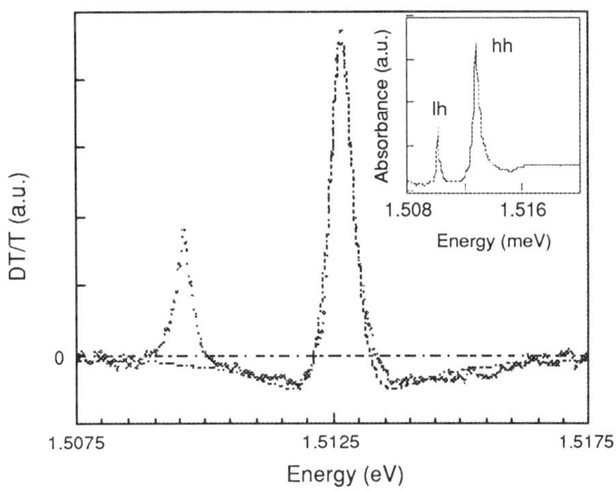

Figure 3. Differential transmission spectrum of the hh exciton. The dashed line is a fit of a collisionally broadened Lorentzian. Inset: Linear absorption spectra.

To confirm this, we first assume that the DT spectrum can be taken as the difference of two Lorentzians distinguished by different linewidths,

$$f\left\{\gamma_2/(\delta^2+\gamma_2^2)-\gamma_1/(\delta^2+\gamma_1^2)\right\} \tag{1}$$

where f is the oscillator strength, δ is the detuning from line center, and $\gamma_{1,2}$ are the dephasing rates without and with the pump beam, respectively. Since the zero crossings in the DT spectrum occur at

$$\delta^2 = \gamma_1\gamma_2, \tag{2}$$

we can determine the density dependent dephasing rate as well as the low excitation density line width. We then fit the hh DT spectrum in Fig. 3 with no adjustable parameter except the peak height. The result is shown as the dashed line. The excellent fit supports the preliminary interpretation of the DT spectra showing that the nonlinearity in this density regime appears to be dominated by line broadening. We designate this behavior as excitation induced dephasing (EID).

Physically, we note that dephasing is the loss of a macroscopic coherence of a coherently excited ensemble: i.e., as discussed in the introduction, a system excited by resonant radiation results in a coherent superposition (referred to as the optical induced coherence) of initial and final states described (in the Schrödinger picture) by a state vector

$$|\psi(t)\rangle = c_i(t)|i\rangle + c_f(t)|f\rangle \tag{3}$$

where in the case of optical experiments, the coherence is detected by the polarization given by the

$$\langle \mu \rangle \propto c_i^* c_f \langle i|\mu|f\rangle + c.c. \tag{4}$$

oscillating at frequency

$$\omega_f - \omega_i \tag{5}$$

The macroscopic polarization is given by the sum over all oscillators.

The individual excitations are not isolated but in fact sit in a reservoir which in the simplest case is just the vacuum radiation field. In the current system, the excitons dynamically couple to each other as well as to phonons, defects, etc. This coupling can result in actual decay of a probability amplitude such as through recombination radiation. However, more important for the current problem is that coupling to the reservoir such as by *elastic scattering* can result in a change of relative phase of the two probability amplitudes. In a simple statistical picture, the effect of these phase jumps is determined by the ensemble average over all of the oscillators. It is easy to show that in the Markoff approximation, this leads to exponential decay of the macroscopic polarization [6]. In the language of nuclear magnetic resonance, this dephasing time is referred to as the transverse relaxation time, T_2. It is important to note that the inverse dephasing time is comprised of the decays of the individual probability amplitudes as well as these so-called pure dephasing processes. In the frequency domain, the dephasing process leads to the homogeneous linewidth. In the case of the above data, the increase in exciton linewidth with increasing exciton density shows that exciton-exciton scattering caused by Coulomb coupling leads to an increase in the dephasing rate, hence the designation of excitation induced dephasing.

In the absence of coherent coupling, DT measurements are an incoherent probe of the nonlinear response. Further insight into the nonlinear response, however, can be had using coherent nonlinear laser spectroscopy techniques. For these experiments, we examined the self-diffracted FWM signal (in the direction of $2\mathbf{k}_2$-$\mathbf{k}_1$). The pulses had an auto correlation width of 3 ps and a spectral width of 0.9 meV to enable the resolution of the hh and lh excitons.

Figure 4a shows the FWM response at the hh resonance as a function of the delay between the two beams when $\mathbf{E}_1$ and $\mathbf{E}_2$ are σ_+ polarized. A prepulse with wave vector -$\mathbf{k}_2$ arriving 3 ps before $\mathbf{E}_1$ is used to generate excitons (N=2x10^{15} /cm^3) which are incoherent with those generated by $\mathbf{E}_1$ and $\mathbf{E}_2$. The density generated by the FWM beams is also 2x10^{15} /cm^3 in this data. The incoherent excitons affect the FWM signal by changing the EID, leading to a large reduction (a factor of 6 in Fig. 4a) of the signal strength. From DT data, the dephasing time in the absence of the prepulse is estimated to be 1.7 ps. The presence of the prepulse further reduces T_2. Since FWM signals decay as $T_2/2$, the measurement is pulse-width limited (FWHM of the signal with no prepulse is 3.1 ps). Hence, the induced change in T_2 is not readily observable in our measurement. As shown in Fig. 4a, the presence of σ_+ and σ_- polarized prepulses result in nearly the same reduction in the hh signal amplitude, indicating the *spin-independent* nature of interactions involved in EID. Spin-flipping of excitons is not important in these measurements since the removal of hh and lh degeneracy significantly decreases the exciton spin-flip rate [7]. In separate measurements we have shown that the combined spin and energy relaxation time of the hh exciton is longer than 20 ps. Density-dependent exciton dephasing has also been reported in earlier DT and FWM measurements in GaAs [8, 9]. The dephasing, however, was attributed to the spin-dependent exchange effects [9].

Figure 4b shows $I_{\|}/I_{\perp}$ at the hh resonance as a function of the exciton density generated by a prepulse arriving 20 ps before $\mathbf{E}_2$, where $I_{\|}$ and $I_{\perp}$ are the FWM signals obtained with $\mathbf{E}_1 \| \mathbf{E}_2$ and $\mathbf{E}_1 \perp \mathbf{E}_2$, respectively ($\mathbf{E}_1$ and $\mathbf{E}_2$ are linearly polarized). At relatively low exciton density where both responses are approaching third order in the applied fields, we find $I_{\|}/I_{\perp} > 10$. The ratio, however, decreases with increasing prepulse intensities.

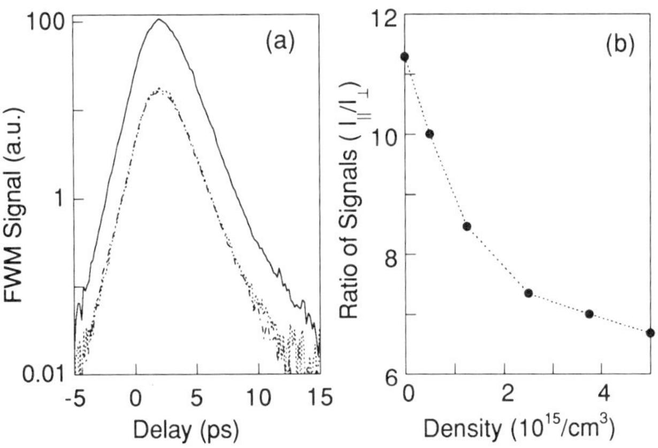

Figure 4. (a) The FWM signal as a function of delay between the two pulses for zero prepulse and for a σ_- and a σ_+ polarized prepulse. The data shows the EID signal is independent of exciton spin. (b) The ratio of $I_{\|}/I_{\perp}$ as a function of prepulse exciton density. At low density, the response is approaching cubic as expected for FWM.

The data clearly shows that in contrast to the more traditional view of optical nonlinearities in semiconductors based on static interactions, the nonlinear response is due to dynamic behavior (i.e., fluctuations in the optical induced coherence) that results from exciton-exciton interactions. The correct analysis of this phenomenon which is needed to fully explain the observations above is based on standard semiconductor Bloch equations which have been modified to include the density dependence in the imaginary part of the exciton self-energy [3]. However, a simple physical but non-rigorous interpretation can be achieved using modified optical Bloch equations [10]. For comparison of qualitative behavior, we include the local field effects using the Lorentz model.

The calculation follows using the standard optical Bloch equations which have been *phenomenologically* modified to describe phase-interrupting excited-state interactions by adding an additional dephasing term. Assuming the excitation dependent dephasing rate is proportional to the excited state population, we have

$$i\hbar\frac{\partial}{\partial t}\rho_{eg} = V_{eg}(\rho_{gg} - \rho_{ee}) + \hbar\omega\rho_{eg} - i\hbar(\gamma + \sigma\rho_{ee})\rho_{eg} \tag{6}$$

The excited state population is determined by

$$i\hbar\frac{\partial}{\partial t}\rho_{ee} = -(V_{ge}\rho_{eg} - c.c.) - i\hbar\gamma_e\rho_{ee} \tag{7}$$

where ρ_{ee} and ρ_{eg} are the standard population density matrix elements (assuming $\rho_{ee}+\rho_{gg}=N$ with N being the density of the two level systems) [6], and $V=-e\mathbf{r}\cdot\mathbf{E}_{loc}$ is the interaction Hamiltonian with $\mathbf{E}_{loc}=\mathbf{E}+L\mathbf{P}$. L is the effective Lorentz local field factor and the electric polarization is given by

$$\mathbf{P}=\mathrm{Tr}(e\mathbf{r}\rho) \tag{8}$$

Note that these equations do not describe the more complicated problem of resonant collisions.

The additional dephasing term not only affects dephasing of the optical polarization but also gives rise to *additional* contributions to the nonlinear optical signal. In a self-diffracting two pulse FWM experiment, the first and second pulses induce optical polarizations in the material propagating in the direction of $\mathbf{k}_1$ and $\mathbf{k}_2$ respectively. The two pulses also interfere in the sample, producing a population grating (for example, an exciton grating). For noninteracting systems, self-diffraction of the second pulse from the grating results in the conventional FWM signal in the direction of $2\mathbf{k}_2$-$\mathbf{k}_1$. In dense media, the optical polarizations can induce electric fields with significant magnitude. Scattering of these electric fields from the population grating also leads to a FWM signal in the direction of $2\mathbf{k}_2$-$\mathbf{k}_1$. Furthermore, the population grating of the excited state can *modulate the optical polarization through the excitation dependent dephasing*, generating an *additional* FWM signal in the direction of $2\mathbf{k}_2$-$\mathbf{k}_1$. (This discussion is valid only when the characteristic length scale of the excited state interaction is small compared with the grating spacing.)

For a homogeneously broadened system, the relevant nonlinear optical polarization can be obtained by solving the population density matrix equation in perturbation. To obtain analytical solutions we assume δ function pulses for the applied optical fields:

$$E = E_1\delta(t - t_1)\exp[i(\mathbf{k}_1 \cdot \mathbf{r} - \Omega t)] + E_2\delta(t - t_2)\exp[i(\mathbf{k}_2 \cdot \mathbf{r} - \Omega t)] + c.c.. \tag{9}$$

The nonlinear optical polarization propagating in the direction of $2\mathbf{k}_2$-$\mathbf{k}_1$ is then given by

$$\begin{aligned} P^{(NL)} = \kappa\exp[i\delta(t - 2t_2 + t_1)]\Big\{&\theta(t - t_2)\theta(t_2 - t_1)e^{-\Gamma_t(t-t_2)-\Gamma_1(t_2-t_1)} + \\ &\frac{N(\sigma + i\eta)}{2\gamma_e}\Big[\theta(t - t_2)\theta(t_2 - t_1)(1 - e^{-\gamma_e(t-t_2)})e^{-\Gamma_t(t-t_2)-\Gamma_1(t_2-t_1)} \\ &+\theta(t - t_1)\theta(t_1 - t_2)(1 - e^{-\gamma_e(t-t_1)})e^{-\Gamma_t(t-t_1)-2\Gamma_2(t_1-t_2)}\Big]\Big\} + c.c. \end{aligned} \tag{10}$$

where $\kappa = -2iN\mu^4\hbar^{-3}E_2^2E_1^*\exp[-i\Omega t + i(2\mathbf{k}_2 - \mathbf{k}_1)\cdot\mathbf{r}]$, $\theta(t)$ is the Heavyside function, μ is the dipole moment between the ground and excited states (assumed to be real), $\eta = 2\mu^2L/\hbar$, and $\delta = \Omega - \omega - \varepsilon$ with $\varepsilon = N\mu^2L/\hbar$ being the static Lorentz shift. Γ_t, Γ_1, and Γ_2 are excitation dependent dephasing rates with $\Gamma_1=\gamma+\sigma N_1$, $\Gamma_2=\gamma+\sigma N_2$, and $\Gamma_t=\gamma+\sigma N_1+\sigma N_2$ where N_i (i=1,2) are the excited state population density due to the excitation of the first and second pulses respectively. To obtain Eq. (10) we have also assumed that decay of the overall excited state population is much slower than dephasing. In general, decay of the nonlinear polarization can become more complicated (nonexponential for example) .

The first term in the curly bracket of Eq. (10) is identical to the *prompt* free polarization decay term for a simple non-interacting two level system except that, as expected, decay of this polarization now depends linearly on the excitation level: Decay of the time integrated signal as a function of t_2-t_1 is determined by $2\Gamma_1$ while decay of the time resolved signal as a function of t-t_2 is determined by $2\Gamma_t$. The second term describes the additional FWM signals due to the local field effects (proportional to ε) and the excitation dependent dephasing (proportional to σ). These additional FWM signals have a rise time (as a function of t) on the order of the total dephasing time as is evident in Eq. (5). Scattering of the polarization induced by E_2 leads to a signal that is *delayed* with respect to the prompt signal from the first term in Eq. (10) when t_2-t_1>0 and to a signal occurring at t_2-t_1<0 ("*negative*" time delay). Similarly, modulating the polarization induced by E_2 through the excitation dependent dephasing also results in a "delayed" signal when t_2-t_1>0 and a signal at "negative" time delay. Figure 5 shows comparison with the above theory. "Negative" time signals are clearly seen due to EID. The modulation is due to quantum beating between the heavy and light hole resonances.

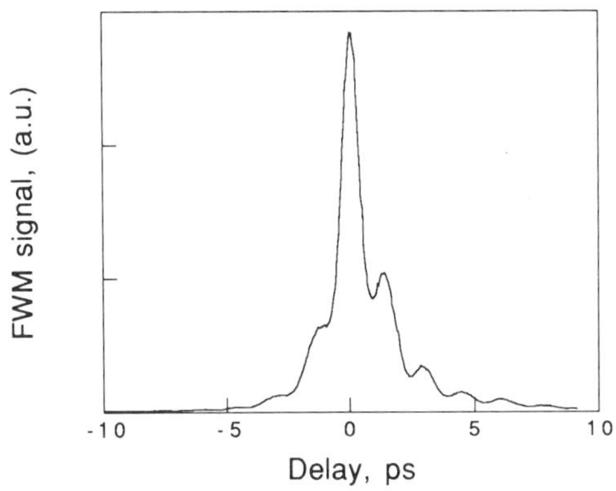

Figure 5. Evidence for the so-call negative time signal due to EID in the hh-lh quantum beats observed in self-diffracted FWM.

For a simple order-of-magnitude estimate of the EID induced contribution to exciton FWM signals in GaAs, we have used the parameters given in [9]:σ=0.8×10^{-4} $cm^3 \cdot s^{-1}$, γ=0.09 meV, exciton Bohr radius a_B=150 Å, and assuming N=$|\phi_{1s}(r=0)|^2=1/(\pi a_B^3)$ where ϕ is the exciton envelope wave function, we estimate that the FWM signal due to the EID can be *as large as 50 times of the normal FWM signal* [represented by the first term in Eq. (10)]. It is interesting to note that recent measurements in a GaAs/AlGaAs quantum well have reported a "delayed" FWM signals 100 times stronger than the normal "prompt" FWM signal [11]. However, the authors attributed the large signal to the local field effects.

Based on this understanding and the data in Fig. 4a, it now becomes possible to understand the polarization properties of transient FWM and the behavior in Fig. 4b. Figure 4a shows that the FWM signal created with pure +1 or -1 hh excitons is affected equally by +1 and -1 excitons (we note that this does not hold for the lh.) Therefore, we take the dephasing rate to depend on the total local exciton density, independent of spin. We now note that when the fields are linearly co-polarized, they are a superposition of σ=+1 and -1 fields. Using the usual selection rules, we see that there are two exciton gratings corresponding to +1 and -1 excitons, but that they are spatially coincident. The nonlinear response is then a sum of the EID response and terms due to other manybody effects. In the case where the two incident fields are cross-polarized, there are again two gratings, however, they are spatially out of phase resulting in no net modulation of the exciton density. Since the dephasing rate is independent of spin, there is no net spatial modulation of the dephasing and hence no EID contribution resulting in a much reduced FWM response at low intensities as seen in Fig. 4b. The decrease of the ratio of the FWM signals shown in Fig. 4b, however, can not be explained by the simple model and is due to the fact that exciton-exciton scattering cross section decreases at higher densities due to screening effects (for a more detailed discussion see [3]).

We conclude here by noting that while these experiments were performed with 3D excitons, the EID effects is expected to be even stronger in ideal 2D systems because of the larger exciton-exciton scattering rate.

COHERENT EXCITATION AND RELAXATION OF EXCITONS IN GaAs QUANTUM WELL STRUCTURES

Reducing the dimensionality from three dimensions to two dimensions results in an increase in the exciton binding energy as well as the energy of exciton transition. The increase in binding energy has the result of increasing the energy separation between the exciton and the bandgap resulting in a relatively sharp excitonic resonance which is clearly observable even at room temperature. However, unlike the three dimensional case, the two dimensional structure is the result of layered crystal growth by various methods, most successfully with molecular beam epitaxy (MBE) techniques. In this approach a lower energy band gap crystal (GaAs) is placed between layers of higher bandgap material ($Al_xGa_{1-x}As$) leading to the formation of a quantum well. The resulting confinement potential results in a quasi-two dimensional region of confinement and excitation in the GaAs. These methods are highly successful but there is strong evidence that even the best material is still characterized by interface roughness due to single atomic layer fluctuations which leads to spatial fluctuations in the electronic confinement potential.

Early insight into the effects of this disorder was obtained in studies of exciton luminescence which confirmed the emission was inhomogeneously broadened due to the corresponding potential fluctuations [12]. These studies were followed by results that indicated that this interface roughness is characterized by a scale length large compared to the Bohr radius of the exciton (see, for example, [13, 14]); however, more recent studies by chemical lattice imaging have shown the presence of monolayer flat islands with a smaller spatial extent of 50-100 Å [15]. While these results are presently somewhat controversial [16, 17], they have led to the proposal that at least two scale lengths for interface roughness are required in order to account for the observations [18]. Further evidence that the island size distribution is bimodal has been recently presented in luminescence and Raman scattering measurements [19]. Details of the interface roughness also depend on specific growth processes, such as interrupted or non-interrupted growth [14], or whether GaAs is grown on $Al_xGa_{1-x}As$ or $Al_xGa_{1-x}As$ is grown on GaAs [20].

Due to these complexities created by nonideal growth in quantum wells , experiments in this section have tended to focus on the nature of the disorder as manifested by the details of relaxation and the nonlinear response. Using a four-wave mixing approach to classical spectral hole burning, ultrahigh resolution spectra can be obtained which reveals the details of inhomogeneous broadening, hole burning and energy migration. As we show in a later section, this technique can also be used to eliminate the effects of spectral diffusion and obtain high resolution spectra of exciton Zeeman splitting. The inhomogeneous broadening also leads to a classical stimulated photon echo which we discuss with polarization properties which are still not fully understood.

Spectral Hole Burning and Spectral Diffusion

Fluctuations in the confinement potential lead to localization of the exciton wave function leading to inhomogeneous broadening of the exciton resonance. These localized excitons couple to other localization sites and, through the emission or absorption of phonons, can migrate between these sites [21, 22]. The strength of the coupling and rate of migration depend on the exciton energy *within* the inhomogeneous width.

To study these effects, we developed a new form of high resolution nonlinear laser spectroscopy. In the case of classical atomic spectroscopy, inhomogeneous broadening is due to Doppler broadening of the resonance which is often two orders of magnitude larger than the atomic resonance. For these systems, high resolution spectroscopy is usually obtained by some form of saturation (or Lamb dip) spectroscopy using *counter propagating* beams. Because the Doppler shift depends on $\boldsymbol{k} \cdot \boldsymbol{v}$ where $\boldsymbol{k}$ is the k-vector of the optical field, the Doppler shift cancels (to first order) in saturation spectroscopy. In specific solids, this has been accomplished through persistent hole burning techniques where a narrow band laser excites a specific frequency group resulting in a spectral hole that lives for minutes to hours. The laser is then scanned over the hole which is detected by an overall change in the absorption.

Since there is no corresponding k-vector dependent broadening nor have we detected any significant persistent hole burning, we developed a new form of real-time spectral hole based on frequency domain four-wave mixing (FWM). A schematic representation of the experimental configuration is shown in Fig. 6 and is closely related to the so-called phase conjugation geometry of importance to optics. All optical beams are supplied by frequency stabilized tunable cw dye lasers pumped by an argon ion laser. Lasers have a peak-to-peak relative frequency jitter on the order of two MHz. Two counter-propagating pump beams $\mathbf{E}_1(\Omega_1,\mathbf{k}_1)$ and $\mathbf{E}_3(\Omega_3,\mathbf{k}_3)$ interact in the sample with a probe beam $\mathbf{E}_2(\Omega_2,\mathbf{k}_2)$ with $\mathbf{E}_3 \perp \mathbf{E}_2 \parallel \mathbf{E}_1$ through the resonant third order susceptibility to generate a signal beam $\mathbf{E}_s(\Omega_s,\mathbf{k}_s)$ proportional to $\tilde{\chi}^{(3)}:\mathbf{E}_1\mathbf{E}_2^*\mathbf{E}_3$. Physically, $\mathbf{E}_1 \cdot \mathbf{E}_2^*$ results in a spatial and temporal modulation of the exciton population, which modifies the optical response of the sample. The coherent nonlinear signal arises from scattering of the backward beam from the modulation [23, 24]. Spectroscopic information can be obtained by studying the dependence of the nonlinear optical response on the relative frequency detuning, the electric field polarization, or the relative angle of the different input beams. For the primary measurements in this paper, information related to relaxation of the system is obtained by measuring the nonlinear response as a function of the frequency of any of the three input beams. (This picture is based on a simple model involving the ground state of the system and an excited state. There are more complex pictures which are required if degeneracy exists as discussed below or if more complex excitations are important such as biexcitons [see for example [25].)

For these studies, samples consisted of 65 periods of 96 Å GaAs wells and 98 Å $Al_{0.3}Ga_{0.7}As$ barriers, grown at 630 °C by MBE on semi-insulating (100) GaAs substrates with interrupted growth. The samples are mounted on a sapphire disk (c axis normal) with

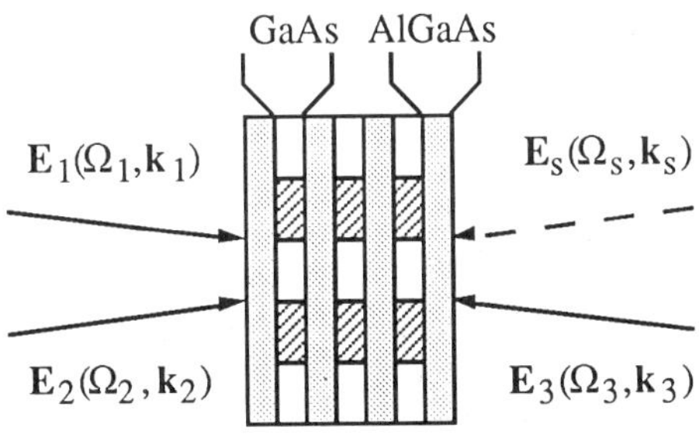

Figure 6. Schematic representation of the experimental configuration for high resolution nonlinear spectroscopy based on frequency domain FWM.

the substrate removed by chemical etching. The data presented in the figures below were obtained on a sample that is characterized by an absorption linewidth of 2.2 meV for the hh1 exciton, and a Stokes shift of 1 meV between the hh1 exciton absorption and emission. The cw FWM measurements are carried out on the hh1 exciton with an exciton density on the order of $10^7/cm^2$.

In Fig. 7, we show the FWM response obtained by scanning the frequency of the third beam (Ω_3) [26, 27]. This enables a direct detection of the spectral hole produced by excitons excited by $\mathbf{E}_1 \cdot \mathbf{E}_2^*$ at energy E (1.5 meV below the absorption line center) as well as detecting any excitons scattered from energy E to E' due to spectral diffusion. The nonlinear response is corrected for sample absorption. The narrow resonance in the response corresponds to exciton spectral hole burning where the effects of the large inhomogeneous broadening have clearly been eliminated. The width of the hole gives an exciton homogeneous line width $2\gamma_{ph} \sim 0.03$ meV, corresponding to a dephasing time of 45 psec. The broad Stokes shifted feature is due to the quasi-equilibrium distribution of the exciton population produced by the migration of excitons from energy E to E'. The response reflects the steady state exciton distribution assuming all excitons in the spectral region concerned give rise to the same cw nonlinear response. The line shape in Fig. 7 can be fit to a simple model which neglects migration to states above the excitation energy. The calculation is based on the nonlinear optical response of a simple two level system and assumes a Gaussian distribution for the quasi-equilibrium population of excitons that have migrated to states below the excitation energy. The spectral profile of the hole-burning resonance is assumed to be Lorentzian. The result is plotted as the solid line in the figure. A small but finite detuning of order 100kHz is set between Ω_1 and Ω_2 to reduce contributions to the data from slower components (discussed elsewhere [28]).

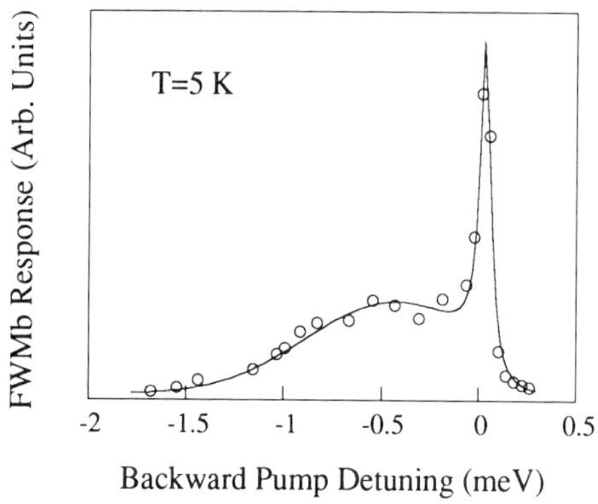

Figure 7. The FWM response obtained by tuning Ω_3 shown in Fig. 6. The narrow peaks is the narrow spectral hole burned by $\mathbf{E}_1 \cdot \mathbf{E}_2^*$ below line center. The broad feature on the left is due to spectral diffusion of excitons to lower energy.

The energy relaxation rate is determined from the line shape of the FWM response obtained by tuning Ω_1 in Fig.6. The results for two different temperatures are shown in Fig. 8. In the simple picture of spectral hole burning in the presence of spectral diffusion, the lifetime of the excitation created by $\mathbf{E}_1 \cdot \mathbf{E}_2^*$ is characterized by two time scales. The initial excitation decays on a short time scale due to spectral diffusion. However, a second time scale is observed due to decay of the equilibrium population; i.e., the recombination time. At 2.5 K, the spectral diffusion rate is slow and the contribution from the spectrally diffused excitons is small. Nevertheless, a weak narrow feature is observed at the top of a broad peak in Fig. 8a. The width of the broad feature corresponds to a spectral diffusion rate of 10's ps. The presence of two time scales is more clearly seen in Fig. 8b at T=10K where the relative contribution of the narrow component has increased reflecting the greater number of spectrally diffused excitons (a direct result of increased spectral diffusion rate). Recall that tuning Ω_1 varies the frequency difference Ω_1-Ω_2 which determines the oscillation time of the grating. The signal decreases when the frequency difference exceeds the relaxation rate (similar to the Fourier transform picture of a time-resolved pump-probe measurement.) The narrow resonance corresponds to a lifetime of 1 nsec and represents the decay time of the quasi-equilibrium exciton distribution (the broad feature in Fig. 7) The wide feature corresponds to a lifetime of 45 psec and is the inverse of the spectral diffusion rate. In addition, setting the detuning of Ω_1-Ω_2 large compared to the width of the narrow feature greatly reduces the contribution of the broad feature in Fig. 7, confirming that the broad feature in Fig. 7 corresponds to the quasi-equilibrium exciton distribution. Eliminating the contribution from this distribution by setting a large value of Ω_1-Ω_2 is critical for Zeeman splitting measurements discussed below.

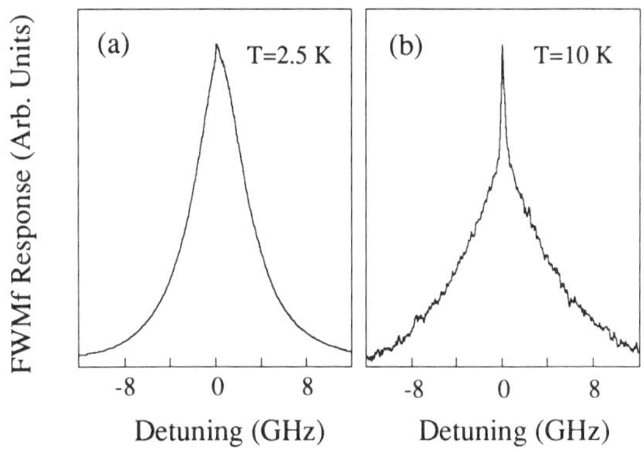

Figure 8. The FWM response obtained by tuning Ω_1 in Fig. 6. The narrow feature at the top has a width determined by the inverse lifetime of the quasi-equilibrium excitons created by spectral diffusion. The width of the broad feature is determined by the spectral diffusion rate.

The mechanism of spectral diffusion which gives rise to the line shape in Fig. 7 and the time scales seen in Fig. 8 have been of considerable interest. A successful discussion of this problem has been presented by Takagahara [21, 22, 29]. This model resembles the Mott model for variable range hopping and considers a localized exciton to couple to another localization site by wave function overlap and dipole-dipole coupling. The energy deficit that occurs in moving from one site to another is accounted for by the absorption or emission of acoustic phonons. At low temperatures, the problem has not been solved analytically. However, numerical calculations show that the functional form for the spectral diffusion rate can be fit by the expression

$$\Gamma = \Gamma_o exp(\beta T^{\alpha}) \tag{11}$$

where β is a parameter that depends on the exciton energy within the inhomogeneous width and α is a parameter between 1.6 and 1.7. Figure 9 shows the excellent agreement with the theoretical prediction. Note that at higher temperatures, the spectral diffusion rate becomes dominated by thermal activation.

The above discussion shows that high resolution FWM spectroscopy provides a wealth of information not readily attainable by other means. This is further high lighted in the section below on magnetic fields. The study of systems characterized by disorder clearly requires spectroscopy capability to eliminate inhomogeneous broadening and high spectral resolution to distinguish one region of the disorder from another. It is also possible to compare these spectroscopic signatures with more detailed understanding of the interface obtained by various microscopy approaches and provide new insight into the nature of the so-called Mott mobility edge which has been reported in these systems [30, 31].

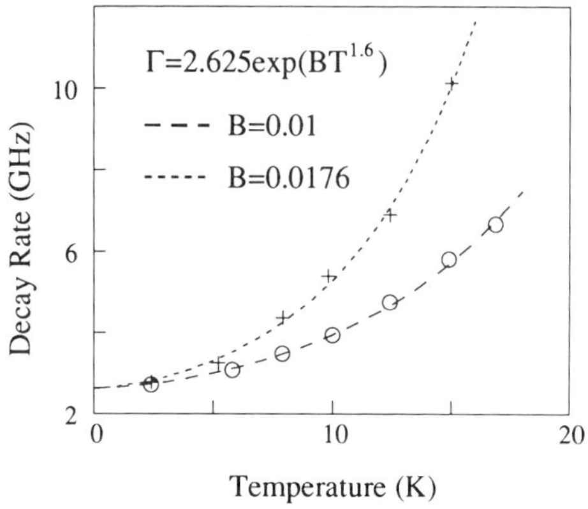

Figure 9. Temperature dependence of the spectral diffusion rates as determined by the linewidth of the broad feature in Fig.8. The dashed lines represent best fits of the phonon assisted migration model developed by Takagahara.

Stimulated Photon Echo and Free Polarization Decay

The phenomena of coherent optical excitations in semiconductors are easily seen in the details of frequency domain nonlinear laser spectroscopy as described above. However, there are some features of coherent phenomena which are more easily appreciated in the time domain and which can provide additional spectroscopic information.

For these experiments, the two pulse geometry in the transient experiments discussed earlier is too limited to permit full exploration of the various relaxation phenomena. Instead, we used the three pulse geometry first developed by Hartmann and colleagues for studies of inhomogeneously broadened ions in crystals . For such systems the emission was designated a stimulated photon echo (SPE) [32]

The experimental geometry is identical to that in Fig. 6 except the fields are pulsed. Relaxation rate information is had as a function of delay between either pulse 1 and 2 or pulse 3 and pulse 1. Causality requires that the time ordering of the fields (for the normal nonlinear response) be pulse 2, followed by pulse 1 and then pulse 3 (note the ordering of pulse 2 and 1 would be opposite for Doppler broadened material.) In the case of homogeneously broadened material, it is well known that the signal emission time with respect to the third pulse is independent of the delays between the incident pulses. For an inhomogeneously broadened system, the emission is delayed with respect to the third pulse by an amount τ, the time between pulse 2 and pulse 1. (For a thorough discussion, see for example [33]). In either case, the presence of a signal even when τ is nonzero is direct evidence of the optical coherence [Eq. (3)] induced by the first pulse and the fact that this coherence exists after the field is removed. To make the unambiguous determination of where the signal arises from a homogeneously or inhomogeneously broadened system, we time resolve the coherent emission by cross correlating the emission with a reference beam from the laser using standard up-conversion techniques.

Using an 8 ps pulse, Fig. 10a shows the time resolved emission where zero corresponds to zero time delay with respect to the third pulse. Figure 10b clearly shows the delay is linear with respect to τ, confirming that in an inhomogeneously broadened quantum well, the emission is a simple stimulated photon echo. Since the dephasing rate (as measured by the line width in spectral hole burning) depends on the relative excitation energy (compared with absorption line center), it is important in these experiments to use narrow band excitation in order to obtain a true measure of the dephasing time. In addition, we will show in further studies discussed below, it is important to operate at low excitation density. Again for the measurements of the dephasing time using the SPE, density *independent* measurements were obtained for densities well below 10^8 excitons/cm^2. The presence of a coherent signal in the direction k_s even when E_1 and E_2 do not over lap in time is direct evidence of the exciton coherence induced by E_2. The delay in the coherent emission shows the loss of phase of the induced coherence and the rephasing induced by E_1.

Using the stimulated photon echo, measurements [34] of the decay rates and corresponding signatures of spectral diffusion and phonon assisted migration are in excellent agreement with the results obtained by frequency domain FWM discussed above.

In the above discussion, the standard treatment of photon echoes has been adequate to explain the data. However, recent measurements have shown that polarization effects lead to behavior which remain unexplained in any unified approach to the nonlinear response. In particular, it has been noticed by several groups [35, 36] that if the polarization of the field which acts in first order (in the sense of perturbation theory, i.e., E_2 in this case) is linearly and orthogonally polarized with respect to the other linearly polarized field(s), then the dephasing rate increases considerably as shown in Fig. 11. Again, as in the above measurements, these observations are at low exciton density.

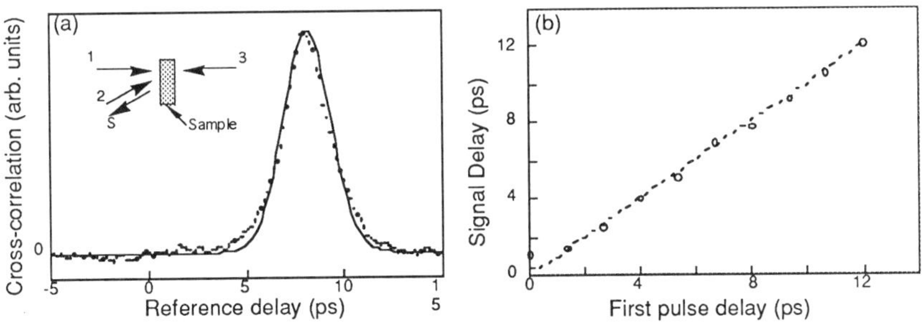

Figure 10. Measurements of the time resolved emission in the transient FWM experiment as a function of delay between the first two pulses. a) For a delay of 8 ps between the first two pulses, the time resolved emission is also delayed 8 ps with respect to the third pulse. b) The delay of the signal as function of delay between the first two pulses showing the linear dependence expected for a stimulated photon echo. The dashed line is fit of theory showing the effects of finite pulse width.

Based on the energy level structure (see Fig. 12) no difference in the FWM signal is expected other than a change in the direction of polarization of the signal. More specifically, for GaAs the conduction band is m=±1/2 character and the heavy-hole valence band is m=±3/2 where in a quantum well the spherical symmetry is lifted by the confinement and the axis of quantization is perpendicular to the plane of the layers. Dipole allowed transitions require $\Delta m=\pm 1$, where in this material transition moments are equal. Hence linearly polarized fields propagating perpendicular to the layers must be taken as a superposition of two circularly polarized fields. We will designate the exciton created by a

σ_+ polarized field as $|+1\rangle$ (consisting of a +3/2 hole and a -1/2 electron) and the exciton created by a σ_- polarized field as $|-1\rangle$ (consisting of a -3/2 hole and a 1/2 electron). The signal is produced by the scattering of E_3 off the grating produced by $\mathbf{E}_1 \cdot \mathbf{E}_2^*$ as discussed above. (This picture is for a homogeneously broadened system but still provides insight for an inhomogeneously broadened system. Although there is no net grating, there is a grating for each frequency group within the inhomogeneous distribution. Each group results in a component of the scattered signal and it is the phase relationship between these components which results in the signal delay.)

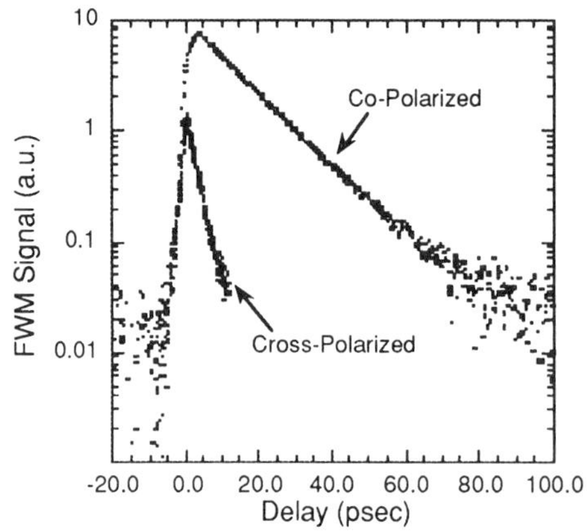

Figure 11. The decay of the signal as a function of delay between E_2 and E_1 for co-polarized and cross-polarized fields.

We see then that for linear co-polarized $\mathbf{E}_1$ and $\mathbf{E}_2$, two spatial gratings consisting of $|+1\rangle$ and $|-1\rangle$ excitons respectively, are created and are spatially coincident. For $\mathbf{E}_1 \perp \mathbf{E}_2$, the two gratings are exactly spatially out of phase. However, $\mathbf{E}_3$ must also be decomposed into circularly polarized components, each of which independently scatters from the gratings. However, the change in decay of the FWM response in Fig. 11 is certainly unexpected.

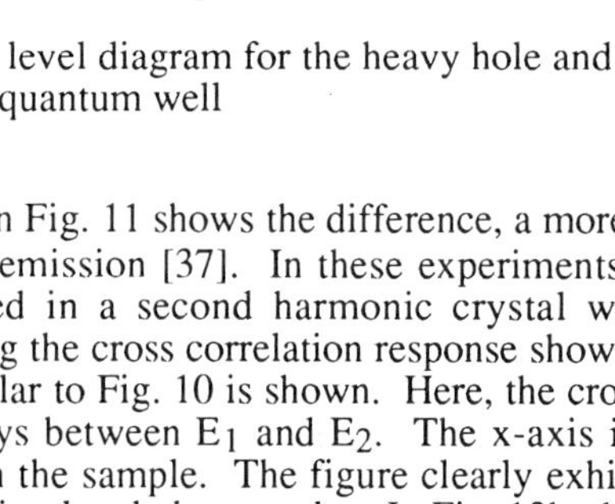

Figure 12. Energy level diagram for the heavy hole and light hole excitons in a GaAs/AlGaAs quantum well

While the response in Fig. 11 shows the difference, a more fundamental difference is seen in the time *resolved* emission [37]. In these experiments, a portion of the original picosecond pulse is mixed in a second harmonic crystal with the signal. The sum frequency is detected giving the cross correlation response shown in Fig. 13. In Fig. 13a, a repeat of the response similar to Fig. 10 is shown. Here, the cross correlation is shown for three increasing time delays between E_1 and E_2. The x-axis is time measured from the arrival of the third pulse in the sample. The figure clearly exhibits the delay discussed in Fig. 10 as expected for a stimulated photon echo. In Fig. 13b, the temporal response of the emitted radiation with $\mathbf{E}_1 \perp \mathbf{E}_2$ is shown for the same three delays. Here there is clearly no delay. (The evidence of a delayed signal is in fact due to an impurity bound exciton with an excessively long dephasing time of 80 ps. The impurity bound excitonic resonance is very strong and is 2.5 meV Stokes shifted with respect to the unbound exciton. It becomes observable in this data because of the long dephasing time compared to the prompt signal.)

The change in temporal behavior as a result of a simple rotation of the polarization reflects a fundamental change in the nature of coherent emission from this system. The fundamental origin of coherent emission in any system such as this is due to the constructive interference of a phased array of radiating dipoles. In the case of the stimulated photon echo such as in Fig. 13a, the pulse that drives the dipoles is E_3. However, because of inhomogeneous broadening, the phases of the different dipoles is determined by the individual frequency of each radiator. Since there is a random distribution of the frequencies corresponding to the origin of the inhomogeneous broadening, the field resulting from the sum of the different dipoles is initially small for the case t_1-t_2>0. However, because of the special properties of the photon echo, these different phases continue to evolve in time until at a time t_1-t_2 after the third pulse, the phases add constructively leading to coherent emission. However, in the case of cross-polarized excitation, Fig. 13b shows only a free polarization decay (in analogy with free induction decay in NMR) and no rephasing at later times. Hence, it seems quite reasonable to conclude that the nonlinear optical response for $\mathbf{E}_1 \| \mathbf{E}_2$ is due to an inhomogeneously broadened distribution of excitonic resonances while for $\mathbf{E}_1 \perp \mathbf{E}_2$ the response arises from a homogeneously broadened system.

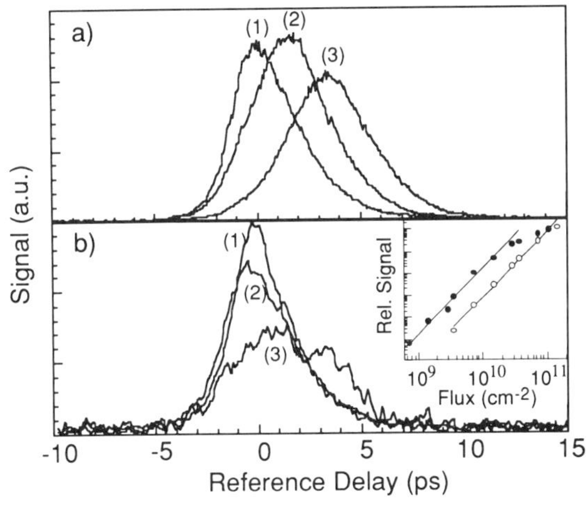

Figure 13. Time resolved emission of the nonlinear optical response as a function of time with respect to the third pulse. Time resolved emission is given for three different delays between the first and second pulse corresponding to 1) 0 ps, 2) 2 ps and 3) 4 ps. (a) co-polarized $\mathbf{E}_1 \| \mathbf{E}_2$. (b) Cross-polarized $\mathbf{E}_1 \perp \mathbf{E}_2$. The inset shows the intensity (photon flux) dependence of the signal for co-polarized (solid) and cross-polarized (open) excitation.

Additional measurements support the picture that these two signals arise from different physical origins. In the inset of Fig. 13, we show the dependence of the signal emission on input intensity. The cross polarized signal is weaker at low intensity but shows no evidence of saturation in contrast to the initially stronger signal for co-polarized excitation. Indeed, at the higher intensities, both signals can be simultaneously observed for co-polarized excitation [38]. Based on the earlier frequency domain work, it has been reasonable to assume the excitons that give rise to the stimulated photon echo are localized and hence inhomogeneously broadened. The saturation of this signal with increasing excitation intensity is simply due to the finite number of localization sites. Figure 14 shows that the energy (i.e., excitation wavelength) and temperature dependence of the two corresponding dephasing rates is also quite different for the two excitations. The co-polarized behavior is again completely consistent with the assumption that this signal arises from strongly localized excitons. The energy and temperature dependence is that predicted for phonon assisted migration and thermal activation discussed above. In contrast, the nearly energy independent dephasing rate for cross-polarized excitation is expected for a *homogeneously* broadened resonance. In addition, while the range of data is not adequate to make a definite determination, the temperature dependence for the dephasing rate for cross-polarized excitation is definitely distinct and appears linear corresponding to single phonon scattering [39].

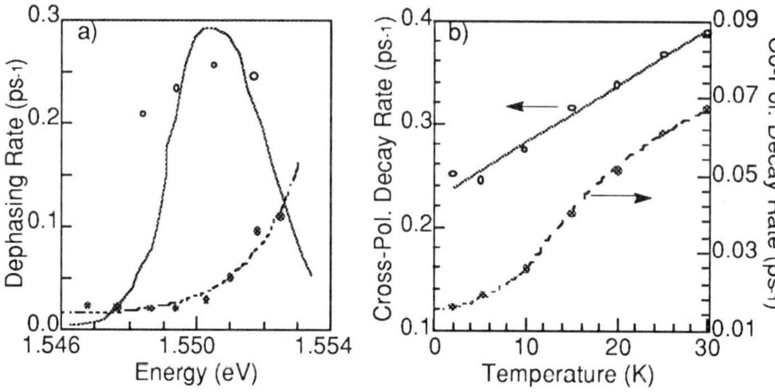

Figure 14. a) Excitation energy dependence of the dephasing rate for co-(closed) and cross-(open) polarized excitation. b) Temperature dependence of the dephasing rate for the two polarization configurations.

The unambiguous interpretation of this data remains unclear though the qualitative features are observed even on high quality samples. A fully consistent explanation remains to be developed. Nevertheless, the relatively simple interpretation is that the radiation arises from two different classes of excitons. The echo arises from strongly localized excitons while the free polarization decay may be due to excitons with more extended wave functions. Such a proposal has some support in the data which shows that the mobility of the excitation observed for $\mathbf{E}_1 \perp \mathbf{E}_2$ is larger than the mobility for $\mathbf{E}_1 \| \mathbf{E}_2$ [37] and may be consistent with recent evidence for multiple scale lengths in interface roughness (see for example [19]). The disappearance of the echo in this case may be associated with the interference effects due to the presence of biexcitons [40]. An alternative explanation may be associated with the spin-independent exciton-exciton dynamic interaction picture described above. In this picture, the echo results from the spin-independent interaction between excitons at the same energy. The disappearance of the echo when $\mathbf{E}_1 \perp \mathbf{E}_2$ is due to the absence of the exciton density grating (for each frequency group). The free polarization decay is due to the spin-dependent exciton interactions (for example, biexcitonic effects). The much faster decay of this signal may be due to the large decay rate of the biexcitons.

EFFECTS OF MAGNETIC FIELDS

Magnetic Freeze Out

The strong Coulomb interaction between electrons and holes leads to the excitonic resonance below the bandgap in direct gap semiconductors which dominates the nonlinear optical response. However, even modest external magnetic fields result in the appearance of Landau resonances. The magnetic field effects are clearly seen in the spectra to dominate the Coulomb coupling. As a result, the nonlinear optical response also changes.

Interestingly enough, studies of magnetic field induced modifications have emphasized the resonant nonlinear optical response of layered systems [41-44] where energy level shifts due to inter-Landau level coupling have been theoretically predicted [41] and observed in quantum wells [42-44] using time domain differential transmission. Other work shows the intramagneto-exciton interaction is quenched in GaAs quantum wells at high fields [42, 44] because of local charge neutrality at a given Landau level [41, 45]. However, much less is known about the nonlinear magneto-optical behavior in bulk semiconductors.

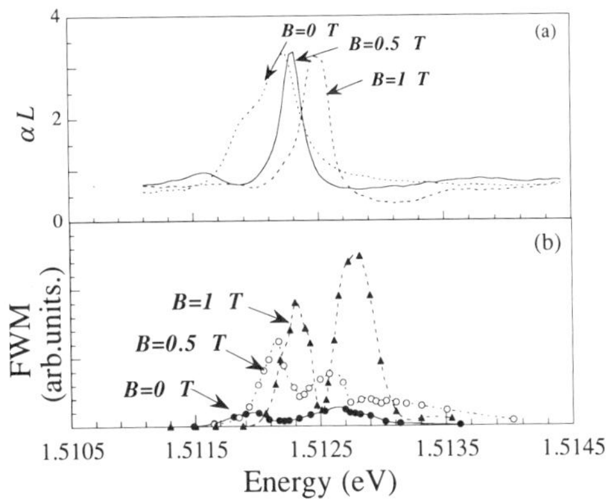

Figure 15. a) Linear absorption spectra and b) Degenerate FWM spectra of the *hh*(1) resonance. Frequency degenerate FWM spectra are not corrected for reabsorption.

Experiments were performed using a high quality bulk GaAs under strain which lifts the hh and lh degeneracy, placing the lh below the hh in energy [46]. The experimental geometry was identical to that used for the other frequency domain experiments described above. The linear absorption spectrum is complicated in the presence of a magnetic field by the presence of so-called forbidden transitions which become weakly allowed due to the effects of strain. However, the degenerate FWM response (where all input frequencies are the same) shows a simple nearly periodic pattern in the presence of the magnetic field due to the Landau resonances.

What is particularly important to this discussion is the effect of the magnetic field on the overall nonlinear response. Figure 15 compares the linear absorption spectrum to the frequency degenerate FWM spectrum of the hh(1). What is most obvious in the data is the enhancement in the FWM response at fields below even 1 T. The only significant effect on the linear absorption response is a shift due to the diamagnetic term, the dipole moment of the transition remains nearly constant. (We note that the dip present in the nonlinear response is due mainly to the strong absorption; however, for $B \geq 1$ T the dip persists after this correction). All measurements were performed at very low cw excitation density ($<10^{14}$ excitons-cm^{-3}) where the response increases linearly with each beam intensity. Hence, the response over the region is clearly given by the third order susceptibility, $\chi^{(3)}$, with no effects due to saturation.

To understand the behavior, we first note that the general linear optical response for a resonance is given by μ^2 / γ_{ph} whereas for the cw frequency degenerate FWM response, the signal varies as $\mu^8 / \gamma_{ph}^4 \gamma_{rel}^2$ [23] where μ is the transition moment, γ_{ph} is the dephasing rate (homogeneous line width), and γ_{rel} is the grating decay rate and includes effects of spatial transport. γ_{rel} is given by $\gamma + 4\pi^2 D / \Lambda^2$, where γ represents the exciton decay, D is the exciton spatial diffusion coefficient and Λ is the grating spacing. Since the linear absorption spectrum shows no change, we consider the possibility that γ_{rel} is reduced by the fields. For a fixed Λ as in the experiments, a change in D would result in a change in the nonlinear response.

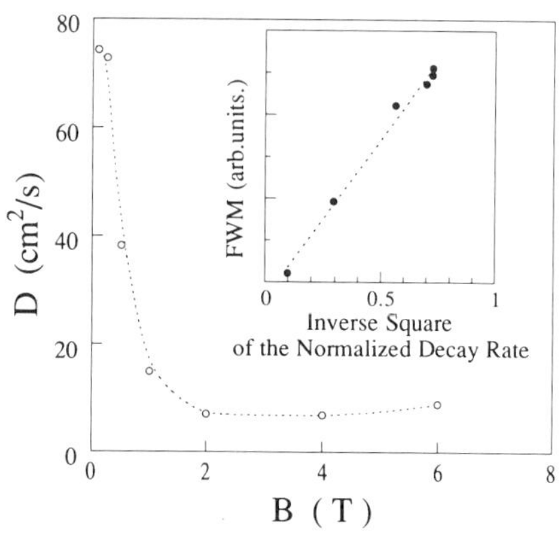

Figure 16. Dependence of diffusion coefficient on magnetic field. Inset: Frequency degenerate FWM versus inverse square of the normalized decay rate $(\gamma_{rel} / \gamma)^{-2} = (1 + 4\pi^2 D / \Lambda^2 \gamma)^{-2}$.

The diffusion coefficient D can be extracted accurately by measuring the dependence of γ_{rel} on Λ which varies as a function of angle between the $\mathbf{E}_1$ and the $\mathbf{E}_2$ beams. Figure 16 shows the measured diffusion coefficient which decreases with increasing B. Such behavior has been predicted in studies of exciton transport in magnetic fields [47]. The increasing field decreases the spatial extent of the exciton wave function. This enhances the likelihood of localization on defects and impurities (i.e., magnetic freeze out) which decreases the mobility. Experimental evidence for this effect was recently reported in germanium [48]. In GaAs, the above measurements show that the diffusion coefficient, D, of $hh(1)$ excitons is reduced by a factor of 5 at fields as low as 1 T. Consequently, a correspondence between the increasing FWM signal and decreasing diffusion is expected. The inset in Fig. 16 shows the effect. The FWM response at a fixed Λ depends linearly on the inverse square of the normalized decay rate, $(\gamma_{rel}/\gamma)^{-2}$, as expected, for magnetic fields ranging from 0 to 6 T. Thus we conclude that the reduction in mobility of magneto-excitons is the origin of the enhancement in the nonlinear response.

Hence, the effects of magnetic fields on these systems show a complex and interesting behavior. The enhancement of the nonlinear response is certainly obvious. More detailed measurements discussed elsewhere [46] also show that nonlinear response is due to Coulomb coupling which leads to strong inter-Landau-level coupling.

Spin Flip Induced Hole Burning

Unlike in bulk material the electronic energy spectrum in a quantum well is fully quantized under a magnetic field parallel to the growth axis. Again, optical absorption reveals a ladder of magnetoexcitons corresponding to transitions between electron and hole Landau levels [49]. Magnetic fields also lift Kramers degeneracy with the resultant Zeeman splitting depending on details of the band structure. In general the energy splitting is determined by the electron and hole g-factors. Early measurements by magnetotransport have shown large exchange-induced enhancement of the electron-*g* [50, 51] and electron spin resonance measurements show the dependence of the electron-*g* on magnetic field for different Landau levels [52, 53] and the importance of nonparabolicity in the conduction band [54]. Determination of the exciton Zeeman splitting has proven to be more elusive [55-57]. A precise determination of the exciton Zeeman splitting in a quantum well using linear optical spectroscopy is difficult since as discussed above interface disorder leads to exciton localization and subsequent inhomogeneous broadening of the absorption profile. The resultant inhomogeneous broadening greatly exceeds the splitting in moderate fields.

In order to directly observe the exciton Zeeman splitting, we have exploited the features of frequency domain spectroscopy based on FWM. In particular, as in spectral hole burning and saturation spectroscopy, frequency domain FWM eliminates the effects of inhomogeneous broadening on spectral measurements as shown above. More importantly, however and unlike earlier cw methods, frequency domain FWM can also reduce the effects of spectral diffusion on the line shape (seen clearly in Fig. 7) which will certainly obscure any effort at high resolution spectroscopy.

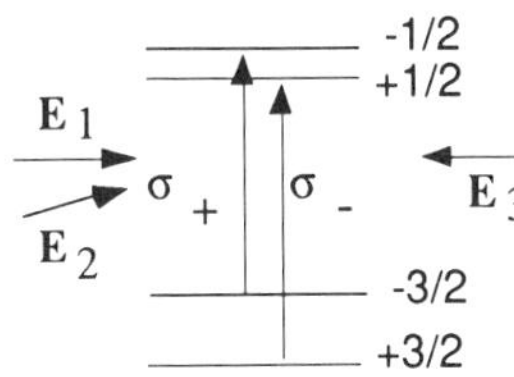

Figure 17. Energy level diagram of the hh exciton in the presence of a magnetic field and FWM configuration .

We consider the case for GaAs/AlGaAs quantum wells. A nearly monochromatic optical beam with σ_- circular polarization can be used to excite a narrow spectral-hole (say at energy E_- within the inhomogeneous absorption profile) of the lowest heavy-hole (hh1) exciton associated with the 3/2 to 1/2 transition as seen in the energy level diagram in a magnetic field shown in Fig. 17 . The width of the spectral-hole is determined by the homogeneous line width. Spin-flips of electrons and holes associated with these excitons generate a spectral-hole at the energy of the -3/2 to -1/2 exciton transition, designated E_+. The *spin-flip-induced* spectral hole at E_+ results from the reduced absorption due to the presence of carriers that have flipped their spins from E_-. This spectral hole can be probed using an optical beam with σ_+ circular polarization. The Zeeman splitting can be obtained by measuring the energy spacing between the spin-flip-induced spectral hole and the original spectral hole.

Unfortunately, the above proposal is complicated by the strong spectral diffusion of the localized excitons shown in Fig.7. In the normal spectral hole burning measurement discussed above, the nonlinear optical signal may be dominated by this distribution. In the limit that the spin relaxation time is long compared with the spectral diffusion time, nearly all spin-flipped excitons have diffused in energy. Hence, the spin-flip-induced spectral hole burning resonance will be overwhelmed by the spectral diffusion process.

To avoid these complications, we developed a new method of spectral hole burning based on FWM [24, 27]. This method can significantly reduce the contribution of the quasi-equilibrium exciton distribution to the nonlinear optical response and, hence, allows us to recover the spin-flip-induced spectral hole burning resonance. The experimental configuration is again based on backward FWM. Nearly degenerate beams $\mathbf{E}_1$ and $\mathbf{E}_2$ interfere in the sample to excite a *traveling* wave grating which oscillates at a frequency equal to the detuning between the two beams $\delta=|\Omega_1-\Omega_2|$ (<<homogeneous line width). (In a normal FWM or spectral hole burning measurement, δ would be 0.) The amplitude of the signal is proportional to $(\delta + i\gamma_{rel})^{-1}$ and the FWM response is probed as a function of Ω_3.

In the presence of spectral diffusion, the spectral-hole diffuses in energy resulting in a spectral redistribution of the excitation. In this case, the FWM response arises from both the spectral-hole and the quasi-equilibrium distribution of the exciton population . The decay of the spectral hole population is determined by the sum of the exciton spectral diffusion, spin relaxation and recombination rates. However, the life time of the quasi-equilibrium distribution is determined by the *recombination time of the exciton* [13]. In the limit that the spectral diffusion rate is much larger than the rate for exciton recombination, setting δ large compared with the recombination rate (but still smaller than or comparable with the spectral diffusion rate) significantly decreases the relative amplitude of the grating associated with the quasi-equilibrium distribution. Hence, the FWM response will be dominated by the hole burning resonance [14].

The results of the experiment with δ=140 MHz and B=4T are shown in Fig. 18. The curve with the open squares corresponds to the FWM response as a function of Ω_3 and all input fields rotating in the same direction. For this choice of polarization, only σ_- excitons

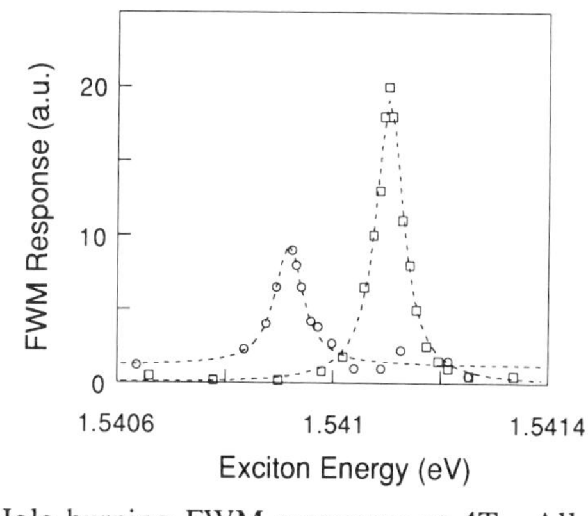

Figure 18. Hole-burning FWM responses at 4T. All three beams are circularly polarized with $\mathbf{E}_1$ and $\mathbf{E}_2$ exciting the σ_- exciton. Squares represent the response when the probe beam $\mathbf{E}_3$ interacts with the σ_- exciton. Circles represent the response when $\mathbf{E}_3$ interacts with the σ_+ exciton. Dashed lines are Lorentzian fits to the response. The inset shows conduction and heavy hole valence band energy levels in GaAs quantum well for a magnetic field parallel to the growth axis.

associated with the 3/2 to 1/2 transition contribute to the response. (Note the absence of the spectrally diffused component seen in Fig. 7). The curve with the open circles represents the FWM response with the field E_3 is rotating opposite the other two fields and hence senses σ_+ excitons. If the electrons or holes associated with the σ_- excitons created by $\mathbf{E}_1 \cdot \mathbf{E}_2^*$ flip their spin at the same localization site, it will produce a spectral hole burning response *at the lower energy of the σ_+ exciton*, as seen in the figure. The energy difference between the two resonances corresponds to the exciton Zeeman splitting: 0.19 meV at 4T.

It now becomes possible to determine the field dependence of the splitting. Recent measurements have shown that at low and intermediate magnetic field, the electron g-factor at the lowest Landau level in GaAs quantum wells is close to the value for bulk GaAs [3,16]. In contrast, the Zeeman splitting of the hh1 exciton obtained above is very small in comparison with that reported for bulk GaAs [58]. Our results seem to be in agreement with the earlier magneto-reflectance measurements where the heavy hole Zeeman doublet was not resolved. The small Zeeman splitting attributed to the strong valence band mixing in quantum well structures has been recently predicted by numerical calculations of magnetoexcitons using the Luttinger Hamiltonian [59]. In particular, the mixing of σ_- excitons with excitons at higher energy pushes the σ_- exciton to lower energy. The above calculation also predicts an eventual sign change of the Zeeman splitting at higher magnetic fields where the band mixing effects overcome those of the Zeeman interaction. However, theoretical determination of the magnetic field at which the zero crossing occurs is difficult since the cancellation of the two competing contributions depends strongly on parameters of the model [59]. The sign change of the splitting has not been observed in our measurements up to 6 T. It is interesting to note that disorder induced localization is expected to enhance the band mixing effects if the localization scale is smaller than the exciton Bohr radius. Using the recombination rate of the localized exciton, we have estimated the localization scale to be comparable to the exciton Bohr radius.

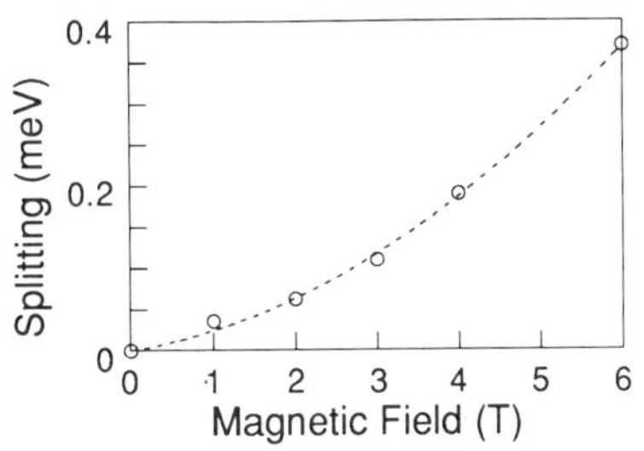

Figure 19. Magnetic field dependence of the exciton Zeeman splitting. Dashed lines are least square fit to a quadratic dependence.

Figure 19 displays the magnetic field dependence of the Zeeman splitting for the hh1 exciton [60]. The dashed line in the figure represents a quadratic dependence. Clearly the quadratic behavior may not extend into the high field region. The observed field dependence is somewhat surprising since the calculations predict a field dependence slower than linear. The observed dependence may be due in part to the nonparabolicity of the conduction band [54], which was not included in the calculations. Note that our results differ considerably from those obtained from nonlinear quantum beats in a 30 Å stepwise GaAs quantum well [56]. Zeeman splittings reported in the quantum beat measurement (0.5 meV at 4T) are proportional to magnetic fields with a field strength ranging from 1 to 5 T, and are very close to those measured for impurity bound excitons in bulk GaAs [61]. The sign of the Zeeman splitting can not be determined in quantum beat measurements.

SUMMARY

The investigation of coherences in semiconductors and their application to the study of fundamental phenomena in these systems is continuing to develop into a rich field of research. Our studies show that they can be observed in either the time domain or the frequency domain. In spite of the rapid developments in this area, there remain many questions and issues that have yet to be resolved which further illustrate the real underlying complexity of these system. Nevertheless, there is no question that these problems will eventually be understood enabling a complete unified picture of these interactions.

ACKNOWLEDGEMENTS

This work is supported by the Air Force Office of Scientific Research and the U.S. Army Research Office. The authors wish to acknowledge helpful discussions with Professor Paul Berman and Stephan Koch and with Dr. Yuan Hu and Rolf Binder.

REFERENCES

1. H. Haug S. Schmitt-Rink, J. Opt. Soc. Am. B **2**, 1135 (1985).
2. D.S. Chemla, S. Schmitt-Rink, and D.A.B. Miller, in *Optical Nonlinearities and Instabilities in Semiconductors,* H. Haug, Editor. (1988, Academic Press, San Diego) p. 83.
3. H. Wang, K. Ferrio, D.G. Steel, Y. Hu, R. Binder, and S. Koch, submitted for publication (1993).
4. F.H. Pollak and M. Cardona, Phys. Rev. **172**, 816 (1968).
5. M. Cardona, Personal Communication. 1992.
6. M. Sargent, M.O. Scully, and W.E. Lamb, *Laser Physics* (Addison-Wesley, Reading, Mass., 1974)
7. G. Lampel, A.N. Titkov, and V.I. Safarov, in *Fourteenth International Conference on the Physics of Semiconductors*, (IOP Conference Proceedings, 1978.)
8. G.W. Fehrenbach W. Schafer, J. Treusch, R.G. Ulbrich, Phys. Rev. Let. **49**, 1281 (1982).
9. L. Schultheis J. Kuhl, A, Honold, C.W. Tu, Phys. Rev. Let. **57**, 1635 (1986).
10. H. Wang, K. Ferrio, D.G. Steel, P. Berman, and S.W. Koch, submitted for publication (1993).
11. D.-S. Kim, J. Shah, T.C. Damen, W. Schafer, F. Jahnke, S. Schmitt-Rink, and K. Kohler, Phys. Rev. Lett. **69**, 2725 (1992).
12. C. Weisbuch, R. Dingle, A.C. Gossard, and W. Wiegmann, Solid State Comm. **38**, 709 (1981).
13. R.C. Miller, C.W. Tu, S.K. Sputz, and R.F. Kopf, Appl. Phys. Lett. **49**, 1245 (1986).
14. C.W. Tu, R.C. Miller, B.A. Wilson, P.M. Petroff, T.D. Harris, R.F. Kopf, S.K. Sputz, and M.G. Lamon, J. Cryst. Growth **81**, 159 (1987).
15. A. Ourmazd, D.W. Taylor, J. Cunningham, and C.W. Tu, PRL **62**, 933 (1989).
16. B. Deveaud, B. Guenais, A. Poudoulec, and A. Regreny, Phys. Rev. Lett. **65**, 2317 (1990).
17. A. Ourmazd and J. Cunningham, Phys. Rev. Lett **65**, 2318 (1990).
18. C.A. Warwick, W.Y. Jan, A. Ourmazd, and T.D. Harris, Appl. Phys. Lett. **56**, 2666 (1990).
19. D. Gammon, B.V. Shanabrook, and D.S. Katzer, Phys. Rev. Lett. **67**, 1547 (1991).
20. T. Tanaka and H. Sakaki, J. Cryst. Growth **81**, 153 (1987).
21. T. Takagahara, Phys. Rev. B **31**, 6552 (1985).
22. T. Takagahara, Jr. Lumin. **44**, 347 (1989).
23. D.G. Steel and J.T. Remillard, Phys. Rev. A **36**, 4330 (1987).
24. H. Wang and D.G. Steel, Phys. Rev. A **43**, 3823 (1991).
25. G. Finkelstein, S. Bar-Ad, O. Carmel, I. Bar-Joseph, and Y. Levinson, Phys. Rev. B **47**, 12964 (1993).

26. H. Wang, M. Jiang, and D.G. Steel, Phys. Rev. Lett. **65**, 1255 (1990).
27. H. Wang and D.G. Steel, App. Phys. **A33**, 514 (1991).
28. M. Jiang, H. Wang, and D.G. Steel, Appl. Phys. Lett. (1992).
29. T. Takagahara, Phys. Rev. B **32**, 7013 (1985).
30. J. Hegarty, L. Goldner, and M.D. Sturge, Phys. Rev. B **30**, 7346 (1984).
31. J. Hegarty and M.D. Sturge, J. Opt. Soc. Am. B **2**, 1143 (1985).
32. R. Kachru, T.W. Mossberg, and S.R. Hartmann, Opt. Comm. **30**, 57 (1979).
33. T. Yajima and Y. Taira, J. Phys. Soc. of Japan **47**, 1620 (1979).
34. M.D. Webb, S.T. Cundiff, and D.G. Steel, Phys. Rev. B **43**, 12658 (1991).
35. H.H. Yaffe, Y. Prior, J.P. Harbison, and L.T. Florez, in Quantum Electronics Laser Science, 1991 *Technical Digest Series* (OSA, 1991.)
36. K. Leo, J. Shah, S. Schmitt-Rink, and K. Köhler, in *VIIth International Symposium on Ultrafast Processes in Spectroscopy*, (IOP, 1992, 1991.)
37. S.T. Cundiff, H. Wang, and D.G. Steel, Phys. Rev. B **46**, 7248 (1992).
38. M.D. Webb, S.T. Cundiff, and D.G. Steel, Phys. Rev. Lett. **66**, 934 (1991).
39. P. Roussignol D. Ricard, C. Flytzanis, N. Neuroth, Phys. Rev. Let. **62**, 312 (1989).
40. G. Finkelstein, S. Bar-Ad, O. Carmel, and I. Bar-Joseph, Phys. Rev. B **47**, 12964 (1993).
41. C. Stafford, S. Schmitt-Rink, and W. Schäfer, Phys. Rev. B **41**, 10000 (1990).
42. J.B. Stark W. H. Knox, D.S. Chemla, W. Schäfer, S. Schmitt-Rink, C. Stafford, Phys. Rev. Lett. **65**, 3033 (1990).
43. T. Rappen, J. Schröder, A. Leisse, M. Wegener, W. Schäfer, N.J. Sauer, and T.Y. Chang, Phys. Rev. B **44**, 13093 (1991).
44. J.B. Stark, W.H. Knox, and D.S. Chemla, Phys. Rev. Lett. **68**, 3080 (1992).
45. G. Röpke and R. Der, Phys. Stat. Sol. (B) **92**, 501 (1979).
46. M. Jiang, H. Wang, R. Merlin, M. Cardona, and D.G. Steel, submitted for publication (1993).
47. V.A. Kharchenko, Sov. Phys. JETP **56**, 1140 (1982).
48. K. Fujii, T. Tomaru, T. Ohyama, and E. Otsuka, *High Magnetic Fields in Semiconductor Physics II* (Springer-Verlag, Berlin, 1989)
49. M. Shinada and K. Tanaka, J. Phys. Soc. Jpn **29**, 1258 (1970).
50. T. Ando and Y. Uemura, J. Phys. Soc. Jpn. **37**, 1044 (1974).
51. T. Englert and K. von Klitzing, Surface Sci. **73**, 70 (1978).
52. D. Stein, K. von Klitzing, and G. Weimann, Phys. Rev. Lett. **51**, 130 (1983).
53. M. Dobers, K. von Klitzing, and G. Weimann, Phys. Rev. **B38**, 5453 (1988).
54. G. Lommer, F. Malcher, and U. Rossler, Phys. Rev. B **32**, 6965 (1985).
55. P. Lefebvre B. Gil, J.P. Lascaray, H. Mathieu, D. Bimberg, T. Fukunaga, H. Nakashima, Phys. Rev. B **37**, 4171 (1988).
56. S. Bar-Ad and I. Bar-Joseph, Phys. Rev. Lett. **66**, 2491 (1991).
57. M.J. Snelling, E. Blackwood, C.J. McDonagh, and R.T. Harley, PR B **45**, 3922 (1992).
58. B. Bimberg, Advan. in Solid State Phys. **XVIII**, 195 (1977).
59. G.E.W. Bauer, *High Magnetic Fields in Semiconductor Physics II*, G. Landwehr, Ed., Springer-Verlag (Berlin 1989).
60. H. Wang, M. Jiang, R. Merlin, and D.G. Steel, Phys. Rev. Lett. **69**, 804 (1992).
61. A.M. White, I. Hinchliffe, and P.J. Dean, Solid State Comm. **10**, 497 (1972).

NON-MARKOVIAN OPTICAL SYSTEMS

Erik T.J. Nibbering, Douwe A. Wiersma and Koos Duppen

Ultrafast Laser and Spectroscopy Laboratory
Department of Chemical Physics, Materials Science Center
University of Groningen, Nijenborgh 4, 9747 AG Groningen
The Netherlands

I. INTRODUCTION

The time resolution of optical experiments has increased dramatically over the last 15 years. With the advent of the CPM- [1] and Ti:sapphire lasers [2] and the development of pulse compressors [3], dynamical processes can now be studied that occur on time scales well below 100 fs. When coherent optical experiments are performed on these time scales, the interpretation of the observed dynamics is preferably done in terms of two-level systems and exponential damping processes.

Of course it is often clear that the two-level approach is insufficient, and in many cases extentions of the theory to multi-level systems are required. On the other hand, it is mostly not realized that even for a well-defined two-level system there are fundamental limits to the applicability of exponential damping parameters. In this paper, we will discuss what the assumptions are that form the basis of the conventional picture of optical dynamics, and sketch how this treatment can be generalized to situations where these basic assumptions do not hold. Especially for very fast dynamics it can be expected that the standard treatment fails, and other, so-called non-Markovian approaches are then required. In this paper, these will be introduced in a phenomenological way, without going into microscopic detail on the mechanisms of the dynamic behaviour.

First, in Section II, the usual treatment of relaxation will be presented, which for a two-level system leads to the optical Bloch equations. In Section III it will be shown how this treatment relies on a separation in time scales between the system relaxation and the heat bath(s) movements that induce the relaxation. When this time scale separation does not hold, a non-Markovian treatment of the optical dynamics is required. Next, in Sections IV and V, two models will be introduced that do not rely on the Markovian approximation. In the stochastic model (Sec. IV), the transition frequency of the system is treated as a stochastically perturbed oscillator, making the transition frequency a stochastic function of time. In this model the propagation of the system and certain bath degrees of freedom are correlated for a finite time. In the Brownian oscillator model (Sec. V), dynamic level shifts

Coherent Optical Interactions in Semiconductors
Edited by R.T. Phillips, Plenum Press, New York, 1994

are closely connected to the non-Markovian relaxation. This arises due to the explicit introduction of potential surfaces for nuclear motions that are randomly perturbed by the heat bath. This model is especially suited to treat optical dephasing and solvation of molecules in liquids. However, when the Brownian oscillatory movements in the potential wells are interpreted in a more abstract way, the model is applicable in more general circumstances as well. The analysis of experimental results on liquids leads to the conclusion that optical dynamics in these systems occurs in a non-Markovian way.

II. THE BLOCH APPROACH

The time evolution of systems in contact with optical fields and some sort of heat bath is usually described in the density matrix formalism:

$$i\hbar\left(\frac{\partial\rho}{\partial t}\right) = [\mathcal{H}_0,\rho] + [\mathcal{H}_i,\rho] + i\hbar\left(\frac{\partial\rho}{\partial t}\right)_{system-bath} \tag{1}$$

where $\mathcal{H}_0$ is the Hamiltonian of the unperturbed system and $\mathcal{H}_i$ describes the coupling of the system with the electromagnetic field(s):

$$\begin{aligned} \mathcal{H}_0 &= \sum_n E_n |n\rangle\langle n| \\ \mathcal{H}_i &= \sum_{n\neq m}\sum_i \frac{1}{2}\boldsymbol{\mu}_{nm}\cdot \boldsymbol{A}_i(r,t)\, e^{-i[\omega_i t - \boldsymbol{k}_i\cdot \boldsymbol{r}+\phi_i]}\, |n\rangle\langle m| \end{aligned} \tag{2}$$

Here, $\boldsymbol{\mu}_{nm}$ is the transition dipole moment between states $|n\rangle$ and $|m\rangle$, and $\boldsymbol{A}_i$ is the amplitude of field i with frequency ω_i, wave vector $\boldsymbol{k}_i$ and phase ϕ_i. The system-bath interaction of Eq. (1) is not written in terms of a Hamiltonian commutator, since the details of this interaction are not available. It is only possible to discuss the influence of a heat bath on the propagation of a system in a statistical, ensemble averaged way. Each individual system of the ensemble will experience a different interaction with the bath at any given time.

Usually, the random interactions in the system-bath coupling are time-averaged to constant parameters, before Eq. (1) is solved [4]. This leads to phenomenological damping terms, which for the diagonal and off-diagonal terms of the density matrix can be written as:

$$\begin{aligned} \left(\frac{\partial\rho_{nn}}{\partial t}\right)_{system-bath} &= \sum_m \gamma_{mn}\rho_{mm} - \sum_m \gamma_{nm}\rho_{nn} \\ \left(\frac{\partial\rho_{nm}}{\partial t}\right)_{system-bath} &= -\Gamma_{nm}\rho_{nm} \end{aligned} \tag{3}$$

The transition probabilities γ_{nm} obey the principle of detailed balancing. When the energy difference between the levels of the system is large compared to kT, only downward transitions occur, corresponding to spontaneous emission of, for instance, photons or phonons. All diagonal elements of the density matrix in principle relax in a combined way, leading to N-1 longitudinal relaxation times when there are N energy levels. Each off-diagonal element decays with a characteristic transverse relaxation time Γ_{nm}, which in general consists of contributions from the finite transition probabilities out of states $|n\rangle$ and $|m\rangle$, and from fluctuations of the energy differences between these levels. For a two-level system and a single electromagnetic field, substitution of Eqs. (2) and (3) in Eq. (1) leads to a form of the well-known Bloch equations, which were first formulated to describe dynamical effects in nuclear magnetic resonance [5]:

$$\frac{\partial \rho_{11}}{\partial t} = -\frac{\partial \rho_{22}}{\partial t} = \frac{\boldsymbol{\mu}_{12}\cdot\boldsymbol{A}_i}{2i\hbar}\left(\tilde{\rho}_{12}e^{+i[\boldsymbol{k}_i\cdot\boldsymbol{r}-\phi_i]} - \tilde{\rho}_{21}e^{-i[\boldsymbol{k}_i\cdot\boldsymbol{r}-\phi_i]}\right) + \frac{\rho_{22}}{T_1}$$

$$\frac{\partial \tilde{\rho}_{12}}{\partial t} = \frac{\boldsymbol{\mu}_{12}\cdot\boldsymbol{A}_i}{2i\hbar}\left(\rho_{11}-\rho_{22}\right)e^{-i[\boldsymbol{k}_i\cdot\boldsymbol{r}-\phi_i]} - \left(\frac{1}{T_2} - i(\omega_{21}-\omega_i)\right)\tilde{\rho}_{12} \tag{4}$$

$$\frac{\partial \tilde{\rho}_{21}}{\partial t} = \frac{\boldsymbol{\mu}_{12}\cdot\boldsymbol{A}_i}{2i\hbar}\left(\rho_{22}-\rho_{11}\right)e^{+i[\boldsymbol{k}_i\cdot\boldsymbol{r}-\phi_i]} - \left(\frac{1}{T_2} + i(\omega_{21}-\omega_i)\right)\tilde{\rho}_{21}$$

Here, the dynamic parameters from Eq. (3) are given their customary two-level names by the substitutions: $(\gamma_{21})^{-1} \equiv T_1$ and $(\Gamma_{21})^{-1} \equiv T_2$, and a transformation was made to the rotating frame: $\tilde{\rho}_{12} \equiv \rho_{12}\exp(-i\omega_i t)$ and $\tilde{\rho}_{21} \equiv \exp(+i\omega_i t)$. Then, all terms oscillating at the double frequencies $\pm 2i\omega_i t$ were neglected (the rotating wave approximation). The presence of inhomogeneous broadening can be easily incorporated by writing a distribution for the system transition frequency ω_{21}. First Eqs. (4) are integrated, and subsequently an integration is performed over this distribution in ω_{21}.

The optical Bloch equations can be used in many circumstances where coherent optical interactions play a role, from cw- to impulsive excitation conditions [6,7]. For instance, for molecular two-level systems it can be shown non-perturbatively how photon echo experiments continuously change into dynamic holography (spatial grating scattering) when the pulse delays are varied in the proper way [8]. In semiconductors, where many-body effects and Coulomb interactions between the charge carriers give rise to excitation dependent phenomena, Eqs. (4) still form the basis of the various more complicated expressions, which are generally known as the semiconductor Bloch equations [9,10,11]. In the following we will concentrate on the last part of Eq. (1), which describes the effect of a heat bath on the propagation of the system. Whenever the expression "system dynamics" is used, it is this part of the total equation of motion that is referred to.

III. NON-MARKOVIAN OPTICAL DYNAMICS

The Bloch treatment discussed above is based upon the assumption of a separation of time scales. In fact, two limiting cases were more or less tacitly introduced: one in the derivation of Eq. (4) from Eq. (1) and one in the introduction of inhomogeneous broadening.

First, it was assumed that the bath moves very fast on the time scale of the dynamics of the system. This means that for the propagation of the system only the time averaged influence of the bath has to be considered. As far as the system dynamics is concerned, the bath is always in thermodynamic equilibrium, and its influence on the evolution of the system can described in terms of constant parameters. The system evolution at a certain time t is then completely determined by the ensemble averaged configuration of the bath at that time, and not by any previous history. The system's future is completely determined by the present and not by its past, since there is no memory of disturbances at earlier times. This is basically the Markovian approximation of statistical physics [12]. In this *fast modulation* or *Markovian limit* the effects of the bath on the dynamics of the system are completely described by phenomenological parameters like T_1 or T_2 for the two-level case.

When the system-bath coupling does not fluctuate infinitely fast compared to the system dynamics, the damping cannot be written in the form of Eqs. (3), and consequently the optical Bloch equations Eq. (4) are not valid as well. In such a case the propagation of the system and (part of) the bath should be treated on equal footing. This will be discussed in the following sections.

The incorporation of inhomogeneous broadening as a distribution of transition frequencies is based upon a separation of time scales as well. This distribution is due to the coupling of the system with its surroundings, but it is assumed now that the time evolution of the system-bath coupling occurs on much longer time scales than the system dynamics. This is the *slow modulation* or *static limit* of system-bath interactions, which leads to an inhomogeneous distribution of spectroscopically distinct transitions in a linear absorption spectrum.

When this interaction is truly static, perfect memory is present. This means that the system remembers, in principle, all previous history of time evolution. Therefore it is possible, by cleverly choosing experimental conditions, to turn back the entire time evolution of the system that is due to this static distribution (photon echo experiments). When the distribution is not truly static, but the time evolution is still much slower than that of the system itself, one speaks of spectral diffusion. When this system-bath coupling evolves on a similar time scale as the system dynamics itself, the slow modulation approximation fails, and again the propagation of the system and (part of) the heat bath should be treated on equal footing.

It is surprising that in so many circumstances a separation occurs of the system-bath coupling into infinitely fast and infinitely slow fluctuations (compared to the system dynamics). However, when either the fast modulation (Markovian) or slow modulation (static) limits fail, the Bloch approach is not valid, and a theory for non-Markovian dynamics should be formulated. The general approach to describe such propagation is the introduction, in a statistical way, of the most relevant bath degrees of freedom in the system equation of motion. This is always possible, since the separation of Eq. (1) into an unperturbed system Hamiltonian, system-field coupling, and system-bath coupling is actually quite arbitrary.

To give an example of the freedom one has in separating the total Hamiltonian, consider the electronic transition of an ensemble of low density gas-phase molecules. Each pure electronic transition is accompanied by a number of vibronic transitions that give rise to a series of sharp lines in the absorption spectrum. It then makes sense to call the relevant electronic and vibrational levels the system. The bath degrees of freedom are the translations of the molecules, which gives rise to collisions among them. So, the nuclear motions of the atoms are divided into system degrees of freedom (the internal modes of the molecule) and bath degrees of freedom (the translations). However, it is quite possible to formulate alternative descriptions, which make sense as well. For instance, all atomic motions can be treated in the same way, as bath degrees of freedom. Then the electronic transition forms a two-level system, and the influence of all nuclear motions on this system is taken into account in a form of a system-bath coupling. Both pictures are in principle equally valid: the accuracy of describing dynamics depends only on technicalities, such as the construction of proper system-bath interactions.

This choise of including or excluding degrees of freedom in the definition of the system is very useful when non-Markovian system-bath interactions are considered. For instance, when the dynamics of a certain two-level system is non-Markovian, it may be possible to model this by incorporating the most relevant bath degrees of freedom into a newly defined, larger system, that may or may not display Markovian dynamics. In this way the extended system, with modified system-bath coupling, acts as a model for the original non-Markovian system. Two such approaches will be presented in the next sections, and used to describe results of a variety of optical experiments on a single system.

IV. THE STOCHASTIC MODEL OF FREQUENCY FLUCTUATIONS

Two pulse degenerate four-wave mixing, which is also known as two-pulse photon echo generation, is well-suited to study ultrafast optical dynamics. The time resolution of such experiments is only limited by the pulse duration. In our case the two pulses were generated with a CPM-laser (45 fs), then amplified with a copper-vapour laser (50 fs), chirped in a monomode optical fiber (1200 fs), and subsequently compressed in time by a sequence of two gratings and four prisms. This yielded a duration for both pulses of about 9 fs. The pulses were crossed at an angle of about 4.5° in a 60 μm thick sample, which consisted of a jet of dimethylsulfoxide (DMSO) with resorufin as dye molecules dissolved in it (10^{-3} mol/liter). The beam geometry and the experimental result are shown in Fig. (1). The nonlinear signals in the phase-matched direction $2\mathbf{k}_2-\mathbf{k}_1$ were detected with a photomultiplier and a lock-in amplifier.

The trace of Fig. (1) displays the distinct asymmetry that is typical for photon echo generation. The fast positive slope is determined by the total spectrum of resorufin that is excited by the pulses, and/or by the finite pulse durations. The much gentler negative slope indicates that the free-induction- (free-polarization-) decay is not irreversible at room temperature. Apparently the dephasing that occurs during time τ after the interaction with the first pulse is (partly) counter balanced by the rephasing that occurs after the interaction with the second pulse. This is only possible when there is memory present of phase disturbing processes. Of course the conventional picture of optical dynamics attributes this memory to inhomogeneous broadening. The decay on the positive delay side is then solely due to homogeneous (exponential) relaxation, which arises as a consequence of system-bath coupling in the Markovian limit. An exponential fit of the observed signal trace gives a decay time of 17 fs (T_2= 68 fs). A weak beat has to be included in the signal analysis since next to the pure electronic transition one vibronic transition of resorufin is excited as well.

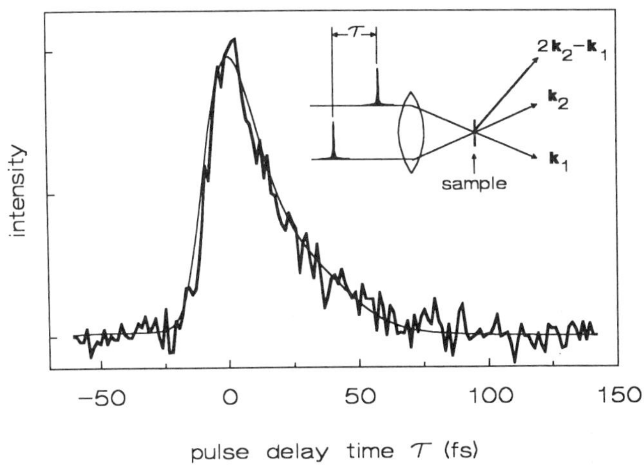

Figure 1 Integrated two-pulse photon echo signal of resorufin in DMSO. The beam geometry is shown in the inset; the smooth curve is theory.

Below we will show that the description of the optical dynamics in terms of homogeneous and inhomogeneous broadening is not consistent with the results of other experiments, such as chirped four-wave mixing. For the moment we just state that it is unlikely that the ultrafast system dynamics of Fig. (1) can be explained by coupling of the system with bath degrees of freedom that move on even much faster time scales. After all, that assumption underlies the Markovian treatment that leads to the T_2 concept. In addition, a true static component in the system-bath interactions is also not expected for this system.

When the Markov approximation fails because system and bath move at similar time scales, the system-bath coupling should be integrated together with the system propagation itself. When very many bath degrees of freedom are important for this dynamics, the explicit introduction of these states in the system Hamiltonian gives rise to a continuum that can be treated in a classical way. The time evolution that occurs in this continuum can be treated as a stochastic process. This approach was adopted by us to treat resonance light scattering and absorption line shape analysis [13,14], and to analyze time resolved experiments such as ultrafast photon echo experiments [15] and chirped four-wave mixing [16]. Later a similar treatment was used by another group to discuss stimulated photon echoes at ultrafast time scales [17].

The stochastic treatment of a quantized system in contact with a classical, non-Markovian bath was discussed extensively in connection with motional narrowing of linear NMR absorption spectra [18,19]. The basic concepts of these 30 year old theories can be directly transferred to linear response in the optical domain, and extensions to nonlinear response are also straigthforward. The formal mathematical theory of stochastic processes is well-documented and can be found in various textbooks [20]. In our case we are dealing with the time dependence of frequency fluctuations. Suppose the frequency of an optical transition between states $|1\rangle$ and $|2\rangle$ has at a certain time a value $\omega(t)=\omega_{21}+\delta\omega(t)$. In the Markovian limit, $\delta\omega(t)$ assumes all values that are possible in an infinitely short time. For a stochastic process the time evolution of the frequency is random (i.e. each time different), but it will follow some characteristic pattern. It now takes some finite average time to sample all available values for $\delta\omega$.

A non-Markovian theory of optical dynamics includes $\delta\omega(t)$ explicitly in its Hamiltonian. The equation of motion for the system must be solved with the time dependence of $\delta\omega(t)$ included. The macroscopic polarization at frequency ω_{21} is proportional to the off-diagonal matrix element $\rho_{21}(t)$. Instead of adopting the Bloch approach, we will calculate this matrix element perturbatively, including possible non-Markovian effects. For linear response after impulsive excitation, it suffices to look at the propagation of a single superposition state from time t=0 to time t=t. If we forget about population relaxation, the integration of the equation of motion has the following general solution:

$$\rho_{21}(t) = \rho_{21}(0)\, e^{-i\omega_{21}t} \left\langle \exp\left\{ -i\int_0^t \delta\omega(t')\,dt' \right\} \right\rangle \qquad (5)$$

The brackets around the exponent indicate averaging over the macroscopic ensemble.

The evaluation of the relaxation in Eq. (5) due to the frequency fluctuations is not trivial. The time dependence of $\delta\omega(t)$ is not known exactly; it will be different for all members of the macroscopic ensemble of systems. However, it is possible to discuss its evolution in statistical terms. The problem is that the ensemble averaging that gives rise to the statistical interpretation is not on the fluctuations itself, but instead on the exponential operator of Eq. (5). The brackets of Eq. (5) have to be worked into the argument of the exponent to be able to express the physics in terms of the statistical properties of $\delta\omega(t)$ itself.

The averaging of Eq. (5) can be performed exactly if $\delta\omega(t)$ describes a stationary process with a Gaussian distribution and a zero mean $\langle\delta\omega(t)\rangle=0$. First the exponent is expanded into a power series:

$$R_1(t) \equiv \langle \exp\left\{-i\int_0^t \delta\omega(t')dt'\right\}\rangle$$
$$= 1 + \sum_{n=1}^{\infty} \frac{(-i)^n}{(n)!} \int_0^t ....\int_0^t \langle \delta\omega(t_1)....\delta\omega(t_n)\rangle \, dt_1....dt_n \tag{6}$$

From the Gaussian property of $\delta\omega(t)$ and its stationarity it follows that all uneven terms are zero [21]:

$$\int_0^t ....\int_0^t \langle \delta\omega(t_1)....\delta\omega(t_{2n-1})\rangle \, dt_1....dt_{2n-1} = 0 \tag{7}$$

while all even terms can be expressed in the two-time correlation function:

$$\int_0^t ....\int_0^t \langle \delta\omega(t_1)....\delta\omega(t_{2n})\rangle \, dt_1....dt_{2n} = \frac{(2n)!}{2^n\, n!}\left[\int_0^t\int_0^t \langle \delta\omega(t_1)\,\delta\omega(t_2)\rangle \, dt_1 dt_2\right]^n \tag{8}$$

From the resulting power series again an exponent can be constructed, but now with the averaging contained in the argument! So, the resulting expression for linear response is:

$$R_1(t) = \exp\left\{-\frac{1}{2}\int_0^t\int_0^t \langle \delta\omega(t_1)\,\delta\omega(t_2)\rangle \, dt_1 dt_2\right\} \tag{9}$$

When the frequency fluctuations are not a Gaussian process, this procedure is not exact. However, in that case it is always possible to expand in terms of cumulants and evaluate the response up to the required accuracy [22].

For the photon echo experiment of Fig. (1), the basic treatment of the stochastic fluctuations is the same as for linear response. Now two periods of coherent superposition states have to be considered, which are inverted compared to each other. The integration of the equation of motion for the system, including the frequency excursions $\delta\omega(t)$, then has the following general solution:

$$\rho_{21}(t,\tau) = \rho_{21}(0,0)\, e^{-i\omega_{21}(t-\tau)} \langle \exp\left\{-i\left(\int_\tau^{\tau+t} \delta\omega(t')dt' - \int_0^\tau \delta\omega(t')dt'\right)\right\}\rangle \tag{10}$$

Notice the different signs of the $\delta\omega(t)$ terms in the exponent, which reflect the fact that the superposition state propagating during time τ is inverted when the second pulse interacts with the system. This enables rephasing of the macroscopic coherence (echo formation) during time t after the second pulse, if there is memory of the manner(s) in which the phase loss occurred.

The procedure to evaluate the statistics of the fluctuations $\delta\omega(t)$ is exactly the same as for linear response: the exponent is expanded, only the even terms are retained, and a new exponent is constructed with the averaging contained in the argument. The result is:

$$\begin{aligned} R_3(t,\tau) &\equiv \left\langle \exp\left\{-i\left(\int_\tau^{\tau+t} \delta\omega(t')dt' - \int_0^\tau \delta\omega(t')dt'\right)\right\}\right\rangle \\ &= \exp\{-\frac{1}{2}\Big(\int_0^\tau\int_0^\tau \langle\delta\omega(t_1)\delta\omega(t_2)\rangle\, dt_1 dt_2 \\ &\qquad + \int_\tau^{\tau+t}\int_\tau^{\tau+t} \langle\delta\omega(t_1)\delta\omega(t_2)\rangle\, dt_1 dt_2 \\ &\qquad - 2\int_\tau^{\tau+t}\int_0^\tau \langle\delta\omega(t_1)\delta\omega(t_2)\rangle\, dt_1 dt_2\Big)\} \end{aligned} \tag{11}$$

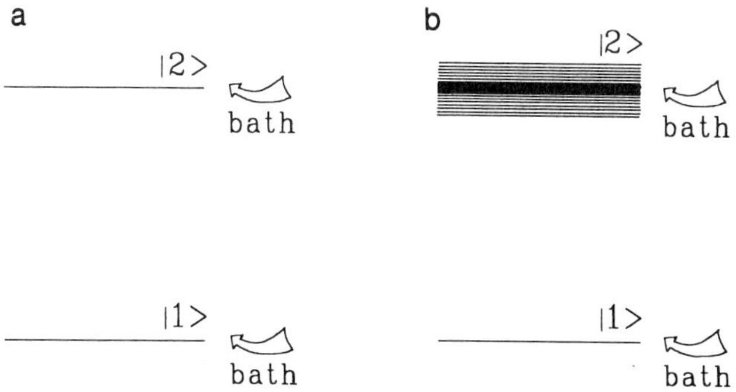

Figure 2 Impression of the stochastic model of non-Markovian dynamics. In Fig. (2b) a Gaussian distribution of states is indicated with possible Markovian dynamics.

In fig (2) is shown how this dynamical model can be interpreted. The time evolution of a two-level system which interacts with some (abstract) heat bath is non-Markovian, because the dynamics of the system and the system-bath coupling occur on the same time scale. This is modelled by replacing the system by a Gaussian distribution of transition frequencies. This distribution also interacts with a (different) heat bath, which induces changes among the states of the Gaussian distribution. The system-heat bath coupling of this extended system may, or may not, occur in the Markovian limit. In Fig. (2b) discrete quantum levels are sketched, but in the model a continuum is assumed.

This Gaussian distribution is *not* the lineshape that is observed in either linear or nonlinear spectroscopy, since motions occur in the continuum: $\delta\omega(t)$ is time dependent. Hence, the line shape is affected by, what can be loosily called, "motional narrowing". It depends on the time scale of the experiment compared to the time scale of the fluctuations what linewidth (or time response) is relevant to the optical experiment.

To quantify this picture, we need to have an analytical expression for the two-time correlation function of Eqs. (9) and (11). This expression will depend on how the frequency fluctuations occur. If the underlying process that leads to the random modulation $\delta\omega(t)$ has a much shorter time scale than the modulation itself, the stationary Gaussian process of frequency modulation is Markovian. In the interpretation of Fig. (2b) this means that the coupling of the Gaussian distribution with the remainder of the heat bath is taken to be Markovian. Doobs theorem [23] states that a Gaussian process is Markovian if and only if its two-time correlation function is exponential:

$$\langle \delta\omega(t_1)\,\delta\omega(t_2)\rangle = \langle \delta\omega(t_1-t_2)\,\delta\omega(0)\rangle = \Delta^2 \exp\{-\Lambda|t_1-t_2|\} \tag{12}$$

Here, Δ is the root mean square amplitude of the frequency excursions (the Gaussian width of fig. 2b), and Λ is the inverse correlation time $(t_c)^{-1}$ of the frequency excursions in the available frequency range.

The stochastic relaxation functions for linear response (Eq. 9) and the photon echo experiment (Eq. 11) can now be evaluated by calculating the two-dimensional integrals with the correlation function Eq. (12). After considerable mathematical manipulation, the results are respectively [15]:

$$R_1(t) = \exp\{-g(t)\} \tag{13}$$

$$R_3(t,\tau) = \exp\{-2g(t)-2g(\tau)+g(t+\tau)\} \tag{14}$$

with the so-called line shape function g(t) given by

$$g(t) = \frac{\Delta^2}{\Lambda^2}[e^{-\Lambda t}+\Lambda t-1] \tag{15}$$

In the long time (Markovian) limit ($t \gg t_c$) this function is exponential, while in the short time (static) limit ($t \ll t_c$) a Gaussian decay results. The ratio of Δ vs. Λ determines whether the decay is predominantly exponential ($\Delta \ll \Lambda$) or Gaussian ($\Delta \gg \Lambda$).

To compare the theoretical expression Eq. (14) with the experimental result of the two-pulse experiment of Fig. (1), the finite pulse width has to be considered. Also, the detector integrates the signal over all times t. Hence, the experimental signal trace $S_{2PE}(\tau)$ has to be compared to the following expression:

$$S_{2PE}(\tau) = \left\{\int_0^\infty [R_3(t,\tau)]^2 dt\right\} \otimes T(\tau) \tag{16}$$

where $T(\tau)$ is a measure of the time resolution, $\otimes$ designates a convolution, and $R_3(t,\tau)$ is given by Eq. (15) substituted in Eq. (14):

$$R_3(t,\tau) = \exp\left\{-\frac{\Delta^2}{\Lambda^2}[2e^{-\Lambda t}+2e^{-\Lambda\tau}-e^{-\Lambda(t+\tau)}+\Lambda(t+\tau)-3]\right\} \tag{17}$$

When Eq. (16) is fitted to the experiment on resorufin in DMSO, the smooth curve of Fig. (1) is obtained. The parameter values of the fit are the root mean square amplitude $\Delta=41$ THz (rad/sec) and $\Lambda=27$ THz (1/sec). Hence, the correlation time of the frequency fluctuations is $t_c=\Lambda^{-1}=37$ fs.

What is the character of the signals? The values found for the parameters Δ and Λ indicate that the dynamics is truly non-Markovian. If Δ was much smaller than Λ, most of the decay (except for very short times) would have occurred in the Markovian limit, while for Δ much larger than Λ the static limit would have applied. Now, with Δ and Λ of the same order of magnitude, the effect of a finite correlation time has to be explicitly considered.

It is instructive to look at the shape of the signal in time. When the derivative of $R_3(t,\tau)$ is taken with respect to t, it is found that the signal has a maximum at:

$$t_{max} = \Lambda^{-1} \ln[2 - e^{-\Lambda\tau}] \qquad \textbf{(18)}$$

From the fact that the signal maximizes at a finite delay from the second pulse, it can be concluded that rephasing of the macroscopic coherence takes place. Hence, the signal is a kind of photon echo. However, the echo does not peak at the conventional time $t\approx\tau$, which is expected for an inhomogeneously broadened system. Here, we have not assumed any inhomogeneous broadening at all; there are only frequency fluctuations on one time scale. Rephasing of the macroscopic coherence is then not due to a static component in the interactions, but is a consequence of interfering fluctuations during times τ and t. The system remembers for about the correlation time how the phase disturbances occurred. Echo formation is only possible for time separations of that order or less.

In Fig. (3) is shown how the maximum of the signal develops as a function of pulse delay for the parameters of Fig. (1). For a conventional echo the maximum occurs (for large inhomogeneous broadening) at a time equal to the pulse separation. It is clear that the maximum in the case of stochastic frequency fluctuations is limited by the correlation time. As a consequence, the integrated intensity of the signal decays faster than exponential until, for times $t \gg t_c$ exponential (homogeneous) decay is observed. In this limit the factor Λ/Δ^2 can be identified as the Markovian parameter T_2.

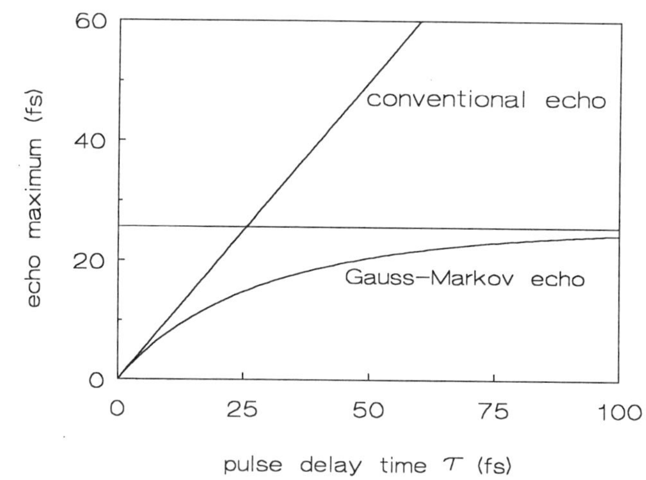

Figure 3 Maximum of the signal amplitude for a conventional echo (Bloch-Hahn) and a stochastic echo (Gauss-Markov).

Now the question has to be answered: how do we prove that the system dynamics is non-Markovian? After all, an exponential decay fits the data of Fig. (1) just as well as the stochastic expression Eq. (17). So far, we have only argued that an explanation of ultrafast system dynamics in terms of much faster bath dynamics is, to our feeling, unphysical.

The main difference between the conventional and the stochastic approach is of course the character of the optical memory. In the conventional picture memory exists for as long as one can speak of inhomogeneous broadening. Thus memory may ultimately disappear because of spectral diffusion. In the stochastic model adopted here, the transition is "homogeneous" when the time scale of observation is longer than the correlation time. Hence, an experiment should be performed that can probe whether memory is present at somewhat longer time scales than observed in the echo experiment. One example of such an experiment is ultrafast transient hole burning [24], another one is chirped four-wave mixing [16,25].

In chirped four-wave mixing two or three chirped optical pulses interact in a sample to generate a nonlinear signal. Although chirped pulses are strongly non-transform limited, their optical phases are well-defined at all times. The carrier frequency is explicitly time dependent, which for a linearly chirped pulse can be written as:

$$\omega(t) = \omega_0 + bt \tag{19}$$

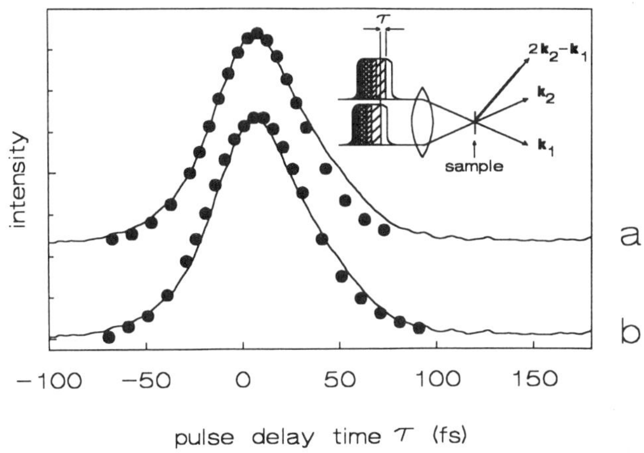

Figure 4 Chirped four-wave mixing signal of resorufin in DMSO. Dots are calculated. Curve (a): stochastic model; curve (b): Brownian oscillators.

Here ω_0 is an off-set frequency and b is the chirp rate.

An experimental result of two-pulse chirped four-wave mixing on resorufin in DMSO is shown in Fig. (4). In this experiment two pulses were used with the beam geometry shown in the inset. These pulses were generated with the same set-up that was used for the photon echo experiment, only the compression stages were left out. The pulse duration was about 1.2 ps with a chirp rate of 2.6 cm^{-1}/fs. As indicated in the inset of Fig. (4), a relative delay τ implies an instantaneous frequency difference between both beams of $\delta\omega = b\tau$ (Eq. 19). If the duration of sweep of the optical field frequency through the half width of the resonance is taken as a typical time for the experiment (just as T_2^{-1} is the HWHM of a homogeneously broadened line), the experiment of Fig. (4) probes the dynamics at ca. 75 fs.

The experimental trace of Fig. (4) clearly shows that the chirped two pulse scattering signal is much more symmetrical than the two pulse photon echo decay. Although qualitative, this is an important signature of the time scale of the experiment compared to that of the optical dynamics. For an optical field with a very fast chirp rate the system dynamics appears to be frozen in, i.e. the resonance appears to be "inhomogeneously broadened" for that field. In that case the chirped signal has the same outlook as a photon echo trace when the relative delay τ is varied (i.e. highly asymmetric) [26].

When the chirp rate is slow compared to the dynamics of the system, the optical transition appears to be "homogeneously broadened" for that field. The chirped signal as a function of relative delay τ is then much more symmetrical than with a fast chirp [16]. In a sense the experiment of Fig. (4) changes from having predominantly time-domain character for fast chirp rates, to having predominantly frequency-domain character for slow chirp rates. Since the measure for fast or slow chirp is the system dynamics, chirped four-wave mixing can be used to study these dynamics from ultrafast time scales up to the lifetime of the excited state.

In the limits of very fast chirp ($b \rightarrow \infty$) and very slow chirp ($b \rightarrow 0$) analytical expressions for the nonlinear polarization can be derived, but in general analytical solutions are not available. However, the nonlinear polarization can be worked out numerically over a wide range of chirp parameters by applying straightforward third-order perturbation theory. This is worked out in detail in ref. [25] for a number of dynamical models. Here we will only use the results of that analysis.

If there was significant memory in the system on the time scale in which the fields sweep through the resonances, the chirp would be fast compared to that dynamical component of the system-bath coupling. The typical asymmetric echo trace would then be observed. Since the experimental result of Fig. (4) does not resemble that of Fig. (1), it is clear that on a 75 fs time scale no memory is present. This can be worked out quantitatively using the expressions for chirped nonlinear response of ref. [25].

The dots in Fig. (4) were calculated using these expressions. In trace (a) the stochastic model of this section was used, in trace (b) the Brownian oscillator model of the next section. The same parameters Δ = 41 THz and Λ = 27 THz were used as in the description of the photon echo. Considering that no fit parameters were involved in this calculation, the agreement between experiment and theory is excellent. Since the conventional models predict an asymmetric echo trace, it must be concluded that all optical memory disappers at ultrafast time scales, (almost) as fast as the coherence decay measured in the photon echo experiments. Hence, we have proven that it makes no sense to separate the system-bath coupling in fast and slow moving components; the dynamics is non-Markovian and characterized by fluctuations on a single time scale, which is expressed in the finite correlation time t_c = 37 fs.

THE BROWNIAN OSCILLATOR MODEL

Although the stochastic model of frequency fluctuations is probably the most developed general non-Markovian theory of spectroscopic transitions, it fails to capture an important aspect of optical excitations, namely ultrafast dynamic level shifts. These also arise due to the coupling of the quantum system with a heat bath. In liquids this is called the dynamic Stokes shift. The basic reason for the failure of the stochastic model to describe such proceses is that the heat bath may influence the system, but the system does not have any effect at all on the heat bath.

This is actually to a large extent true for NMR transitions, for which the stochastic theory was originally developed [18,19]. At room temperature the transition frequency of the propagating spin system is much smaller than kT. Therefore the heat bath hardly notices when the system is excited. In these circumstances only fluctuations of the transition frequency occur; there is no net-energy transfer between the system and the heat bath. However, this high temperature limit does not hold for optical excitations at room temperature, so usually energy will be transferred between the system and the bath when the state of the system is changed. Fluctuations (dephasing) and energy relaxation are thus intimately linked. Both influence the (non-)linear optical response.

An elegant way to take a finite correlation time into account, and incorporate shifts of energy levels as well, is the multi-mode Brownian oscillator (MBO-) model for optical dynamics. We recently started to use this model to analyse data on ultrafast processes in liquids [27,28]. In the MBO-model, that was developed by Yan and Mukamel [29,30], some specific nuclear solvent degrees of freedom are assumed to dominate the effect of nuclear motion on the electronic transitions. These modes, that are projected out of the heat bath, are treated explicitly and experience in their propagation the influence of all remaining degrees of freedom (which in this model is now called the "bath"). Due to the coupling with the electronic two-level system the dynamics of the nuclear modes occur within potential wells, while the interactions with the bath cause the modes to behave in a damped random fashion. The time evolution of these effective nuclear modes is similar to that of a Brownian particle in a potential well; therefore the nuclear modes are called Brownian oscillators.

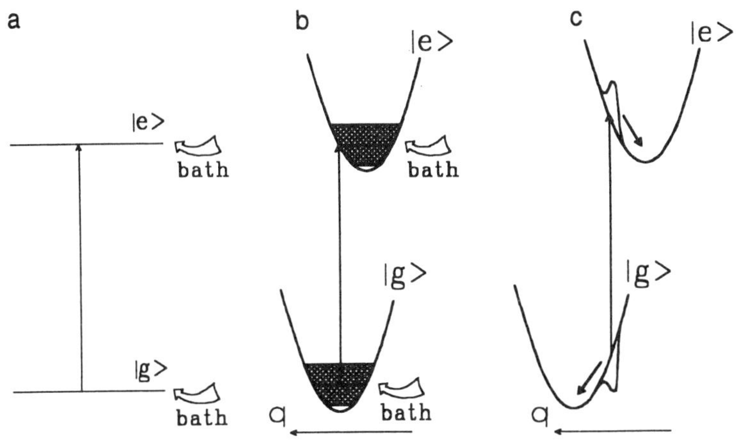

Figure 5 Impression of the Brownian oscillator model for non-Markovian optical dynamics. Both damping of the two-level quantum coherence (b) and relaxation (c) are described.

In Fig. (5) is shown how such displaced Brownian oscillators affect the optical response. Due to the fluctuations (damping) of the oscillatory movement, the width of the electronic transition increases. Due to the displacement of the potential surfaces, a nonequilibrium situation is created upon excitation with respect to the nuclear coordinate. When an optical pulse excites a fraction of the solute molecules from the ground state to the excited state, a "hole" is created in the ground state distribution of the oscillator, and a "particle" in the excited state. The subsequent evolution of the "particle"and the "hole" leads to a red or a blue shift of the electronic transition frequency, depending on their position relative to the equilibrium minimum of the potential wells.

Since the coupling between the molecular system and the heat bath is mediated now by the Brownian nuclear mode, all dynamical optical properties can be evaluated from the motions of this mode, which is subject to random forces from the heat bath. Just as for the stochastic model, this is treated in a classical, non-quantum mechanical way. The equation of motion for the oscillator can be expressed in terms of generalized Langevin equations for coordinate q and momentum p, which in turn can be rewritten as an equation of motion for a correlation function $M(t)$. Optical properties such as the shift of the electronic transition frequency after optical excitation or dephasing of an electronic coherent superposition state can then be evaluated in terms of this dynamical (normalized and real) correlation function, with $M(0)=1$ and $M(\infty)=0$.

The generalized Langevin equation for $M(t)$ is [29,30]:

$$\frac{d^2M(t)}{dt^2} + \int_0^t d\tau\, \gamma(t-\tau)\frac{dM(\tau)}{d\tau} + \omega^2 M(t) = 0 \qquad \textbf{(20)}$$

where ω is the frequency of the oscillator and $\gamma(t)$ is a non-Markovian friction kernel. This kernel is related to the correlation function of the random force $f(t)$ that is exerted by the heat bath on the oscillator, which can be expressed in a fluctuation-dissipation relation:

$$\langle f(t)f(0)\rangle = \gamma(t)\,[n(\omega)+\frac{1}{2}] \qquad \textbf{(21)}$$

where $n(\omega)$ is the Bose-Einstein thermally averaged occupation number of the oscillator. If the thermal motions of the bath are assumed to be fast compared to the Brownian oscillator dynamics, the friction of Eq. (20) is instantaneous, and the Markovian limit of the MBO model is obtained: $\gamma(t)=2\gamma\delta(t)$. In this limit Eq. (20) can be solved analytically.

Interesting is the case of a strongly overdamped oscillator ($\gamma \gg \omega$). The correlation function then has a particularly simple form [29,30]:

$$M(t) = \exp(-\Lambda t) \qquad \textbf{(22)}$$

Here Λ is the inverse correlation time of the oscillator, which is defined as $\Lambda=\omega^2/\gamma$. Just as for the stochastic model, the optical response can be described in terms of a line shape function g(t). When the energy of the oscillator is smaller than the thermal energy of the heat bath ($\hbar\omega \ll$ kT), this line shape function is:

$$g(t) = i\frac{\hbar\Delta^2}{2kT\Lambda}[1-e^{-\Lambda t}] + \frac{\Delta^2}{\Lambda^2}[e^{-\Lambda t}+\Lambda t-1] \qquad \textbf{(23)}$$

where the Δ is a measure of the frequency range over which transition probability is allowed between the two potential curves that are displaced by an amount d:

$$\Delta^2 = \omega^2 d^2 [n(\omega)+\tfrac{1}{2}] \qquad \textbf{(24)}$$

This is treated more extensively in Refs. 28-30.

The result of the Brownian oscillator model in the overdamped limit reads as an straightforward extension of the stochastic model, although the basic concepts are completely different. The line shape function Eq. (23) is identical to the stochastic line shape function Eq. (15), except for the imaginary part. We will now finish this paper by discussing how that affects the (non-)linear response compared to the analysis of the previous section.

The photon echo intensity relaxes as:

$$[R_3(t,\tau)]^2 = \exp[-2\,Re\{2g(t)+2g(\tau)-g(t+\tau)\}] \tag{25}$$

This is identical to the expression that occurs in Eq. (16)! Apparently the imaginary part of the lineshape function, which is connected to dynamic level shifts, is unimportant for the photon echo decay. The reason for this behaviour is that only coherent superposition states are involved in the generation of the echo signal, and that an inversion of phase occurs when the second pulse is applied. Only fluctuations (damping) are then important for the decay of the signal.

So, the fit of the echo trace of Fig. (1) applies to the Brownian oscillator model as well, with the same values for the parameters Δ and Λ. For the chirped experiment this is not completely true. The optical fields interact with the sample for several tens of femtoseconds, and populations in the ground and excited states are propagated for finite times. As a consequence, the calculated result of the Brownian oscillator model is slightly different from that of the stochastic model. In fact, it fits the data somewhat better. This is shown in Fig. (4) as traces (a) and (b).

However, the main difference between both models is in linear response. In the stochastic treatment, where only fluctuations occur, the emission spectrum is identical to the absorption spectrum. In liquids this is usually not observed, the emission is red shifted compared to the absorption. We will now calculate the absorption and emission spectra without any kind of additional fitting of the dynamics, with the parameters from the femtosecond photon echo experiment.

The room temperature absorption and emission spectra of resorufin in DMSO are displayed in Fig. (6). It is clear that many vibronic transitions contribute to these spectra. Optical dephasing leads to a line width of about 400 cm^{-1} for the individual absorption and emission lines in the vibronic manifold. The influence of the dynamic level shifts is evident from the observed difference between the absorption and emission maxima. The steady-state Stokes shift is 114 cm^{-1} for this particular system. The absorption spectrum $S_A(\omega)$ can be written as:

$$S_A(\omega) = 2\,\omega\, Re\{\int_0^\infty dt \exp[i(\omega-\omega_{21})t]\exp[-g(t)]\} \tag{26}$$

where ω is the frequency of the absorbed light and ω_{21} is the transition frequency of the electronic two-level system, which is centered in between the absorption and the emission maxima. The line shape function $g(t)$ is given by Eq. (15) for the stochastic model and by Eq. (23) for the Brownian oscillator model. The steady state emission spectrum $S_F(\omega)$ is:

$$S_F(\omega) = 2(\omega)^3\, Re\{\int_0^\infty dt \exp[i(\omega-\omega_{21})t]\exp[-g^*(t)]\} \tag{27}$$

where ω is the frequency of the emitted light and $g^*(t)$ is the complex conjugate of $g(t)$. In the stochastic model, where g(t) is real, the absorption and emission spectra are practically identical. In the Brownian oscillator model they are mirror images of each other around the center frequency ω_{21}.

The dashed lines in Fig. (6) show the calculated absorption and emission spectra of resorufin in DMSO at room temperature. Again, just as in the calculation of the photon echo decay and the chirped four-wave mixing signal, a single overdamped oscillator was assumed to determine the optical response, with parameter values Δ = 41 THz and Λ = 27 THz. The different vibronic components were considered as separate transitions, with absorption intensities as indicated by the structure underneath the spectra. Both the positions of these vibronic lines and their relative intensities are known from a report on resorufin in a glass [31].

The enlarged part of Fig. (6) shows the origin region of the electronic transition of resorufin in DMSO in greater detail. The one overdamped Brownian oscillator does not only describe the absorption and emission spectrum very well, but also, within 10% accuracy, the steady-state Stokes shift. The oscillator model further predicts that this time-dependent shift occurs typically in about the correlation time Λ^{-1}. This has yet to be tested.

CONCLUSIONS

We have shown that the ultrafast optical dynamics in a liquid system should be treated in a non-Markovian way. In particular the combination of ultrafast photon echo experiments and chirped four-wave mixing clearly demonstrates that the conventional picture of homogeneous and inhomogeneous broadening fails. The stochastic model describes most optical features very well. When the influence of dynamic level shifts has to be considered the Brownian oscillator model is better suited, since complex line shape functions are then required. With only two parameters we were able to fit optical experiments that probed these molecular systems on time scales from femtoseconds up to steady-state.

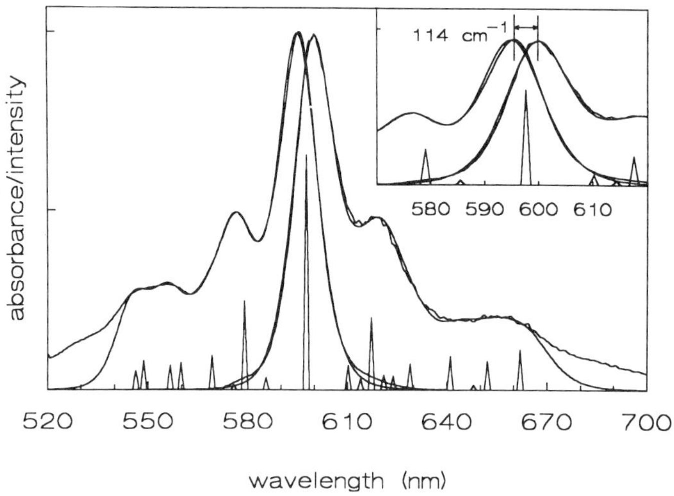

Figure 6 Absorption and emission spectra of resorufin in DMSO. Solid lines are measured; dotted lines are calculated. The inset shows an enlarged view of the origin region and the Stokes shift.

REFERENCES

1. J.A. Valdmanis and R.L. Fork, IEEE J. Quant. El. **QE-22**, 112 (1986).
2. P.F. Curley, C. Spielmann, T. Brabec, F. Krausz, E. Wintner and A.J. Schmidt, Opt. Lett. **18**, 54 (1993).
3. J.-M. Halbout and D. Grischkowsky, Appl. Phys. Lett. **45**, 1281 (1984).
4. N. Bloembergen, *Nonlinear Optics*, Benjamin, New York (1965).
5. F. Bloch, Phys. Rev. **70**, 460 (1946).
6. L. Allen and J.H. Eberly, *Optical resonance and two-level atoms*, John Wiley, New York (1975).
7. R. Loudon, *The quantum theory of light*, Clarendon Press, Oxford (1983).
8. K. Duppen and D.A. Wiersma, J. Opt. Soc. Am. B **3**, 614 (1986).
9. S. Schmitt-Rink, D.S. Chemla and H. Haug, Phys. Rev. B **37**, 941 (1988).
10. M. Lindberg and S.W. Koch, Phys. Rev. B **38**, 3342 (1988).
11. J.R. Kuklinski and S. Mukamel, Phys. Rev. B **44**, 11253 (1991).
12. N. Wax, ed., *Selected Papers on Noise and Stochastic Processes*, Dover, New York (1954).
13. E.T.J. Nibbering, K. Duppen and D.A. Wiersma, J. Chem. Phys. **93**, 5477 (1990).
14. E.T.J. Nibbering, K. Duppen and D.A. Wiersma, p. 471 in *Ultrafast Phenomena VII*, C.B. Harris, E.P. Ippen, G.A. Mourou and A.H. Zewail, eds., Springer Verlag, Berlin (1990).
15. E.T.J. Nibbering, D.A. Wiersma and K. Duppen, Phys. Rev. Lett. **66**, 2464 (1991).
16. E.T.J. Nibbering, D.A. Wiersma and K. Duppen, Phys. Rev. Lett. **68**, 514 (1992).
17. J.-Y. Bigot, M.T. Portella, R.W. Schoenlein, C.J. Bardeen, A. Migus and C.V. Shank, Phys. Rev. Lett. **66**, 1138 (1991).
18. R. Kubo, p. 23 in *Fluctuation, Relaxation and Resonance in Magnetic Systems*, D. ter Haar, ed., Oliver & Boyd, Edinburgh (1962).
19. J.R. Klauder and P.W. Anderson, Phys. Rev. **125**, 912 (1962).
20. N.G. van Kampen, *Stochastic Processes in Physics and Chemistry*, North Holland (1981).
21. H.J. Risken, *The Fokker-Planck Equation*, Springer Ser. Synergetics **18**, H. Haken, ed., Springer, Berlin (1984).
22. R. Kubo, M. Toda and N. Hashitsume, *Statistical Physics II, Nonequilibrium Statistical Mechanics*, Springer, Berlin (1985).
23. M.C. Wang and G.E. Uhlenbeck, Rev. Mod. Phys. **17**, 323 (1945).
24. C.H. Brito-Cruz, R.L. Fork, W.H. Knox and C.V. Shank, Chem. Phys. Lett. **132**, 341 (1986).
25. K. Duppen, F. de Haan, E.T.J. Nibbering and D.A. Wiersma, Phys. Rev. A **47**, 5120 (1993).
26. T. Tokizaki, Y. Ishida and T. Yajima, Opt. Commun. **71**, 355 (1989).
27. E.T.J. Nibbering, K. Duppen and D.A. Wiersma, J. Photochem. Photobiol. A: Chem. **62**, 347 (1992).
28. E.T.J. Nibbering, D.A. Wiersma and K. Duppen, Chem. Phys., submitted for publication.
29. Y.J. Yan and S. Mukamel, J. Chem. Phys. **89**, 5160 (1988).
30. Y.J. Yan and S. Mukamel, Phys. Rev. A **41**, 6485 (1990).
31. R. van den Berg and S. Völker, Chem. Phys. **128**, 257 (1988).

ULTRASHORT COHERENT EXCITATIONS IN SEMICONDUCTORS

Walter E. Bron

Department of Physics
University of California, Irvine
Irvine, California 92717 USA

I. INTRODUCTION

The origin of this line of research was marked by the realization that optical four-wave mixing techniques can be used to generate coherent excitations in matter[1]. The four-wave mixing technique of particular interest here is "Coherent Anti-Stokes Raman Scattering" (CARS) and its time-resolved (TRCARS) version[2]. A semiclassical theoretical description of these nonlinear processes has been presented by Placzek[3].

In this treatment the optical polarizability, α is expanded, such that the leading term in a Raman active medium is written as

$$\alpha = \alpha_0 + (\partial\alpha/\partial Q_v)Q_v + \cdots, \tag{1}$$

in which Q_v is the coordinate corresponding to some excitation of the solid, and α_0 refers to a static polarizability if one is present.

An electromagnetic field, E, can interact with the polarization. The pertinent term in the interaction Hamiltonian, H_1, corresponding to the lowest-order Raman process, is

$$H_1 = \frac{1}{2}\frac{\partial\alpha}{\partial Q_v}Q_v E^2. \tag{2}$$

It follows that as a result a force,

$$F = -\frac{\partial H_1}{\partial Q_v} = \frac{1}{2}\frac{\partial\alpha}{\partial Q_v}E^2, \tag{3}$$

acts on the medium. Moreover, under the action of both the excitation and the electromagnetic field, a polarization, P, is induced, namely,

$$P = -\frac{\partial H_1}{\partial E} = N\frac{\partial\alpha}{\partial Q_v}Q_v E, \tag{4}$$

Coherent Optical Interactions in Semiconductors
Edited by R.T. Phillips, Plenum Press, New York, 1994

in which N is the number density of the excited modes. We shall demonstrate shortly that a part of the CARS technique generates coherent states[4] which most closely approximate the classical limit. For example, if a coherent phonon state is generated, it will most closely approximate a harmonic oscillator. In a number of the experiments discussed below, it is observed that laser generated coherent phonon and polariton states, decay at some rate, Γ, even after the laser excitation is turned off. We arrive, therefore, at a picture of the excitation of the solid as a series of damped, driven oscillators, Q_v, of mass m, and write an equation of motion

$$\ddot{Q}_v + 2\Gamma\dot{Q}_v + \omega_v^2 Q_v = \frac{1}{2m}\frac{\partial\alpha}{\partial Q_v}E^2. \tag{5}$$

As we shall discover below, under coherent excitation, $\langle Q\rangle$ becomes the coherent amplitude, $1/\Gamma$ is related to the dephasing time T_2, and ω_v is the resonant frequency of the oscillator.

In the actual experiment, an electromagnetic field is chosen which contains two frequency components, ω_ℓ and ω_s, produced by two well-defined synchronously pumped coherent laser beams, propagating with wavevector $\vec{k}_\ell$ and $\vec{k}_s$. The em field amplitude due to these beams is

$$E(x,t) = \frac{1}{2}E_\ell \exp[i(\omega_\ell t - \vec{k}_\ell \cdot \vec{x})] + \frac{1}{2}E_s \exp[i(\omega_s t - \vec{k}_s \cdot \vec{x})] + c.c., \tag{6}$$

in which $\vec{x}$ is the position vector and t is the temporal coordinate. It is often the case that the available ultrashort pulsed lasers have output frequencies limited to near, or in, the visible spectrum. Clearly if the goal is to produce an effective excitation in the far infrared, then $\omega << \omega_\ell, \omega_s$. In contrast, excitations in the ultraviolet require that $\omega >> \omega_\ell, \omega_s$.

In order to solve Eq. 5 for Q_v, we use the trial function,

$$Q_v = \frac{1}{2}q_v \exp(i\omega t) + c.c., \tag{7}$$

and pick out the terms which are consistent with the inequalities in ω_ℓ, ω_s and ω, i.e.,

$$(\omega_v^2 - \omega^2 - i2\Gamma\omega)q_v \exp(i\omega t) = \frac{1}{2m}\frac{\partial\alpha}{\partial Q_v}E_\ell E_s \exp[i(\omega_\ell \pm \omega_s)t - (\vec{k}_\ell - \vec{k}_s)\cdot\vec{x}]. \tag{8}$$

Note that the temporal terms in Eq. 8 requires that $\omega = \omega_\ell \pm \omega_s$. If $\omega_\ell = \omega_s$ and $\vec{k}_\ell = \vec{k}_s$, the result is the well known case of two-photon absorption (TPA) to which we shall return from time to time. For the present purposes we limit discussion to the case $\omega_\ell \neq \omega_s$, but $\omega_\ell - \omega_s = \omega_v$ with ω_v corresponding to Raman active phonon (polariton) modes.

Solving Eq. 8 for Q_v yields

$$Q_v = \frac{1}{2m}\frac{\partial\alpha/\partial Q_v}{[\omega_v^2 - (\omega_\ell - \omega_s)^2 + i2\Gamma(\omega_\ell - \omega_s)]}E_\ell E_s \exp[-i(\omega_\ell - \omega_s)t - (\vec{k}_\ell - \vec{k}_s)\cdot\vec{x}]. \tag{9}$$

It is clear from Eq. 9 that a strong resonant excitation can occur if $\omega_\ell - \omega_s$ equals the frequency of a Raman active mode, ω_v, providing that the phase-matching condition $\vec{k}_v = \vec{k}_\ell - \vec{k}_s$ can be met. This nonlinear process is termed Coherent Raman Excitation (CRE).

It is most important to note that the range of frequencies and wavevectors allowed for the excited mode is highly limited by the wavevectors of the incident beams, the degree of focus, and the uncertainty of ω_ℓ and ω_s, etc. But these effects are very small. Thus, unlike incoherent spontaneous Raman or stimulated Raman scattering, the phase space available to CRE is severely limited. Consequently, the occupation probability of the excited states may be quite large even with modest laser intensities. The resultant "coherent phonon state" does indeed resemble most closely a classical harmonic oscillator and is the basis for Eq. 5.

It is this coherent phonon state which is of particular interest in the investigation reviewed here. In real media there always exists the possibility that the coherent state somehow scatters from some component of the medium. For example, a vibrational mode may scatter at crystal surfaces, bulk imperfections, electronic carriers, or through phonon-phonon interactions, etc. During the scattering process the wavevector of a component of the coherent state may *change*. The excitation is said to "dephase" with a characteristic rate of $2/T_2$.

The discussion so far has neglected to explicitly include the fact that the laser output is not cw but rather consists of trains of ultrashort (femto-to picosecond) duration pulses. If the durations of the exciting laser pulses Δt_ℓ and Δt_s are short compared to the relaxation time of the coherent excitation and to the time between successive laser pulses, then the coherent intensity $|\langle Q_\nu \rangle|^2$ can be observed to decay, through a coherent Raman interaction between the excitation and a time delayed third (pulsed) "probe" laser field of amplitude E_3. This process is called TRCARS, for time resolved coherent anti-Stokes Raman scattering.

Specifically, the presence of a coherent excited state causes an oscillatory polarizability at frequency ω_v, as indicated in Eqs. 1-5 which can, in turn, interact with a synchronized probe laser beam at frequency ω_3 to yield a non-linear polarization P^{NL}

$$P^{NL} = \chi^{(3)} E_\ell E_s E_3, \tag{10}$$

where $\chi^{(3)}$ is the total third-order non-linear susceptibility which, in general, contains contributions from the response of the electronic, $\chi_E^{(3)}$, and vibrational $\chi_R^{(3)}$ components. The polarization acts as a source term in Maxwell's equations to produce a strong coherent output beam with $\omega_4 = \omega_3 + (\omega_\ell - \omega_s)$ provided that $\vec{k}_4 = \vec{k}_\ell - \vec{k}_s + \vec{k}_3$ can be satisfied. It is usual to set $\omega_3 = \omega_\ell$ by taking a small component of the incident "ℓ" laser beam. Therefore, in practice, $\omega_4 = 2\omega_\ell - \omega_s$. The corresponding energy and wavevector diagrams are shown in Figs. 1 and 2. It is clear from the figure that the interaction leads to CARS. Accordingly, we identify ω_{AS} with ω_4 and $\vec{k}_{AS}$ with $\vec{k}_4$, and $\Gamma \propto 2/T_2$.

Figure 2 illustrates yet another property of the CARS technique. In this figure, the resonant coherent excitation is taken to be a lattice vibration (phonon, polariton) with energy $\hbar\omega_v = \hbar\omega_\ell - \hbar\omega_s$, and wavevector $\vec{q}_v$. It is, however, not necessary that the coherent excitation be resonantly excited. As Fig. 1 clearly illustrates, four wave mixing can be satisfied by transitions to virtual states. Polarization of bound electrons due to the mixing of ω_ℓ and ω_s is a case in point. This excitation is always present whether $\omega_\ell - \omega_s = \omega_v$ or not.

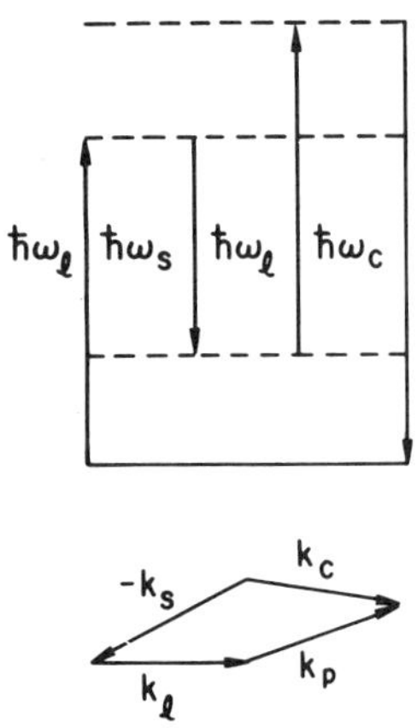

Fig. 1 Energy levels and wavevector system for a non-resonant "virtual" Raman polarization of bound carriers.

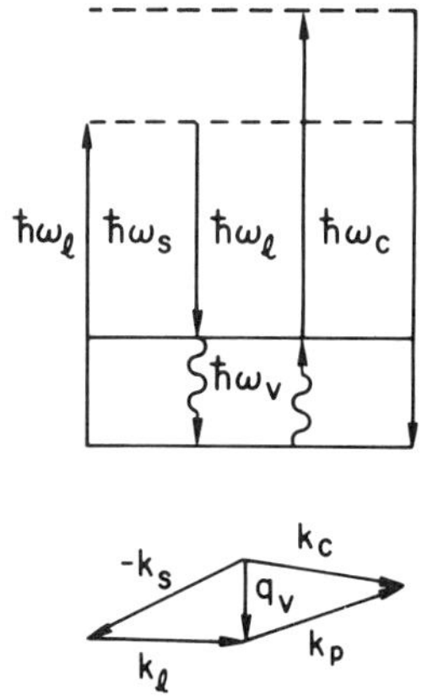

Fig. 2 Energy levels and wavevector system for a resonant excitation of a low frequency Raman active vibrational mode. In both Fig. 1 and 2 energy and momentum (k-vector) conservation is obeyed. Both cases are called coherent anti-Stokes Raman scattering (CARS).

In an illustrative case, to be discussed in some detail below, the nonresonant component is represented by the polarization of bound electrons and the resonant component is represented by the generation, and subsequent decay, of longitudinal optical (LO) phonons in simple compound semiconductors such as GaP. If we add to this picture the coherent, synchronous, time delayed, probe beam E_p, then the total nonlinear polarization of Eq. 10 becomes

$$P^{NL} = \chi^{(3)} E_\ell E_s E_p = (\chi_E^{(3)} + \chi_R^{(3)}) E_\ell E_s E_p \tag{11}$$

It follows from the above discussion and Eqs. 1, 3, 4, and 5, that the intensity of the TR-CARS signal, as a function of the delay time between the pump and the probe laser pulses, can be written as[5]

$$I(\Delta t) = AS(\Delta k) \int_{-\infty}^{+\infty} dt |E_p(t - \Delta t)[NR_a Q(t) + 3\chi_E^{(3)} E_\ell(t) E_s^*(t)]|^2, \tag{12}$$

where

$$S(\Delta k) = \frac{\sin^2(\Delta k L/2)}{(\Delta k L/2)^2}, \tag{13}$$

and

$$\Delta k = k_4 - [k_\ell + k_p - k_s], \tag{14}$$

and

$$\ddot{Q} + 2\Gamma\dot{Q} + \omega_v^2 Q = \frac{1}{2m}\frac{\partial\alpha}{\partial Q_v}E^2 = R_a m^{-1} E_\ell(t)E_s^*(t) - \frac{4\pi}{\epsilon_x}\chi^{(2)}E_\ell(t)E_s^*(t). \qquad (15)$$

In these expressions R_a is the Raman tensor which is related to $\chi_R^{(2)}$, and the coherent amplitude $\langle Q \rangle$ of the excited phonon packet is given by Eq. 15. The wavevector mismatch is given by Δk; it is set experimentally to approximately zero (hence, $S(\Delta k) \approx 1$). Finally, Δt is the temporal delay between the pulses of the probe laser, p, and those of the lasers ℓ and s; m is the reduced lattice mass, N is the number of primitive cells per unit volume and $A = 2\pi\omega_4^2 L^2/c\epsilon_4$. In the expression for A, c is the speed of light, L is the effective length within the medium over which spatial overlap of the three-wave mixing exists, and ϵ_4 is the dielectric constant of the medium at ω_4.

It is perhaps worthwhile to emphasize immediately the special properties of the coherent phonon (and polariton) state. This state produced through coherent Raman excitation (CRE) of phonon (polariton) distributions which is essentially monochromatic, possesses known wavevectors (magnitude and direction), belongs to specified polarization branches, and (at least initially) possesses fixed relative phase relationships, and exist at a point in a solid specified by the overlap volume of the two laser beams which are required for CRE. Clearly, such unique specificity is rarely available in other experimental techniques.

That brings to a close a brief discussion of the components of a theoretical basis which describes the experimental observables; namely, the response of condensed matter which has been coherently excited by ultrashort duration optical pulses. The application of the experimental techniques to the excitation and dephasing of various physical phenomena is the major remaining task of this review. However, before proceeding to the physics of optical excitations in solids, it is necessary to at least introduce the very special laser system used to carry out the experimentation.

II. EXPERIMENTAL BACKGROUND

With but one exception the experimental techniques of importance here involve anti-Stokes Raman scattering using pulsed laser sources. These sources are applied to produce ultrashort coherent pulses with durations in the picosecond and/or femtosecond regime. Again, the literature which has accrued on various applicable laser sources is very large, and even a casual perusal of current scientific journals in the field finds ever more new and more sophisticated methods in every new issue of the journals. I, therefore, limit the following to a very brief review of mostly that apparatus developed in my research group, to accomplish the research reviewed here.

The basic experimental apparatus necessary to conduct TR-CARS experiments is illustrated in Fig. 3. Mode-locking is a technique to phase order the various oscillating modes of a laser. This technique together with inhomogeneous broadening of certain gain media, permits the formation of orderly pulse trains of very short duration pulses.(6) The duration of the output pulses is about 100 ps. The two tunable dye lasers, are passively mode locked using cavity lengths which are matched to that of the "mode-locked laser". The output of the Nd:YAG mode-locked laser is doubled to 532 nm and is split in half so as to synchronously pump two dye lasers.

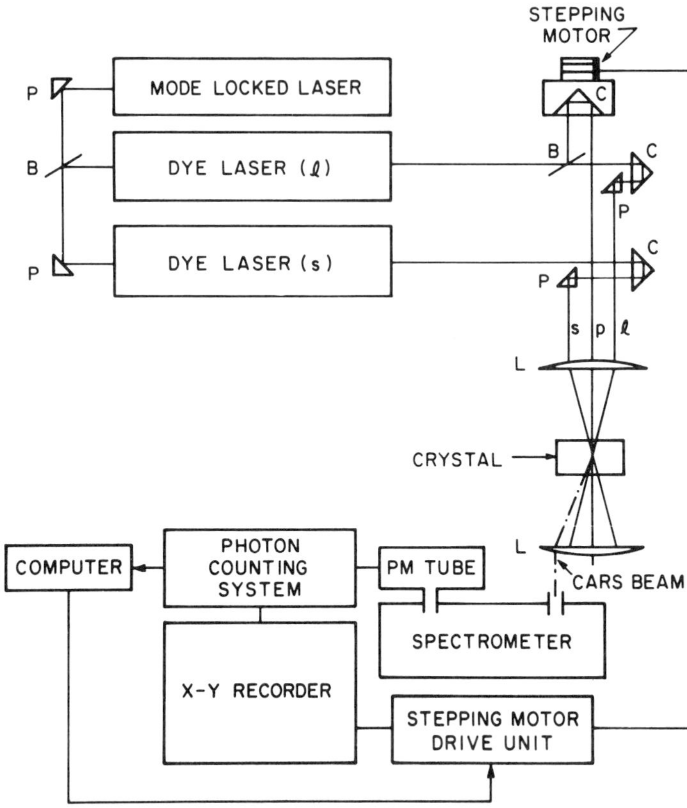

Fig. 3. Basic apparatus for TRCARS four-wave mixing of the incident laser beams in the overlap volume inside some solid. B denotes a beam splitter, C a retroprism, L a lens, and P a turning prism.

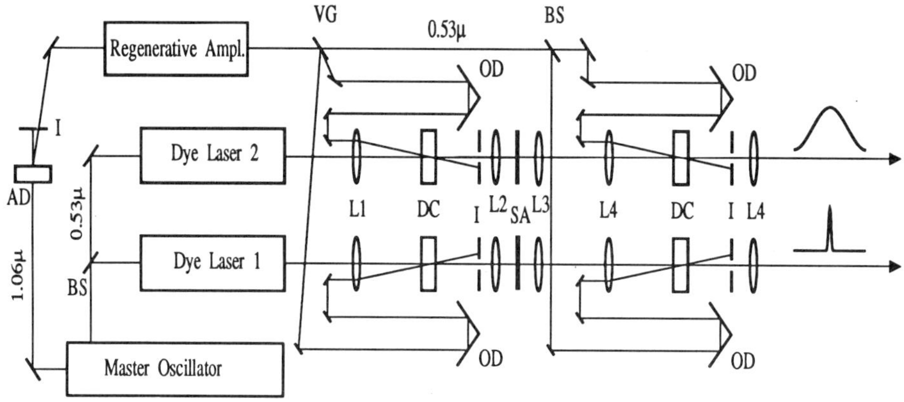

Figure 4. Sketch of a dual, synchronous, tunable amplifiable ultrashort pulsed laser system. The master oscillator (ML) is a mode-locked Nd:YAG laser, DL are two tunable dye lasers, RA is a regenerative Nd:YAG amplifier, AD is an acousto-optic deflector, BS are beam splitters, L are lenses, DC are flow through dye laser quvettes, SA saturable absorber, OD optical delays, I are irises and VG glass wedges

The pulse repetition rate is 76 MHz, and pulse duration of the output of the dye lasers is $\sim$ 7ps with an average output power of $\sim$ 100 mW. The output beams of the two dye lasers are tuned such that $\omega_\ell - \omega_s$ is equal to some Raman active excitation of the solid. In order to achieve mixing, the pulses which had been separated at the entrance of the two dye lasers, must be reunited such that the two pulses achieve maximum spatial and temporal overlap.

In a number of experiments the output power of the pulse laser needs to be increased by several orders of magnitude. For these purposes a regenerative amplifier (RA) was added to the basic laser system.(7) The amplifier output is doubled with a beta barium borate (BBO) crystal and reaches an energy per pulse of 0.8 mJ at 532 nm (45% conversion efficiency).

The doubled output of the amplifier is used to longitudinally pump dual, two-stage, dye amplifier chains. In this manner an amplification factor of 6×10^4 is reached in each of the two dye laser output beams.

For a number of experiments it is necessary to obtain trains of 50 to 100fs duration pulses. For this purpose we have constructed a synchronized femtosecond laser which can be substituted for one of the synchronized dye lasers discussed above. It should be understood that multi-wave mixing reaches its maximum output only when the maximum of all pulses overlap temporally and spatially. Therefore as we mix one pico- and one-femtosecond pulse, nonlinear excitation of the medium occurs only as long as the shortest duration laser pulse is present. Thus, for the generation of femtosecond excitation, one femtosecond laser is substituted for one of the picosecond dye lasers.

The femtosecond dye laser(8) is synchronously pumped by the doubled output of the Nd:YAG mode-locked master oscillator. The femtosecond laser consists of a standard six-mirror linear cavity dye laser. The laser delivers 60-80 fs pulses at approximately 630 nm wavelength using Rhodamine 590 tetrafluoroborate gain, and

DODCI absorber, dyes. The output power is 50 mW and the amplitude fluctuation of the output pulses is less than 8 percent. As with the case of the picosecond lasers discussed above, the output of the femtosecond laser can be synchronously amplified with the output of the regenerative amplifier.

III. COHERENT EXCITATION OF LO PHONONS AND THEIR SUBSEQUENT DEPHASING

As a first example of the dynamics of a coherent phonon state, we discuss the coherent longitudinal optic (LO) phonon in GaP. (Other examples of compound semiconductors are cited in the literature.) In this example, two laser pulse trains (ℓ, s) are tuned so that $\omega_\ell - \omega_s$ is equal to that of the near zone-center LO phonon (or polariton) in compound semiconductors. The excitation yields a coherent phonon state with wavevector $\vec{q}_v$, which dephases over periods of a few picoseconds,[9] or hundreds of femtoseconds.[10] After the laser pulse ceases, the excitation is observed to decay (more correctly dephase).

The subsequent temporal variation of the coherent LO phonon packet is measured with the TR-CARS technique. For this purpose a small part of the ω_ℓ beam is extracted and delayed by an optical delay line, so that pulses from it appear at the overlap volume at any delayed time, Δt, of between 0 and 0.3 ns, with a resolution of fractions of picoseconds. As described above, the delayed beam scatters from that component of the LO phonon state which remains coherent after Δt, producing thereby a coherent beam propagating along $\vec{k}_4$.

Figure 5 is an example of the typical temporal variation of the TRCARS signal intensity, $I(\Delta t)$, for near zone-center LO phonons with ω_v = 403 cm^{-1} and at an ambient temperature of 5 K for a GaP crystal with a carrier concentration of about 10^{16} cm^{-3}. The component of the signal from $\approx$ - 40 to $\approx$ +20 ps is primarily the result of non-linear excitation of the bound electronic system, i.e., through $\chi_E^{(3)}$ as described in section II. The full shape of this signal (dashed curve) can be obtained by slightly detuning $\omega_\ell - \omega_s$ away from ω_v.

The signal beyond 20 ps is a measure of the dephasing of the coherent intensity $\langle Q\rangle^2 \propto \exp[-2t/T_2]$, where $\langle Q\rangle$ is the coherent amplitude of the excited LO mode. The signal intensity is seen from Fig. 5 to decay exponentially over several orders of magnitude with a time constant $\tau = T_2/2$. Here $T_2/2$ is the traditional "dephasing time". The temperature dependence of τ, as observed over a range from 5 to 300 K, is shown in Fig. 6 (solid circles). It is generally accepted that the linewidth δv (in cm^{-1}) of the same Raman mode, ω_v, as observed from spontaneous Raman scattering is related to T_2 by $\delta v = (\pi c T_2)^{-1}$ provided that the Raman line is homogeneously broadened[11]. Linewidth measurements of the LO phonon (at $k \approx 0$) have been reported[12] through high-resolution Fabry-Pérot interferometry. The corresponding dephasing times are designated as BPTU in Fig. 6 (open circles). We have repeated these measurements using our samples and a standard double monochromator providing 0.15 cm^{-1} resolution. Phonon dephasing times are obtained by deconvoluting the spectral function of the instrument from the measured Raman spectrum assuming a Gaussian profile for the spectrometer function and a Lorentzian profile for the Raman line. The results are displayed as crosses in Fig. 6.

The time T_2 includes contributions from all dephasing phenomena, such as elastic scattering from surfaces, from impurities, from imperfections, and from electronic carriers, etc., and also includes phonon-phonon scattering which leads to phonon

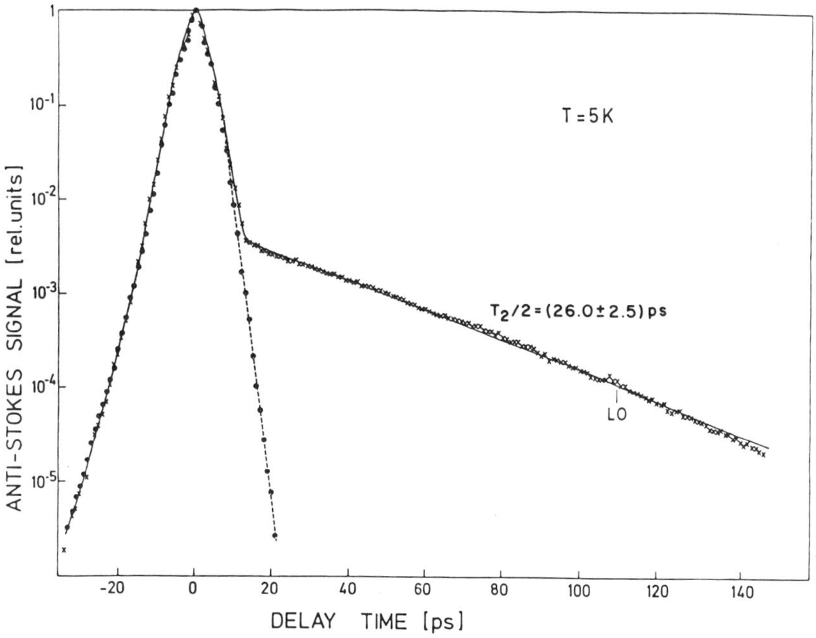

Figure 5. Coherent anti-Stokes Raman scattering signal as a function of the delay time. These results were obtained for a very high purity sample of GaP held at 5K. The measured dephasing time is 26.0 ± 2.5ps.

population decay. The population decay time is conventionally referred to as T_1. Since T_2 contains T_1 it follows that $T_2 \leq T_1$.

In the absence of any detailed knowledge of the strength of the various scattering processes it is not possible to calculate the magnitude of T_2, although it is possible to predict theoretically its temperature dependence. Contributions to the temperature dependence of T_2 can arise from scattering from thermally activated carriers and from various depopulation processes. We concentrate here on the latter processes. At low ambient temperature, such that $\hbar\omega_v \gg k_B T$, the spontaneous three-phonon decay of $k \approx 0$ optical phonons into two acoustic phonons (plus the corresponding recombination process) should dominate all other phonon-phonon interactions.

The governing equation of motion for a zone centered LO phonon, with resonant frequency ω_{LO}, nonlinearly coupled to two acoustic phonons gj and $g'j'$, and being driven by a pulsed laser with temporal profile, $F(t)$ can be written as

$$Q_{LO}(t) + \omega_{LO}^2 Q_{LO}(t) + \omega_{LO}\Sigma_{gjj'} V_3 \begin{pmatrix} 0 & q & q' \\ LO & j & j \end{pmatrix} Q_{gj}(t) Q_{g'j}(t) = F(t) \qquad (16)$$

in which V_3 is the (cubic) anharmonic coupling coefficient, Q_{LO} is the LO phonon normal coordinate and Q_{gj} is that of the acoustic phonons. Here g refers to the mode's wavevector, and j to the mode's dispersion branch.

Eq. 16 can be rewritten in terms of the self energy, $\tilde{\Pi}_{LO}(\tau)$,

$$Q_{LO}(t) + \omega_0^2 Q_{LO}(t) + 2\omega_{LO} \int_{-\infty}^{\infty} \Pi_{LO}(t-t') Q_{LO}(t') dt' = F(t) \qquad (17)$$

with the Fourier Transform of $\tilde{\Pi}_{LO}(\tau)$ being:

$$\tilde{\Pi}_{LO}(\omega) = \Delta_{LO}(\omega) + i\Gamma_{LO}(\omega) \qquad (18)$$

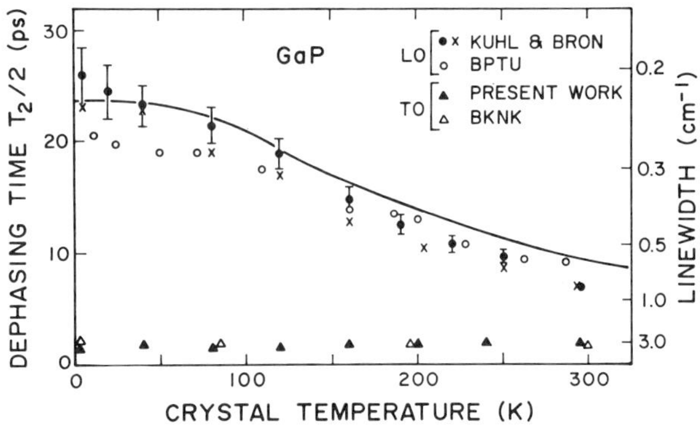

Figure 6. Temperature dependence of the dephasing time and the bandwidth of LO phonons in GaP. The solid line is a theoretical comparison to the data.

and $\Delta_{LO}(\omega)$ is the Kramers-Kronig transform of $\Gamma_{LO}(\omega)$. This is the well known result that the anharmonicity leads to a change of the LO phonon frequency (Δ_{LO}) and a broadening of its spectral profile (Γ_{LO}).

The solution for the damping rate is found to be

$$\Gamma_{LO}(\omega) = \frac{\pi}{2}\sum_{\vec{q}}\left|V_3\begin{pmatrix} 0 & \vec{q} & -\vec{q} \\ LO & j & j' \end{pmatrix}\right|^2 \delta(\omega - \omega_{\vec{q}j} - \omega_{\vec{q}j'})[n_{\vec{q}j} + n_{\vec{q}'j'} + 1]. \tag{19}$$

In Eq. 19 $\delta(\omega)$ is the joint density of states and the n_{qi} are the occupation numbers of the acoustic phonon states into which LO phonon decays. Equation 19 may be simplified to

$$\Gamma_{LO}(\omega) = \Gamma^0_{LO}[n_{qj} + n_{q'j'} + 1] \tag{20}$$

where Γ^0_{LO} is the damping rate of the LO phonons at T=0 K.

The topography of the dispersion relations for GaP[14] suggest that the zone-center LO phonon can only decay into two LA phonons each with energy $\hbar\omega_{LO}/2$ and wavevector $\vec{q}$ and $-\vec{q}$ so to maintain energy and momentum conservation. As we shall see in section IV on polariton decay, this restriction to the "half energy" LA phonons is not the general case. It is, however, the simplest illustrative case. Symbolically, we write the decay process as

$$LO(\omega, 0) \rightarrow LA(\omega/2, q) + LA(\omega/2, -q). \tag{21}$$

Thus in Eq. 19 $\omega_{\vec{q}j} = \omega_{\vec{q}'j'}$ and under thermal equilibrium at temperature, T,

$$n_{qj} = n_{q'j'} = n^T_{\vec{q},j} \tag{22}$$

where $n^T_{\vec{q}j}$ is determined from Bose-Einstein statistics to be

$$n_{\vec{q}j}(\omega, T) = [\exp(\frac{\hbar\omega}{k_B T}) - 1]^{-1}. \tag{23}$$

Expression 20, with its magnitude fit to the low-temperature data, appears as the solid line in Fig. 6. Comparison with the data shows indeed that the low-temperature dependence of T_2 is dominated by three-phonon anharmonic interactions, i.e., by T_1. Beyond this temperature additional processes, including higher order phonon-phonon interactions and scattering from thermally activated carriers, must also be considered. We conclude, therefore, that it is possible to measure the temperature dependence of the dephasing time of optical phonons and to account for the observation in terms of Bose-Einstein statistics.

Finally we note, that in these experiments the coherent optical phonons were purposely generated well within the bulk of the sample. As a consequence of the low group velocity of optical phonons,[14] it becomes possible to neglect boundary scattering during the limited experimental observation time ($\sim$ 200 ps). It is easy to show that elastic scattering at point imperfections is, unlike anharmonic decay, essentially independent of temperature. Thus, the effective dephasing time $T_2/2$ in the presence of both elastic and anharmonic decay becomes

$$\frac{2}{T_2} = \frac{1}{T_{imp}} + \frac{(1 + n' + n'')}{\tau_0}, \tag{24}$$

where T_{imp} is the dephasing time due to elastic scattering at point imperfections. The best fit to the data of Fig. 6 results in values for $T_{imp} = \infty$ and $\tau_0 = 24.6$ ps. Thus, impurity scattering does not play a role in optical phonon dephasing for samples with impurity content of $\leq 10^{16}$ cm^{-3}.

IV. COHERENT EXCITATION OF POLARITONS AND THEIR SUBSEQUENT DEPHASING

We shall return to the interaction between optical phonons and electronic carriers in section VI. Instead we discuss next the interaction of polaritons with thermal phonons, in the same sense as we have discussed the interaction of LO phonons with a thermal bath. Polaritons are excitations of coupled transverse optical (TO) phonons and the polarization from electromagnetic excitations traveling in a medium. The net result of this coupling is illustrated in Fig. 7 which is the dispersion relations for polaritons in GaP. For the smallest values of $\hbar ck$ (a measure of the crystal momentum) the properties of the polariton approach that of pure photons, whereas the properties approach those of TO phonon modes at high values of $\hbar ck$. The most interesting section of the polariton dispersion is clearly in the intermediate region, in which the polaritons can be considered to be a mixture of photons and phonons. We have determined the polariton lifetime in the coupled region in much the same way as the lifetime of LO phonons discussed above.

It should be noted from the outset that the restrictions imposed by the dispersion relation for the lifetime of LO phonons in GaP are somewhat relaxed for polaritons in GaP. Namely, the restriction on LO phonon decay into two LA phonons, each of half the energy of the LO phonon (and wavevector $\vec{k}$ and $-\vec{k}$) is lifted for polariton decay in GaP. Depending on the frequency of the polariton, ω_π, energy and momentum may be conserved, during polariton (π) decay into one LA phonon with wavevector near the X-point and one TA phonon near the minus X-point of the Brillouin zone, in addition to decay into two LA phonons each with energy $\omega_\pi/2$ and equal, but oppositely signed wavevector.

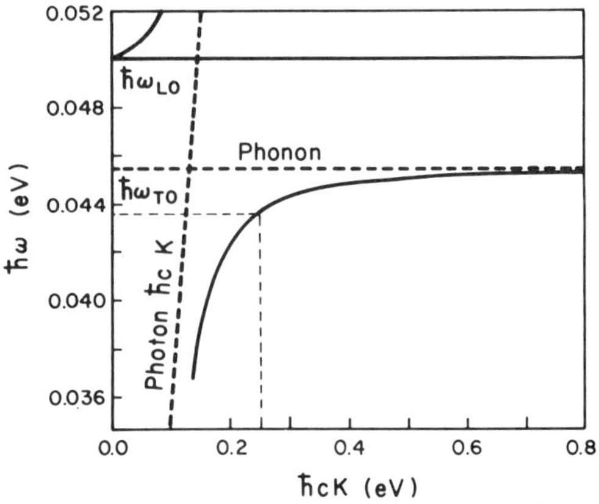

Figure 7. Dispersion relations for photons and LO, TO phonons and polaritons for very small values of the k-vector.

The experiment used to evaluate the more complicated dephasing of polaritons as compared to LO phonons, is described below. The existence of polariton decay into the TA(X) state has also played a major role in the discovery of anamously long-lived phonons to be discussed in section V.

The femtosecond laser system discussed in section II is adapted for the coherent generation of polaritons, and their lifetime is determined through the TR-CARS technique. In this case, the laser system contains one synchronously pumped tunable picosecond and one synchronously pumped linear cavity femtosecond laser.

A number of factors influence the effective spectral distribution of polaritons generated through CRE. Among these are the forward scattering geometry which limits the wavevector of the coherent polaritons (see insert to Fig. 8), and the narrow spectral bandwidth of the picosecond laser (~ 5 cm^{-1}) compared to that of the femtosecond laser ($\sim$45 cm^{-1}) means that the latter laser dominates the effective bandwidth of the CRE. In this experiment, the femtosecond laser peak frequency was fixed at 624 nm, and that of the picosecond laser is mixed with the femtosecond laser to produce various excitation frequencies. Once the distribution of polariton wavevectors is determined, the corresponding spectral distribution can be obtained through the polariton dispersion relations of GaP.[14,15] A typical result is illustrated in the insert of Fig. 9. In this example the central polariton wavevector is $q \sim 2 \times 10^3$ cm^{-1}, which corresponds to a spectral distribution peaked at $\sim$354 cm^{-1}. However, a 120 mm focal length lens is used to focus the ℓ, s and p beams into the sample, which results in an angular spread in the wavevector of the laser beams and, hence, in a spread in polariton wavevector of between $\sim 1.40 \times 10^3$ to $\sim 2.6 \times 10^3$ cm^{-1}. The corresponding range of polariton frequencies is from $\sim$330 cm^{-1} to $\sim$360 cm^{-1} (see insert to Fig. 9).

In order to avoid the generation of LO phonons (ω_{LO} $\sim$403 cm^{-1}) in addition to the polaritons, the bandwidth of the femtosecond laser was limited to $\sim$45 cm^{-1} which corresponds to a pulse duration of $\sim$300 fs. This is achieved through adjustments of two intracavity prisms in the femtosecond laser, through which it is also possible to

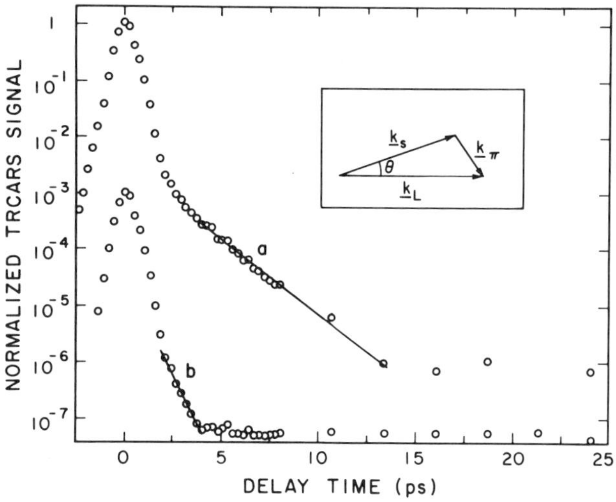

Figure 8. Plot a: Normalized TRCARS signal as a function of the delay of the probe pulse. Plot b: The residual TRCARS signal after the linear portion of plot "a" has been analytically subtracted. The inset depicts wavevector conservation between the ℓ and s incident laser beams and the polariton wavevector k_π. The ambient temperature is ~ 5K.

produce pulse durations of less than 70 fs (see section II above). The energy per pulse in the output of each laser was ~0.6nJ.

Figure 8 illustrates the temporal dependence of the TR-CARS signal amplitude at an ambient temperature of 5K. The TR-CARS signal shows three different components. The first component from ~0 ps $< \Delta t <$~2.5 ps is the response of the bound electrons in GaP through the third order nonlinear electronic susceptibility, $\chi_E^{(3)}$, (see section I above). The second component of the signal (~1.5 ps $< \Delta t <$~ 5ps), which occurs after the CRE is turned off, is nonlinear on a semilog plot, i.e., the dephasing of the polaritons in this time interval does not correspond to a single decay rate. The third component (~5 ps $< \Delta t <$ 12.5 ps) illustrates a clearly exponential decay corresponding to a polariton dephasing time $T_2/2 = 1.72 \pm 0.16$ ps. Analytically subtracting this third component from the total signal, uncovers a second exponential decay with a dephasing time $T_2/2 = 620 \pm 60$ fs. In order to investigate the temperature dependence of the dephasing of the polariton, these measurements are repeated at 50K intervals from 5K to 300K. As a result of the limiting (300 fs) temporal resolution inherent in the measurements it is possible to observe the dephasing of the faster (shorter $T_2/2$) decay channel only up to 200K. Figure 9 displays typical experimental results at 300K. At this temperature the dephasing time of the slower decay component is $T_2/2 = 630 \pm 50$ fs, whereas the dephasing time of the faster component is shorter than the limiting temporal resolution (see Fig. 9). The temperature dependence of $T_2/2$ and that of the polariton dephasing rate (in cm^{-1})$[\Gamma_\pi = (2\pi c T_2/2)^{-1}]$, for both decay channels, is indicated in Fig. 10. As noted earlier, at low ambient temperatures, such that the polariton energy $\hbar\omega_\pi >> k_B T$, three particle decay should dominate all other polariton-

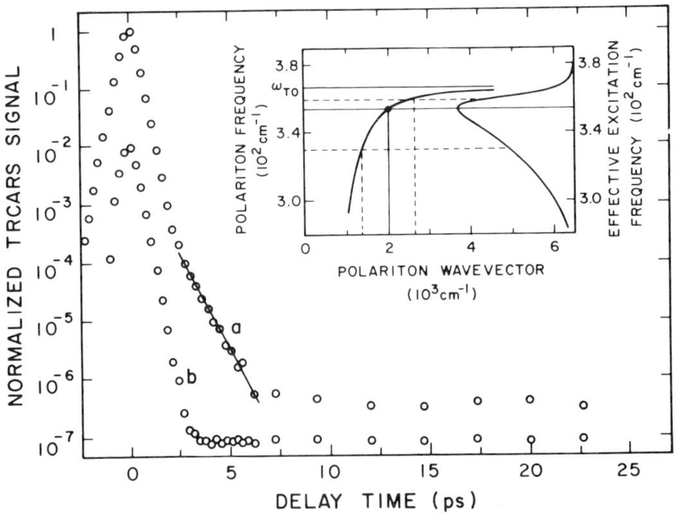

Figure 9. Plot a: Normalized TRCARS as a function of the probe delay. The ambient temperature is 300K. Plot b: Residual signal after the linear part of plot a has been analytically subtracted.

phonon interactions. We now set out to prove that in the present case the two possible decay channels, mentioned at the beginning of this section, are the source of the more complicated experimental observations. The two possible decay channels are written symbolically as $\pi(\omega_\pi, q \sim 0) \rightleftarrows LA(\omega_\pi/2, q') + LA(\omega_\pi/2, -q')$ and $\pi(\omega_\pi, q \sim 0) \rightleftarrows LA(\sim X) + TA(\sim -X)$, in which π refers to polaritons, and X refers to the X-point of the Brillouin zone. For simplicity, we refer to the first as the "half-energy" channel, and to the second as the "X-point" channel.

The governing equation of motion is the same as that for LO phonons (see Eq. 19) with ω_π substituted for ω_{LO}. The formal result is the same; namely

$$\Gamma(\omega_\pi) = \frac{\pi}{2}\sum_{\vec{q}} \left| V_3' \begin{pmatrix} 0 & \vec{q} & \vec{q}' \\ \pi & j & j' \end{pmatrix} \right|^2 \delta(\omega - \omega_{\vec{q}j} - \omega_{\vec{q}'j'})\{n_{\vec{q}j} + n_{\vec{q}'j'} + 1\}, \tag{25}$$

We performed two-parameter, least χ^2, fits of the data of Fig. 10 to Eq. 25. With n_{qj} obtained from Eq. 23 above. The two fitting parameters are Γ_0 and ω_1 with the constraint that $\omega_\pi = \omega_1 + \omega_2$. The fits to the data appear as the solid lines in Fig. 10. For the longer $T_2/2$ with ω_π = 354 cm^{-1}, the best fit is $\omega_1 = (118 \pm 16)$ cm^{-1}, $\Gamma_0 = 2.96 \pm 0.4$ cm $^{-1}$, $\omega_2 = (236 \pm 16)$ cm^{-1} with χ^2 =0.9. This result, when compared to the phonon frequencies at the X-point as obtained from neutron scattering $\{\omega_{TA(X)} = (107 \pm 3)$ cm^{-1} and $\omega_{LA(X)} = (249 \pm 4)$ cm$^{-1}\}$ identifies this as the "X-point" decay channel. Values of $\chi^2 \leq 1$ are obtained for ω_π over a restricted range of $\pm$ 6 cm^{-1} about 354 cm^{-1}.

For the shorter $T_2/2$ (Fig. 10) a best fit is obtained with ω_π = 344 cm^{-1}, ω_1 = (172 ± 26) cm^{-1}, $\Gamma_0 = (8.2 \pm 0.3)$ cm^{-1}, $\omega_2 = (172 \pm 26)$ cm^{-1}, and χ^2 = 0.5. For all values of $330 \lesssim \omega_\pi \lesssim\sim 348 cm^{-1}, \omega_1 = \omega_2 = \omega_\pi/2$ with $\chi^2 \lesssim 0.8$. This identifies the "half-energy" decay channel. The dispersion relations for GaP[14] permit only excitation of LA phonons for these values of $\omega_\pi/2$. It is obvious from these results

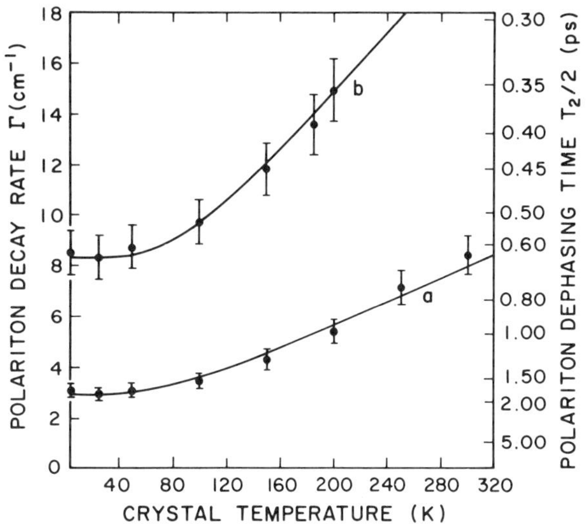

Figure 10. Temperature dependence of the polariton decay rate for the "half-energy" decay channel (b), and the "x-point" decay (a).

that the general case of the decay of excitations need not be limited to single decay channels. On the contrary, the general case has multi-channel decay. However, the analytical tools presented here can be used to ferret out the various contributing decay channels.

V. LONG-LIVED TRANSVERSE ACOUSTIC PHONONS

In the foregoing sections we have identified the decay channels of LO phonons and polaritons in GaP by analyzing the temperature dependence of the dephasing of the excitation. Although the TR-CARS technique used in these analyses is very accurate, it would be helpful to have available an independent experimental observation with which to check the decay channels. The obvious auxillary measurement is the spectral profiles of the acoustic phonon generated through LO phonon or polariton decay.

Unfortunately, the acoustic phonons are not Raman active (in first order) and thus, cannot be detected by TR-CARS. Vibronic sideband phonon spectroscopy (VSPS), which the author developed some time ago for the detection of heat pulses, was adapted to the present experiment in order to monitor the population of polaritons, and the LA and TA acoustic phonons formed as the result of their decay. In VSPS, electronic transitions (zero phonon luminescence), localized at some optically active impurity, are modulated by phonons or polaritons which can couple to the transition. As a result Stokes and anti-Stokes sidebands arise in much the same way as its radio frequency, FM, counterpart. In the presence of phonons an anti-Stokes sideband appears which reflects the phonon spectral distribution. In this way the concentration of polaritons and TA phonons were monitored as a function of a delayable probe laser pulse which triggered the zero phonon luminescence.

The first observations along these lines were reported in reference 16 and demonstrated that indeed, the excitation in GaP of LO phonons at 403 cm^{-1} lead to vibronic sidebands only at 201.5 cm^{-1}, i.e. only through the "half energy" decay. These early experiments are relatively crude, but nevertheless demonstrate the point. More recent experimentation, at better signal-to-noise ratios, on the decay of polaritons leads to more detailed results. Most importantly, temporal resolution on the nanosecond scale has been accomplished and is incorporated into the VSPS measurements of the polaritons in GaP.

As before, polaritons were coherently generated through CRE. However, in place of the probe laser of TR-CARS, another pulse laser is substituted to excite the vibronic luminescence via VSPS. The laser system contains two pulsed nitrogen lasers, one of which, together with two Rodamine 6G tunable dye lasers, is used to produce two laser beams with linewidth of 3 cm^{-1}, pulse duration of 5 ns, and 70 μJ energy per pulse. The two "yellow" output beams are tuned such that their frequency difference equals 354 cm^{-1} or 341 cm^{-1} and the beam directions are oriented to conserve crystal momentum when a polariton is generated through CRE. A second pulsed nitrogen laser is used to pump a dye laser containing coumarin 485 to excite a "green" VSPS zero-phonon luminescence at 535 nm. The source of the luminescence is known to result from an excitation of and subsequent decay of a bound exciton associated with nitrogen impurities in GaP.[17] The luminescence can be delayed electronically over a range of $\sim$100 ns with a resolution of $\sim$5 ns. The diameter of the focal spot of the various laser beams inside the sample was 100 μm. The output of the VSPS signal is collected and digitally stored in an optical multichannel analyser, OMA. The delay between the "yellow" lasers and the "green" lasers is separately monitored through a digital oscilloscope. The range of detection of the OMA was set to cover the frequency from that of polaritons to that of the TA phonons and to block out the very strong zero-phonon luminescence. The entire system is calibrated using standard reference lines. The sample is held to 4 to 5K, in order to obtain a sideband signal with the best signal to noise ratio. The multiple beam overlap volume should be as near as possible to the entrance surface of the crystal because of the strong absorption at the "green" frequency.

From the earlier work on polariton decay cited above, we expect $\omega_\pi \sim 341$ cm^{-1} polaritons to decay into two LA phonons with frequency $\omega_{LA} = \omega_\pi/2$. This result is indeed observed and is demonstrated in the top panel of Fig. 11 which shows sidebands only at the polariton frequency and at the $\omega_\pi/2$ frequency. In the current context we are more interested in the formation and decay of the TA phonon at the X-point of the Brillouin zone which is formed by the decay of the $\omega_\pi \sim 354$ cm^{-1} polariton. The corresponding experimental results are indicated in Fig. 14 which is a display of the VSPS signal amplitude vs. the phonon (and polariton) frequencies for various delays between the "yellow" pump lasers and the "green" probe laser pulses. The second highest panel of Fig. 11 corresponds to $\Delta t = 0$ and indeed indicates the presence of $\omega_\pi \sim 354 cm^{-1}$, the LA(X) phonon at $\sim$ 236 cm^{-1} and the TA(X) phonon at $\sim$ 118 cm^{-1}. Note, within a delay of < 10 ns, that both the polaritons and the LA phonons disappear which is consistent with their expected short lifetimes. All of these observations support the predictions of the earlier results on polariton decay channels.[10]

Of some particular note is the decay of the $\omega_\pi \sim 354$ cm^{-1} polariton into an LA phonon near the X-point of the Brillouin zone and on TA phonons near the minus X-point or vice versa. All of the phonons and polaritons, with one important exception, possess lifetimes in the picosecond or femtosecond regime. The important exception is the TA phonon near the X-point for which Figs. 11 and 12 lead to an observed decay of 66.5 ± 5 ns.

The existence of such long-lived TA phonon has been a source of speculation for nearly three decades. We have now observed a TA phonon near the minus (or positive) X-point which exhibits a lifetime which is 4 to 6 orders of magnitude longer than other phonons and polaritons in GaP. One possible reason for this difference in decay times may stem from variations in the density of states into which these various phonons and polaritons decay. However calculations of the single and joint density of phonon states(17) are not consistent with sharply different decay times. Another possibility is strong luminescence from the miriad excitonic transitions in GaP(17) which happen to be in the 118 cm^{-1} range and possess a recombination lifetime near 66 ns. The strong luminescence would have to be generated during the VSPS excitation, and would need to be present at all times, including the decay of the 341 cm^{-1} polaritons for which no such luminescence is observed (see top panel of figure 11). This luminescence was also not observed if we slightly detuned the excitation lasers away from 354 cm^{-1}.

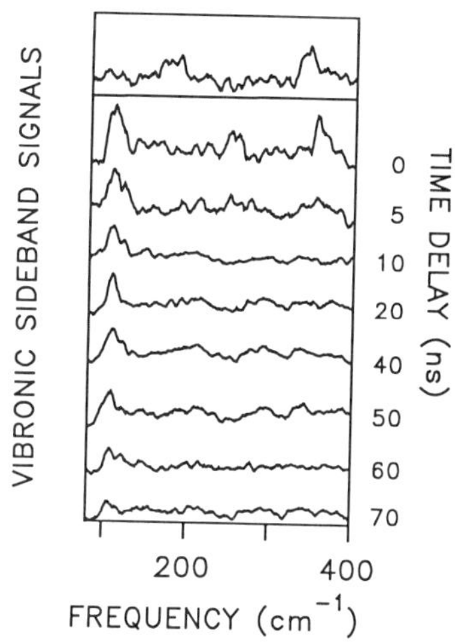

Figure 11. Vibronic sideband signal as a function of particle energy and the delay time of the probe laser pulse. The uppermost panel is the sideband obtained for $\Delta t = 0$ and polariton peak (341 cm^{-1}) and the "half energy" peaks. The remaining panels depict the decay of the original distribution of "X-point" formed through the original decay of 354 cm^{-1} polaritons.

As an alternative to the above explanation, Orbach and Vredevoe(18) have proposed that, in dispersive media, the reason for the anomalously long lifetime is that the TA phonons cannot readily decay by three-particle interactions and simultaneously obey energy and (crystal) momentum conservation. As a consequence, these phonons should possess anomalously long lifetimes.

At the time of Orbach and Vredevoe's paper the existence of such long-lived

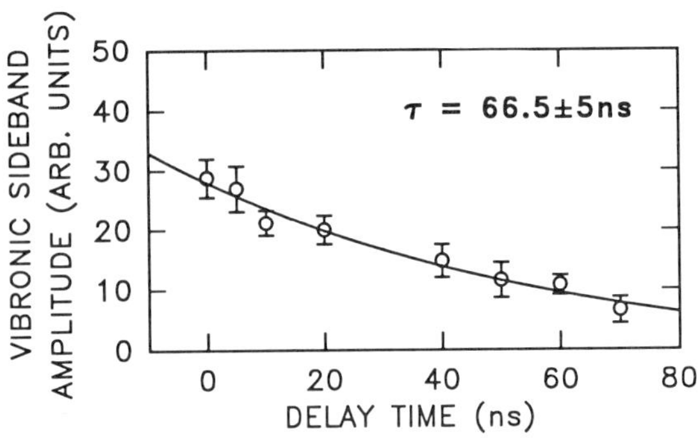

Figure 12. Compilation of the vibronic sideband signal following the slow decay of the TA phonons at the X-point.

phonons would have verified the simple, but critical, tenet that the phonon branches are in general dispersive and, in the absence of other scattering mechanisms, that the lowest order phonon-phonon anharmonic interaction, plus energy and momentum conservation, govern phonon lifetimes. Although most of these tenets have since been verified by other means, none of these means have unambiguously measured the lifetime of the long-lived phonons. This missing link in phonon physics has attracted, over time, the imagination of a number of investigators. Since long-lived phonons would also be useful as probes of various interactions in condensed matter, the continued hunt for these phonons has, at times, been referred to as the "Holy Grail" of phonon physics.

Thus it has been demonstrated that a long-lived species of TA phonons exists with wavevector near the X-point of GaP. The lifetime of these phonons is 66.5 $\pm$ 5 ns. This lifetime is much longer than that of LO ($\sim$ 26 ps), polaritons ($\sim$ 0.1 ps), and LA phonons. The relatively long life results from restrictions imposed by energy and momentum conservation.

VI. INTERACTION OF COHERENT LO PHONONS WITH A TWO-COMPONENT PLASMA

We return now to the TR-CARS technique to determine the effect of a two-component (electron-hole) plasma on the dephasing of near-zone-center LO phonons in GaP[20]. In this case, an electron-hole plasma is formed in a pure crystal through the decay of excited carriers after they had been produced through two-photon absorption (TPA) of incident pulsed laser radiation. This method of photoexcitation may produce a plasma which is nonstationary (i.e., varies in time) during the experimental observation period. It is, therefore, appropriate to conduct temporaly resolved observations to provide information on the dynamics of the interaction between the two-component plasma and the phonons. As has been described above, time-resolved coherent anti-Stokes Raman scattering (TR-CARS) is an outstanding tool for measurements of the dephasing rate of LO phonons and its dependence on ambient temperature. We have also shown that the dynamics of the LO phonon dephasing is influenced by the presence of high concentrations of acoustic phonons.[21]

Recently, we have applied TR-CARS to the study of the temporal evolution of the interaction of a two-component plasma and LO phonons in GaP.[20] The investigation provided, for the first time, direct observation of the transient dynamics of the plasma-phonon interaction. We observed an increase in the instantaneous dephasing rate of the LO phonons during the first 150 ps of the interaction. The dephasing rate increases as the incident laser irradiance increases (i.e., with increasing plasma density). Furthermore, we observe that the increase in the dephasing rate becomes negligible after 600 ps which suggests plasma transport out of the laser interaction volume.

The phonons and the two-component plasma are excited using the dual synchronously amplified picosecond laser system operating at 1 kHz (see section II) with outputs at the frequencies ω_ℓ and $\omega_s(\omega_\ell = 2.137$ eV and $\omega_s = 2.087$ eV). We add to this laser system another laser beam used to inject the two-component plasma into the mutual interaction volume of all the other laser beams. The difference in output frequencies of the two laser beams $(\omega_\ell - \omega_s)$ is resonantly tuned to the LO phonon frequency and a variably time delayed part of the ω_ℓ beam is used to probe the dephasing of the LO phonons; note that as usual the probe frequency $\omega_p = \omega_\ell$. An additional synchronized laser pulse with frequency ω_s is used to "inject" a nonstationary electron-hole plasma (NEHP) via two-photon absorption, TPA, $(2\omega_s)$ with an initial excess energy of 1.05 eV above the direct gap of 2.7 eV. (Nonstationary in the present context means that the effect of the plasma on the phonons varies with time.) The injection laser pulse arrives at the interaction volume at a fixed time after the excitation of the LO phonons. The NEHP density is varied by changing the incident irradiance of the injection laser pulse. The intensities of the ω_ℓ, ω_s, and ω_p beams are kept sufficiently weak to avoid additional TPA and intense LO phonon generation that could cause nonexponential dephasing of the LO phonons;[21] thus pulse energies for phonon excitation and probing are kept below 1 nJ, whereas the injection pulse energies are varied from 5 to 500 nJ. The four beams are focused into a high purity ($< 10^{16}N$ impurities per cm^3) GaP crystal maintained at 5 K. The experiment is carried out by alternately measuring the TR-CARS signal with and without the presence of the NEHP at a set of fixed probe delay times in the interval between - 50 and 200 ps, in order to minimize long-term laser fluctuations.

Figure 13 displays typical experimental results on the intensity of the TR-CARS signal as a function of the probe delay time for three values of the irradiance of the injection laser pulse. The upper curve, which represents the LO phonon dephasing without NEHP injection, clearly shows exponential dephasing. The lower two curves indicate LO phonon dephasing in the presence of the NEHP. The dip at ~26 ps delay is consistent with the timing of the injection pulse, but its origin is not yet understood. An increase in the dephasing rate, due to the presence of the NEHP, follows the injection pulse. The inset of Fig. 13 illustrates the difference between the natural logarithm of the TR-CARS signal observed with the NEHP and that observed without the NEHP as a function of delay time, and indicates the additional dephasing resulting from the plasma-phonon interaction. The additional dephasing, depicted in the inset of Fig. 13, represents a time-dependent interaction between the LO phonons and the NEHP.

We now present a synopsis of a theoretical model which we apply to the experimental observations. The interaction of the LO phonons and the NEHP can be described in terms of a coupled vibrational system consisting of the LO phonons and the electron-hole plasma interacting by way of a local polarization field. The equations of motion of such a system are given by

$$\ddot{W} + \omega_t^2 W + \Gamma \dot{W} = (e^*/M)E, \tag{26}$$

$$m_e \ddot{X}_e = -(1/\tau_e) m_e \dot{X}_e - \eta(\dot{X}_e - \dot{X}_h) + eE, \tag{27}$$

$$m_h \ddot{X}_h = -(1/\tau_h) m_h \dot{X}_h - \eta(\dot{X}_h - \dot{X}_e) - eE, \tag{28}$$

$$E + 4\pi \left(\frac{e^*}{M} E + \frac{e^*}{V_0} W + ne(X_e - X_h) \right) = 0, \tag{29}$$

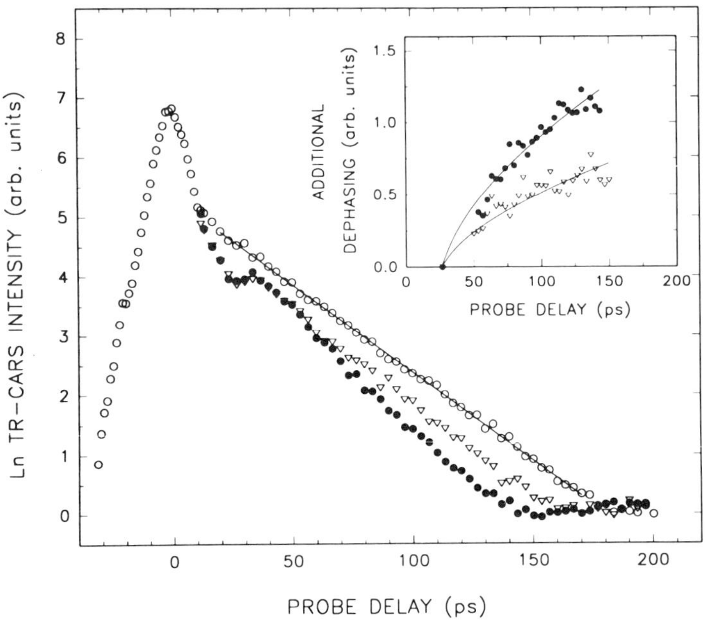

Figure 13. TRCARS signal as a function of probe delay for high purity GaP with no plasma present (open circles) and increasing presence of a nonstationary electron-hole plasma (open triangles and closed circles). The inset depicts the additional dephasing. The solid curves represent the theoretical prediction of the additional dephasing.

where W is the amplitude of the relative interatomic lattice displacement; ω_t is the TO phonon frequency, Γ is the spontaneous anharmonic phonon decay rate; E is the local electric field; M and e^* are the reduced mass and effective charge of Ga and P, respectively; $\dot{X}_3, \dot{X}_h$ and m_e, m_h are the drift velocities and effective masses of the electrons and holes, respectively; τ_e and τ_h are the momentum relaxation times of the electrons and holes due to scattering by acoustical phonons; and η is due to electron-hole Coulomb scattering.

In order to solve Eqs. 26 - 29 we adopt the following assumptions: (a) The plasma frequency $\omega_p << \omega_{LO}$; this corresponds to a low plasma density under which the optical lattice vibrations are only slightly perturbed. The plasma density is estimated in the present experiment as $n \sim 10^{17}$ cm^{-3}, thus $\omega_p \sim 10^{13}s^{-1}$ and $\omega_{LO} \sim 10^{14}s^{-1}$; (b) $1/\omega_{LO} << \tau_{e-e}, \tau_{e-h}, \tau_{h-h}$; this indicates slow carrier-carrier scattering compared to the period of an LO phonon oscillation. For electrons and holes in GaP, $\tau_{e-e}, \tau_{e-h}, \tau_{h-h} \sim 10^{-13}s^{-1}$. Under these conditions, the coupled system yields a solution for the temporal evolution of the coherent phonon intensity $|A(t)|^2$ in the presence of the plasma. The result can be expressed in the following form:

$$|A(t)|^2 = |A(0)|^2 \exp[-\Gamma t - \beta \int_0^t \gamma(t')dt'], \tag{30}$$

where $A(0)$ is the coherent phonon amplitude at $t = 0, \beta = (1 - \epsilon_\infty/\epsilon_0)\omega_p/\omega_\ell$ is the coupling constant between the LO phonons and the NEHP, and $\gamma(t)$ is the time-dependent damping of the plasma oscillations. In the absence of a NEHP, the solution reduces to the standard exponential LO phonon dephasing with decay rate Γ. In order to arrive at an expression for $\gamma(t)$, it is necessary to determine the temporal dependence of the carrier temperature. After an initial carrier cascade due to the Fröhlich interaction (<2 ps) the carriers form a plasma at energies below that of the $k \cong 0$, LO phonon and the phonons generated as a result of inter-X-valley electron scattering ("XX scattering"). Starting at this point of the cooling of the plasma, we have calculated the further cooling as the carriers undergo additional LO phonon and XX scattering assisted by carrier-carrier scattering, deformation potential (DA) scattering, and piezoelectric (PA) scattering from acoustical phonons. The result of this calculation appears in Fig. 14. It can be seen from Fig. 14 that the plasma cools from a starting temperature of approximately 150 K to a temperature of 30 K within the first 25 ps after injection. The initial rapid cooling is due to LO phonon emission and XX scattering. The cooling rate then decreases as DA and PA scattering become the dominant mechanisms of energy relaxation. In order to determine the plasma damping, it is necessary to consider only those carrier interactions which do not conserve current. Assuming a nearly parabolic energy dispersion for the electrons and neglecting contributions due to the anharmonicity of the hole band, the expression for the plasma damping reduces to

$$\gamma(t) = \frac{1}{\tau_{e-h}} = n\sigma v = n(4\pi r_D^2)\left(\frac{3k_B T_c}{m^*}\right)^{\frac{1}{2}}, \tag{31}$$

where n is the plasma density, σ is the scattering cross section, v is the thermally controlled drift velocity of the carriers, $T_c(t)$ is the plasma temperature as given in Fig. 14, $r_d(\text{Å})$ is the Debye screening radius at temperature T_c, k_B is the Boltzmann constant, and m^* is the reduced mass for electrons and holes. Substituting Eq. 31 into 30 and using the derived carrier temperature T_c from Fig. 15 yields an expression for the dependence of the additional phonon dephasing on the carrier density n.[21] A best fit is then made to the additional dephasing using the theoretical expression to determine the carrier density n. The resulting additional dephasing is shown in the inset to Fig. 13. A one-parameter fit to the data indicates good agreement with the data and yields $n = 2.1 \times 10^{17}$ and 1.2×10^{17} cm^{-3} for the illustrative cases.

Whereas the temporal evolution of the additional dephasing indicates good agreement between the theoretical model and the data, it is necessary to determine the reliability of the fitted parameter, i.e., the carrier density n. An alternate evaluation of the carrier density n can be obtained through a calculation of the number of carriers generated through TPA in the mutual interaction volume of the three laser beams.

Only carriers excited in the interaction volume are available for the interaction, since transport of carriers out of the volume is negligible during the 150-ps interaction time.

It is clear that we have observed additional phonon dephasing due to the interaction of lattice vibrations and plasma oscillations. The model is in good agreement with the observed dephasing of the LO phonons, which depends on the plasma density and plasma damping. Moreover, the plasma density as obtained from the model and that calculated independently from TPA are also in good agreement. Since ballistic transport at the thermal velocities does not result in significant carrier transport out

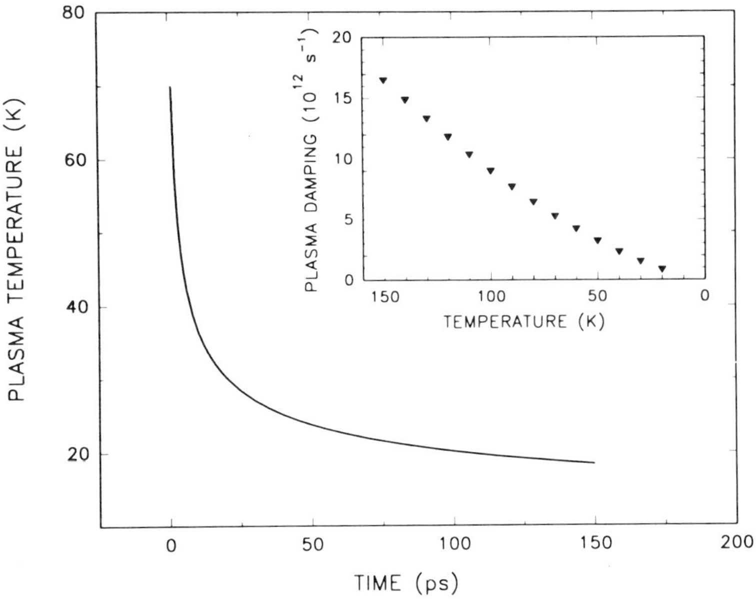

Figure 14. Calculated plasma temperature as a function of time. Inset: Plasma damping as a function of the ambient temperature.

of the interaction volume within the first 200 ps of the observation, we eliminate changes in the plasma density as a source of the change in the magnitude of the additional phonon dephasing. The plasma damping results from electron-hole scattering as described in Eqs. 30 and 31. The plasma oscillations are damped by any carrier-scattering event which does not conserve current. The assumption that electron-hole scattering dominates the damping appears valid for the conditions of the experiment. The temporal dependence of the magnitude of the scattering is due to the cooling of the plasma which decreases the plasma velocity and thus the scattering cross section. For the convenience of the reader we include, in the lower inset to Fig. 14, the temperature dependence of the plasma damping as obtained from Eq. 31 and the fitted values of the carrier density.

In conclusion, we have observed, an interaction of an optically induced two-component electron-hole plasma with LO phonons in GaP using the TR-CARS technique. A time-dependent increase in the dephasing rate of the LO phonons is observed to arise from the coupling of the vibrational modes of the plasma and those of the phonons. It is demonstrated, on experimental and theoretical grounds, that the increase of the dephasing rate depends on the density and on the damping of the plasma. The temporal dependence of the increased dephasing arises from a non-stationary plasma temperature. Plasma damping is readily observable for temporal resolution of the order of picoseconds. The present experimental technique makes it possible to observe, directly in the time domain, the interaction of a plasma with optical phonons.

VII. SUMMARY REMARKS

I have discussed, in some detail, a set of investigations on the ultrashort dynamics of various processes in condensed matter. Experimentation is conducted (for the most part) directly in the time domain. For this purpose a unique set of lasers have been developed that permit temporal resolution down to tens of femtoseconds. Moreover, coherent excitations and coherent detection are brought to bear which permits detection over many orders of magnitude. The experimental arrangement permits investigation over a wide variety of physical phenomena. Among those which have so far been investigated are:

- Dephasing of LO phonons
- Dephasing of polaritons
- Generation and detection of long lived phonons
- Nonlinear coupling between LO and LA phonons
- Dephasing of phonon interacting with a two component plasma

Not covered in this review, but just as interesting, is our applications of ultrashort laser pulses to

- Thermomodulation of thin normal metal films
- Ultrashort switching of Pb and high Tc superconducting strip lines
- Carrier dynamics in C_{60} (Buckyballs) thin films
- Optically induced ultrashort laser damage

Many other phenomena await investigation. This field of inquiry is far from complete.

ACKNOWLEDGMENT

The author acknowledges support through NSF DMR88-3888, NSF DMR 89-13289, and ARO DAAL-03-89-K-0600. He also thanks the many co-workers who have contributed to this research effort.

REFERENCES

1. See e.g. Levenson, M. D., and Song, J. J. in *Coherent Nonlinear Optics*, ed. Feld, M. S. and Letokhov, V. S. (Springer: Berlin) 1980.
2. See e.g. Demtröder, W. D. in *Laser Spectroscopy*, (Springer: Berlin) 1981.
3. Placzek, G., *Handbuck der Radiologie*, ed. Marx. E., (Akademische Verlagsgesellochaft: Leipsig) 1934.
4. Loudon, R., in *The Quantum Theory of Light* (Clarendon Press, Oxford) 1983.
5. Zinth, W., Lauberau, A. and Kaiser, W., Opt. Commun. **49**, 935 (1978).
6. See e.g. Verdeyen, J. T., in *Laser Electronics*, 2nd edition (Prentice Hall, Englewood Cliffs, NJ) 1989.
7. Juhasz, T., Kuhl, J., and Bron, W. E., Opt. Lett. **13**, 577 (1988).
8. Juhasz, T., Smith, G. O., Mehta, S. M., Harris, K. and Bron, W. E., IEEE J. Quantum Electronics, **25**, 1704 (1989).
9. Kuhl, J. and Bron, W. E., Sol. State Commun. **49**, 935 (1984), Bron, W. E., Kuhl, J., and Rhee, B. K., Phys. Rev. **34**, 6961 (1986).
10. Juhasz, T. and Bron, W. E., Phys. Rev. Lett. **63**, 2385 (1989).
11. Lauberau, A., and Kaiser, W., Res. Mod. Phys. **50**, 607 (1978).
12. See e.g. Bairamov, B. Kh., Kitaev, Yu. E., Negoduiko, N. and Kashkotgev, M. K., Sov. Phys. Solid State **16**, 1323 (1974).
13. Bairamov, B. H., Parshin D. A., Toporov, V. V. and Ubbaillaev, S. B., Sov. JTF Letters **5**, 1116 (1979).
14. See e.g. Bilz, H. and Kress, W., in *Phonon Dispersion Relations in Insulators* (Springer: Berlin) 1979.
15. Ushioda, S. and Muller, J. D., Sol. State Commun. **11**, 299 (1972).
16. Rhee, B. K. and Bron, W. E., Phys. Rev. **34**, 7107 (1986).
17. Thomas, D. G., Gershenson, M. and Hopfield, J. J., Phys. Rev. **34**, 7107 (1986).
18. Kunc, K., Ann. Phys. (Paris) **8**, 319 (1973).
19. Orbach, R. and Vredevoe, L. A., Phys. **1**, 91 (1964).
20. Smith, G. O., Juhasz, T., Bron, W. E., and Levinson, V. B., Phys. Rev. Lett. **68**, 2366 (1992).
21. Bron, W. E., Juhasz, T., Mehta, S., Phys. Rev. Lett. **62**, 1655 (1989), Mehta, S., Juhasz, T., and Bron, W. E., Phys. Rev. B**45**, 209 (1992).

VERTICAL TRANSPORT STUDIED BY SUB-PICOSECOND FOUR-WAVE MIXING EXPERIMENTS

J. Feldmann[1], G. von Plessen[1], T. Meier[1], P. Thomas[1], E.O. Göbel[1], K.W. Goossen[2], D.A.B. Miller[2], and J.E. Cunningham[2]

[1] Department of Physics and Materials Sciences Center
Philipps-University of Marburg, D-35032 Marburg, Germany
[2] AT&T Bell Laboratories, Holmdel, New Jersey 07733, USA

INTRODUCTION

Since the realization of semiconductor heterostructures vertical transport of electrically injected carriers has been one of the most interesting topics in semiconductor physics.[1,2] Precise engineering of semiconductor layers and thus electronic energy levels allows the tailoring of transport properties over a wide range and has even led to the invention of semiconductor devices relying on ballistic electron transport.[1,3] In addition, negative differential resistance (NDR) can be realized by using tunneling diodes containing double-barrier heterostructures[4] or a superlattice structure.[5,6] Actually, the proposal of Esaki and Tsu[5] to use electronic Bloch oscillations in the miniband of a semiconductor superlattice to realize NDR marked the starting point for the physics and applications of semiconductor heterostructures.

The study of vertical carrier transport in electrically biased semiconductor heterostructures by means of time-resolved photoluminescence[7–12] and transient pump and probe spectroscopy[13–19] has gained considerable interest during the last years. These time-resolved incoherent optical studies have led to a much better understanding of the dynamical aspects of electron and hole tunneling in double quantum well, multiple quantum well (MQW) and superlattice structures. In particular, transient pump and probe experiments for studying the dynamics of field-induced carrier sweep-out in MQW structures are of great practical importance.[15] Fast carrier escape leads to short charging and discharging times and to high saturation intensities of quantum well modulators and thus improves the switching speed of self-electro-optic effect devices (SEED).[20] In particular, shallow MQWs are promising device structures, because they combine high electroabsorptive contrast ratios with ultrafast carrier sweep-out.[16,21,22]

The mentioned time-resolved optical techniques are not sensitive to the coherent

Coherent Optical Interactions in Semiconductors
Edited by R.T. Phillips, Plenum Press, New York, 1994

aspects of electron and hole tunneling. In recent years, techniques like four-wave mixing (FWM)[23–27] and detection of terahertz emission[28,29] have been used to study the field-induced coherent dynamics of optically excited electronic wavepackets in semiconductor heterostructures. Back and forth tunneling of an electronic wavepacket in an asymmetric double quantum well,[23,28] Bloch oscillations of an electronic wavepacket in a semiconductor superlattice,[24–26,29] and field-induced exciton ionisation by electron escape tunneling out of shallow multiple quantum wells[27] were investigated. Here, we review incoherent pump and probe experiments and coherent FWM experiments both performed on shallow MQW and superlattice structures. We show that coherent nonlinear spectroscopy provides direct access to transport phenomena taking place on a nanometer spatial scale and femtosecond time scale. The knowledge of the initial field-induced carrier dynamics obtained by FWM experiments together with the knowledge of the carrier sweep-out rates obtained by incoherent pump and probe experiments then allows a more complete analysis of the transport dynamics.

SHALLOW MULTIPLE QUANTUM WELLS

Strong and well-resolved excitons have been observed in the absorption spectra of shallow $GaAs/Al_xGa_{1-x}As$ MQWs for values of x as low as 0.02.[21] In addition, these shallow quantum wells exhibit strong electroabsorption at low electric fields applied perpendicular to the layers. This pronounced electroabsorption effect at low biases is not only caused by the red shift of the exciton as observed for deeper quantum wells,[30] but also is ascribed to the fact that the low $Al_xGa_{1-x}As$ barriers cannot prevent the field ionization of the exciton.[21]

Pump and Probe Experiments: Ultrafast Carrier Sweep-out

A pump and probe experiment on a p-i-n diode under electric field differs from conventional pump and probe experiments performed on field-free semiconductor samples, since phase space filling or screening of the electron-hole Coulomb attraction are not the main sources for the differential absorption signal. Instead, space charge effects are rather governing the pump-induced change in absorption. The nonlinear optical signal is caused by the field-induced spatial separation of photocreated electrons and holes. Carriers accumulate in the doped regions and reduce the effective voltage between the electrodes. Since the absorption coefficient depends on the electric field via the electroabsorptive effect (quantum confined Stark effect, Wannier-Stark ladder effect,...), the absorption is changed by the partial screening of the electric field. Thus the differential absorption signal directly monitors the arrival of carriers at the contact layers. Accordingly, the rise time of the signal is equal to the carrier sweep-out time and can be measured as a function of the applied electric field.[13–18]

We have investigated the carrier sweep-out dynamics for three different shallow MQW p-i-n diodes with barrier Al-concentration of 2%, 4%, and 6% by performing transient pump and probe experiments using laser pulses of about $0.7ps$.[16] In Fig.1, the experimentally determined sweep-out times at $T = 10K$ are plotted as a function of the applied voltage. We find a decrease of the sweep-out times with increasing electric field and a subsequent saturation at a value of about $5ps$. The electric field strength where saturation occurs depends on the barrier height. At low fields the carrier transport is hampered by the AlGaAs barriers , i.e., the time for carriers to tunnel through the barriers is the limiting factor for the sweep-out at low temperatures.[16] This also explains the fact that the sweep-out times are longer for samples with higher AlGaAs barriers. For each sample the tunneling rate increases with increasing electric field, since the effective barrier height of the quantum wells decreases leading to an enhanced leakage of the quantum well wave function into the barrier material.[31] When the applied voltages become larger than $0V, 2V$, and $4V$ for the $x = 2\%, 4\%$, and 6% samples, respectively, the sweep-out times are independent of the applied field and are approximately $5ps$ for all samples. At these high fields the tunneling process out of the quantum wells must be efficient enough that the carrier sweep-out is mainly limited by saturated drift to the contact layers. We can estimate the saturated drift velocity to be about $10^7 cm/s$, which is reasonable for GaAs.

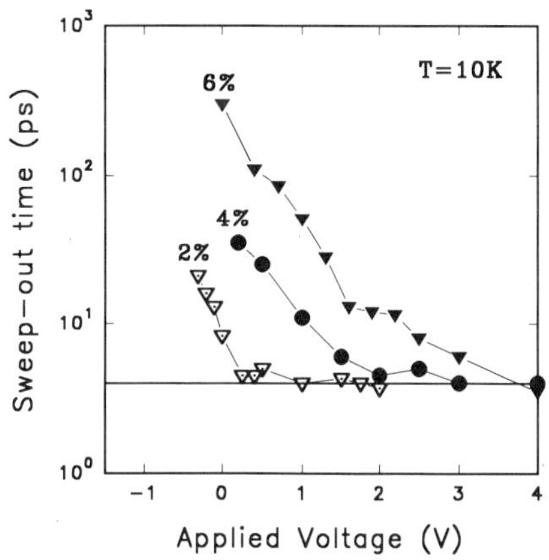

Figure 1. Experimentally determined sweep-out times versus applied bias voltage for three shallow MQW samples at $T = 10K$ (positive applied voltages correspond to reverse bias voltages).

However, the knowledge of the sweep-out rates as shown in Fig.1 is not sufficient to analyse the transport dynamics. The experimentally determined sweep-out times represent complex average values, since each excited quantum well has a different distance from the contact layer. In addition, it is not clear whether retrapping of already escaped carriers into adjacent quantum wells contribute to the long sweep-out times at low fields. For a complete analysis the time-constant for the escape tunneling of the carriers out of the *individual* quantum wells within the MQW structure is needed. This escape tunneling should lead to a broadening of the homogeneous linewidth of the exciton transition.

In Fig.2, the linewidth (HWHM) of the lowest heavy-hole exciton resonance taken from photocurrent measurements is plotted as filled dots as a function of the applied voltage for the $x = 6\%$ sample. An increase of the linewidth is observable only for voltages larger than $U = 1V$. In addition, contributions from inhomogeneous and homogeneous broadening of the linewidth cannot be distinguished from each other by using linear optical methods. As shown in the next section, transient FWM experiments allow the determination of the field-induced broadening of the excitonic *homogeneous* linewidth also at low fields and thus directly monitor the escape tunneling out of the *individual* wells, i.e., the field ionization of the exciton as illustrated in the upper part of Fig.3. It is shown that carrier retrapping indeed plays an important role in the low field regime.

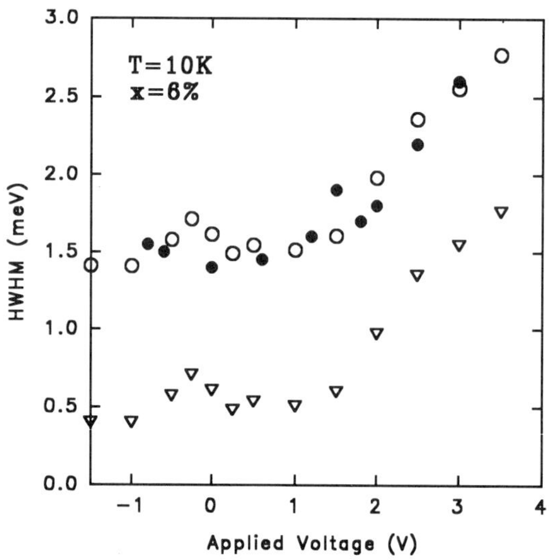

Figure 2. Homogeneous linewidth (triangles) and total linewidth (full and hollow circles) of the $n = 1$ heavy-hole excitonic transition as a function of the applied voltage.

Four-Wave Mixing Experiments: Field-Ionization of Excitons

We use a two-pulse self-diffraction technique, where the sample is excited by two subsequent $110fs$ laser pulses with wave vectors $\vec{k_1}$ and $\vec{k_2}$ at times $t = 0$ and $t = \tau$, respectively. A diffracted FWM signal is detected in the phase-matched direction $2\vec{k_2} - \vec{k_1}$ by using a slow photodetector. The decay of this signal is determined by the loss of coherence of the optical transition. The corresponding decay time allows the determination of the dephasing time T_2 and thus the homogeneous linewidth $\Gamma = 2\hbar/T_2$ irrespective of inhomogeneous broadening effects arising from crystal imperfections and from inhomogeneities of the electric field.

In the lower part of Fig.3, the diffracted FWM signals are plotted versus the time delay between the two laser pulses for the $x = 6\%$ shallow MQW sample for two distinct voltages corresponding to the flatband condition ($U = -1.5V$) and high electric field ($U = 4V$). The central laser photon energy is located at the low energy part of the $n = 1$ heavy-hole exciton line. The data of Fig.3 show that application of an electric field leads to a shortening of the signal decay time τ_d, i.e., of the dephasing time $T_2 = 4 \cdot \tau_d$ (as valid for inhomogeneously broadened transitions). From the T_2 data we can now calculate the field-dependent homogeneous linewidth $\Gamma_{hom}(F) = \hbar/T_2$ (HWHM). The $\Gamma_{hom}(F)$ values are plotted as hollow triangles in Fig.2. We now assume a field-independent inhomogeneous broadening of $\Gamma_{inh} = 1meV$ and obtain the hollow circles for the total linewidth $\Gamma_{tot}(F) = \Gamma_{hom}(F) + \Gamma_{inh}$. The excellent agreement with the total linewidth taken from the photocurrent spectrum shows that the field-induced broadening of the total linewidth at high fields is solely due to a broadening of the homogeneous linewidth and not due to inhomogeneities of the electric field.

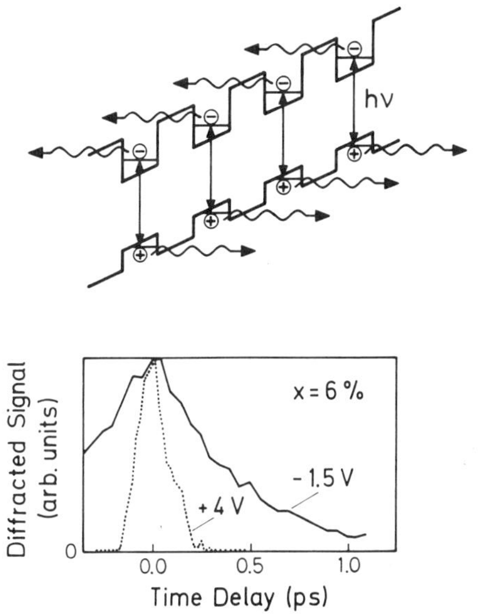

Figure 3. Upper part: Schematic illustration of carrier escape tunneling leading to enhanced dephasing rates. Lower part: Diffracted FWM signal versus time delay for the $x = 6\%$ shallow MQW sample.

The field-dependence of the dephasing-rate is caused by the field-ionization of the optically excited excitons, i.e. by the carrier escape tunneling out of the quantum wells. It is reasonable to assume that the measured dephasing rate $T_2^{-1}(F)$ at non-zero electric field F is given by the sum of the dephasing rate at flat band $T_2^{-1}(F = 0)$ caused by scattering with phonons and crystal imperfections and of the field-induced tunneling rate $T_{tun}^{-1}(F)$:

$$\frac{1}{T_2(F)} = \frac{1}{T_2(F = 0)} + \frac{1}{T_{tun}(F)} \qquad (1)$$

Thus, we can extract the tunneling times $T_{tun}(F)$ from the dephasing times $T_2(F)$. On the left side of Fig.4, the experimentally determined tunneling times are plotted as dots versus the voltage applied to the respective diode. In the case of the $x = 4\%$ sample, the tunneling times decrease monotonically with increasing field, whereas in the case of the $x = 6\%$ sample the tunneling time exhibits a pronounced minimum at about $-0.2V$.

We have performed transfer matrix (TM) calculations for the shallow MQW structures in the presence of an electric field, since they provide not only energy levels of confined electronic states, but also the electron lifetime in the particular quantum confined state, i.e., the electron escape tunneling time.[11,27,32] The transmission coefficient of an electronic wave incident on the MQW structure peaks whenever the energy of the incident wave corresponds to the energy of a quasi-bound $n = 1$ quantum well state. The energetic width Γ_T (FWHM) of the transmission peak in the presence of an electric field is related to the lifetime of the electron (hole) in the $n = 1$ quantum well state, i.e., it determines the tunneling time of the carriers through the so-called Breit-Wigner formula:

$$T_{tun}(F) = \frac{\hbar}{\Gamma_T(F)} \tag{2}$$

For both sample structures, it turns out that the calculated tunneling times of the heavy-holes are much longer than those of electrons. Accordingly, electron escape tun-

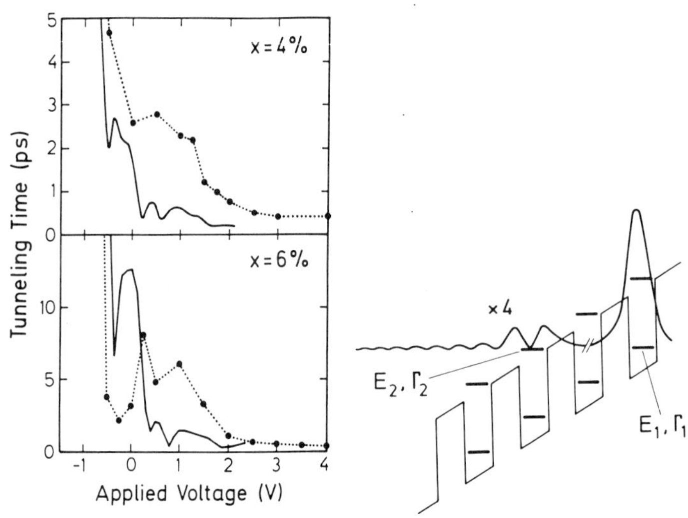

Figure 4. Left side: The experimentally determined tunneling times are plotted as dots versus the applied voltage for both MQW samples. The solid lines represent theoretically expected electron tunneling times calculated according to Eq.(2). Right side: Calculated squared wavefunction for the quasi-bound $n = 1$ quantum well state in the conduction-band when it is resonant with the unbound $n = 2$ state of the second-nearest quantum well.

neling in the conduction-band is the governing field-induced exciton ionization or dephasing mechanism. The calculated electron tunneling times are shown as solid lines in Fig.4. A comparison with the experimental data shows that in the case of the $x = 4\%$ sample, the calculations overestimate the carrier escape rate. The same is true for the $x = 6\%$ sample for voltages higher than $0.2V$. The additional tunneling barrier provided by the electron-hole Coulomb attraction, enhanced localization of the quasi-bound states by alloy disorder in the shallow *AlGaAs* barriers (coherent backscattering), and fast carrier relaxation in the continuum[33] may explain the reduced experimental escape rate.

The tunneling times calculated by the TM method for the $x = 6\%$ sample also show a resonantly enhanced tunneling rate at about $-0.5V$. The lifetime shortening arises from resonant mixing of lifetimes[34] of the quasi-bound $n = 1$ quantum well state and the resonant $n = 2$ continuum state of the second-nearest quantum well. As shown on the right side of Fig.4, the probability of finding the electron outside of the confining quantum well potential is enhanced when the two states come into resonance.

A comparison of the data shown in Fig.1 and Fig.4 now allows a distinction to be made as to whether or not retrapping of already escaped carriers in adjacent wells plays a significant role. At high electric fields, where the sweep-out times in Fig.1 have a constant value of $5ps$, the respective tunneling times in Fig.4 are on a sub-picosecond time-scale. This is in agreement with the above mentioned assumption that carrier sweep-out at high electric fields is limited by the saturated drift velocity and not by tunneling through barriers. At lower electric fields, the sweep-out times shown in Fig.1 are much longer than the sum of the respective tunneling times taken from Fig.3 and the $5ps$ for the drift (the drift should be saturated for voltages larger than $U = 0V$). This definitely shows that retrapping of already escaped carriers into adjacent quantum wells is an important contribution to the dynamics of carrier sweep-out.

The shallow MQW system allowed us to study field-induced escape tunneling from a quasi-bound quantum well state into a continuum of states. Now, we turn to the problem of incoherent and coherent carrier dynamics *within* a band continuum of states in the presence of an applied electric field.

SUPERLATTICES

Electrons in a band of a periodic crystal are described by Bloch's eigenfunctions

$$\psi(\vec{k},\vec{r},t) = u(\vec{k},\vec{r}) \cdot exp(i\vec{k} \cdot \vec{r} - i \int^{t} \frac{E(\vec{k})}{\hbar} dt'), \tag{3}$$

where $E(\vec{k})$ is the band dispersion. In the presence of an electric field $\vec{F}$, the equation of motion for an electronic Bloch state is given by the so-called acceleration theorem[35–39]

$$\hbar\dot{\vec{k}} = e\vec{F}. \tag{4}$$

A Bloch electron composed of Bloch states moves through $\vec{k}$-space with a field-dependent velocity as depicted on the right side of Fig.5. Due to Bragg-reflections at the Brillouin-zone (BZ) boundary the electron executes so-called Bloch oscillations within a single band with a time period

$$T_B = \frac{h}{eFd}, \tag{5}$$

where d is the lattice constant. To a good approximation, the electronic wavefunctions in the presence of an electric field are still given by Eq.(3), if $\vec{k}$ in Eq.(3) is made time-dependent according to the acceleration theorem given by Eq.(4). However, these so-called Houston-functions[40] neglect the 'leakage' of electrons from one band into another and thus the possibility of field-induced electron tunneling into higher bands (Zener tunneling).[41] In addition, in real crystals electron scattering due to phonons, impurities, etc. limit the coherent motion described by the acceleration theorem to a finite time τ_{sc}.[42] We note that the acceleration theorem is quantum mechanically correct even in the presence of Zener tunneling, whereas the concept of Bloch oscillations within a single band is only valid for times less than the Zener tunneling time.[43]

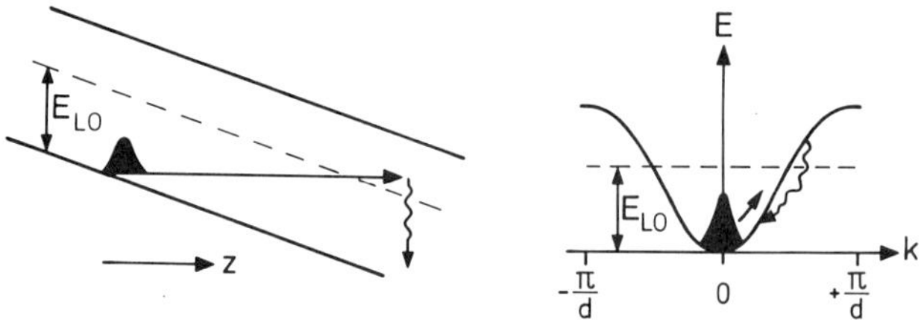

Figure 5. Schematic illustration for the field-induced coherent motion of an electronic wavepacket initially created at the bottom of a miniband having a width exceeding the LO-phonon energy E_{LO}. Scattering by LO-phonons is illustrated in the real space picture (left side) and the k-space picture.

The Houston-functions can be used to calculate the matrix-elements for the optical transitions from the valence-band $E_v(\vec{k})$ to the conduction-band $E_c(\vec{k})$ in the presence of an electric field.[44,45] The resulting one-particle matrix-elements then contain the non-sinusoidal phase-factor[44]

$$exp(i \int^t \frac{E_c(\vec{k}(t')) - E_v(\vec{k}(t'))}{\hbar} dt'). \tag{6}$$

Due to the periodic traversal of the electron through the BZ the phase-factor in Eq.(6) is periodically modulated with a time-period equal to the period $T_B = h/eFd$ expected for Bloch oscillations. The corresponding Fourier component

$$\frac{1}{2\pi} \int_{-\infty}^{+\infty} dt \quad exp(i \int^t \frac{E_c(\vec{k}(t')) - E_v(\vec{k}(t')) - \hbar\omega}{\hbar} dt') \tag{7}$$

then exhibits sidebands with energies

$$E_n = E_0 + n \cdot eFd, \qquad n = 0, \pm1, \pm2, ... \tag{8}$$

giving rise to the well-known Wannier-Stark (WS) ladder transitions as depicted in the band scheme of Fig.6.

In real space, the field-induced oscillatory motion of the electron causes a WS localization of the eigenfunction. As a result, the spatial extent of the wavefunction is restricted to a localization length

$$\lambda = \frac{\Delta}{eF}, \tag{9}$$

where Δ is the energetic width of the electronic band. As shown in Fig.6, the continuous spectrum of band states changes into a series of equally spaced WS energy levels.[46–53] The requirement for both the formation of the WS ladder and the occurrence of Bloch oscillations is the same, namely that the period T_B for Bloch oscillations is less than the scattering time τ_{sc}.[42,54] The electron must be able to make at least one oscillation through the BZ before being deflected. In the case that for all reasonable electric fields the scattering time τ_{sc} is shorter than the Bloch oscillation period T_B no WS transitions but Franz-Keldysh oscillations (FKO) can be observed in the absorption spectrum.[55,56] For the calculation of FKO only $E(\vec{k})$-states next to $\vec{k} = 0$, where the parabolic band approximation applies, are considered in Eq.(6) and (7),[44,45] since electrons are assumed not to reach large $\vec{k}$-states due to the strong scattering. The transition from the *band regime* to the *WS regime* occurs at an applied electric field where electrons are fast enough in $\vec{k}$-space to beat the scattering and to reach the BZ boundary coherently.[54]

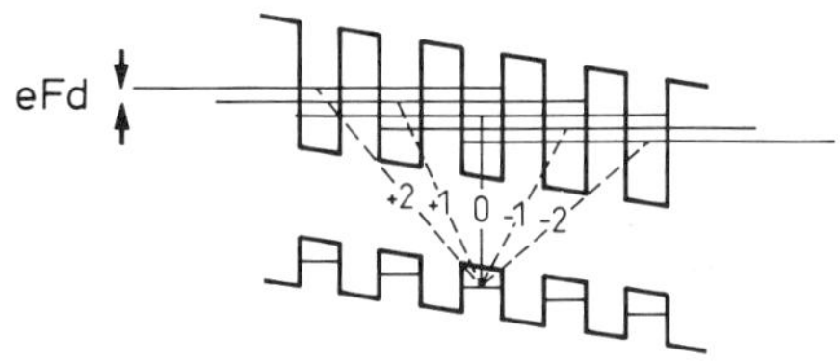

Figure 6. Bandscheme of the conduction- and valence-band potential profiles of a superlattice in the presence of an electric field F.

In particular, electron scattering by optical phonons is expected to efficiently damp Bloch oscillations and broaden the homogeneous linewidth of WS transitions, since electron-LO-phonon scattering times are less than $180fs$ in the $GaAs/AlGaAs$ system.[57] Semiconductor superlattices offer the unique opportunity to tailor the lattice constant d and the miniband width Δ. As depicted in Fig.5, we expect an efficient damping of Bloch oscillations when the miniband width Δ exceeds the LO-phonon energy. Then the transition from the *(mini)band regime* to the *WS regime* occurs at high electric fields when the time-period T_B for Bloch oscillations becomes comparable to $\tau_{sc} = 180fs$, i.e., when the WS spacing eFd is larger than the corresponding homogeneous linewidth (FWHM) of about $\Gamma_{hom} = 2\hbar/\tau_{sc} = 7.5meV$. Indeed, to the best of our knowledge, WS ladder spacings less than $7.5meV$ were only observed for superlattices having electronic miniband widths of less than the LO-phonon energy. Once the WS ladder levels have formed ($T_B \leq \tau_{sc}$) the scattering rate between different WS states decreases with increasing electric field.[58,59] This is due to the fact that the spatial overlap of initial and final WS wavefunctions decreases with increasing Stark localization.

In the following, we first review transient pump and probe experiments for electric fields in the *WS regime*.[17] The experimental results clearly show that Stark localization hampers carrier sweep-out from superlattices and represent an independent proof for the occurrence of NDR. Secondly, we present transient FWM experiments on superlattices with different miniband widths. Using $110fs$ laser pulses, Bloch oscillations for electric fields in the *WS regime* can only be observed in superlattices with miniband widths less than a LO-phonon energy. In superlattices with wide minibands the application of an electric field leads to a dramatic shortening of the FWM decay already in the *miniband regime*. Model calculations also show this characteristic transient behaviour when electron-LO-phonon scattering is included in the calculations.

Pump and Probe Experiments: Stark Localization and NDR

Cohen and Bar-Joseph[17] performed picosecond pump and probe experiments on a $GaAs/Al_{0.3}Ga_{0.7}As$ superlattice sample in the presence of an applied electric field. The conduction miniband width of their superlattice sample is approximately $70meV$ and is thus larger than the LO-phonon energy. In Fig.7, low-temperature photocurrent spectra at $0V/cm$ (bottom), $1.7 \cdot 10^4 V/cm$, $5 \cdot 10^4 V/cm$, and $8.7 \cdot 10^4 V/cm$ (top) are shown. At zero electric field, a heavy-hole exciton resonance related to the lower edges of the conduction- and valence-minibands (M_0 exciton) is observed. The exciton resonance disappears at a field-strength of about $5 \cdot 10^3 V/cm$. The authors[17] attribute this broadening to field-ionization of excitons within the lowest miniband. In this *miniband regime*, electrons and holes move in opposite directions without performing Bloch oscillations, i.e., no WS ladder transitions can be resolved for these low fields. As can be seen in Fig.7, WS ladder transitions having a spectral spacing of $eFd \approx 28meV$ are observable at a field strength of $5 \cdot 10^4 V/cm$ but not at a field-strength of $1.7 \cdot 10^4 V/cm$, where a WS ladder spacing $eFd \approx 9.5meV$ is expected. The narrowest eFd spacing which is observed by the authors is approximately $13meV$ at about $2.6 \cdot 10^4 V/cm$.[17] Obviously, the above statement that for miniband widths exceeding $36meV$ efficient LO-phonon scattering only allows the observation of WS ladder spacings for eFd-values larger than $7.5meV$ holds for this superlattice sample.

Cohen and Bar-Joseph[17] performed transient pump and probe experiments on the superlattice sample for eFd spacings larger than about $22meV$, i.e., only in the *WS regime*. The rise times of the temporal differential absorption signal, which represent the carrier sweep-out times, are plotted as a function of the applied electric field in Fig.8. The electron sweep-out time from the superlattice is a nonmonotonic function of the applied electric field showing a maximum at about $7 \cdot 10^4 V/cm$ corresponding to $eFd \approx 40meV$. The increase of the sweep-out times with increasing electric field is a consequence of field-induced Stark localization, whereas their decrease at higher fields is due to tunneling into higher bands (Zener tunneling).[17] Altogether, these transient optical measurements provide an independent proof for the occurrence of NDR induced by Stark localization of electronic wavefunctions. However, the low-field carrier transport in the *miniband regime* was not investigated. In that regime, the sweep-out should be dominated by drift within the intact miniband leading to decreasing sweep-out times with increasing electric field as expected for conventional transport in bulk semiconductors (mobility regime).

We now show that transient FWM experiments provide additional information on the initial carrier dynamics in the *miniband regime* and the *WS regime*.

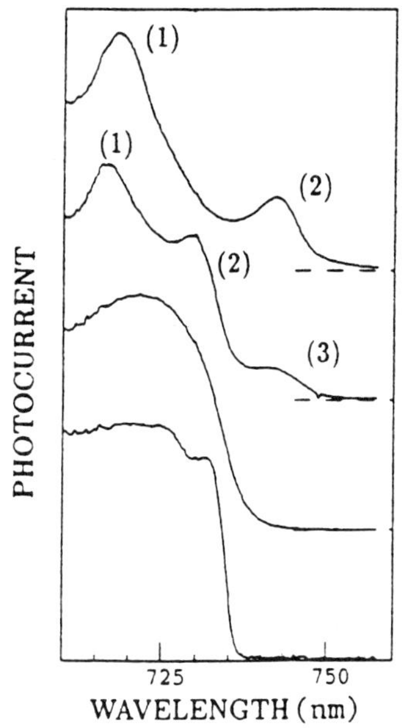

Figure 7. Photocurrent spectra at applied fields of $0V/cm$ (bottom), $1.7 \cdot 10^4 V/cm$ (middle lower), $5 \cdot 10^4 V/cm$ (middle up), and $8.7 \cdot 10^4 V/cm$ (top) for a superlattice with a conduction-miniband width of about $70meV$ (taken from Ref. 17).

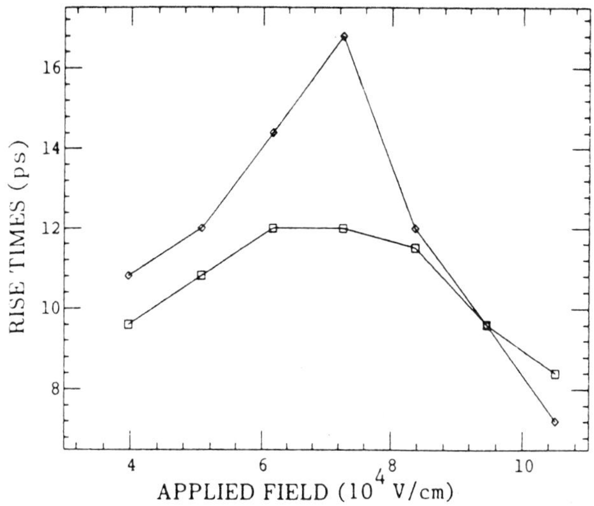

Figure 8. Rise time of the differential absorption signal as a function of the applied fields at two illumination powers (taken from Ref.17).

Four-Wave Mixing Experiments

In 1973, Zakharov and Manykin[60] theoretically analyzed the effect of a constant electric field on the diffracted photon echo in a transient FWM experiment for a bulk semiconductor. These authors found that the photon echo signal should reflect the temporal development of the wavevector $\vec{k}(t)$. However, due to the small lattice constant in bulk crystals and due to efficient carrier scattering Bloch oscillations or WS ladders have not been observed in bulk semiconductors. In a more specific theoretical treatment, von Plessen and Thomas[61] proposed to detect Bloch oscillations directly in the time domain by performing FWM experiments on an electrically biased superlattice. Here, we summarize their main results. In k-space, the ensemble of vertical optical transitions ($\Delta k = 0$) between the valence-miniband and the conduction-miniband states can be viewed as an inhomogeneously broadened ensemble of two-level systems with transition frequencies $\omega(k)$. When a spectrally broad (short) laser pulse excites all of these two-level systems at $t = 0$, their dipole moments are initially in phase leading to a macroscopic (first order) polarization. Even without an applied electric field, the phases $\omega(k) \cdot t$ of the individual dipole moments evolve differently in time due to the different (but fixed) phase velocities $\omega(k)$. As a consequence, the macroscopic polarization vanishes. However, for uncoupled two-level systems the second pulse at $t = \tau$ leads to a zero phase shift at $t = 2\tau$. The recovered macroscopic (third order) polarization then leads to the emission of a photon echo.[62] With an applied electric field, the temporal variation of the quasi-momentum $k(t)$ [Eq.(4)] implies that the phases $\int \omega(k(t))dt$ of the dipole moments now evolve according to the temporally varying phase velocities $\omega(k(t))$. As a result, the nonlinear interaction due to the second laser pulse generally does not lead to a perfect zero phase shift at the instant the photon echo should arise ($t = 2\tau$). Only for specific values of the field, such that the electron is able to traverse the entire BZ and return to the initial state during the time between the pulses, i.e., for $t = T_B$, the radiators will again be in phase at $t = 2\tau$. Altogether, the transient FWM signal for a biased superlattice should exhibit periodic modulations as a function of the time interval τ between the pulses. In Fig.9, the calculated echo amplitude for excitation with δ-pulses and without any damping is plotted as a function of the time delay τ between the laser pulses for a WS ladder spacing of $eFd = 10meV$. Pronounced modulations with time period T_B can be seen as a consequence of Bloch oscillations.

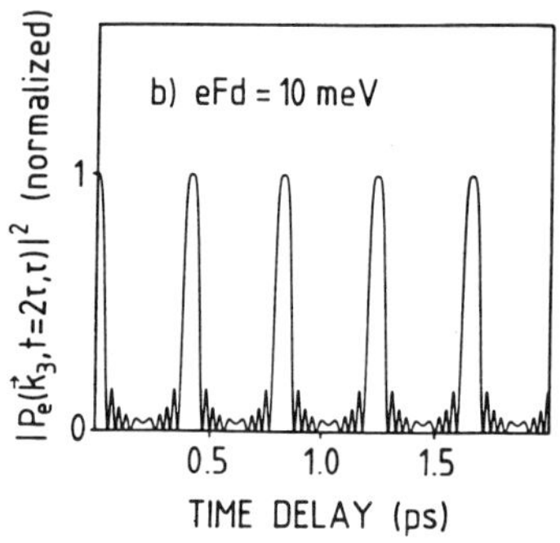

Figure 9. Calculated FWM signal for a WS ladder spacing $eFd = 10meV$ as a function of the time delay τ between the δ-function laser pulses and in the absence of damping.

Bloch Oscillations in Narrow Minibands. We now review and discuss results of sub-picosecond FWM experiments performed on a $GaAs/Al_xGa_{1-x}As$ superlattice sample. The intrinsic superlattice structure consists of 91 periods of 95Å GaAs and $15\AA Al_{0.3}Ga_{0.7}As$ embedded in a p-i-n diode.[24,25] The conduction miniband width of this superlattice sample is approximately $21meV$ and is thus less than the LO-phonon energy. Fox et al.[52] studied the linear optical properties of this superlattice sample in detail. Fig.10 shows low-temperature photocurrent spectra for electric fields up to $2.5 \cdot 10^4 V/cm$ ($-1V$ reverse bias). A well-defined heavy-hole Stark-ladder, i.e., a fanlike behavior of the state energies as a function of field according to Eq.(8) is evident. The *WS regime* already starts at fields lower than $4 \cdot 10^3 V/cm$, where WS ladder transitions with eFd spacings of less than $4meV$ are still resolved in the photocurrent spectrum. We argue that the absence of LO-phonon scattering within the $21meV$ wide miniband prolongs τ_{sc} and allows electrons to perform 'slow' Bloch oscillations at low electric fields.

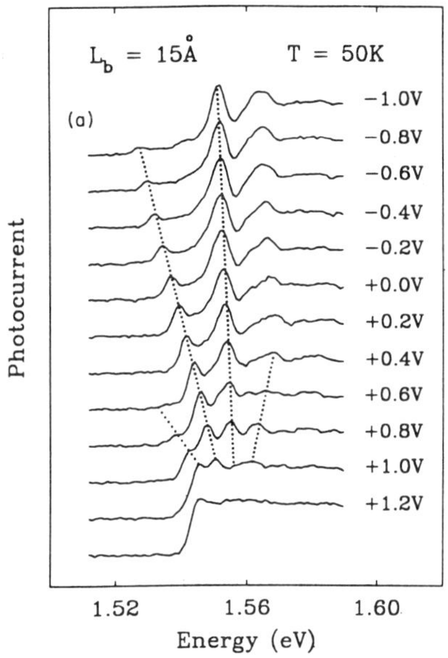

Figure 10. Photocurrent spectra for a superlattice with a conduction-miniband width of about $21meV$ (taken from Ref. 52).

In Fig.11, the diffracted FWM signal is shown as a function of the time delay τ between the $110fs$ laser pulses for a WS ladder spacing of $6.2meV$. A pronounced beating with a time period of $0.7ps$ is observed in good agreement with the time period $T_B = h/eFd = 0.67ps$ expected for Bloch oscillations. By lowering the electric field T_B-values as long as $1.4ps$ corresponding to $eFd = 3meV$ were observed.[24,25] For a very similar superlattice structure, also with a miniband width less than the LO phonon energy, T_B-values as long as $1.7ps$ corresponding to WS ladder spacings as low as $eFd = 2.4meV$ were measured by Leo et al.[26] We conclude that the scattering times τ_{sc} for electrons and holes must be on a picosecond time scale for all k-states throughout the mini-BZ. Otherwise, electrons with such low velocities could not coherently move in k-space for such long time intervals. Probably, electron-acoustic phonon scattering is the limiting scattering mechanism for these narrow miniband superlattices.[63]

In a theoretical paper, Bouchard and Luban[64] showed that dipole radiation in the terahertz range should be detectable as a consequence of optically generated Bloch oscillations as initially proposed by Esaki and Tsu.[5] Very recently, coherent electromagnetic radiation originating from Bloch oscillations in a superlattice structure with a narrow miniband width was indeed detected by Waschke et al.[29]

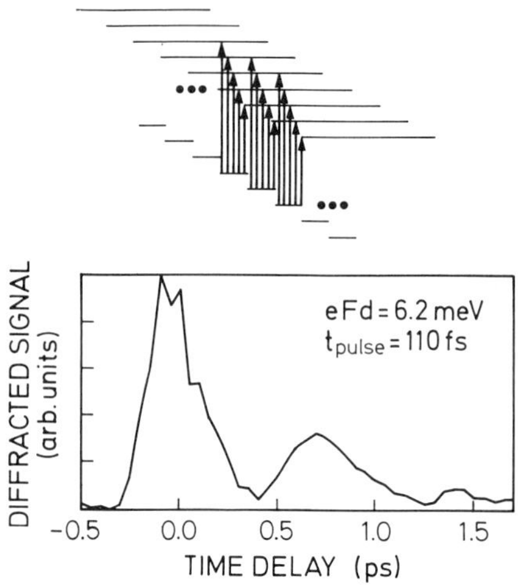

Figure 11. FWM signal versus time delay for $eFd = 6.2meV$ using $110fs$ laser pulses.

So far, Coulomb effects have been neglected in the analysis of Bloch oscillations. Dignam and Sipe[65] calculated how excitonic effects alter the fanlike behaviour of the WS ladder transitions. For a particular field-range and for particular miniband widths excitonic effects become important and lead to level repulsion in the absorption spectra of electrically biased superlattices. Fox et al.[52] experimentally confirmed that excitonic effects may considerably change the fanlike absorption spectrum. In addition, these authors found that for the superlattice sample which showed Bloch oscillations[24,25] (see Fig.11) excitonic mixing effects are negligible, supporting the validity of the simple one-particle analysis for this superlattice sample as described above. However, the contribution of excitonic effects to the temporal dynamics of optically excited wavepackets in electrically biased superlattices is an important issue.[66]

Altogether, the above described transient FWM experiments[24–26] and THz experiments[29] give convincing experimental evidence for the existence of Bloch oscillations in semiconductor superlattices. These experimental facts should help end the historical controversy about the theoretical validity of the concept of Bloch oscillations.

LO-Phonon Scattering in Wide Minibands. As already discussed above electron scattering by LO-phonons has to be considered in order to understand the low-temperature photocurrent spectra of superlattices with miniband widths larger than the LO-phonon energy. In the presence of an electric field, electrons move in the $E(k)$-superlattice dispersion as depicted on the right side of Fig.5. When they reach a kinetic energy higher than the LO-phonon energy they might be scattered by emission of a LO-phonon. Whether or not the electron performs Bloch oscillations by undergoing Bragg reflections at the BZ boundary now depends on the electron's k-space velocity. In an equivalent real space scheme as shown on the left side of Fig.5, a standing wave within the miniband is only formed when the electron coherently reaches the upper edge of the miniband and becomes reflected. Again, whether or not a standing wave (WS wavefunction) is formed depends on the field-induced velocity of the electron.

We now present theoretical and experimental results on the influence of LO-phonon scattering on transient FWM experiments.[67] The calculations are performed in k-space, where the optically coupled states between the valence-miniband and the conduction-miniband can be viewed as an ensemble of two-level systems.[25,61]

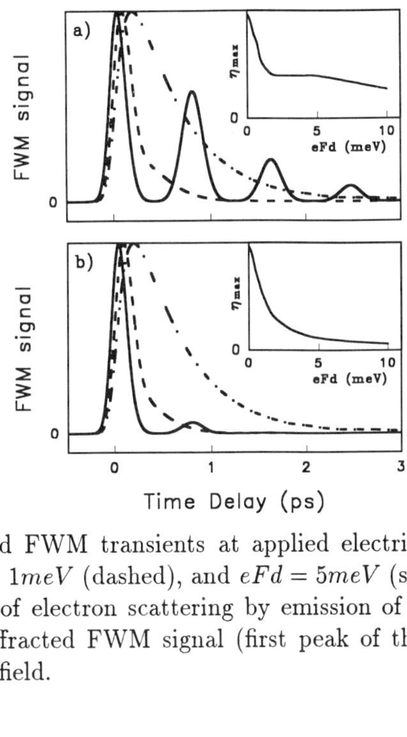

Figure 12. Calculated FWM transients at applied electric fields of $eFd = 0meV$ (dashed-dotted), $eFd = 1meV$ (dashed), and $eFd = 5meV$ (solid line) (a) without and (b) with consideration of electron scattering by emission of LO-phonons. The insets show the maximum diffracted FWM signal (first peak of the FWM transients) as a function of the applied field.

In the presence of an electric field, each two-level system moves in k-space according to the acceleration theorem and thus changes its transition frequency $\omega(k(t))$ as discussed above. We have calculated the FWM signal as a function of the delay time τ for a superlattice with miniband width exceeding the LO-phonon energy of $36meV$.

Laser pulses with a temporal width of $110fs$ and a spectral width of $22meV$ centered at the optical band gap are assumed, which generate electronic wavepackets around $k = 0$ (k-space), i.e., at the bottom of the wide miniband (real space) as depicted in Fig.5. We now investigate two cases. First, a k-independent constant dephasing time $T_2 = 1.7ps$ is assumed for all two-level systems, i.e., fast scattering by emission of LO-phonons for $E(k)$-states high in the miniband is not included. In Fig.12a, the calculated transients of the FWM signal are shown for three values of the applied electric field, namely $eFd = 0meV$ (dashed-dotted), $1meV$ (dashed), and $5meV$ (solid line). With no applied field, the signal decay is given by the assumed dephasing time $T_2 = 1.7ps$. Application of a field of $eFd = 1meV$ corresponding to a time period of $T_B = 4.1ps$ for Bloch oscillations leads to a shortening of the signal decay. This shortening is not caused by dissipative dephasing processes but by the field-induced coherent motion of the electronic wavepacket initially excited at the bottom of the miniband. For $eFd = 5meV$, corresponding to $T_B = 0.83ps$, electronic Bloch oscillations cause pronounced periodic modulations of the signal damped by the dephasing time $T_2 = 1.7ps$.

The set of curves in Fig.12b is obtained from calculations using the same parameters as for Fig.12a, however, a k-dependent dephasing rate is now assumed. For $E(k)$-states below $E_{LO} = 36meV$, the dephasing time is again $T_2 = 1.7ps$, whereas we assume a dephasing time of $T_2 = 180fs$ for $E(k)$-states higher in energy than $E_{LO} = 36meV$. As in Fig.12a, the shortening of the decay of the FWM signal is obtained when a field of $eFd = 1meV$ is applied. The initial motion of the electronic wavepacket at the bottom of the miniband is not affected by the short dephasing time of $T_2 = 180fs$ for $E(k)$-states above $E_{LO} = 36meV$. However, a comparison of the solid curves in Fig.12a and Fig.12b shows that for $eFd = 5meV$ the modulations due to Bloch oscillations are drastically damped when LO-phonon scattering for $E(k)$-states above $E_{LO} = 36meV$ is taken into account. For $eFd = 5meV$, the time period $T_B = 830fs$ for Bloch oscillations is too slow compared to the LO-phonon scattering time of $180fs$ to allow several coherent circulations of the electron within the mini-BZ.

The insets of Fig.12a and b show the maximum FWM signal as a function of the applied electric field. In both cases, it is seen that the signal intensity drops with increa-

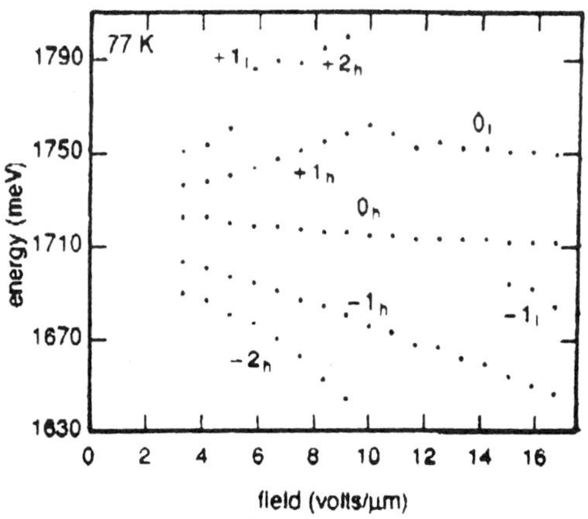

Figure 13. 'Fan diagram' showing peak positions versus field for the superlattice sample with a conduction-miniband width of about $70meV$.

sing electric field. This drop in intensity is a consequence of the coherent motion of the wavepacket away from the bottom of the miniband, where the wavepacket is initially excited.[67]

We have performed transient FWM experiments using $110fs$ laser pulses on a p-i-n $GaAs/Al_{0.3}Ga_{0.7}As$ superlattice structure having a miniband width of approximately $70meV$. The parameters of the superlattice structure happen to be exactly the same as the one used by Cohen and Bar-Joseph[17]. In Fig.13, the results of photocurrent measurements are shown. The energetic positions of the photocurrent peaks are plotted versus the applied field and clearly show the fan of the heavy-hole WS ladder transitions. A minimum WS ladder spacing of $eFd \approx 14meV$ is observed at $F \approx 3.3 \cdot 10^4 V/cm$ close to the value of $\approx 13meV$ obtained by Cohen and Bar-Joseph[17]. The FWM experiments are performed for applied electric fields below $F \approx 3.3 \cdot 10^4 V/cm$, i.e., in the *miniband regime* where WS ladder transitions are not resolved. The central laser photon energy is tuned to $1.692eV$ to excite the electronic wavepacket at the bottom of the miniband, as assumed in the calculations described above. In Fig.14, the experimentally determined FWM transients are shown for applied electric fields of $F = 0V/cm$ (dashed-dotted), $F = 1.2 \cdot 10^4 V/cm$ (dashed), and $F = 2.8 \cdot 10^4 V/cm$ (solid line). A continuous shortening of the decay time of the FWM signal is observed with increasing electric field in agreement with the set of calculated FWM curves shown in Fig.12b. We thus conclude that the shortening is caused by the field-induced coherent motion of the electronic wavepacket initially excited at the bottom of the miniband. The observed field-induced decrease of the maximum diffracted FWM signal shown in the inset of Fig.14 supports this interpretation. Using $110fs$ laser pulses, periodic modulations of the FWM signal due to Bloch oscillations are not observed at any applied field. The theoretical results shown in Fig.12 show that the LO-phonon scattering time of $180fs$ does not allow to observe Bloch oscillations in superlattices with miniband widths larger than

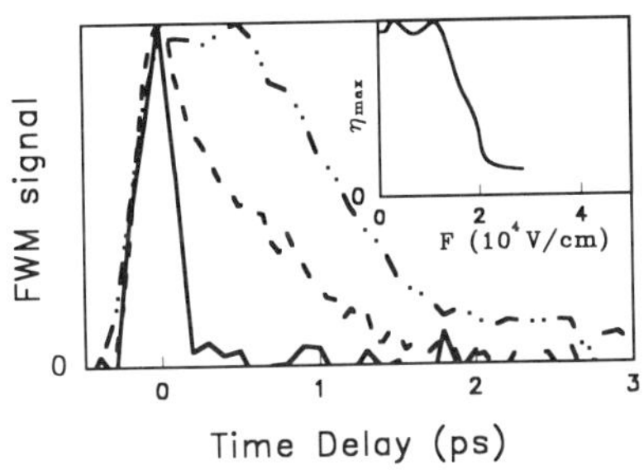

Figure 14. Experimentally determined FWM transients at applied electric fields of $0V/cm$ (dashed-dotted), $1.2 \cdot 10^4 V/cm$ (dashed), and $2.8 \cdot 10^4 V/cm$ (solid line). The inset shows the maximum diffracted FWM signal (first peak of the FWM transients) as a function of the applied field.

$36meV$ when laser pulses of $110fs$ pulse width are used. For the same reason, the WS ladder transitions can only be resolved in the photocurrent spectrum for large enough eFd-values. However, transient FWM experiments using laser pulses much shorter than $110fs$ should allow the observation of Bloch oscillations in the *WS regime* of superlattices with wide minibands.

CONCLUSIONS

We have reviewed incoherent and coherent time-resolved optical studies on electrically biased shallow multiple quantum well and superlattice structures. Incoherent pump and probe experiments on p-i-n structures allow to determine sweep-out times of photoinjected carriers, i.e., the time it takes for carriers to reach the p- and n-contact layers. However, this technique does not provide direct determination of the initial carrier escape times out of the individual quantum wells. In contrast, sub-picosecond four-wave mixing experiments provide direct access to the initial carrier dynamics. It thus becomes possible to study the coherent dynamics of optically created wavepackets and field-induced broadening of homogeneous linewidths of excitonic transitions, i.e., exciton ionization by field-induced carrier escape. We have discussed resonantly enhanced carrier escape out of shallow multiple quantum wells, Bloch oscillations of electrons in superlattices and their damping caused by LO phonon scattering.

ACKNOWLEDGEMENTS

We thank J. Shah, J. Sipe, S.L. Chuang, A.M. Fox, K. Leo, A. Bouchard, H. Grahn, M. Koch, B. Schlichtherle, and G. Weiser for stimulating discussions and M. Preis for excellent technical assistance. The work at Marburg University is financially supported by the Deutsche Forschungsgemeinschaft through the Leibniz-Förderpreis.

REFERENCES

1. C. Weisbuch and B. Vinter, 'Quantum Semiconductor Structures (Fundamentals and Applications)', Academic Press, San Diego (1991).
2. 'Physics of Quantum Electron Devices', F. Capasso, ed., Springer-Verlag, Berlin (1990).
3. M. Heiblum and M.V. Fischetti, 'Ballistic Electron Transport in Hot Electron Transistors', in: 'Physics of Quantum Electron Devices', F. Capasso, ed., Springer-Verlag, Berlin (1990).
4. T.C.L.G. Sollner, W.D. Goodhue, P.E. Tannenwald, C.D. Parker, and D.D. Peck, Appl. Phys. Lett. 43:588 (1983).
5. L. Esaki and R. Tsu, IBM J. Dev. 14:61 (1970).
6. A. Sibille, J.F. Palmier, H. Wang, and F. Mollot, Phys. Rev. Lett. 64:52 (1990).

7. Y. Masumoto, S. Tarucha, and H. Okamato, Phys. Rev. B33:5961 (1986).

8. M. Tsuchiya, T. Matsusue, and H. Sakaki, Phys. Rev. Lett. 59:2356 (1987).

9. D.Y. Oberli, J. Shah, T.C. Damen, C.W. Tu, T.Y. Chang, D.A.B. Miller, J.E. Henry, R.F. Kopf, N. Sauer, and A.E. DiGiovanni, Phys. Rev. B40:3028 (1989).

10. M.G.W. Alexander, W.W. Rühle, R. Sauer, and W.T. Tsang, Appl. Phys. Lett. 55:885 (1989).

11. T. Matsusue, M. Tsuchiya, J.N. Schulman, and H. Sakaki, Phys. Rev. B42:5719 (1990).

12. H.T. Grahn, H. Schneider, W.W. Rühle, K. von Klitzing, and K. Ploog, Phys. Rev. Lett. 64:2426 (1990).

13. R.J. Manning, P.J. Bradley, A. Miller, J.S. Roberts, P. Mistry, and M. Pate, Electron. Lett. 24:854 (1988).

14. G. Livescu, D.A.B. Miller, T. Sizer, D.J. Burrows, J.E. Cunningham, A.C. Gossard, and J.H. English, Appl. Phys. Lett. 54:748 (1989).

15. A.M. Fox, D.A.B. Miller, G. Livescu, J.E. Cunningham, and W.Y. Jan, IEEE J. Quantum Electron. 27:2281 (1991).

16. J. Feldmann, K.W. Goossen, D.A.B. Miller, A.M. Fox, J.E. Cunningham, and W.Y. Jan, Appl. Phys. Lett. 59:66 (1991).

17. G. Cohen and I. Bar-Joseph, Phys. Rev. B46:9857 (1992).

18. A. Miller, C.B. Park, and P. Li Kam Wa, Appl. Phys. Lett. 60:97 (1992).

19. J.A. Cavailles, D.A.B. Miller, J.E. Cunningham, P. Li Kam Wa, and A. Miller, IEEE J. Quantum Electron. 28:2486 (1992).

20. For a recent review, see D.A.B. Miller, Opt. and Quantum Electron. 22:61 (1990).

21. K.W. Goossen, J.E. Cunningham, and W.Y. Jan, Appl. Phys. Lett. 57:2582 (1990).

22. K.W. Goossen, J.E. Cunningham, and W.Y. Jan, Appl. Phys. Lett. 59:3622 (1991).

23. K. Leo, J. Shah, E.O. Göbel, J.P. Gordon, and S. Schmitt-Rink, Semicond. Sci. Technol. 7:B394 (1992).

24. J. Feldmann, K.Leo, J. Shah, D.A.B. Miller, J.E. Cunningham, T. Meier, G. von Plessen, A. Schulze, P. Thomas, and S. Schmitt-Rink, Phys. Rev. B46:7252 (1992).

25. J. Feldmann, in: 'Advances in Solid State Physics', U. Rössler, ed., Vieweg, Braunschweig (1992).

26. K. Leo, P.H. Bolivar, F. Brüggemann, R. Schwedler, and K. Köhler, Solid State Commun. 84:943 (1992).

27. G. von Plessen, J. Feldmann, E.O. Göbel, K.W. Goossen, D.A.B. Miller, and J.E. Cunningham, Appl. Phys. Lett., in press.

28. H.G. Roskos, M.C. Nuss, J. Shah, K. Leo, D.A.B. Miller, A.M. Fox, S. Schmitt-Rink, and K. Köhler, Phys. Rev. Lett. 68:2216 (1992).

29. C. Waschke, H.G. Roskos, R. Schwedler, K. Leo, H. Kurz, and K. Köhler, Phys. Rev. Lett. 70:3319 (1993).

30. D.A.B. Miller, D.S. Chemla, T.C. Damen, A.C. Gossard, W. Wiegmann, T.H. Wood, and C.A. Burrus, Phys. Rev. B32:1043 (1985).

31. A. Larsson, P.A. Andrekson, S.T. Eng, and A. Yariv, IEEE J. Quantum Electron. 24:787 (1988).

32. P.J. Price, Appl. Phys. Lett. 62:289 (1993).

33. J. Shah, in: 'Optics of Semiconductor Nanostructures', F. Henneberger, S. Schmitt-Rink, and E.O. Göbel, eds., Akademie-Verlag, Berlin (1993).

34. J.E. Avron, Annals of Physics 143:33 (1982).

35. F. Bloch, Z. Phys. 52:555 (1928).

36. C. Kittel, in: 'Quantum Theory of Solids', Wiley, New York (1963).

37. J. Zak, in: 'Solid State Physics', H. Ehrenreich, F. Seitz, and D. Turnbull, eds., Academic, New York (1972).

38. A. Nenciu and G. Nenciu, Phys. Lett. 78:101 (1980).

39. J.B. Krieger and G.J. Iafrate, Phys. Rev. B33:5494 (1986).

40. V.W. Houston, Phys. Rev. 57:184 (1940).

41. C. Zener, Proc. R. Soc. A145:523 (1934).

42. P. Roblin and M.W. Muller, Semicond. Sci. Technol. 1:218 (1986).

43. G. Nenciu, Rev. Mod. Phys. 63:91 (1991).

44. W. Franz, Z. Naturforschg. 13a:484 (1958).

45. L.V. Keldysh, Sov. Phys. JETP 34:788 (1958).

46. G.H. Wannier, Phys. Rev. B117:432 (1960) and Rev. Mod. Phys. 34:645 (1962).

47. J. Bleuse, G. Bastard, and P. Voisin, Phys. Rev. Lett. 60, 220 (1988).

48. E.E. Mendez, F. Agullo-Rueda, and J.M. Hong, Phys. Rev. Lett. 60:2426 (1988).

49. P. Voisin, J. Bleuse, C. Bouche, S. Gaillard, C. Alibert, and A. Regreny, Phys. Rev. Lett. 61:1639 (1988).

50. I. Bar-Joseph, J.M. Kuo, R.F. Kopf, D.A.B. Miller, and D.S. Chemla, Appl. Phys. Lett. 55:340 (1989).

51. H. Schneider, K. Fujiwara, H.T. Grahn, K. von Klitzing, and K. Ploog, Appl. Phys. Lett. 56:605 (1990).

52. A.M. Fox, D.A.B. Miller, J.E. Cunningham, W.Y. Jan, C.Y.P. Chao, and S.L. Chuang, Phys. Rev. B46:15365 (1992).

53. E.E. Mendez and G. Bastard, Physics Today, 34 (June 1993).

54. F. Beltram, F. Capasso, D.L. Sivco, A.L. Hutchinson, S.N.G. Chu, and A.Y. Cho, Phys. Rev. Lett. 64:3167 (1990).

55. D.E. Aspnes and A.A. Studna, Phys. Rev. B7:4605 (1973).

56. C. Coriasso, D. Campi, C. Cacciatore, C. Alibert, S. Gaillard, B. Lambert, and A. Regreny, Europhys. Lett. 16:591 (1991).

57. J.A. Kash, R.G. Ulbrich, and J.C. Tsang, Solid State Electron. 32:1277 (1989).

58. R. Tsu and G. Döhler, Phys. Rev. B12:680 (1975).

59. R. Ferreira and G. Bastard, Surf. Sci. 229:424 (1990).

60. S.M. Zakharov and E.A. Manykin, Izv. Akad. Nauk SSSR 37:2171 (1973).

61. G. von Plessen and P. Thomas, Phys. Rev. B45:9185 (1992).

62. L. Allen and J.H. Eberly, in: 'Optical Resonance and Two-Level Systems', Dover Publications, New York (1975).

63. L. Schultheis, A. Honold, J. Kuhl, K. Köhler, and C.W. Tu, Phys. Rev. B34:9027 (1986).

64. A.M. Bouchard and M. Luban, Phys. Rev. B47:6815 (1993).

65. M.M. Dignam and J.E. Sipe, Phys. Rev. B43:4097 (1991).

66. M. Dignam, J.E. Sipe, and J. Shah, submitted for publication.

67. G. von Plessen, T. Meier, J. Feldmann, E.O. Göbel, P. Thomas, K. Goossen, D.A.B. Miller, and J.E. Cunningham, submitted for publication.

AMPLITUDE DECAY AND INSTANTANEOUS FREQUENCY DYNAMICS OF EXCITONIC POLARIZATION IN SEMICONDUCTOR QUANTUM WELLS

J.-Y. Bigot[1], M.-A. Mycek[2], S. Weiss[2], R.G. Ulbrich[3], and D.S. Chemla[2]

[1] Institut de Physique et Chimie des Matériaux de Strasbourg, Unité Mixte 380046 CNRS - ULP - EHICS, 5, rue de l'Université, 67084 Strasbourg Cedex, France

[2] Physics Department, University of California at Berkeley and Materials Sciences Division, Lawrence Berkeley Laboratory, Berkeley, California 94720

[3] Department of Physics, University of Göttingen, Göttingen, Germany

INTRODUCTION

Coherent light scattering is an ideal tool to investigate the microscopic interactions that take place in various diluted or condensed phase systems[1]. It is of particular interest in semiconductors where the Coulomb interaction between electrons and holes brings new features in the dynamics of the polarization established by ultrashort optical pulses[2]. Due to this Coulomb coupling the coherent emission observed in different wave-mixing geometries behaves in a non-trivial fashion. The usual description of the material relaxation with a simple transverse time T_2, as is done for atomic-like systems[3], is no longer sufficient. Although the underlying physics is more complex, one may picture these different regimes by analogy with oscillators: the exponential T_2 behavior corresponds to a damped harmonic oscillator, while the Coulomb interacting system leads to a non-exponential damping, as with nonlinear coupled oscillators.

A spectacular manifestation of this non-exponential decay regime occurs near the band gap of semiconductors where excitons dominate the nonlinear refractive and absorptive optical properties[4]. Each exciton, in spite of its Boson character, "*feels*" the Pauli exclusion principle and the Coulomb interaction between the extended electronic states from which it is built. While the Pauli exclusion principle is a nonlinear density dependent effect, whose analogous effect in atomic systems corresponds to a population saturation, the Coulomb coupling between excitons is large at low exciton densities and has no counterpart in non-interacting two-level systems[5]. This Coulomb coupling leads to a non-exponential decay and to a dynamical frequency shift of the polarization. It is the aim of this review article to describe some experimental procedures that allow such behaviors to be revealed for excitons in GaAs quantum well structures (QWS).

In the first part we describe the different techniques which can be used to determine the nonlinear decay and frequency shift of a polarization wave. The study of the amplitude decay of Four-wave mixing (FWM) emission from GaAs QWS is reported in detail in the second part. Finally, in the third part, we will report recent results concerning instantaneous frequency variations of this FWM emission. These experiments, together with the amplitude decay study, represent a first attempt to fully characterize experimentally the coherent emission from Coulomb interacting systems.

I. Amplitude and Phase Characterization of Four-Wave Mixing Signals: definitions and techniques

The full characterization of an electromagnetic field requires the knowledge of its amplitude and its phase. From an experimental point of view, it is a difficult task to determine both of them. One usually uses cross-correlation techniques, where the signal to be measured is mixed with a known reference field in a nonlinear optical correlator which, in most cases, is a transparent nonlinear crystal combined with a precise optical delay line[6]. An alternative method is to perform an auto-correlation of the signal in the same type of crystal. These techniques have various limitations. First, the knowledge of the reference pulse itself is incomplete, particularly in short pulse spectroscopy where obtaining a perfectly Fourier-transformed pulse also requires its full characterization. Second, there is a limitation in the nonlinear process order obtained with the particular crystal used, which prevents high order auto-correlations. Third, another purely technical difficulty is the low conversion efficiency of the available nonlinear crystals, which becomes a major drawback for the detection of very weak signals such as those obtained in low excitation FWM experiments reported here. In spite of these limitations, we will show that a combination of intensity and interferometric correlations and power spectrum measurements give a reasonable understanding of the temporal evolution of both the amplitude and phase of the coherent emission from a dense system.

We consider a self-diffracted FWM experiment in which two pulsed beams, with wave vectors k_1 and k_2, spatially coincide on a sample at time t and $t - \Delta t$ to generate a signal with wave vector $k_s = 2k_2 - k_1$. The corresponding third order nonlinear polarization $P^{(3)}(t, \Delta t)$ induced contains information on the interactions which take place in the material. For a set of homogeneously broadened and independent two-level systems, all the information can be retrieved from the intensity of the time integrated signal $S_{ti}(\Delta t) \propto \int_{-\infty}^{\infty} \left|P^{(3)}(t, \Delta t)\right|^2 dt$ which, in the case of excitation with $\delta -$ function pulses, is known to decay[3] with a time constant $T_2/2$.

In the case of interacting excitons, it has been shown[7] that the polarization field $P^{(3)}(t, \Delta t)$ is a complicated function of the intrinsic exciton damping, phase space filling (due to the Pauli exclusion principle), and Coulomb interaction. For a single resonance, these effects reduce to three parameters T_2, Ψ_s, and V, respectively (see section II). Moreover, the theory demonstrates that a single measurement of $S_{ti}(\Delta t)$ is **not** sufficient to separate the influence of each mechanism on the dephasing dynamics. Intuitively, this can be understood from the fact that a population saturation is an instantaneous process (on the timescale of an optical cycle) while the Coulomb interaction between the $k-$states responsible for the exciton creation is a nonlocal effect in real space, which therefore takes time to build up. Although the time integrated technique allowed for the demonstration of the existence of the two contributions[8], it is inadequate to temporally separate each one.

Further improvements have been made by time resolving the FWM signal[9]. These measurements are based on an intensity cross-correlation of the emitted FWM field with the incident reference pulse intensity $|E|^2$. The observed quantity is given by : $S_{tr}(t, \Delta t) \propto \int_{-\infty}^{\infty} \left|P^{(3)}(t-u, \Delta t)\right|^2 |E(u)|^2 du$. This new step, which requires powerful reliable pulses, has been made possible by the use of self-modelocked Ti:Sapphire femtosecond lasers[10].

Another step beyond time resolved measurements was performed by addressing the experimental problem of complete phase retrieval of the excitonic polarization wave[11]. This was made by power spectra and interferometric correlation measurements. The measured quantities are the FWM power spectrum: $S_{PS}(\omega, \Delta t)$, the first order auto-correlation: $S_a(\tau, \Delta t) \propto \int_{-\infty}^{\infty} \left|P^{(3)}(u, \Delta t) + P^{(3)}(\tau - u, \Delta t)\right|^2 du$ and the first order cross-correlation: $S_c(\tau, \Delta t) \propto \int_{-\infty}^{\infty} \left|P^{(3)}(u, \Delta t) + E(\tau - u)\right|^2 du$. All three quantities can be measured for a fixed time delay Δt. Ideally one would also like to measure the second order interferometric auto- and cross-correlations to obtain a closer evaluation of the amount of chirp contained in the emitted FWM signal. This however appears as a very difficult experimental task for the particular case of excitonic polarization at low densities, owing to the few femtojoules available in the generated FWM signal.

The three quantities defined above already allow a further insight in the phase dynamics. A lineshape analysis of the power spectrum (PS) of the FWM signal gives two types of information. Its peak wavelength corresponds to the excited resonance while its width and symmetry account for the dynamics. For unperturbed two-level systems with T_2 exponential decay, the PS is a Lorentzian centered at the frequency Ω_0 of the *oscillating dipoles* and width $2/T_2$. Any departure from this functional form allows one to distinguish between independent or interacting dipoles. Such analysis is formally equivalent to the lineshape analysis introduced by R. Kubo for stochastic systems[12]. As discussed below (part III), in the context of excitons, it allows one to show three features: the exciton screening at high densities, the interplay of light and heavy hole bands for different excitation wavelengths, and an asymmetric shape at low excitation densities due to the exciton-exciton coupling. Although in theory $S_{PS}(\omega, \Delta t)$ and $S_a(\tau, \Delta t)$ are Fourier transforms of each other through Parseval's theorem, in practice they are measured over finite frequency or time windows and with a finite number of sampling points. Therefore depending on the information desired one of the techniques is better suited. Peak positions or long temporal delays are better resolved spectrally while large off-resonance asymmetries or *early* dynamics are easier to determine temporally. This consideration naturally depends on the particular system under study and more specifically on the temporal range of investigation. We believe that in the femtosecond domain first order interferometry becomes particularly relevant.

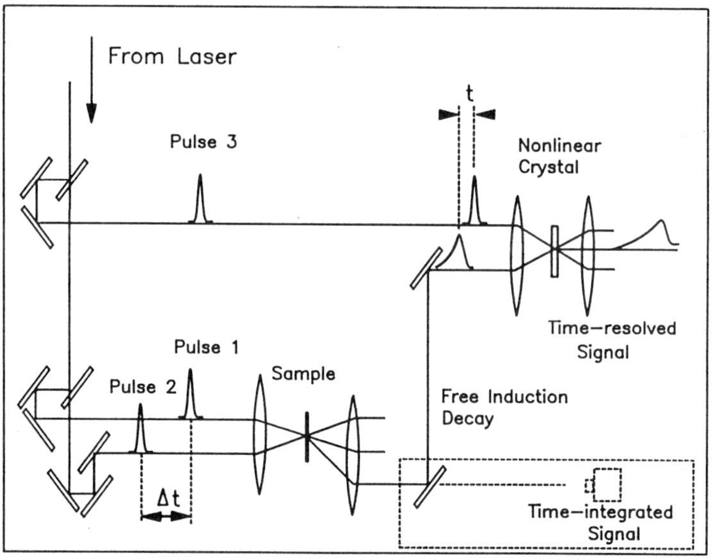

Figure 1. Experimental set-up used to measure the temporal profile of the FWM signal. For power spectra and interferometric auto- or cross-correlations measurements, the dotted inset is replaced respectively by an OMA, a Michelson or a Mach-Zehnder interferometer.

Additional interesting information can be retrieved from $S_a(\tau, \Delta t)$ by a careful analysis of the interferogram which we name Differential Fringe Spacing. This analysis is developped in section III where we show that, combined with cross-correlation interferometric and time resoved amplitude measurements, it gives a reasonable understanding of the instantaneous frequency dynamics.

A sketch of the experimental set-up showing the five types of measurements (S_{ti}, S_{tr}, S_{PS}, S_a, S_c) is displayed in Fig. 1. The laser beam from a tunable Ti : Al_2O_3 cavity delivering Fourier transformed pulses of about 100 fs duration at 88 MHz repetition rate is split into three parts. Two beams are used to generate the FWM signal. This signal can be either time integrated (TIS) and recorded as function of delay Δt between the two pulses, or time resolved (TRS) by cross-correlation with the third beam. The cross-correlation is performed by up-conversion in a 300 μm-thick KDP crystal. This configuration allows one to determine the time profile of the FWM signal for different delays Δt. Both delays Δt and t are produced by computer controlled stepper motors and the detection system is composed of photomultipliers and standard lock-in techniques. The power spectra measurements have been performed with a spectrometer and an OMA. The interferometric correlations are performed in Michelson (for auto-correlation) and modified Mach-Zehnder (for cross-correlation) interferometers whose delays τ are also mechanically varied with stepper motors.

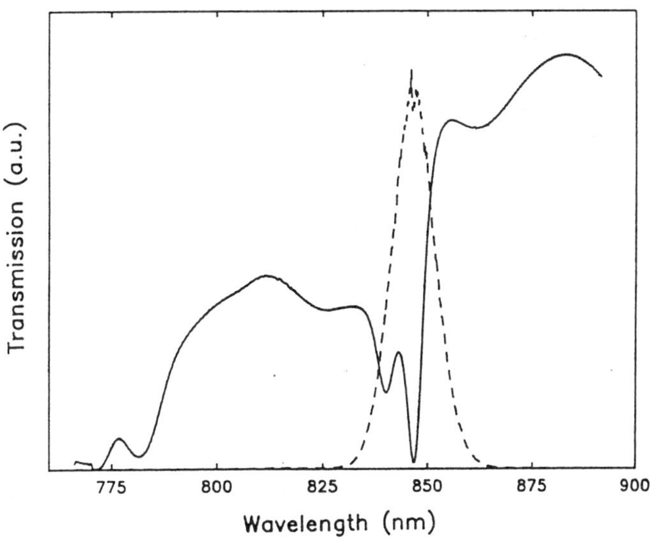

Figure 2. Transmission spectrum of the multiple quantum well structure (solid line) and spectrum of the laser (dashed line).

II. Amplitude Decay: time resolved and time integrated Four-Wave Mixing signals from resonantly excited excitons

In this section, we describe the effects of many-body interactions on the amplitude decay of the excitonic polarization of a GaAs quantum well structure.

The sample under study consists of 47 periods of 98 Å GaAs QW with 96 Å $Al_{0.3}Ga_{0.7}As$ barrier layers, the substrate having been removed by etching. The sample was antireflection coated to prevent spurious linear and nonlinear Fabry-Pérot effects and mounted on a sapphire substrate. The sample's transmission is shown in Fig. 2 together with the incident pulse spectrum tuned to the excitonic resonance. The experiments were made at room temperature.

Figure 3 represents the amplitude decays of the FWM signal for two different exciton densities N_x of $4 \times 10^{11}\text{cm}^{-2}$ (left) and 10^{11}cm^{-2} (right). Curves (b) correspond to the TI signals as function of Δt and curves (a) to the TR signals as function of t for $\Delta t = 0$. All traces have been normalized and time shifted in order to better show the different dynamics. A clear broader nonexponential decay is observed on the TR signal, while the TI traces display a slight exponential tail.

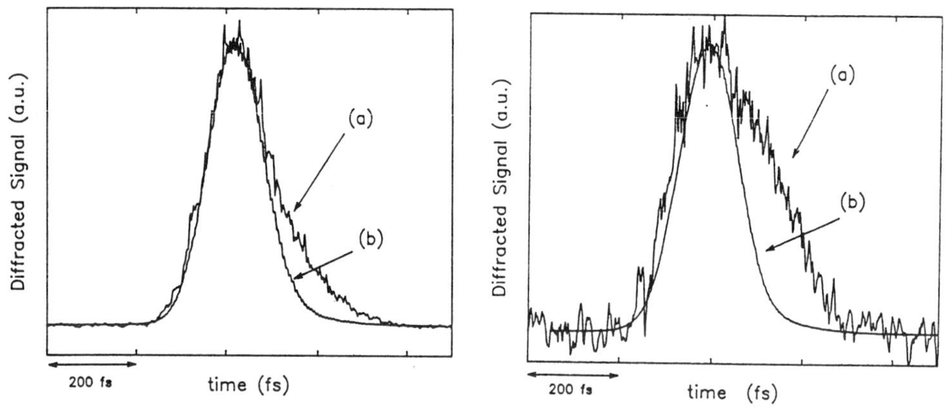

Figure 3. Time resolved (a) with $\Delta t = 0$ and time integrated (b) signals for $N_x = 4 \times 10^{11}\text{cm}^{-2}$ (left) and $N_x = 10^{11}\text{cm}^{-2}$ (right).

Before interpreting these results, let us consider the problem of lineshape homogeneity for excitons. At room temperature, collisional broadening with thermal phonons homogenizes the excitonic lineshape. This is in contrast with low temperature measurements, where the inhomogeneous broadening due to layer thickness fluctuations of the QWS has to be considered. This homogeneous nature of the system was checked several times in our experiment by measuring the temporal position of the intensity maximum of the TR signal for different Δt.

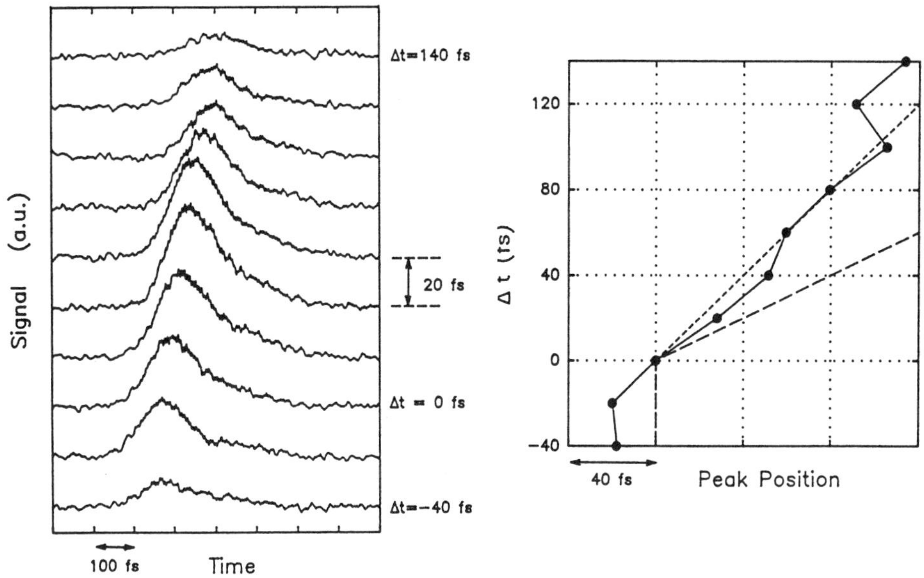

Figure 4. Time resolved signals for $N_x = 3 \times 10^{11}\text{cm}^{-2}$ and varying Δt (left) and relative position of the maximum time resolved signal (right). The dotted (dashed) line represents a homogeneous (inhomogeneous) system.

Fig. 4 (a) shows such a set of TR signals. The corresponding evolution of their maxima in $(t, \Delta t)$ coordinates is displayed in Fig. 4 (b) (black-circles). The dotted line, with slope 1, represents the expected behavior of an ideal free decay signal (homogeneous system) while the dashed one, with slope 2, represents a photon echo signal (inhomogeneous system). A first attempt in fitting the TI signals of Fig. 3 within the usual independent two-level system formalism gives a value $T_2 = 135$ fs. Obviously one cannot explain the TR signals with this single constant. Indeed, these results can be well understood and fitted by using the excitonic wave function formalism. In this theory, within the Bloch sphere approximation, the exciton amplitude $\Psi(t)$ can be shown to satisfy a nonlinear Schrödinger equation[13] :

$$\frac{\partial \Psi(t)}{\partial t} = -i(\Omega - i/T_2)\Psi(t) + i\mu E(t) - i\mu E(t)\frac{|\Psi(t)|^2}{\Psi_s^2} - iV|\Psi'(t)|^2\Psi(t)$$

where μ is the interband dipole moment, Ω the exciton resonance frequency and the other parameters were defined previously. A numerical solution of this equation for the polarization emitting in the direction k_s, using Gaussian pulses having the duration used in the experiment, gives the TR signals represented in Fig. 5 (full curve). The fitting parameters $V\Psi_s^2$ and T_2 are, respectively, 0.01 and 130 fs (0.01 and 190 fs) for the two exciton densities corresponding to Fig. 3.

The two nonlinear terms, proportional to $|\Psi(t)|^2$ in the above equation, contribute to the observed behavior. The first one originates from the Pauli exclusion principle between the k-space extended states which constitute the exciton. It takes into account phase space filling (PSF) effects in the semiconductor. Its contribution to the signal is represented in dashed dotted line in Fig. 5. The second one, proportional to V, accounts for exciton-exciton interactions (XXI). It is represented as a dashed line and is seen to prevail at low densities of excitation (right curve), while PSF is predominant at high densities (left curve). Both contributions are delayed with respect to the temporal origin $t = 0$. For the PSF term, this delay originates from the finite pulse duration. It is due to a convolution between the building signal which follows the pulse arrival and the relaxation driven by the damping constant T_2, which starts as soon as dipoles are created. In the limit of infinitely short pulses, this PSF contribution is instantaneous. In contrast, the delayed XXI contribution originates from the building of the polarization wave itself since it is proportional to $\Psi(t)$. Importantly, it remains delayed even for δ – function pulses.

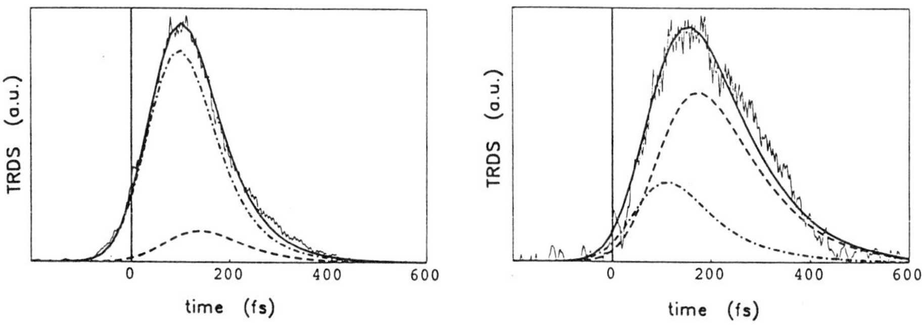

Figure 5. Fit of the time resolved signals corresponding to Fig. 3. The dashed-dotted (dashed) line corresponds to the PSF (XXI) interaction terms.

These differences are at the origin of the main features observed in the non-exponential exciton dephasing. The two source terms act differently. First, the Rabi frequency term μE establishes coherence into the excitonic system; a process which is completely driven by the applied field, and which may be reduced when population saturation (or PSF, in the present case) occurs for high applied field. Second, the created dipoles also establish coherence, but with a source term which is the polarization wave itself. This leads to some delay in its contribution. This second source term has a formal analogy with a local field effect in the case of a single excitonic resonance, in the sense of an electromagnetic field being modified by the induced polarization. Strictly "*speaking*" this terminology is incorrect and may lead to confusion. One must remember that for excitons the coupling term V contains all the information on the extended states of the semiconductor. In this sense it is a completely non-local interaction.

The fit of Fig. 5 is made for $\Delta t = 0$. For low densities ($N_x \lesssim 5 \times 10^{11}\mathrm{cm}^{-2}$) it was found that the entire set of TR traces, obtained with varying Δt, can be reproduced numerically with the same constant $V\Psi_s^2 = 0.01$. The relaxation time T_2, however, had to be reduced from 190 fs to 160 fs for negative Δt. This effect is possibly due to an additional dephasing source term brought by free carriers created by exciton ionisation through exciton-phonon collisions. In our experiment pulse 1 is more intense than pulse 2. Therefore, the amount of free carriers generated by this exciton ionisation mechanism is higher when pulse 1 arrives before pulse 2, i.e. for $\Delta t < 0$, leading to a slightly shorter dephasing time T_2.

Figure 6 shows different TR traces obtained for increasing exciton densities at $\Delta t = 0$. In the high density regime ($N_x \geq 10^{12}\mathrm{cm}^{-2}$), the effect of XXI becomes negligible. The fitting process with the nonlinear Schrödinger equation then requires shorter dephasing times and lower parameters $V\Psi_s^2$. At such high densities the band gap renormalization becomes important and leads to the dissociation of excitons and the creation of free carriers. This in turn perturbs the phase and reduces the coupling by screening the Coulomb interaction[14,15]. This last effect acts as the reverse of what occurs in the process of Dicke superradiance for molecular systems[16]. While superradiance is a collective effect which shows up at high densities, excitonic coupling is a low density effect which originates from the large coherence range of an exciton.

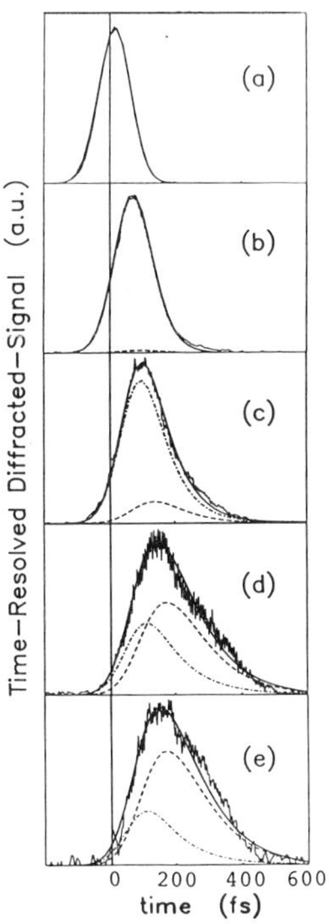

Figure 6. Time resolved signals measured at $\Delta t = 0$ for exciton densities of (a) 10^{13}, (b) 10^{12}, (c) 4×10^{11}, (d) 2×10^{11} and (e) $10^{11}\mathrm{cm}^{-2}$. The solid line corresponds to a fit including both XXI (dashed line) and PSF (dashed-dotted line) contributions.

For the highest intensities, the above model is no longer adequate to describe the dephasing dynamics and higher order many body interactions have to be considered. The excitonic resonance is then completely screened, as shown unambiguously in the power spectra measurements reported in the next section.

III. Instantaneous Frequency Dynamics: an attempt to retrieve the phase of the excitonic polarization wave

In this section, we concentrate on the frequency dynamics of the FWM signal for different excitation wavelengths of the incident laser beam tuned close to the excitonic resonance.

A lineshape analysis of the FWM signal generated from excitons allows a first insight into the phase dynamics. The degeneracy of the excitonic resonance in GaAs QW structures is lifted at $k = 0$ into heavy-hole (hh) and light-hole (lh) excitons with respective resonances $\Omega_{\mathrm{hh}} = 1.467$ eV and $\Omega_{\mathrm{lh}} = 1.482$ eV. Since their oscillator strengths are different, three experimental situations, corresponding to different laser wavelengths Ω_{L}, are distinguished in the following. They are referenced with respect to the hh exciton by the detuning parameter $\delta\omega = \Omega_{\mathrm{hh}} - \Omega_{\mathrm{L}}$. The first case corresponds to the excitation of a single resonance, the hh exciton, obtained with $\delta\omega = 14$ meV. In the second one, the detuning is set to $\delta\omega = -\ 9$ meV and both resonances are excited with, however, more emphasis on the hh exciton. The third situation corresponds to an excitation even further above resonance, so that equal weights are given to both excitons, with a detuning set to - 13 meV.

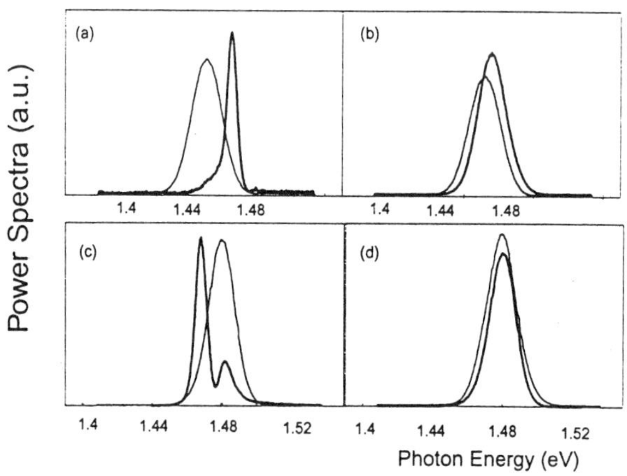

Figure 7. Power spectra of the FWM signal (thick lines) and laser pulse (thin lines) for different exciton densities N_x and detunings $\delta\omega$: $N_x = 3 \times 10^9 \mathrm{cm}^{-2}$ (a,c), $3 \times 10^{11} \mathrm{cm}^{-2}$(b,d) and $\delta\omega = 14$ meV (a,b), -9 meV (c,d).

FWM power spectra $S_{PS}(\omega, \Delta t = 0)$ are shown in Fig. 7 for the first two cases and for two different exciton densities. Note that the exciton density is calculated by considering the total absorption of the pulse without any lineshape considerations. The pulse spectrum is also displayed in the thin line. For a very low density $N_x = 3 \times 10^9 \text{cm}^{-2}$ (curves (a) and (c)), $S_{PS}(\omega, \Delta t = 0)$ is maximum near the hh resonance, showing the resonant character of the FWM emission. For a higher density $N_x = 3 \times 10^{11} \text{cm}^{-2}$ (curves (b) and (d)), the maximum of S_{PS} almost coincides with the one of the laser. For intermediate intensities (not shown here) the maximum of the FWM power spectrum gradually shifts from Ω_{hh} to Ω_L.

The behavior of Fig. 7, although recorded for different incident intensities, suggests a dynamical frequency shift of a single FWM emission. This is reinforced by the asymmetric shape of S_{PS} which is clear in the case of single resonance excitation (Fig. 7 (a)). From the lineshape discussion in section I, this is the signature of a non-exponential decay. It is interesting to notice that this asymmetry disappears at high densities. This is consistent with the fact that XXI, which is at the origin of the non-exponential decay, dominates at low densities and vanishes at high densities when PSF prevails. It is also worth noticing that the excitonic resonance is almost completely bleached for densities of the order of $5 \times 10^{11} \text{cm}^{-2}$. For this density regime, dephasing by free carriers becomes predominant as discussed in the preceeding section. This result is in complete agreement with previous measurements on two-dimensional carrier-carrier screening performed with 8 fs pulses at room temperature on a similar system[17].

The dynamic frequency shift suggested by the FWM power spectra can be further exhibited by interferometric correlations. A complete characterisation of the phase is not feasible experimentally for the reasons given in part I. However, we found that a data analysis of first order auto-correlations $S_a(\tau, \Delta t)$ in terms of the dynamic fringe spacing, taking into account the amplitude decay measurements, gives results close to the temporal evolution of the instantaneous frequency of the FWM emission. The analysis is performed in the following way. For each delay Δt, we have first measured the dynamic fringe spacing (DFS(τ)), defined as the number of fringes during the delay τ, for $S_a(\tau, \Delta t)$ and the laser pulse. Both measurements are made passing through the same optical elements, including the sapphire holder of the sample, so that any phase distortion is compensated. We then compute the differential fringe spacing defined as $\Delta F(\tau) = \text{DFS}_{\text{laser}}(\tau) - \text{DFS}_{\text{FWM}}(\tau)$. For a linear temporal dependence of the phase, ΔF corresponds exactly to the phase. For more complicated temporal dependence, there is in general no simple relation between the two. However, for *well behaved* amplitude signals, i.e. without any abrupt change over a few optical cycles, we have noticed a very good agreement between the temporal dependence of the phase and ΔF. This dependence was verified numerically on a large number of trial functions having a smooth temporal dependence. We should stress that, at the present time, we have no mathematical proof of this behavior and although the criterion *well behaved* is vague, we believe that it applies to a number of experimental situations. Naturally, this analysis relies on the knowledge of both the temporal amplitude variation and *the sign* of the phase change relative to the laser. This sign was determined in our case by cross-correlation of the FWM signal with the laser.

In order to fully characterize our set-up, we first applied the above technique to the laser beam itself, the reference being in this case the same laser but running CW. The differential fringe spacing indicates no chirp in the pulse. This result was confirmed by the usual second order interferometric autocorrelations (which are possible in this case due to the high power available in the pulse). We also made a careful investigation of what may bring distortion in these kinds of measurements if one does not proceed in a differential way. This can be done easily with a CW laser. We found that the most significant distortions are produced by the stepper motors. They can be of the order of $\lambda/4$ over a few hundreds of femtoseconds for the 0.1 μm-step motors commonly available.

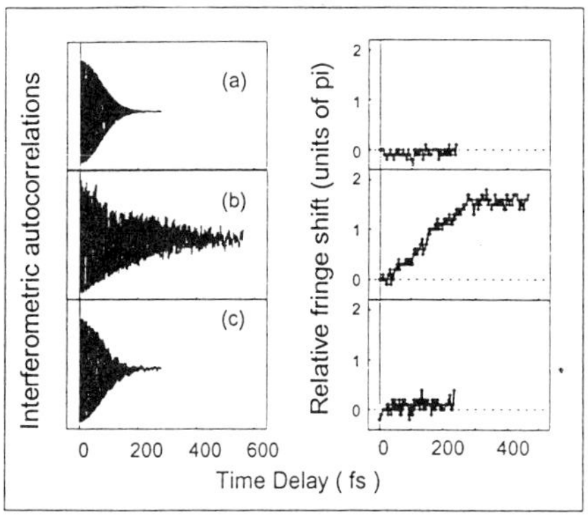

Figure 8. Interferometric autocorrelation and differential fringe spacing for the laser (a), the low density FWM signal (b) and the high density FWM signal (c). The corresponding power spectra of (b) and (c) are displayed in Figs. 7(a) and (b).

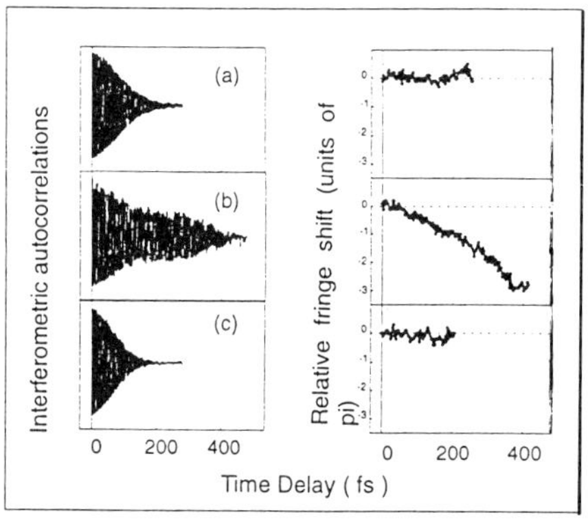

Figure 9. same as Fig. 8 with $\delta\omega$ = - 9 meV. The corresponding power spectra are displayed in Figs. 7(c) and (d).

Figure 8 shows $S_a(\tau, \Delta t)$ and $\Delta F(\tau)$ for: the reference laser pulse (a), the low density FWM signal (b), the high density FWM signal (c). Curves (b) and (c) are obtained for the detuning $\delta\omega = 14$ meV and therefore correspond to the power spectra (a) and (b) in Fig. 7. Figure 9 shows the same type of curves but for the detuning $\delta\omega = -9$ meV (spectra (c) and (d) in Fig. 7).

In both cases the high intensity results show no particular change in the frequency dynamics. This is expected in the model developed in section II where we have seen that, in the high density regime, XXI become negligible. The additional information obtained here is that, at higher density, the system is completely driven by the laser and the FWM signal is emitted at the laser frequency Ω_L.

For low excitation density and $\delta\omega > 0$, an initial linear increase in the $\Delta F(\tau)$ curve simply reproduces the chosen detuning. After about 200 fs it saturates, indicating a *"low energy shift"* of the resonance. For $\delta\omega < 0$, the same type of shift is observed in addition to the trivial linear increase which now has a negative slope due the sign of $\delta\omega$.

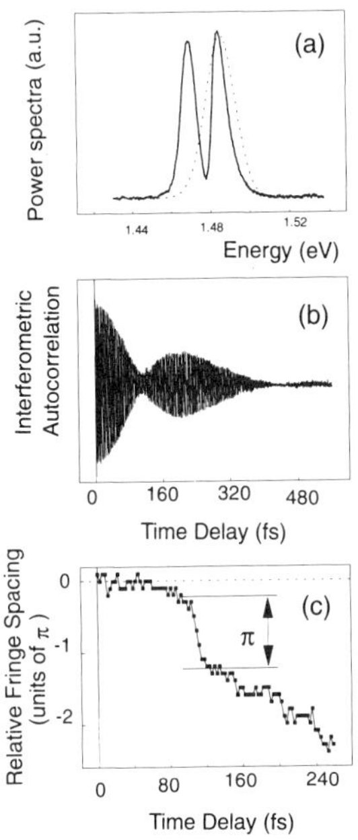

Figure 10. Power spectrum (a), interferometric autocorrelation (b) and differential fringe spacing of the FWM signal at low exciton densities and equal contributions of the hh and lh excitonic resonances.

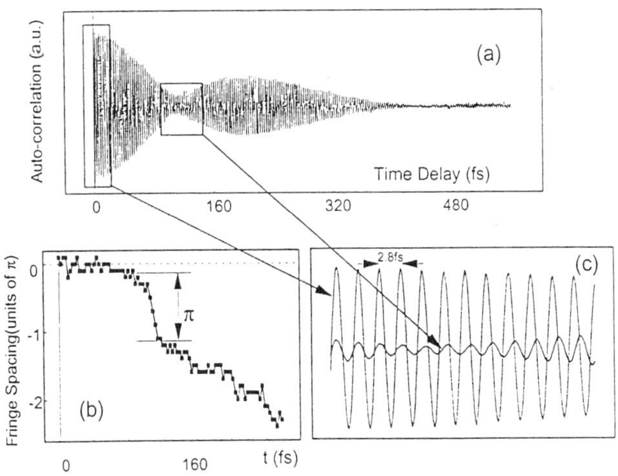

Figure 11. Detail around the beginning and the first node of the interferogram shown in Fig. 10 (b). The π shift occurs within about 15 optical cycles.

There is at the present time no complete theory that models the results shown in figures 8 and 9. The model discussed in section II gives only a qualitative understanding of the instantaneous frequency dynamics. For a full description, one has to take into account the two excitonic resonances and possibly go beyond a mean field description of the XXI, which was assumed in the numerical solution of the nonlinear Schrödinger equation to fit the TR amplitude decay.

Another set of differential fringe spacing measurements was taken for the detuning $\delta\omega = -13$ meV. Figure 10 (a) shows the FWM power spectrum obtained at low density ($N_x = 9 \times 10^9 \text{cm}^{-2}$) for $\Delta t = 0$. $S_a(\tau, \Delta t = 0)$ and $\Delta F(\tau)$ are represented in Fig. 10 (b) and (c), respectively. An equal contribution from the hh and lh excitons being optimized for this detuning, quantum beats clearly show up with a period $\Delta T = 230 \pm 10$ fs. This time period corresponds to the energy splitting $\Delta E_{\text{hh-lh}} = 15$ meV between the two excitons. Such quantum beats have been observed recently in the modulation of the FWM intensity profile[18]. The new information obtained here is the dynamical frequency change during the occurence of the beating process. This is shown in Fig. 10 (c) where a corresponding π shift is obtained in the $\Delta F(\tau)$ curve around the first node. A detailed view of this effect is shown in Fig. 11. Close-ups of the fringe dynamics are shown at the beginning and around the first node of $S_{\text{a}}(\tau, \Delta t = 0)$. It is seen that the π shift occurs within an elapsed time of about 15 optical cycles. These interferograms are unsmoothed and correspond to an experimental accuracy of 0.13 fs.

We believe that, apart from the quantum beat phenomenon observed directly in the frequency dynamics, this type of measurement opens new possible investigations in the domain of femtosecond spectroscopy.

CONCLUSION

Time resolved studies of the polarization decay induced by resonant excitation of excitons in semiconductor quantum well structures have revealed new aspects of coherent emission in condensed matter systems. We have shown that exciton-exciton Coulomb coupling leads to non-exponential decays of Four-Wave Mixing signals emitted by excitons. This decay is the manifestation of a coherent polarization wave that builds up in the material with a certain delay as compared to the incident laser pulse. The decay properties of this polarization wave contrast with the usual polarization free decay observed in atom-like two-level systems. In particular, it leads to an instantaneous frequency shift of the emission, an effect which disappears for high exciton densities when phase space filling and/or exciton screening occur. The ultrafast spectroscopy techniques employed to study the time resolved profile and the instantaneous frequency dynamics of the coherent signal become very reliable and accurate. It is very likely that development and improvements of these techniques, in particular for phase measurements as reported here, will allow a better understanding of the processes that take place during the interaction of a coherent pulse with a condensed phase material.

REFERENCES

1. See, for example, L. Allen and J.H. Eberly, "Optical Resonances and Two Level Atoms" Wiley, New York (1975).
2. See, for example, E.O. Göbel in : "Festkörperprobleme: Advances in Solid State Physics" Vieweg, Braunschweig, Vol. 30, p. 269 (1990).
3. See, for example, T. Yajima and Y. Taira, *J. Phys. Soc. Jpn.* 47:1620 (1979).
4. See, for example, "Excitons", E.I. Rashba and M.D. Sturge eds., North Holland, Amsterdam, (1982); Optical Nonlinearities and Instabilities in Semiconductors, H. Haug ed., Academic, New York (1988).
5. S. Schmitt-Rink, D.S. Chemla, and D.A.B. Miller, *Adv. Phys.* 38:89 (1989)
6. See, for example, W. Rudolph and B. Wilhelmi, "Light Pulse Compression", V.S. Letokhov, C.V. Shank, Y.R. Shen, and H. Walther eds. Harwood Academic Publishers, London, Chapter 6, p. 108 (1999).
7. S. Schmitt-Rink and D.S. Chemla, *Phys. Rev. Lett.* 57:2752 (1986)
   S. Schmitt-Rink, D.S. Chemla, and H. Haug, *Phys. Rev. B* 37:941 (1988)
   A.V. Kuznetsov, *Phys. Rev. B* 44:8721 (1991).
8. M. Wegener, D.S. Chemla, S. Schmitt-Rink, and W. Schäfer, *Phys. Rev. A* 42:5665 (1990); K. Leo, M. Wegener, J. Shah, D.S. Chemla, E.O. Göbel, T.C. Damen, S. Schmitt-Rink and W. Schäfer, *Phys. Rev. Lett.* 65:1340 (1990).
9. S. Weiss, M.-A. Mycek, J.-Y. Bigot, S. Schmitt-Rink, and D.S. Chemla, *Phys. Rev. Lett.* 69:2685 (1992); D.S. Kim, J. Shah, T.C. Damen, W. Schäfer, F. Jahnke, S. Schmitt-Rink, and K. Kohler, *Phys. Rev. Lett.* 69:2725 (1992).
10. D.E. Spence, P. N. Kean, and W. Sibbett, *Opt. Lett.* 16:42 (1991).
11. J.-Y. Bigot, M.-A. Mycek, S. Weiss, R.G. Ulbrich, and D.S. Chemla, *Phys. Rev. Lett.* 70:3307 (1993).
12. R. Kubo, M. Toda, and N. Hashitsume, "Statistical Physics II: Nonequilibrium Statistical Mechanics" Springer Series in Solid State Science, Vol. 31, Springer-Verlag, Berlin (1985).
13. M.-A. Mycek, J.-Y. Bigot, and D.S. Chemla (to be published).
14. L. Schultheis, J. Kuhl, A. Honold, and C.W. Tu, *Phys. Rev. Lett.* 57:1635 (1986); A. Honold, L. Schultheis, J. Kuhl, and C.W. Tu, *Phys. Rev. B* 40:6442 (1989).
15. M. Lindberg, R. Binder, and S.W. Koch, *Phys. Rev. A* 45:1865 (1992).
16. R.H. Dicke, *Phys. Rev.* 93:99 (1954); A.M. Ponte Gonçalves and A. Tallet *Phys. Rev. A* 1:1472 (1970).

17. J.-Y. Bigot, M.T. Portella, R.W. Schoenlein, J.E. Cunningham, and C.V. Shank, *Phys. Rev. Lett.* 67:636 (1991).
18. E.O. Göbel, K. Leo, T.C. Damen, J. Shah, S. Schmitt-Rink, W. Schäfer, J.F. Müller, and K. Köhler, *Phys. Rev. Lett.* 64:1801 (1990); K. Leo, J. Shah, E.O. Göbel, T.C. Damen, S. Schmitt-Rink, W. Schäfer, J.F. Müller, and K. Köhler, *Phys. Rev. Lett.* 66:201 (1991).

DEPHASING TIME MEASUREMENTS IN QUANTUM DOTS BY NON–DEGENERATE FOUR WAVE MIXING

A. Uhrig, U. Woggon, M. Portuné, V. Sperling, and C. Klingshirn

Fachbereich Physik, Universität Kaiserslautern
E.-Schrödinger-Str., D-67663 Kaiserslautern
Germany

INTRODUCTION

Studies on the relaxation process of coherently excited polarizations associated with excited states are interesting to reveal the mechanisms of dephasing in quantum confined semiconductor systems. In two-dimensional systems like quantum wells (QW) or superlattices (SL), the most important mechanisms considered are the scattering of the excited electron-hole pairs with phonons (LO-phonons or confined acoustic phonons), with interface potential fluctuations or localized states or, at higher excitation densities, the scattering by carrier–carrier interaction. In QD's, at present, the relative contribution of these mechanisms to the dephasing process is an open question and one is trying to identify the different dephasing processes from the experiment. To determine ultrafast relaxation times in the range below 100 fs, various nonlinear spectroscopic methods in the frequency domain have been applied, like hole-burning spectroscopy [1–3] or coherent transient spectroscopy using broadband incoherent light [4,5]. In the present paper the method of nondegenerate four-wave mixing (NDFWM) has been used to derive the dephasing time T_2. The samples investigated are CdS and $CdS_{0.7}Se_{0.3}$ quantum dots in glass with radii 2 to 3 times larger than the Bohr radius. The large sizes were chosen to minimize the interface influence and by this to investigate the electron-hole pair states before trapping. The minor importance of trap states have been deduced from the dominance of a sharp band-edge related luminescence for these large dot sizes [6]. The dephasing times determined are consistent with results from differential absorption spectroscopy [7,8] at comparable excitation densities. As a new phenomenon we discuss the influence of carrier–carrier scattering in the weak confinement range and determine the dephasing time of this process by increasing gradually the excitation intensity.

EXPERIMENTAL METHOD

Two tunable laser beams with frequencies $\omega_1 \neq \omega_2$ and different wave vectors

$\vec{k}_1 \neq \vec{k}_2$ intersect in a sample to produce a nonstationary intensity distribution with a temporal modulation of the maximum intensity with a frequency $\omega_3 = 2\omega_1 - \omega_2$ ("moving grating" (see Fig. 1)). For the measurements of the diffraction efficiency, the beam of the laser 1 at ω_1 was usually fixed in frequency and the output light intensity at ω_3 was measured during the stepwise scanning of laser 2 with ω_2. The strength of the decrease of the diffraction efficiency with increasing detuning between the both lasers is a measure for the dephasing time T_2. The theoretical description of the NDFWM

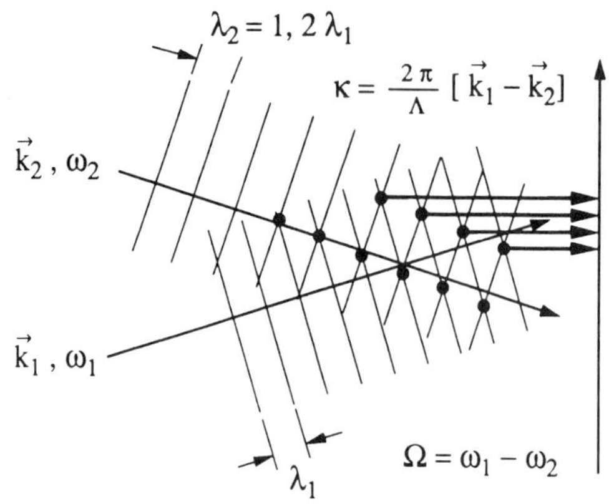

Fig. 1. *Nondegenerate four wave mixing, scheme of the experiment showing the formation of the moving grating by interference of two lasers with different frequencies* ω_1 *and* ω_2.

for resonant excitation of an inhomogeneously broadened optical transition has been developed in [9]. In the present stage of application on quantum dots the two-level model has been used keeping in mind that this is only a first approximation. For the output light intensity I at $\omega_3 = 2\omega_1 - \omega_2$ holds $I \sim |\chi^{(3)}|^2$ with

$$|\chi^{(3)}|^2 = \frac{|K_3|^2}{(1+\Omega^2 T_1^2)(1+\Omega^2 T_2^2)} \qquad \text{with} \qquad \Omega = \omega_2 - \omega_1 \qquad (1)$$

where $K_3 = -2\pi\ |d_{21}|^4\ N\rho_0\ g(\Omega)\, T_1\ T_2\, \hbar^{-3}$ with d_{21} the dipole matrix element, N the density of two-level atoms, and $g(\Omega)$ the size distribution function normalized in energy. The change around $\Omega = 0$ is caused by the value of the lifetime T_1. A sharp peak is expected if T_1 is large compared to T_2. The tails in the detuning curves were determined by the dephasing time T_2. The ratio of the signal intensity at zero detuning to the signal intensity for large values of Ω is determined by the ratio in the times T_1 and T_2. For fast dephasing times T_2 the absolute value of the diffracted beam is generally decreased and the detection of signals needs high sensitivity. If the T_1 time is short and in the same order of magnitude as the T_2 time, then the contributions arising from the population grating and the polarization grating can no longer be clearly distinguished and the analysis of T_2 requires a wide dynamic range in detection of $|\chi^{(3)}|^2$.

RESULTS AND DISCUSSION

Using a high-quality sample containing CdS quantum dots with well known radii of $R = 7.5$ nm and a size deviation of $\pm 15\%$ the intensity dependence of the dephasing time has been investigated. The lowest quantum confined states have been identified by differential absorption spectroscopy (DAS) at T = 20 K (see [8]) at an energy of 2.65 eV. The DAS at low excitation gives for the homogeneous linewidth Γ an estimated value of 20 meV.

The measurements of NDFWM have been carried out at T = 300 K at an excitation energy at 2.6 eV, which corresponds to the maximum of the spectral change in the refraction coefficient. The lifetime $T_1 = 40$ ps applied in the fitting procedure has been derived from the luminescence decay curve. The results are shown in Fig. 2. The symmetric change for positive and negative detuning supports the approximation

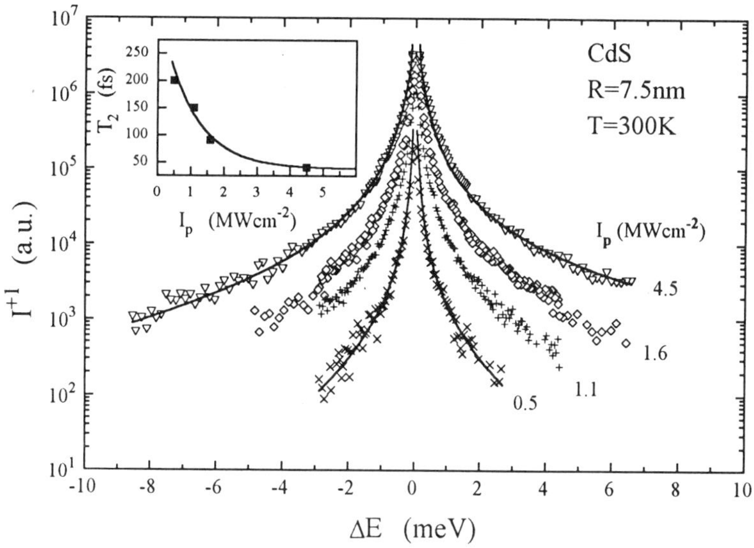

Fig. 2. *Diffraction efficiency measured for CdS quantum dots of $R = 7.5$ nm. The inset shows the dephasing times obtained from eq.(1) as a function of excitation intensity.*

based on the two-level model without intensity-dependent modification of the lifetime as well as the neglection of nonresonant contributions in the fitting process. As calculated in [10] the consideration of trap states results for the diffraction efficiency in an asymmetric lineshape on the high and low energy side around zero detuning.

At low excitation intensities of 500 kW/cm^2 implying less than one electron-hole pair per dot, the dephasing time has been determined to $T_2 = 200$ fs. This is one of the largest values reported up to now for the dephasing time in CdS quantum dots and ascribed to the very low excitation densities used in the NDFWM measurements. In this range of excitation the dephasing is mainly attributed to the interaction of the excited electron–hole pairs with phonons (see below).

For excitation intensities of 1 MW/cm^2 and more, a significant increase in the wings of the detuning curves can be seen. The excitation of a few electron-hole pairs inside the dot is accompanied by the increase of carrier – carrier scattering overwhelming all other dephasing channels. The characteristic time constant for this scattering mechanism is 20 fs, derived from the limit at highest excitation densities. This is in good agreement with values known from the bulk, e.g. in electron–hole plasmas, for this scattering process resulting in a loss of phase coherence.

The influence of the temperature has been investigated at a $CdS_{0.7}\,Se_{0.3}$ commercial glass (OG 550) with quantum dots of $R \approx 7$ nm corresponding approximately to

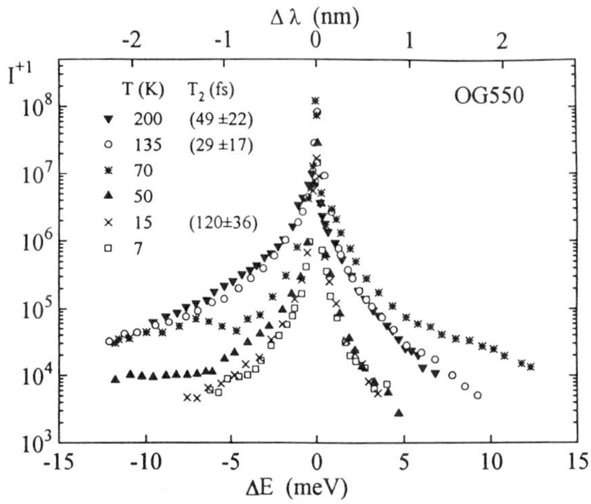

Fig. 3. *Temperaturdependence of the dephasing for an OG 550 filter for pump intensity of* 30 MW/cm^2

two times the Bohr radius (Fig. 3). The excitation energy has been chosen resonantly to the quantum confinement state to exclude excitation of the localised states in the tail of the absorption spectrum typically found in mixed crystals [12].

For temperatures of 7 and 15 K the dephasing time is nearly constant at values of $\geq$ 120 fs. With increasing temperature the phase relaxation time decreases down to a value around 30 fs for 200 K.

This behaviour is in good agreement with the temperature dependence measured for the homogeneous linewidth in hole–burning experiments for pure CdSe dots [7]. In these experiments only a gradual increase of Γ below 50 K has been found. However a considerable change occured above this threshold while in [10,11] a linear dependence was found.

On the one hand as reason for the temperature dependence of the phase relaxation in [7] the change of the electron – phonon coupling caused by the differences in the wavefunctions for electron and holes due to the influence of Coulomb interaction were discussed. On the other hand for temperatures above 50 K an additional contribution appears by the increasing coupling of the carriers with LO phonons. The temperature dependent dephasing yields a modification of the homogeneous linewidth of a factor of more than 2 between 7 and 200 K.

For low temperatures the contribution to the dephasing presumably will be a remainder scattering at surface states and, also, at this high pump intensity of 30 MW/cm^2 a further dominance of the carrier – carrier scattering.

CONCLUSION

NDFWM is a useful tool for investigating isolated quantum dots in glasses because the diffusion process is of minor importance, the expected T_2 times lie in the range below 1 ps and are also small compared to the lifetime T_1.
Investigations depending on the excitation intensity and temperature allow us to separate the different contributions to the total dephasing rate. The set in of carrier – carrier scattering with a mimimum dephasing time of 20 fs as the main contribution to the homogeneous linewith in the intensity dependent experiments hints to the excitation of more than one electron–hole pair in one dot. The temperature dependent part of the phase relaxation time shows the same behaviour as measurements of the homogeneous linewidth in DAS. Further investigations especially under low excitation and

low temperature should show a more pronounced picture of the dephasing mechanism in this regions.

Acknowledgement

The authors are grateful to B. Hönerlage, Strasbourg, for discussion and M. Müller, Jena, for the support for growing CdS samples. The work has been supported by the Deutsche Forschungsgemeinschaft and by the Materialforschungsschwerpunkt der Universität Kaiserslautern.

References

[1] A.P. Alivisatos, A.L. Harris, N.J. Levinos, M.L. Steigerwald, and L.E. Brus, J. Chem. Phys. **89**, 4001 (1988).

[2] N. Peyghambarian, B. Fluegel, D. Hulin, A. Migus, M. Joffre, A. Antonetti, S.W. Koch, and M. Lindberg, IEEE J. Quant. Electr. **QE-25** ,2516 (1989).

[3] P. Roussignol, D. Ricard, and C. Flytzanis, Phys. Rev. Lett. **62**, 312 (1989).

[4] N. Morita and T. Yajima, Phys. Rev. **A 30**, 2525 (1984).

[5] G.L. Huang and H.S. Kwok, J. Opt. Soc. Am. B **9**, 2019 (1992).

[6] A. Uhrig, L. Banyai, Y.Z. Hu, S.W. Koch, C. Klingshirn, N. Neuroth, Z. Phys. B **81**, 385 (1990).

[7] U. Woggon, S. Gaponenko, W. Langbein, A. Uhrig, C. Klinghshirn, Phys. Rev. **B 47**, 3684 (1993).

[8] U. Woggon, A. Uhrig, M. Saleh, and C. Klingshirn, Proc. II–VI conference Rhode Island/USA 1993, to be publ. J. Crystal Growth.

[9] T. Yajima and H. Souma, Phys. Rev. A **17**, 309 and 324 (1978).

[10] E. Hanamura, Phys. Rev. **B 46**, 4718 (1992).

[11] R. W. Schoenlein, D.M. Mittleman, J.J. Shiang, A.P. Alivisatos, and C.V. Shank, Phys. Rev. Lett. **70**, 1014 (1993).

[12] U. Siegner, D. Weber, E.O. Göbel, D. Bennhardt, V. Heuckeroth, R. Saleh, S.D. Baranovskii, P. Thomas, H. Schwab, C. Klingshirn, J.M. Hvam, V.G. Lyssenko, Phys. Rev. **B 46**, 4564 (1992).

COHERENCE EFFECTS ON THE EXCITON RADIATIVE RECOMBINATION IN QUANTUM WELLS

B. Deveaud[1]*, B. Sermage[2], and D.S. Katzer[3]

[1]Centre National d'Etudes des Télécommunications
22300 Lannion, France
[2]Centre National d'Etudes de Télécommunications
92120 Bagneux, France
[3]Naval Research Laboratory
Washington, DC 20375, USA

Introduction

Low temperature luminescence of quantum wells is strongly dominated by excitonic effects [1]. The first reason for this abrupt change, from 3D to 2D systems, lies in the breakdown of the translational symmetry which modifies the usual picture of polariton states (see Hopfield [2]). In 3D systems, coupling of the exciton to photons does not lead directly to absorption or to light emission but to stable states of the system. On the contrary, for excitons confined in two dimensions, coupling to the photon field leads to a finite lifetime of those excitons with a wavevector $|k|<k_o=n\omega_o/c$ [3] (here ω_o is the frequency of the photon at the exciton energy, n the material refractive index and c the speed of light). On the contrary, excitons with k above $n\omega_o/c$ should not recombine at all; this is because conservation of wavevector cannot be obtained in such a case. When computed in the absence of collisions, the decay rate of near k = 0 2D excitons integrates individual dipole elements over the whole quantum well volume. The radiative lifetime of 2D excitons is then very short of the order of 10ps [3-6].

As the region where radiative recombination is allowed is very narrow (k_o corresponds to an exciton state less than 0.1 meV above the ground state), the observation of such a very short radiative lifetime requires the use of a sample where interface roughness is

* Present address: Ecole Polytechnique Fédérale de Lausanne
Institut de Micro-optoélectronique
CE ECUBLENS
CH1015 Lausanne, Switzerland

minimized, with excitation exactly at the exciton energy, and the sample being held at very low temperature. In such a case, very short lifetimes of the luminescence have indeed observed [7].

Different interpretations for this very short decay (which has been observed by different authors) have been discussed in the literature. We shall discuss here the possible explanations, and we shall show that, in particular the dependence of the observed decay behaviour on coherence of the exciton, greatly favors our interpretation of the very short decay times as a radiative lifetime.

Exciton in a Quantum Well

Let us first begin by a short overview of the theoretical description of excitons in quantum wells. If we follow Andreani et al. [6] and Citrin [8], we get an exciton lifetime Γ_R which can be given, if the homogeneous lifetime is longer than the radiative lifetime, by :

$$\Gamma_R = \frac{2E_o}{3K_BT}\Gamma_o \qquad (1)$$

where Γ_o is the radiative rate of k=0 spin allowed excitons :

$$\Gamma_o = \frac{\pi e^2}{m_o c_o \sqrt{\varepsilon_\infty}} f_o \text{ with}$$

$$f_o = 2m_o E_{ex}(0)|F_{ex}(0)|^2 \sum_s \left|\langle (cv)_s |R.n_j| 0\rangle\right|^2 \frac{\mathrm{Re}(P)}{\hbar^2} \qquad (2)$$

and, if on the contrary, the dephasing rate is larger than the radiative rate :

$$\Gamma_R = \frac{4\pi E_o}{3\hbar\Gamma_h}\left(1-\exp(\frac{\hbar\Gamma_h}{2\pi k_B T})\right)\Gamma_o \qquad (3)$$

Let us note however, that the discussion and precise computation of the different effects is very difficult as none of the involved times can be neglected against the others : the radiative rate, dephasing rate and thermalization rate are all in the range of 10 to 30 ps at low temperature and low density. The different models use assumptions which are not well verified in the practical case of quantum wells. For example, Eq.1 assumes a thermalized distribution which cannot be obtained if the collision dephasing time is longer than the radiative decay time.

If we try to summarize the main properties of excitons, we get the following picture. Two classes of excitons are present in quantum wells, the spin allowed and spin forbidden excitons. The exchange splitting is not yet determined precisely but is expected to be rather small (see for example [9]). The spin allowed excitons can only recombine radiatively if their energy is less than 0.1 meV. In such a case, their recombination should be very fast (from 10 to 30 ps depending on the parameters of the quantum well). If created resonantly, the spin allowed excitons will have a tendency to take the temperature of the lattice (usually larger than the radiative width of 0.1 meV = 1K), and to equilibrate with spin forbidden excitons. The thermalization and spin reversal times can be different in particular because the spin flip times for excitons, electrons and holes can be quite different. It is expected that the thermalization time will be of the order of a few times the homogeneous lifetime (several collisions are needed to establish a thermal distribution). A non thermal as well as non spin equilibrium regime will then exist at short time. This distribution will have a tendency to transform into an equilibrium regime at longer times.

Experiments

We have studied at low temperature (from 2K to 50K), with a resolution of about 5 ps, the dynamics of the exciton luminescence in very high quality quantum wells. In particular, the density of photo excited excitons has been varied in order to study the influence of the dephasing time on the exciton dynamics. The sample is a GaAs / AlAs quantum well obtained with growth interruption at each interface. The sample shows a well defined splitting of the exciton peak, with a negligible Stokes shift (see [10] for details). This allows us to study a sample where both interface roughness and Rayleigh scattering may be neglected in first order. A typical series of time variations of the exciton luminescence intensity is shown on Fig. 1 for a lattice temperature of 10 K, and for different exciton densities. In general, the behaviour is not a single exponential, our interpretation is that the long time decay behaviour corresponds more or less to a thermalized/spin averaged distribution of excitons, the initial decay being due both to the fast radiative recombination of the non-thermal distribution of spin allowed excitons ($k \approx 0$), and to the modifications of the initial distribution (spin averaging and obtention of a thermalized distribution).

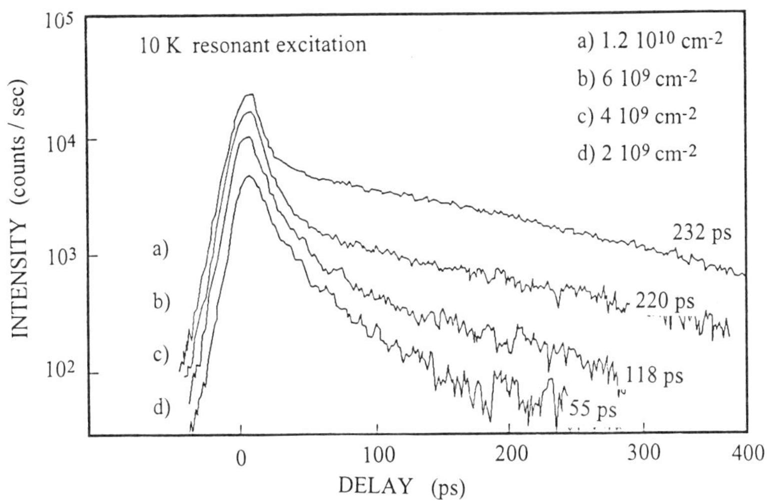

Fig. 1. Exciton luminescence decays for different excitation densities. Note the increase of the long time decay time constant, and the decrease of the short time constant around the laser pulse.

This was modeled in ref.7 by a very simple rate equation model which would need to be improved to take into account the different features of Eqs. 1 and 3. The short time behaviour becomes faster and faster as the dephasing rate becomes larger (larger excitation density). In a rough approximation the short time decay τ_{decay} is given by the radiative τ_{rad} and the dephasing times τ_{deph} as :

$$\frac{1}{\tau_{decay}} = \frac{1}{\tau_{rad}} + \frac{1}{\tau_{deph}} \tag{4}$$

The long time behaviour becomes longer and longer as given by Eq.3. The particular structure of our samples imposes a limit to the observable times due to tunneling of the excitons out of the double barrier tunnel structure of the samples.

Let us now discuss the relations of our interpretation to other possible explanations of the fast decay of the exciton luminescence. This decay has been attributed by us to radiative recombination [7], by Oberhauser et al [11] to trapping by defects, and by Stolz et al [12] to Rayleigh scattering.

Defect trapping

Oberhauser et al [11] proposed that the short decay could be due to trapping by defects or by localized excitons. Such an explanation can be discarded in our case for several reasons. First, we do observe free excitons : this has been discussed at length in [10]. The main reason is that the line shape is indeed a Lorentzian line shape.

The second explanation (trapping by defects) can be ruled out for two main reasons; The first one, discussed in [7] is that the variation of the initial intensity upon variation of the excitation power is clearly sublinear. The second reason comes from the variation with excitation energy. A small detuning of the excitation energy is not expected to affect in any significant manner the trapping on impurities. In our samples, we do observe a drastic change in the temporal behaviour of the exciton luminescence as soon as we detune the laser energy from the exciton resonance. Such a behaviour is on the contrary clearly in agreement with the model of short radiative recombination : a small detuning should increase the exciton temperature and slow down the radiative recombination as observed.

Rayleigh scattering

It has also been argued by different authors [12,13] that the rapid transient at short times might be explained by Rayleigh scattering. Indeed, when the laser energy is brought to resonance with the exciton, a strong enhancement of the diffracted signal is observed, due to enhancement of the Rayleigh scattering [14]. It is then difficult, a priori, to rule out a possible contribution of the Rayleigh scattering to the fast short time decay. One has to note however first that a strong enhancement of the luminescence intensity is also obtained due to the increased absorption at the exciton energy, which is further increased in time resolved measurements by the decrease of the decay time. We report time decays in conditions where the scattering from the laser (in off resonance conditions) is much weaker than the luminescence signal. When the two signals are enhanced at resonance, luminescence stays stronger.

When we bring the laser closer and closer to the exciton resonance (see Fig. 6 of Ref.10), the decay of the signal at the exciton energy becomes continuously shorter and shorter. For example, for a detuning of 1.5 meV, a clear difference can be made between the Rayleigh scattering and the exciton luminescence. In such a case, decay times much shorter than 100 ps are observed (down to 60 ps about). This sets an upper limit to the radiative lifetime of a 2D exciton.

We have also carefully studied the temporal behaviour close to resonance, and found that the luminescence signal that we observe is very slightly delayed with respect to the laser pulse. This delay is very difficult to quantify precisely, but it is of the order of 10 ps at low excitation densities ($\leq 10^9$ cm^{-2}), and becomes very short when the excitation density get higher (above $5x10^9$ cm^{-2}). The observation of a delay between the excitation pulse and the signal, as well as the reduction of this delay when the homogeneous lifetime is shortened are very strong arguments against the interpretation in terms of Rayleigh scattering.

Also, the importance of Rayleigh scattering strongly depends on the imperfections of the sample [15] and we precisely study a sample of the highest available quality, where any effect of this type should be minimized.

Conclusions

We have studied the effects of coherence of the excitonic states in a quantum well on the radiative recombination. We vary the dephasing time by the changes in the excitation density. We find a clear increase of the observed decay time, in very good agreement with theoretical estimations.

Acknowledgments

The authors would like to acknowledge useful discussions with : I. Abram, I. Bar Joseph, G. Bastard, F. Clérot and C. Weisbuch.

References

[1] R.J. Elliot, *Phys. Rev.*, **108**, 1384 (1957)
[2] J.J. Hopfield, *Phys. Rev., **112**, 1555 (1958)*
[3] V.M. Agranovitch, O.A. Dubovskii, *JETP Lett.*, **3**, 223 (1966)
[4] M. Orrit, C. Aslangul, P. Kottis, *Phys. Rev.*, **B25**, 7263 (1982)
[5] E. Hanamura, *Phys. Rev.*, **B38**, 1228 (1988)
[6] L.C. Andreani, F. Tassone, F. Bassani, *Solid State Commun.*, **77**, 641 (1990)

[7] B. Deveaud, F. Clérot, N. Roy, K. Satzke, B. Sermage, D.S. Katzer, *Phys. Rev. Lett.*, **67**, 2355 (1991)

[8] D.S. Citrin, *Phys. Rev.,* **B47**, 3832 (1993)

[9] R. Bauer, D. Bimberg, J. Christen, D. Oertel, D. Mars, J.N. Miller, D. Miller. T. Miller, T. Fukunaga, H. Nakashima, *Proccedings of the 18th Conf. on the Phusics of semmiconducttors,* Ed. O Engström, World Sci.(1979), p.525

[10] B. Deveaud, F. Clérot, B. Sermage, C. Dumas, D.S. Katzer, in *Optical phenomena in semiconductors of reduced dimensions.* Eds. D:J: Loockwood, A. Pinczuk, Kluwer Academic Press (1993) p. 129

[11] D. Oberhauser, H. Kalt, W. Schlapp, H. Nickel, C. Klingshirn, *J. Lumin.*, **48 & 49**, 717 (1991)

[12] H. Stolz, D: Schwartze, W. Von der Osten, G: Weimann, *Phys. Rev.*, **B47**, 9669 (1993)

[13] R. Eccleston, R. Strobel, W.W. Rühle, J. Kuhl, B.F. Feuerbach, K. Ploog, *Phys. Rev.,* **63**, 1395 (1992)

[14] J. Hegarty, M.D. Sturge, C. Weisbuch, A.C. Gossard, W. Wiegmann, *Phys. Rev. Lett.*, **49**, 930 (1982)

THE RELATIONSHIP BETWEEN REAL AND VIRTUAL EXCITATION MECHANISMS FOR NONLINEAR REFRACTION

M.G. Burt

BT Laboratories
Martlesham Heath
Ipswich
IP5 7RE
United Kingdom

INTRODUCTION

It is well known that the nonlinear refraction that arises from real incoherent excitation in semiconductors is several orders of magnitude larger than that arising from virtual excitation.In a semiquantitative way this can be understood using a two level model.Under CW illumination in the tail of the absorption peak the model predicts[1,2,3] that the refractive index change for real excitation is a factor $2T_1/T_2$ larger than that for virtual excitation where T_1 is the energy relaxation time (lifetime) and T_2 is the dephasing time.Because T_1 is typically in the order of nanoseconds and T_2 in the order of picoseconds or femtoseconds the ratio $2T_1/T_2$. can account,at least in part,for the relative efficacy of real and virtual excitation.The main purpose of this paper is to highlight the physical origin of the $2T_1/T_2$ enhancement factor.

PHYSICAL ORIGIN OF THE ENHANCEMENT FACTOR

The key[3,4] to understanding the origin of the $2T_1/T_2$ enhancement factor lies in the appreciation that $T_2/2$ is the time constant for the conversion of coherent excitation into incoherent excitation because the probability of finding an initially coherent excitation still in a coherent state after time t decays as $\exp(-2t/T_2)$.Hence $T_2/2$ is the time taken to create real incoherent excitation and $2T_1/T_2$ represents the ratio of lifetime to creation time for real excitation.This ratio represents a figure of merit for real excitation,a measure of how long the excitation is available for blocking transitions,and hence altering the refractive index,compared with the time it takes for the light to create it.On the other hand it is reasonable to suppose that this ratio is unity for virtual excitation because the latter corresponds to adiabatic following;the excitation rises and falls faithfully with the pulse intensity,a point supported by the algebraic result described in the next section.It therefore

Coherent Optical Interactions in Semiconductors
Edited by R.T. Phillips, Plenum Press, New York, 1994

follows that one would expect real incoherent excitation to provide a factor $2T_1/T_2$ greater refractive index change than virtual excitation provided the overall excitation level remains small and saturation effects avoided.It has,of course,been assumed that the blocking of transitions depends only on the occupation probabilities alone and is independent of whether the occupation is due to real or virtual excitation or a mixture thereof.This can be readily verified for the two level model[4].It also turns out that for virtual excitation of a three level system with equally spaced levels that the nonlinear polarisation depends on the occupation probabilities alone[5].

TIME DEPENDENT ILLUMINATION

Up to now the discussion has been restricted to CW illumination.It is useful to consider the case of time dependent illumination because it tests the assertion that the ratio of lifetime to creation time for virtual excitation is unity.It also shows what time constant takes the place of T_1 and $T_2/2$ individually when one has virtual as opposed to real excitation.

We consider a two level system illuminated near resonance with exponentially rising and falling edges.For $t < 0$ the optical intensity rises as $I_0\exp(\gamma^{(-)}t)$ and for $t > 0$ falls off as $I_0\exp(-\gamma^{(+)}t)$ with both $\gamma^{(+)}$ and $\gamma^{(-)}$ small compared to the detuning so that virtual excitation as opposed to real coherent excitation occurs in the absence of dephasing.The refractive index change,Δn,for an ensemble of identical two level systems can be found from a straightforward application of the results in reference 4 (Note that in reference 4 $\gamma^{(+)}$ and $\gamma^{(-)}$ are rise and fall rates of the electric field not the intensity as in this paper.) For $t < 0$ it is given by

$$\frac{\Delta n}{(\Delta n)_{virtual}} = \frac{2\Gamma_2 + \gamma^{(-)}}{\Gamma_1 + \gamma^{(-)}}\exp(\gamma^{(-)}t)$$

and for $t > 0$ by

$$\frac{\Delta n}{(\Delta n)_{virtual}} = \frac{2\Gamma_2 - \gamma^{(+)}}{\Gamma_1 - \gamma^{(+)}}\exp\left(\gamma^{(+)}t\right) + \left(\frac{2\Gamma_2 + \gamma^{(-)}}{\Gamma_1 + \gamma^{(-)}} - \frac{2\Gamma_2 - \gamma^{(+)}}{\Gamma_1 - \gamma^{(+)}}\right)\exp(-\Gamma_1 t),$$

where Γ_1 and Γ_2 are $1/T_1$ and $1/T_2$ respectively.$(\Delta n)_{virtual}$ is the refractive index change for CW illumination with intensity I_0.Both Δn and $(\Delta n)_{virtual}$ are proportional to I_0 .We see that we easily regain the CW illumination result

$$\frac{\Delta n}{(\Delta n)_{virtual}} = \frac{2\Gamma_2}{\Gamma_1} = \frac{2T_1}{T_2}$$

by setting both $\gamma^{(+)}$ and $\gamma^{(-)}$ equal to zero.We can also illustrate the adiabatic following feature of virtual excitation by setting both Γ_1 and Γ_2 equal to zero to obtain

$$\frac{\Delta n}{(\Delta n)_{virtual}} = \exp(\pm\gamma^{(\mp)}t)$$

where the upper sign applies for $t < 0$ and the lower sign for $t > 0$.We also get this result,of course,in the coherent transient regime with both $\gamma^{(+)}$ and $\gamma^{(-)}$ much larger than both Γ_1 and

Γ_2 provided the detuning is large enough to allow this limit to be taken.(The detuning must be much larger than both $\gamma^{(+)}$ and $\gamma^{(-)}$ for the excitation to remain virtual).

What is interesting about the general expression for Δn is that the rate of creation,$2\Gamma_2$, and the rate of decay,Γ_1,of real incoherent excitation are always associated with the rise and fall rates,$\gamma^{(-)}$ and $\gamma^{(+)}$,of the optical intensity.The lifetime and creation time of incoherent real excitation are both replaced by the characteristic time constant of the pulse in the virtual excitation regime.

COMMENT ON THE VIRTUAL EXCITATION LIFETIME CONCEPT

A standard physical picture of the mechanism by which a system interacts with a light beam in the virtual excitation regime uses the virtual excitation lifetime concept derived from the uncertainty principle.In the context of the two level model,the electron makes a transition from the ground state up to the excited state,but can only remain there for a time,equal to the inverse of the detuning,consistent with the energy time uncertainty principle.One might then argue that in the virtual excitation regime one would expect the inverse detuning to take the place of T_1.It would also appear reasonable to replace the creation time,$T_2/2$,by the inverse detuning in the virtual excitation regime because this is the minimum pulse duration that will result in virtual excitation;any shorter pulse in the absence of dephasing will have a sufficiently broad spectrum to produce real coherent excitation and the system will not return to the ground state when the optical intensity has returned to zero.With these assignments the ratio of lifetime to creation time for virtual excitation would be unity in keeping with the adiabatic following property of virtual excitation.

However,there is no sign of the inverse detuning playing such a role in the transition from real to virtual excitation in the time dependence of Δn.It would appear,therefore,that we have pushed this virtual particle lifetime concept too far.This is not that surprising.The concept of a lifetime is one that has its roots in the theory of incoherent excitation.One describes the time evolution of the system in terms of the time dependence of the state occupation probabilities alone because the dephasing processes are fast compared with the time constants of the external perturbation.To describe energy absorption in the virtual excitation case as mediated by transitions from the ground state to the excited state where the electron resides (on average) for a time equal to the inverse detuning and then returns to the ground state ignores the essential coherent nature of virtual excitation,that the electron state needs more than an occupation probability to describe it; it needs a wavefunction that is a superposition of the ground state and excited states of the unperturbed system.

SUMMARY

The $2T_1/T_2$ enhancement factor obtained for the refractive index change due to real incoherent excitation compared with that for virtual excitation for CW illumination near resonance in the absence of saturation effects can be understood in terms of the differences in the ratio of lifetime to creation time for the two types of excitation.$T_2/2$ is the mean creation time for real incoherent excitation,which has life time T_1.In the virtual excitation regime,at least for a model pulse with exponentially rising and falling edges,the appropriate pulse time constant takes on the role of both the creation time and the lifetime.This reflects the adiabatic nature of virtual excitation;its existence,creation and decay,is determined entirely by the optical pulse.The analysis finds no support for the idea that virtual excitation has a lifetime equal to the inverse of the detuning.It is suggested that this is because the lifetime concept essentially ignores the coherent nature of virtual excitation.

REFERENCES

1. Wherrett B S Proc Roy Soc A **390** 373 (1983);Phil Trans R Soc A **313** (1984)
2. Burt M G Proc SPIE __ Int Soc Opt Eng (USA) **1280** 35 (1990)
3. Burt M G Semicond Sci Technol **5** 1215 (1990)
4. Burt M G Semicond Sci Technol **8** 1393 (1993)
5. Walmsley M , Abram R A and Burt M G Semicond Sci Technol **8** 268 (1993)

RESONANT RAYLEIGH SCATTERING IN EPITAXIALLY GROWN $ZnSe_{1-x}S_x$ LAYERS

M. Jütte[1], H. Stolz[1], W. von der Osten[1]
J. Söllner[2], K.-P. Geyzers[2], M. Heuken[2], and K. Heime[2]

[1]Fachbereich Physik, Universität-GH, 33095 Paderborn, Germany
[2]Institut für Halbleitertechnik I, RWTH, 52056 Aachen, Germany

INTRODUCTION

In recent years the development of blue light emitting devices led to an increasing interest in ZnSe/ZnS layer systems epitaxially grown on a GaAs substrate. Most important to understand the optical processes in these materials are the excitonic states. Contrary to bulk crystals, they are affected by intrinsic biaxial strain which occurs due to lattice mismatch and different thermal expansion between layer and substrate and which reduces the crystal symmetry from T_d to D_{2d} (for literature see e.g. ref.[1]). As a consequence, the fourfold degenerate valence band and, hence, the excitonic states are strain-split into heavy (hh) and light (lh) hole states, each twofold degenerate. In this contribution we present a study of the exciton properties by Resonant Rayleigh Scattering (RRayS). Originally introduced by Hegarty *et al.*[2], RRayS represents a linear coherent spectroscopic method that hitherto was rarely applied but can provide information beyond that obtained by other optical techniques[3].

EXPERIMENTAL DETAILS

All experiments were carried out at $T \approx 7K$ with $ZnSe_{1-x}S_x$ layers grown by MOVPE on (001) GaAs substrate. The epilayers had low sulphur concentrations ($x < 2\%$) and thicknesses of 60 nm ("thin" sample) and 2-9 μm ("thick" sample). For excitation an argon ion laser pumped dye laser system was used operating with Stilbene 3. The light emission of the sample is analyzed by a Spex double monochromator and registered by a photomultiplier with single photon counting. A special feature of our setup is the possibility to vary the photon energy of the exciting laser and the monochromator automatically in identical steps. The wavelength of the dye laser is continuously measured by a wavemeter with an absolute accuracy of < 0.05 meV.

Coherent Optical Interactions in Semiconductors
Edited by R.T. Phillips, Plenum Press, New York, 1994

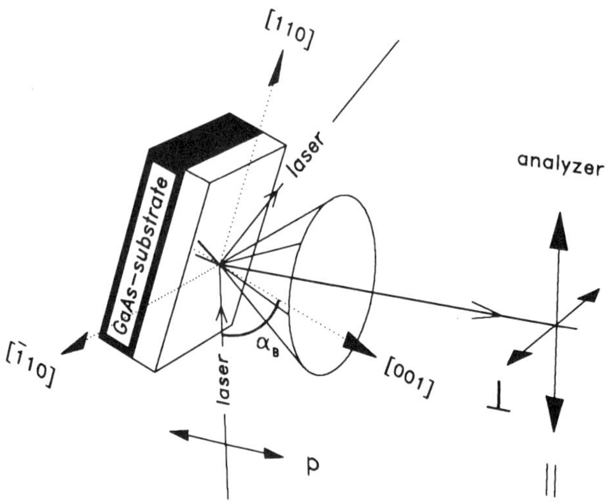

Fig. 1. *Brewster geometry for scattering. The polarisation of the incoming laser is parallel (p) to the plane of incidence, the analyzer positions are correspondingly designated by* $||$ *and* $\perp$. *The cone indicates the solid angle of light emission which is coupled into the monochromator.* α_B : *Brewster angle.*

The samples were arranged in Brewster angle geometry as illustrated in Fig. 1. Due to the high refractive index of ZnSe, within the sample the directions of propagation of the exciting light and the detected part of scattered light are nearly parallel to the normal of the surface ($||$ [001]). This arrangement simplifies the selection rules for the exciton states. As only p-polarised light is entering the sample, the undesired specular reflection is minimized improving the signal-to-noise ratio for RRayS.

PRINCIPLE AND WAYS OF MEASUREMENT

RRayS is an elastic process that requires the existence of any static disorder giving rise to spatial fluctuations of the susceptibility[2,3]. Being usually superimposed by strong non-resonant elastic scattering, in high quality samples it becomes resonant in the spectral region of excitons leading to enhancement of intensity up to several orders of magnitude and allowing to characterize the states involved.

As discussed in some detail previously[3], strictly speaking RRayS refers to the coherent part of the elastically scattered light, for which the information on the phase of the incident light is retained. Besides, incoherent Hot Luminescence (HL) occurs at the same energy for which phase memory is lost.

The criterion to distinguish between RRayS and HL is based on the degeneracy and polarisation behaviour of the exciton states. In accordance with the D_{2d} symmetry of the strained crystal, the exciton is doubly degenerate having transition moments along [110] and [$\bar{1}$10] (see Fig. 1). Following resonant excitation with linearly p-polarised light, only the [110] component of the exciton state is being occupied, the perpendicular [$\bar{1}$10] component becoming exclusively populated by elastic scattering between the degenerate states. As a result the measured intensity $I_\perp$ corresponds to (incoherent) HL while the intensity $I_{||} - I_\perp$ remaining in parallel polarisation is (coherent) RRayS.

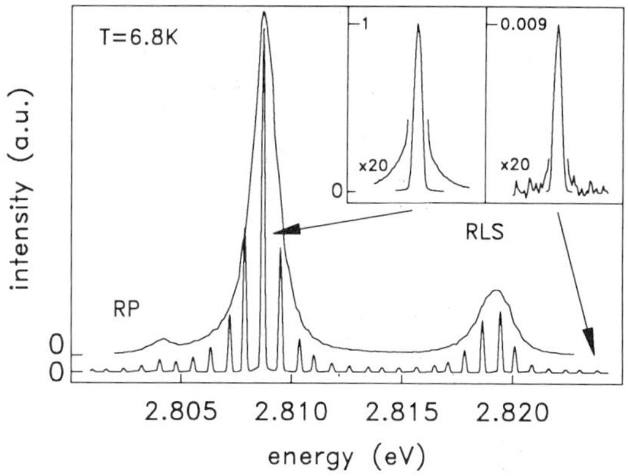

Fig. 2. *RRayS spectra in the excitonic region of a thin $ZnSe_{1-x}S_x$-sample (60 nm). The resonance line spectrum (RLS) is shown together with the resonance profile (RP). Only the dominant intensity component is displayed occurring for incident and scattered light polarised in the plane of incidence (p,||). Inset: Rayleigh line in (left) and off (right) resonance with line pedestal in an enlarged scale.*

Experimentally we used two ways to measure RRayS (Fig. 2). One was to change the spectral position of the dye laser stepwise and, for each position, take a spectrum around the laser line ("Resonance Line Spectrum": RLS). In resonance with an excitonic state the scattering line has a spectral width equal to that of the laser ($\delta E \simeq 0.15$ meV). It sits on top of a broader background due to luminescence or resonant inelastic scattering processes which is absent off resonance (cf. inset). This enables rightaway to distinguish the elastic scattering from other radiative processes. In the other type of measurement the laser and monochromator energies are varied synchronously and in parallel (step width 0.1 meV) and only the peaks of the scattering lines are detected ("Resonance Profile": RP). Here the elastic scattered (coherent) part is obtained from the difference $I_{\|} - I_{\perp}$, in which any depolarised luminescence component is eliminated.

RESULTS AND ANALYSIS

Based on a recently published theory[3], for cw excitation the time-integrated intensities of RRayS and HL are given by

$$
\begin{aligned}
I_{RRayS} &= I_{\|} - I_{\perp} \\
&\propto \left| \sum_{\alpha=1}^{2} (\vec{M}_\alpha \cdot \vec{e}_L)(\vec{M}_\alpha \cdot \vec{e}_S) \right|^2 \frac{\tau_{coh}}{2}
\end{aligned}
\tag{1}
$$

$$
\begin{aligned}
I_{HL} &= I_{\perp} \\
&\propto \frac{1}{2} \sum_{\alpha,\alpha'=1}^{2} \left| (\vec{M}_\alpha \cdot \vec{e}_L)(\vec{M}_{\alpha'} \cdot \vec{e}_S) \right|^2 \left(T_1 - \frac{\tau_{coh}}{2} \right)
\end{aligned}
\tag{2}
$$

where α, α' denote the (twofold) degenerate exciton states. According to Eqs.(1,2),

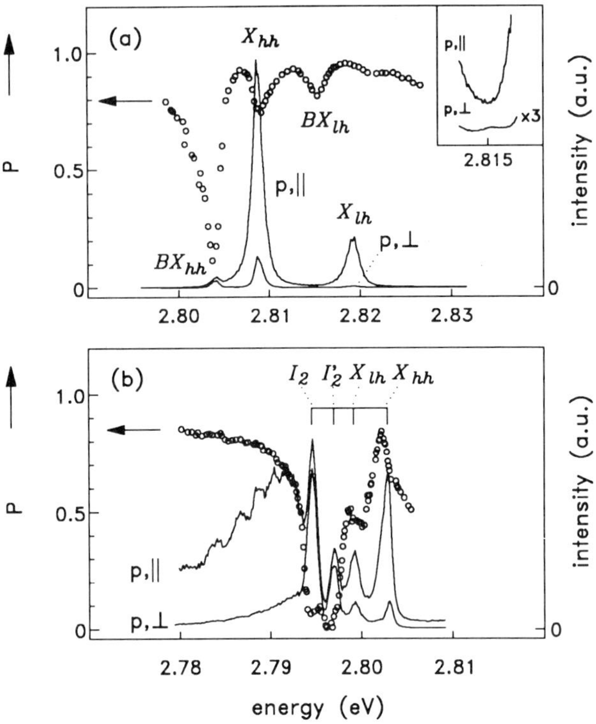

Fig. 3. *Resonance profiles $I_{p,\parallel}$ and $I_{p,\perp}$ and degree of polarisation P measured for two $ZnSe_{1-x}S_x$ epilayers: a) thin (60nm) sample of Fig. 2, b) thick (8.35μm) sample (x low). Inset: Detail of the resonance profiles in the region of the bound lh exciton in an enlarged scale.*

the scattered intensities depend on the dipole matrix elements ($\vec{M}_{hh}$ and $\vec{M}_{lh}$ for the hh and lh exciton transitions) whereby $\vec{e}_L$,$\vec{e}_S$ are the polarisation unit vectors for the incident (laser) and scattered light. Moreover they are determined by two relaxation times with the corresponding rates related to each other by $1/\tau_{coh} = 1/2T_1 + 1/T_2'$. Here T_1 and T_2' are the population lifetime and pure dephasing time, respectively. They account for the inelastic and the elastic scattering processes that both cause dephasing of the intermediate state described by the coherence time τ_{coh}. For the used scattering geometry and the mentioned selection rules it is straightforward to derive from Eqs.(1,2) the linear degree of polarisation

$$P = (I_{\parallel} - I_{\perp})/(I_{\parallel} + I_{\perp}) = 1/(1 + 2T_1/T_2') \quad . \tag{3}$$

P is determined by the ratio of the time constants but does not provide information on their absolute values.

Figure 3a,b displays resonance profiles $I_{\parallel}$ and $I_{\perp}$ representative for a thin and thick epilayer, respectively. The hh and lh *free exciton states* (X_{hh}, X_{lh}) are characterized by strong resonances of Rayleigh scattered intensity (given by the difference $I_{\parallel} - I_{\perp}$) as well as by a high degree of polarisation. The assignment of X_{hh} and X_{lh} is based on the difference in magnitude of the transition matrix elements which enter the expression for the Rayleigh intensity according to $I_{RRayS} \propto |M|^4 \cdot \tau_{coh}$ (Eq.(1)). Since for the thin sample (Fig. 3a) the strain splitting of the exciton states (10.5 meV) is much larger than the estimated exchange splitting of ≈ 0.5meV[4], the ratio of squared matrix elements is $|M_{hh}|^2/|M_{lh}|^2 = 3$ predicting a much stronger resonance enhancement for X_{hh} than for X_{lh}. Based on that, X_{hh} is lower in energy than X_{lh} which, in view of the known stress dependence of energies, implies that compressive strain is effective. Distinctly different from that, in the thick sample (Fig. 3b) the exciton splitting is reverse. This would suggest tensile strain, however, as the exchange and strain splitting energies are of comparable magnitude a definite determination of the strain requires a more elaborate analysis. In addition, from the degree of polarisation it is concluded that T_2' is larger by about one order of magnitude compared with T_1 indicating that the coherence is lost by depopulation of the excited states rather than by pure dephasing.

For the *bound exciton states* ($BX_{hh,lh}$, I_2, I_2') the Rayleigh scattered intensity ($I_{\|} - I_{\perp}$) exhibits only weak resonances and low degree of polarisation. As seen from Eq.(3) this implies a ratio of relaxation times $T_1/T_2' \gg 1$. As an exception to this the bound exciton transition BX_{lh} (Fig. 3a) seems to have a relatively high degree of polarisation which, however, is caused by the influence of the nearby free exciton line X_{lh} (see inset).

The structure on the low energy side of the I_2 line in Fig. 3(b) are Fabry-Perot interferences due to multiple reflection of elastically scattered light between the $ZnSe_{1-x}S_x$ layer boundaries, which have to be of good quality to show this effect. These oscillations are only visible if the absorption is sufficiently weak but strong enough for light scattering to be observable. Therefore, the interference appears only in a small energy region below the bound excitons. Reflectivity spectra of the thick samples taken in quasi-back reflection show an interference pattern with peaks at the same energy positions. As expected, the thin epilayers show no such oscillation structure in this energy region because the free spectral range of a very small Fabry-Perot resonator is very large exceeding the depicted energy region.

In conclusion, our measurements demonstrate that RRayS in systems with spatial variations of susceptibility is a useful tool for exciton spectroscopy that allows characterization of exciton states. Especially in layer structures in which direct absorption measurements are usually not possible because of the opaque substrate, RRayS can be used alternatively.

ACKNOWLEDGMENTS

The authors are indebted to the Deutsche Forschungsgemeinschaft for support of this project.

REFERENCES

[1]J. Gutowski, N. Presser, G. Kudlek, phys. stat. sol. (a) **120**, 11 (1990).

[2]J. Hegarty, M.D. Sturge, C. Weisbuch, A.C. Gossard, W. Wiegmann, Phys. Rev. Lett. **49**, 930 (1982).

[3]H. Stolz, D. Schwarze, W. von der Osten, G. Weimann, Superlatt. & Microstruct. **9**, 511 (1991); Phys. Rev. B, **47**, 9669 (1993).

[4]M. Jütte, H. Stolz, W. von der Osten, to be published.

SPECTRALLY-RESOLVED FEMTOSECOND-FOUR-WAVE MIXING ON SEMICONDUCTORS

T. Rappen[1], U. Peter[1], W. Schäfer[2], and M. Wegener[1]

[1]Institut für Physik, Universität Dortmund, D-44221 Dortmund
[2]HLRZ, Forschungszentrum Jülich, D-52428 Jülich
Federal Republic of Germany

INTRODUCTION

Femtosecond-four-wave mixing is a powerful tool to study the loss of coherence (dephasing) of elementary excitations in solids[1]. Usually the spectrally integrated, diffracted wave is detected as a function of time delay. It is, however, in many situations a priori not clear, which states actually contribute to the coherent diffracted wave. In order to get a more detailed understanding of the coherent dynamics of continuum states and their interplay with excitons, we spectrally resolve the diffracted wave in femtosecond-four-wave mixing as a function of time delay for different excitation conditions[2]. We also investigate the dependence of the four-wave-mixing spectra on the polarization of the two incident beams. These low-density results deliver information on the density dependence of the dephasing time of both, continuum and exciton states.

The experimental setup has been discussed in detail in Ref.2. We use 140-fs transform-limited pulses for the four-wave-mixing experiment in the self-diffraction geometry. By means of a 0.25-m grating spectrometer and an optical multichannel analyzer spectra of the diffracted wave are taken as a function of time delay[2]. We are exciting within the low-density limit (carrier density $\approx 10^{15} cm^{-3}$) to be within the third order perturbation regime. The high quality 3-μm-thick germanium single-crystal sample (antireflection coated on both sides) is immersed in liquid helium at a temperature of 1.8K. The sample does not exhibit any linear birefringence.

CIRCULAR POLARIZATION

We start by discussing the results for co-circular polarization of the two incident beams with wavevectors $\vec{q}_1$ and $\vec{q}_2$, respectively. The experimental results for excitation within the continuum of states [Fig. 1(a)] are compared with solutions of the semiconductor Bloch-equations, assuming a constant and equal dephasing rate for all states (for details see Ref.2). Two clearly separated maxima, one narrow line at the exciton and a much broader contribution from the continuum states are observed.

Coherent Optical Interactions in Semiconductors
Edited by R.T. Phillips, Plenum Press, New York, 1994

The exciton peak is surprisingly large considering the small overlap between the laser spectrum and the exciton as observed in the absorption spectrum. Furthermore, the exciton contribution vanishes at earlier time delay than the continuum. One might in fact expect the opposite behavior, arguing that the exciton should have a longer dephasing time. These astonishing effects can be traced back to the interaction (strength of the exciton) and interference (temporal evolution) of the coherent polarization of excitons and continuum states. The general scenario is well described by the semiconductor Bloch-equations [Fig. 1(b)]. For a more detailed discussion see Ref.2.

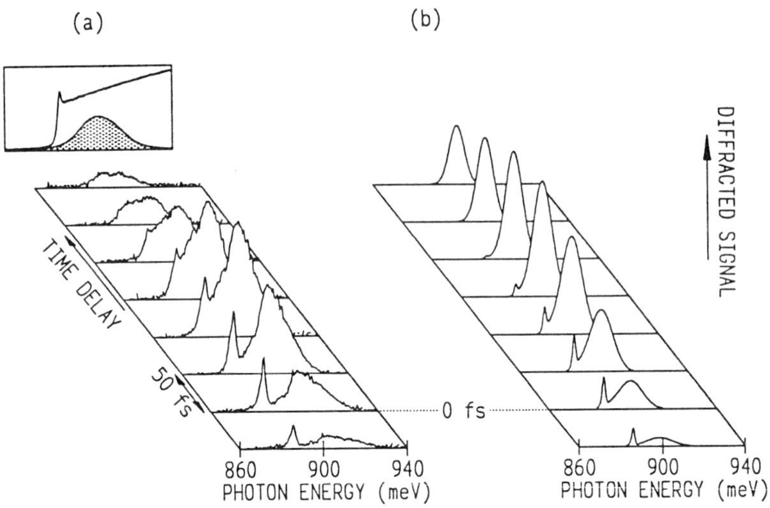

Figure 1. Spectrally-resolved femtosecond-four-wave mixing as a function of time delay (in steps of 50 fs). The lattice temperature is 1.8K. The absorption spectrum and the laser spectrum (shaded) are depicted in the background. (a) experiment, (b) theory

LINEAR POLARIZATION

When using linear polarization of the two incident beams and an analyzer for the diffracted beam in direction $\vec{q}_3 = 2\vec{q}_2 - \vec{q}_1$ there are eight possible configurations if each of the two beams is polarized either in the plane of incidence (x) or the orthogonal direction (y). From the experiment we find the two independent and nonvanishing polarization combinations (x,x,x) and (y,x,y) for the beams with wavevectors ($\vec{q}_1$, $\vec{q}_2$, $\vec{q}_3$). In summary: (x,x,x) = (y,y,y); (y,x,y) = (x,y,x); (x,x,y) = (y,y,x) = 0 and (x,y,y) = (y,x,x) = 0. These results are consistent with symmetry considerations. For the (x,x,x) configuration [Fig. 2(a)] we find a similar scenario as for co-circular excitation (not shown), whereas for (y,x,y) the signal (Fig. 2(b), note the scaling factor of 10) is about a factor 20 weaker. This fact clearly resembles earlier results of Ref.3 on excitons. For both cases the spectrally-integrated traces exhibit a closely exponential decay with an identical decay time of about 160 fs.

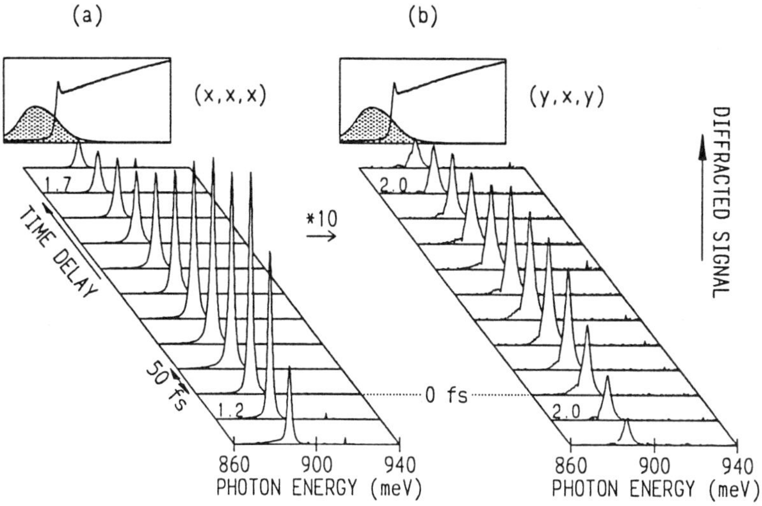

Figure 2. Experiment: Spectrally-resolved femtosecond-four-wave mixing for exclusive excitation of the exciton. a) (x,x,x), (b) (y,x,y).

Furthermore, the polarization of the involved beams influences the evolution of the exciton linewidth in the four-wave-mixing spectra. In the (x,x,x) configuration the half width at half maximum (HWHM) (indicated at two spectra in each series in units of meV) monotonously increases from 1.2 meV at early time delays to 1.7 meV HWHM at larger time delays, similar to the (+,+,+) behavior[2]. In contrast to this, in the (y,x,y) configuration the HWHM of 2 meV is significantly broader and constant for all time delays.

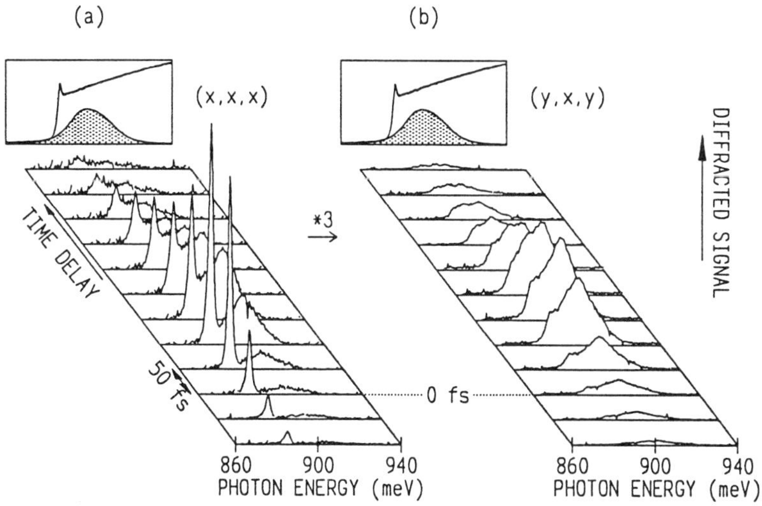

Figure 3. Experiment: Spectrally-resolved femtosecond-four-wave mixing for excitation well above the direct gap. a) (x,x,x) (b) (y,x,y).

We observe an even more dramatic polarization dependence for excitation within the continuum states as depicted in Fig. 3. For the (x,x,x) configuration [Fig. 3(a)] the data are rather similar to the case of co-circular excitation [Fig. 1(a)]. For the (y,x,y) configuration the spectra are strikingly different [Fig. 3(b)]. The continuum contribution decreases by less than a factor of three with respect to (x,x,x), whereas the exciton almost completely vanishes. Some minor hints of the exciton can still be seen from the low-energy shoulder in the spectra. Let us note at this point, that in a spectrally integrated experiment one observes mainly the excitonic dynamics for (x,x,x) and the dynamics of continuum states for the (y,x,y) configuration. This has to be considered for the interpretation of such data (for excitation within the continuum of states).

The semiconductor Bloch-equations (with a constant dephasing time), which were employed for circular polarization (Fig. 1), predict no polarization dependence, i.e. (x,x,x) = (y,x,y). The experimental results, however, can be understood qualitatively when accounting for the two degenerate optical transitions with different spin and also accounting for a density-dependent dephasing rate $T_2^{-1} = \gamma_2 = \gamma_2^0 + (\partial\gamma_2/\partial n)\, n$ as previously discussed for excitons[3]. This introduces a nonlinear coupling between the two transitions. As usual we employ a third order perturbation approach.

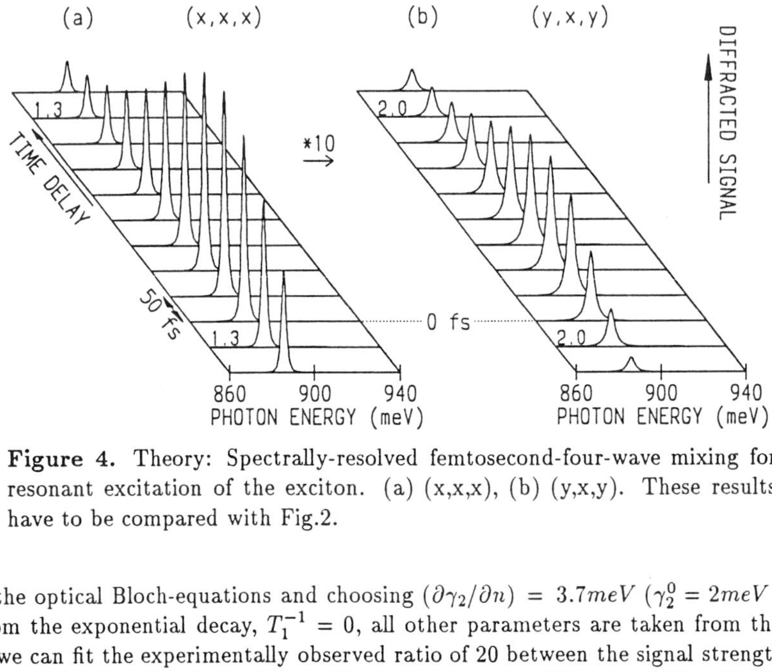

Figure 4. Theory: Spectrally-resolved femtosecond-four-wave mixing for resonant excitation of the exciton. (a) (x,x,x), (b) (y,x,y). These results have to be compared with Fig.2.

Using the optical Bloch-equations and choosing $(\partial\gamma_2/\partial n) = 3.7meV$ ($\gamma_2^0 = 2meV$ is obtained from the exponential decay, $T_1^{-1} = 0$, all other parameters are taken from the experiment) we can fit the experimentally observed ratio of 20 between the signal strengths in the (x,x,x) and the (y,x,y) polarization configuration (compare Fig. 4 and Fig. 2). Moreover, we simultaneously reproduce the different observed linewidths for (x,x,x) and (y,x,y). The density-induced dephasing leads to mathematical terms similar to the local field model[4]. At negative time delay one obtains a square of a Lorentzian in the spectrum and a Lorentzian at positive time delay. For (y,x,y) the additional terms cancel and the linewidth is simply given by the inverse of the observed decay time (160-fs decay time ($T_2 = 320fs$) corresponding to about 2meV).

In order to obtain a calculable model for excitation within the continuum of states we completely neglect the Coulomb interaction in the semiconductor Bloch-equations. This model can be interpreted as an inhomogeneously broadened system following a square-root density of states. We again use $(\partial\gamma_2/\partial n) = 3.7meV$. In Fig. 5 these results are superimposed with those of Fig. 4. Clearly the continuum gains relative strength in the (y,x,y) configuration as compared to the (x,x,x) configuration.

In a microscopic theory the phenomenological dephasing rate in the semiconductor Bloch-equations has to be replaced by the electron scattering terms exceeding Hartree-Fock. It turns out that the experimental results cannot be understood without correct inclusion of all contributions within the second Born-approximation, especially vertex corrections. Respective calculations are under progress, the results will be published soon.

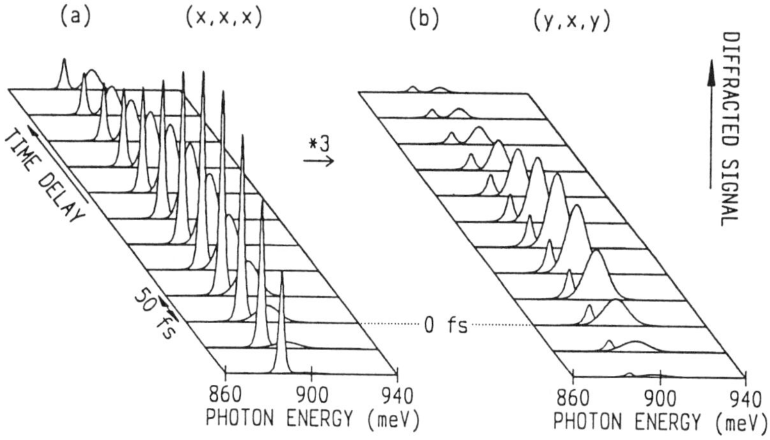

Figure 5. Theory: Spectrally-resolved femtosecond-four-wave mixing including the continuum of states. (a) (x,x,x), (b) (y,x,y).

CONCLUSION

We have studied the simultaneous coherent dynamics of continuum states and excitons by spectrally-resolved femtosecond-four-wave mixing. The experimental spectra of the diffracted wave show a distinct maximum at the exciton even for excitation many (10-15) Rydberg energies above the gap and a much broader structure associated with the continuum states. Surprisingly, the exciton vanishes at earlier time delays than the continuum contribution. This is understood by the interference and interaction of the polarizations of the exciton and continuum states. The experimentally observed polarization dependences of the spectra of four-wave mixing on continuum states and excitons are interpreted in terms of a density-dependent dephasing rate.

REFERENCES

1. L. Schultheis, J. Kuhl, A. Honold, and C. W. Tu, Phys. Rev. Lett. **57**, 1635 (1986), P.C. Becker, H. L. Fragnito, C. H. Brito-Cruz, R. Fork, J. E. Cunningham, J. E. Henry, and C. V. Shank, Phys. Rev. Lett. **61**, 1647 (1988)
2. T. Rappen, U. Peter, M. Wegener, and W. Schäfer, Phys. Rev. B (August 1993), in press
3. see contributions of S.W. Koch and D.G. Steel in this book
4. M. Wegener, D. S. Chemla, S. Schmitt-Rink, and W. Schäfer, Phys. Rev. A **42**, 5675 (1990)
5. M. Lindberg, R. Binder, S. W. Koch, Phys. Rev. A **45**, 1865 (1992)
6. R. Binder, S. W. Koch, M. Lindberg, W. Schäfer, and F. Jahnke, Phys. Rev. B **43**, 6520 (1991)

QUANTUM BEATS AND POLARIZATION INTERFERENCE FROM QUANTUM WELL EXCITONS

[1]G. von Plessen, M. Koch, J. Feldmann, E.O. Göbel, P. Thomas,
[2]J. Shah, and [3]K. Köhler

[1]Department of Physics and Materials Sciences Center
Philipps University, Renthof 5, D-35032 Marburg, Germany
[2]AT&T Bell Laboratories, Holmdel
New Jersey 07733, USA
[3]Fraunhofer Institute for Applied Solid State Physics
D–7800 Freiburg, Germany

1. INTRODUCTION

One of the most interesting recent developments in coherent laser spectroscopy on semiconductors is the introduction of quantum beat spectroscopy [1], which has been used for determining energy splittings, and measuring the dynamics of excitonic wavepackets. When two or more closely spaced optical transitions are simultaneously excited by a short (i.e. spectrally broad) laser pulse, interference phenomena will be observed during a time interval of the order of the dephasing time, which are generally referred to as quantum beats. These beats may be of quite different nature, depending on whether the transitions involved are independent, or are coupled with each other.

To illustrate this difference, let us consider two optical transitions, which may be either uncoupled, as in the case of two independent two-level absorbers, or coupled with each other, as, e.g., in a three-level system. In the first case, which is schematically depicted in Fig. 1(a), the polarization is a superposition of the individual temporal responses of the two-level systems and interferes externally in the detector (in the following referred to as polarization interference (PI)). In the second case, which is shown in Fig. 1(b), the beats are caused by a quantum mechanical interference due to the coupling of the transitions via the common state (in the following referred to as quantum beats (QBs)). Unfortunately, an experimental distinction between PI and QB turns out to be very difficult. In particular, this applies to transient four-wave mixing (FWM) experiments with time integrated detection of the nonlinear signal, where a qualitatively very similar beat structure will be found in either case. This experimental ambiguity posed a severe problem for the interpretation of results presented in Ref. 2, where FWM experiments were performed on a quantum well (QW) structure with spatially separated interface islands, which differ in thickness by one monolayer. The simultaneous excitation of excitons associated with the different interface islands led to a beating behaviour of the transient FWM signal, which was classified as quantum beats. Subsequently, it was argued [1,3] that the observed oscillatory structure should not be called QBs, since it

Coherent Optical Interactions in Semiconductors
Edited by R.T. Phillips, Plenum Press, New York, 1994

originates from two spatially separated oscillators, i.e. from two uncoupled two-level systems (PI situation).

In the present paper, we solve this problem by demonstrating that QBs can be umambiguously distinguished from PI in a *time-resolved* FWM experiment. Using this method, we are able to identify the beats originating from different QW island excitons as QBs.

2. QB VERSUS PI: THEORETICAL APPROACH

In the following, we will summarize the results of a simple model calculation [4], which will help to elucidate our approach. Let us consider the two examples for PI and QBs shown in Fig. 1(a) and (b), respectively. It is assumed that in either case, both optical transitions are excited simultaneously by sufficiently short (i.e. spectrally broad) laser pulses, and are separated in energy by a value ΔE. For the sake of clarity, we consider here purely homogeneous transitions and neglect many-body effects.

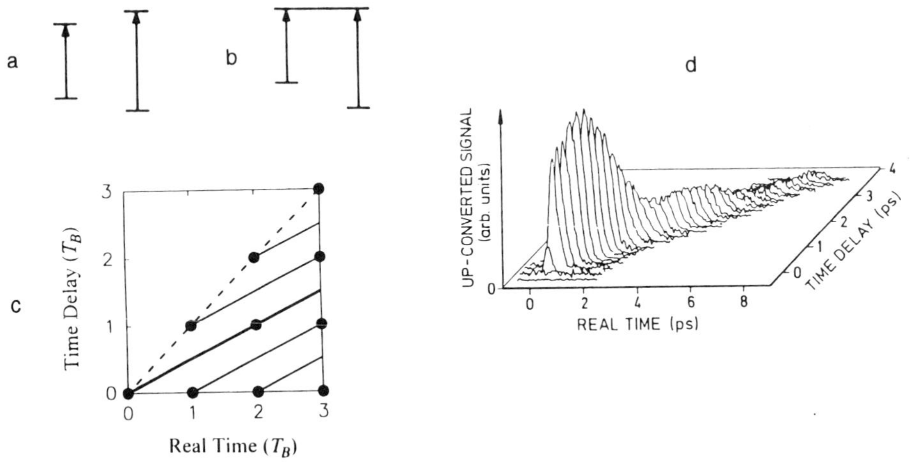

Figure 1. (a) (2 × 2)–level system as an example for a polarization interference system. (b) Three–level system as an example for a quantum beat system. (c) Solid lines: temporal positions of FWM signal maxima for a (2 × 2)–level system as discussed in Section 2. The thick line corresponds to $t = 2\tau$. Dots: temporal positions of the signal maxima for a three–level system. (d) Intensity of the up–converted signal as a function of the real time t and the time delay τ between the pump laser pulses for the sample described in Section 4.

A single pulse (wave vector $\vec{k}_1$) incident on the sample at time $t = 0$ will set up a coherent first-order polarization, which for both PI and QBs subsequently develops in time. The interference effect manifests itself in a periodic amplitude modulation of the polarization, i.e. a beating. Since the modulation is the same in the two cases, it is, in principle, impossible to experimentally distinguish QBs from PI using linear optical methods. However, in a transient FWM experiment in the photon echo configuration, the situation is different. In such an experiment a second laser pulse (wave vector $\vec{k}_2$) is incident on the sample at a delay time $t = \tau$ after the first pulse. The nonlinear interaction of the coherent first-order polarization with the second pulse generates a third–order polarization, $P^{(3)}$, which gives rise to a signal with intensity $I = \mid P^{(3)} \mid^2$ diffracted into the phase–matched direction $2\vec{k}_2 - \vec{k}_1$ for times $t \geq \tau$.

For both PI and QB, $P^{(3)}$ will be a periodic function of the 'real' time t as well as of the time delay τ between the two pulses. However, there will be a marked difference between

PI and QBs with regard to the kind of periodicities involved. It follows from Eq.1 and 3 of Ref. 4 that in the PI case, the FWM signal intensity as a function of t and τ will be peaked along lines $t = 2\tau \pm nT_B$, where $T_B = h/\Delta E$ is the beat period and n is an integer (with the restriction $t \geq \tau$). These lines are schematically shown in Fig. 1(c). The difference between PI and QB lies in the fact that in the PI case, only transitions within each two–level system occur, whereas in the QB situation, the coupling of the transitions via the common excited state gives rise to additional interference terms ('QB–terms') in the nonlinear polarization. In other words, $P^{(3)}_{QB}(t,\tau) = P^{(3)}_{PI}(t,\tau)+$ 'QB-terms', as can be seen from Eq.2 and 3 of Ref.4. This means that in the QB situation, an additional interference pattern is superimposed on that indicated by the solid lines in Fig. 1(c). A brief calculation shows that the signal intensity peaked along these lines will now be periodically modulated as a function of τ and $t-\tau$ so that it exhibits maxima (shown as dots in Fig. 1(c)) wherever the two additional conditions $\tau = lT_B$ and $t-\tau = \pm mT_B$ $(l,m = 0,1,2,\ldots)$ are fulfilled, and minima in between. In particular, there will be a periodic modulation along the line $t = 2\tau$, whereas in the PI case, no such modulation is to be expected. This is of particular relevance for the case of strong inhomogeneous broadening, where the polarization is considerable only along the line $t = 2\tau$, and is suppressed anywhere aside from that line by the inhomogeneous broadening (photon echo situation, [5]).

3. QB VERSUS PI: EXPERIMENTAL VERIFICATION

To verify the criterion described in the previous section experimentally, we have performed subpicosecond four–wave–mixing experiments with time–resolved (as opposed to time–integrated) detection of the nonlinear signal. As model systems we use two different GaAs/AlAs quantum well structures which have been designed to represent the cases of PI and QBs, respectively. To realize the conditions for two uncoupled two–level absorbers (PI situation), we have designed a sample containing spatially separated GaAs/AlAs QWs of 80 and 90Å width (Fig. 2(a)). This results in two quantum–mechanically independent and spectrally adjacent heavy–hole excitonic transitions ($\Delta E_{hh1-hh2} = 13.4meV$), which can be excited separately from the corresponding light–hole exciton transitions. For the demonstration of QBs, the heavy and light hole transitions in a quantum well of 150Å width are utilized, where the two excitons share the electron states while different hole states are involved (Fig. 2(b)). The energy splitting between the two transitions is measured in a transmission experiment to be $\Delta E_{lh-hh} = 5.4meV$.

The experiments are performed using a tunable Kerr–lens mode locked Ti:sapphire laser as the excitation source. This laser system generates bandwidth–limited pulses of typically $110fs$ width, and $21meV$ spectral width (FWHM); the pulses can be tuned into resonance with the respective exciton transitions. The experiments are performed at a sample temperature of $T = 5K$. We use an optical sampling technique in order to time-resolve the FWM signal. The diffracted signal is up-converted in a $1mm$ LBO crystal using a third laser pulse. Accordingly, we can detect the real–time behaviour of the diffracted signal by scanning the delay t between the FWM signal and the third reference pulse. More details about the experimental set–up and the samples can be found elsewhere [4,6,7].

The experimental results are shown in Fig. 2, where the up-converted signal is plotted as a function of both the real time t and the time delay τ. In Fig. 2(c), a beating of the signal as a function of t can be seen clearly; along the line $t = 2\tau$ a rise of the signal and a subsequent slow decay is observed, but no periodic modulation. This is essentially the behaviour expected for PI. In contrast, in Fig. 2(d), a pronounced modulation of the signal along $t = 2\tau$ is observed in addition to the beating as a function of t. We also note that in both cases, all observed beats, with few exceptions, show a time period that agrees well with the expected values $T_B = h/\Delta E$. We conclude that the two samples indeed show the behaviour predicted for PI and QBs. Thus, we will henceforth be justified in using the difference in the FWM interference patterns as a criterion for distinguishing QBs from PI.

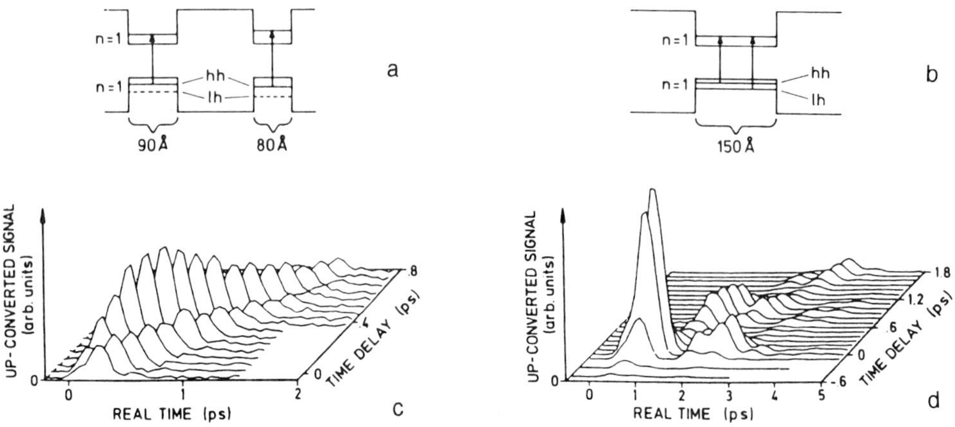

Figure 2. (a) Experimental realization of a (2 × 2)–system by two spatially and spectrally separated heavy–hole excitonic transitions belonging to different GaAs/AlAs quantum wells of 80 and 90Å width, respectively. (b) Experimental realization of a three–level system by the light– and heavy–hole excitonic transitions of a 150Å GaAs/AlAs quantum well. (c) Intensity of the up–converted signal as a function of the real time t and the time delay τ between the pump laser pulses for the sample of Fig. 2(a). (d) Same as Fig. 2(c) for the sample of Fig. 2(b).

4. EXCITONS ON DIFFERENT QW INTERFACE ISLANDS

In the following, we will apply the criterion developed in the previous two sections to the problem of beats from excitons on different QW interface islands [7]. The sample investigated is the same as that studied in Ref. 2. In this sample, the occurence of spatially separated interface islands in the 70Å quantum well leads to a difference in the confinement energies of the corresponding excitons, which results in beats in the time–integrated FWM signal with a beat period $T_B = 1.33ps$ [2]. Our experimental set–up is identical to that described in the previous section.

The experimental findings are shown in Fig. 1(d). For each time delay τ, the time resolved signal exhibits a peaked, nearly symmetric shape as a function of the real time t. The temporal positions of the maxima folllow the line $t = 2\tau$ as expected for sufficient inhomogeneous broadening (photon echo case, [5]).

The temporal halfwidth of the signal as a function of t is about $800fs$; from this, we determine the inhomogeneous broadening to be approximately $3.3meV$. The signal maximum shows a beating as a function of τ; the beat period is 1.33 ps and hence exactly the same as observed previously in the time integrated FWM experiment [2]. Since the beat period is larger than the temporal width of the signal peak, it is clear that no periodic modulation in the signal amplitude as a function of t can be observed.

An important consequence of the observed beating along the line $t = 2\tau$ in Fig. 1(d) is that the beating cannot result from PI, since otherwise the signal would be unmodulated along $t = 2\tau$, as we have seen in the previous sections. This means that the two excitonic transitions associated with spatially separated islands having different quantum well thicknesses must somehow be coupled with each other. In other words, the observed signal beats do not arise from polarization interference but are due to quantum beats as claimed by Göbel et al. [2].

There are several possible mechanisms which might lead to a coupling of the two transitions. The first possibility is that the excitation of one excitonic transition might lead to a bleaching of the other transition, since both exciton states might use the same one–particle k–states. Secondly, screening of the Coulomb interaction might lead to a coupling of the two exciton transitions. The coupling may also be caused by Coulomb–exchange interaction. Further experimental and theoretical work is clearly needed to determine the nature of the coupling mechanism.

5. CONCLUSIONS

In summary, we have developed a criterion for distinguishing quantum beats from polarization interference. The criterion is based on the difference in the temporal patterns of the signal beats in time-resolved FWM. We have verified the validity of the criterion by performing subpicosecond FWM experiments with time-resolved signal detection on especially designed quantum well structures. We have then applied this method to the problem of beats from excitons on spatially separate QW interface islands with different confinement energies. We have shown that the optical transitions associated with these excitons are coupled with each other and thus lead to quantum beats. We propose bleaching and screening as possible coupling mechanisms.

Acknowledgements

We thank S.T. Cundiff, D. Bennhardt, T. Meier, and D. Weber for helpful discussions and M. Preis for excellent technical assistance. The work at Marburg University is financially supported by the Deutsche Forschungsgemeinschaft through the Leibniz-Förderpreis. One of us (J.S.) is grateful to the Alexander von Humboldt foundation for a Humboldt senior scientist research award.

REFERENCES

1. K. Leo in: F. Henneberger, S. Schmitt-Rink, and E.O. Göbel eds., "Optics of Semiconductor Nanostructures", Akademie Verlag, Berlin (1993), p. 127, and references therein.
2. E.O. Göbel, K. Leo, T.C. Damen, J. Shah, T.C. Damen, J. Shah, S. Schmitt-Rink, W. Schäfer, J.F. Müller, and K. Köhler, "Quantum beats of excitons in quantum wells", *Phys. Rev. Lett.* 64:1801 (1990).
3. B.F. Feuerbacher, J. Kuhl. R. Eccleston, and K. Ploog, "Quantum beats between the light and heavy hole excitons in a quantum well", *Sol. Stat. Comm.* 74:1279 (1990).
4. M. Koch, J. Feldmann, G. von Plessen, E.O. Göbel, and P. Thomas, "Quantum beats versus polarization interference: An experimental distinction", *Phys. Rev. Lett.* 69:3631 (1992).
5. M.D. Webb, S.T. Cundiff, and D.G. Steel, "Observation of time-resolved picosecond stimulated photon echoes and free polarization decay in GaAs/AlGaAs Multiple Quantum Wells", *Phys. Rev. Lett.* 66:934 (1991).
6. M. Koch, D. Weber, J. Feldmann, E. O. Göbel, T. Meier, A. Schulze, P. Thomas, S. Schmitt-Rink, and K. Ploog, "Subpicosecond photon-echo spectroscopy on GaAs/AlAs short-period superlattices", *Phys. Rev. B* 47:1532 (1993).
7. M. Koch, J. Feldmann, E.O. Göbel, P. Thomas, J. Shah, and K. Köhler, "Coupling of exciton transitions associated with different quantum well islands", *Phys. Rev. B* 48:11480 (1993).

NATURE OF COHERENT FOUR-WAVE MIXING BEATS IN SEMICONDUCTORS

J. Erland,[a] I. Balslev,[a] and J.M. Hvam[a,b]

[a]Fysisk Institut, Odense Universitet
Campusvej 55, DK-5230 Odense M
[b]Mikroelektronik Centret
Technical University of Denmark, DK-2800 Lyngby
Denmark

INTRODUCTION

Coherent quantum beat spectroscopy has recently gained importance for the investigation of semiconducting materials and structures. The variety of excitonic resonances in bulk and low-dimensional semiconductors, and the advent of ultrafast lasers with pulse lengths shorter than typical dephasing times of these excitons, have made it possible to observe quantum beats between close-lying excitonic transitions.[1,2,3] In particular, the beat phenomena have been observed in four-wave mixing (FWM) experiments on exciton complexes in GaAs multiple quantum wells,[4,5,6,7,8] and in bulk semiconductors.[3]

In these experiments, the laser beam with ultrafast pulses is split into two beams incident on the sample in the directions $\mathbf{k}_1$ and $\mathbf{k}_2$, and the nonlinear FWM signal is emitted in the direction $2\mathbf{k}_2 - \mathbf{k}_1$, and detected time integrated as a function of the delay τ between the pulses in the two beams.[9] These correlation traces will for two-level systems show an exponential decay $I_{FWM}(\tau) \propto I(0)e^{-c\gamma\tau}$, from which the dephasing time $T_2 = 1/\gamma$ can be determined if the coefficient c is known. In a purely homogenously (inhomogeneously) broadened two-level system, c=2 (c=4).[9] In semiconductors, exciton interactions may imply that a description by two-level atoms is no longer adequate. Then, the temporal behaviour of the nonlinear signal is neither as a free induction decay (c=2), nor as a photon echo (c=4).[10] Then, the integrated intensity of the FWM signal also has a contribution for negative delay (pulse #2 arriving before #1).[11] In the following, we shall consider only low exciton densities, where the FWM process is often well described by a two-level model.[12,13]

QUANTUM BEATS AND POLARIZATION INTERFERENCES

In nearly degenerate multilevel systems, that are simultaneously and coherently excited by the same ultrashort laser pulses, the correlation trace may additionally show modulations due to quantum interferences, e.g. in the simplest form[14]

$$I_{FWM}(\tau) = I_{ave}(\tau)[1+I_m(\tau)\cos(\Delta\omega\tau+\phi)] \quad (1)$$

where $\Delta\omega = \omega_{20} - \omega_{10}$ is the splitting of the transition frequencies. The decays of $I_{ave}(\tau)$ and of the modulation amplitude $I_m(\tau)$ are governed by the dephasing rates γ_{j0} of the two transitions. The phase of the modulation is specified by ϕ. Such modulations have been observed in GaAs quantum wells due to beats between heavy-hole excitons (hhx) and light-hole excitons (lhx),[3,4] free and bound excitons,[6] excitons and biexcitons[7,8] and between Landau split magnetoexcitons.[1] Beats have also been observed between different free and bound exciton levels and complexes in bulk CdSe crystals.[3]

It is an interesting question, whether the observed modulations are due to true quantum beats between transitions with one state in common, or they are due to polarization interferences between independent, but coherently excited, two-level systems.[14,15] Koch et al.[16] have recently reported an experiment distinguishing between the two cases in a specially prepared GaAs quantum well sample. They time-resolve the FWM signal by cross correlation with a third, delayed laser beam and observe the modulation in real time for different delays between the two interfering incident beams. We have found an alternative way to distinguish experimentally between true quantum beats and polarization interferences, by spectrally resolving the time-integrated FWM signal.[17] As we shall see, the information is embedded in the amplitude and the phase of the modulation in Eq. (1), that are functions of the optical frequency ω of the nonlinear signal.[18]

To gain insight in the temporal and spectral dependences of the FWM signal in the two cases, we have solved simple Bloch equations for two-level and three-level systems and calculated the third order nonlinear polarization using Greens functions techniques.[9,13,18]

Time Dependence

For two independent two-level systems with the complex transition frequencies $\Omega_{j0} = \omega_{j0} - i\gamma_{j0}$, the solution is straight forward. For $\tau>0$ (no signal for $\tau<0$), the total third order polarization has the form

$$P^{(3)}(t,\tau>0) = \Theta(t-\tau)[M_{10}^4 N_1 e^{-i\Omega_{10}(t-2\tau)} + M_{20}^4 N_2 e^{-i\Omega_{20}(t-2\tau)}] \quad (2)$$

where t is the time after the arrival of pulse #1, M_{j0} is the dipole matrix elements and $\Theta(t-\tau)$ is the step function. N_j is a complicated function of the time-integrated pulse amplitude, its spectral overlap with the resonance number j and the concentration of two-level centers of this type. The FWM signal intensity is proportional to the absolute square of this polarization, as shown in Fig. 1(a). The temporal interference results from the cross term which has a phase that varies as $t-2\tau$ in agreement with the experiments by Koch et al..[16]

For a three-level system with a common ground state of the two nearly degenerate transitions, the solution is to the same approximations for central and coherent excitation by the same laser pulse[18]

$$\begin{aligned} P^{(3)}(t,\tau>0) \propto \Theta(t-\tau)[&2M_{10}^4 e^{-i\Omega_{10}(t-2\tau)} + 2M_{20}^4 e^{-i\Omega_{20}(t-2\tau)} \\ &+ M_{10}^2 M_{20}^2 (e^{-i\Omega_{10}(t-\tau)+i\Omega_{20}\tau} + e^{-i\Omega_{20}(t-\tau)+i\Omega_{10}\tau})] \end{aligned} \quad (3)$$

In this case, the intensity of the FWM signal, as shown in Fig. 1b, behaves in a more complicated fashion, than previously predicted,[16] with modulations, quantum beats, varying as τ, t, $t-\tau$ and $t-2\tau$. There are indications in the previously reported experiments of such a complex behaviour, although the term varying as $t-\tau$ was emphasized.[16]

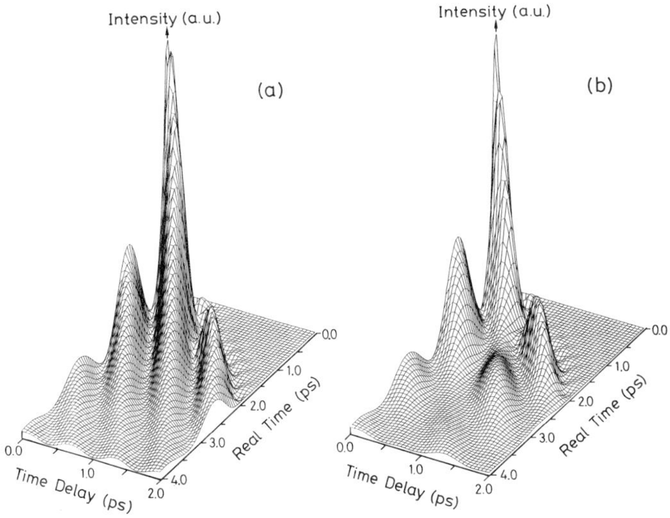

Figure 1. Calculated intensity of the FWM signal in the real-time, time delay domain for two independent two-level systems (a) and for a three-level system (b) with $\Delta\omega = 4.84\times10^{12}s^{-1}$ and $\gamma_1 = \gamma_2 = 5\times10^{11}s^{-1}$.

Spectral Dependence

For two independent two-level systems, the spectral dependence is the Fourier transform of Eq. (2)

$$P^{(3)}(\omega,\tau>0) = \left[\frac{M_{10}^4 N_1 e^{i\Omega_{10}^*\tau}}{\Omega_{10} - \omega} + \frac{M_{20}^4 N_2 e^{i\Omega_{20}^*\tau}}{\Omega_{20} - \omega}\right] \tag{4}$$

The absolute square of Eq. (4), yields an expression as in Eq. (1), where the beats originate from the cross product involving both resonance denominators. This gives a variation in the amplitude and phase of the modulation when ω passes through one of the transition frequencies, as shown in Fig. 2(a). In resonance ($\delta\omega=\omega-\omega_{20}\approx0$), the modulation amplitude is minimum, and the phase undergoes a shift of about π (see vertical line in Fig. 2). To observe a substantial modulation of the nonlinear signal, the two terms in Eq. (4) have to be of the same order of magnitude. For comparable matrix elements of the two transitions, $M_1 \approx M_2$, this can be obtained by exciting near one resonance, $N_1 > N_2$, and detecting near the other resonance $\omega \approx \omega_{20}$.

In Fig. 2(b) is shown the spectrally resolved FWM signal around the I_2 bound exciton in CdSe when the excitation is centered close to the I_1 bound exciton.[17] The observed beats have a period of 900 fs, corresponding to the I_1 - I_2 splitting of 4.8 meV. The qualitative agreement between the experiments and the calculations in Fig. 2, for a representative set of parameters, verifies that the beating bound exciton transitions can be considered non-interacting, as expected for low donor and acceptor concentrations.

In a three-level system, the spectral dependence of the FWM signal is the Fourier transform of Eq. (3)

$$P^{(3)}(\omega,\tau>0) \propto \left[\frac{2M_{10}^4 e^{i\Omega_{10}^*\tau} + M_{10}^2 M_{20}^2 e^{i\Omega_{20}^*\tau}}{\Omega_{10} - \omega} + \frac{2M_{20}^4 e^{i\Omega_{20}^*\tau} + M_{10}^2 M_{20}^2 e^{i\Omega_{10}^*\tau}}{\Omega_{20} - \omega}\right] \tag{5}$$

Detecting the signal near one resonance, $\omega \approx \omega_{10}$, the first term in Eq. (5) will dominate.

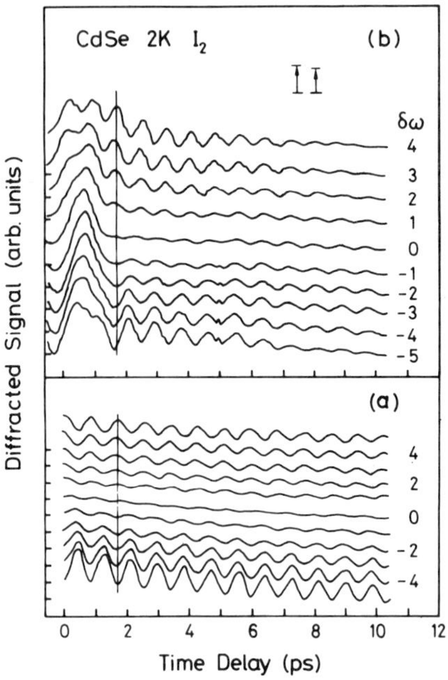

Figure 2. Diffracted FWM signal for different detunings $\delta\omega = \hbar(\omega - \omega_2)/0.08\text{meV}$ around the I_2 bound exciton resonance in CdSe. All the curves are normalized to their maximum value, and they are shifted along the ordinate axis as indicated by the zero-level marks. The lower curves (a) are calculated from Eq. (4) with $\hbar(\omega_{20} - \omega_{10}) = 4.8$ meV, $\gamma_{10} = 0$, $\hbar\gamma_{20} = 0.03$ meV, $M_{10} = M_{20}$ and $N_1/N_2 = 1.6$.The upper curves (b) are experimental with an excitation level of 0.8 kW/cm^2.

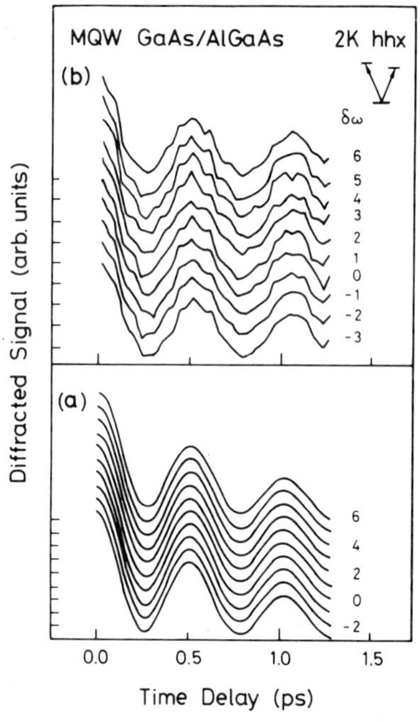

Figure 3. Diffracted FWM signal for different detunings $\delta\omega = \hbar(\omega - \omega_1)/0.08\text{meV}$ around the hhx resonance (ω_1) in a GaAs MQW. All the curves are normalized to their maximum value, and they are shifted along the ordinate axis as indicated. The lower curves (a) are calculated from Eq. (5) with $\hbar(\omega_{20} - \omega_{10}) = 7.7$ meV, $\hbar\gamma_{10} = \hbar\gamma_{20} = 0.3$ meV, and $M_{10} = M_{20}$ .The upper curves (b) are experimental with an excitation level of 0.2 kW/cm^2.

For comparable matrix elements of the two transitions, a strong modulation of the signal will appear, now as a true quantum beat. In this case, the amplitude and phase of the modulation are nearly constant through the resonance, as shown in Fig. 3(a) with a set of parameters matching the experimental quantum beats between the hhx and the lhx resonances in 116 Å wide GaAs multiple quantum wells (see Fig. 3(b)[17]). The agreement between the spectral dependences of the experimental and the calculated quantum beats is rather convincing, confirming the simple nature of the observed modulations.

CONCLUSIONS

We have described the nature of the observed beats in a time-integrated four-wave mixing experiment in different coherently excited multi-resonance systems. In an impurity-bound exciton system in CdSe, we observe a phase shift of the beat signal, when we move through one of the resonances. This is a signature of polarization interference between two independent two-level systems. In GaAs quantum wells, where the hhx and lhx share a common ground state, no such phase shift is observed, as predicted for quantum beats in a three-level system. In view of the general nature of the calculations, spectral resolution of the time-integrated FWM signal thus seems to provide distinct and unambiguous information about the nature of the observed beats in systems where level schemes and system interactions are not known in advance.

Acknowledgements

The authors want to thank V.G. Lyssenko, K.-H. Pantke, B.S. Razbirin, D. Oberhauser and G. Weimann for their contributions to this work, which was supported by the Danish Natural Science Research Council.

REFERENCES

1. V. Langer, H. Stolz, and W. von der Osten, Phys. Rev. Lett. **64**, 854 (1990).
2. W.A.J.A. van der Poel, A.L.G.J. Severens, and C.T. Foxon, Optics Commun. **76**, 116 (1990).
3. K.-H. Pantke, V.G. Lyssenko, B.S. Razbirin, J. Erland, and J.M. Hvam, *Proc. 21st International Conference on the Physics of Semiconductors*, Beijing 1992, edited by P. Jiang and H.-Z. Zheng (World Scientific, Singapore 1992) p. 129.
4. K. Leo, T.C. Damen, J. Shah, E.O. Göbel, and K. Köhler, Appl. Phys. Lett. **57**, 19 (1990).
5. B.F. Feuerbacher, J. Kuhl, R. Eccleston, and K. Ploog, Solid State Commun. **74**, 1279 (1990).
6. K. Leo, T.C. Damen, J. Shah, and K. Köhler, Phys. Rev. B **42**, 11359 (1990).
7. D.J. Lovering, R.T. Phillips, G.J. Denton, and G.W. Smith, Phys. Rev. Lett. **68**, 1880 (1992).
8. K.-H. Pantke, D. Oberhauser, V.G. Lyssenko, and J.M. Hvam, Phys. Rev. B. **47**, 2413 (1993).
9. T. Yajima and Y. Taira, J. Phys. Soc. Japan **47**, 1620 (1979).
10. D.-S. Kim, J. Shah, T.C. Damen, W. Schäfer, F. Jahnke, S. Schmitt-Rink, and K. Köhler, Phys. Rev. Lett. **69**, 2725 (1992).
11. K. Leo, M. Wegener, J. Shah, D.S. Chemla, E.O. Göbel, T.C. Damen, S. Schmitt-Rink, and W. Schäfer, Phys. Rev. Lett. **65**, 1340 (1990).
12. L. Schultheis, J. Kuhl, A. Honold, and C.W. Tu, Phys. Rev. Lett. **57**, 1635 (1986).
13. C. Dörnfeld and J.M. Hvam, IEEE J. Quantum Electron. **25**, 904 (1989).
14. L.Q. Lampert, A. Compaan, and I.D. Abella, Phys. Rev. A **4**, 2022 (1971).
15. H. Stolz, V. Langer, E. Schreiber, S. Permogorov, and W. von der Osten, Phys. Rev. Lett. **67**, 679 (1991).
16. M. Koch, J. Feldmann, G. von Plessen, E.O. Göbel, and P. Thomas, Phys.Rev. Lett. **69**, 2631 (1993).
17. V.G. Lyssenko, J. Erland, I. Balslev, K.-H. Pantke, B.S. Razbirin, and J.M. Hvam, Phys. Rev. B **48**, (1993).
18. J. Erland and I. Balslev, Phys. Rev. A **48**, (1993).

ANALOGIES BETWEEN COHERENT OPTICAL INTERACTIONS AND QUANTUM TRANSPORT IN SEMICONDUCTOR DEVICES

W. Quade[1], F. Rossi[2], and E. Schöll[1]

[1]Institut für Theoretische Physik
Technische Universität Berlin
Hardenbergstr.36
D-10623 Berlin, Germany

[2]Fachbereich Physik
Universität Marburg
Renthof 5
D-35037 Marburg, Germany

INTRODUCTION

The new generations of ultrasmall semiconductor devices are characterized both by very short transit times of the carriers from source to drain of less than a picosecond and by very high electric fields of up to some $10^6 V/cm$. Under such conditions, carriers reach very high energies and are subjected to time scales (for example imposed by phonon scattering) of the order of 100 femtoseconds. Energy nonconserving and collisionally broadened transitions and the influence of the field during the collision (intracollisional field effect) strongly influence the transport behaviour [1]. It is hard to treat such problems in a full spacedependent quantum mechanical theory. The necessary approximations become very serious and at the very end it is hard to recognize whether the final result still contains characteristic information or is dominated by the approximations. Furthermore specific experiments in devices are very hard to accomplish and measurements often allow for different interpretations.

Under these circumstances it may help to look for similar phenomena in neighbouring fields which can be better accessed and studied by experiments. One can, for example, draw links between generation of carriers by a laser and the injection of carriers from a contact into the high–field region of a semiconductor device, or between coherent states in laser experiments and non–completed collisions in transport. Furthermore a common feature in these two fields is the presence of an electric field, ac in the one and (approximately) dc in the other case.

The high electric field in devices leads to the Zener effect [2]. A valence electron tunnels across the forbidden gap into the conduction band. In the language of optics this means that even a static electric field can induce polarization. Another important generation process in devices is impact ionization [3] where a high–energetic conduction electron collides with a valence electron and, as a consequence, this latter electron undergoes a transition from the valence to the conduction band. The role of impact ionization in ultrasmall semiconductor devices is particularly pronounced since this is an autocatalytic process, which may lead to the eventual destruction of the device. In the following we will concentrate on the Zener effect and especially on impact ionization [4] to emphazise the analogies between transport in devices and coherent optical interactions.

PHYSICAL SYSTEM AND THEORETICAL APPROACH

Our semiconductor is modeled by the following two-band (valence band v and conduction band c) Hamiltonian

$$H = H_0 + H_{\mathcal{E}} + H_{ii}\,, \tag{1}$$

where

$$H_0 = \sum_k E_c(k)\, c_k^+ c_k + \sum_k E_v(k)\, d_k^+ d_k \tag{2}$$

describes the free evolution of electrons in the conduction band $E_c(k)$ and of holes in the valence band $E_v(k)$. The operators c_k^+, c_k and d_k^+, d_k are the creation and annihilation operators of electrons and of holes, respectively, and k is the wavevector. The scalar–potential Hamiltonian $H_{\mathcal{E}}$ describes the action of the applied uniform electric field $\mathcal{E}$. It is given in the first quantization scheme (and coordinate representation) by $H_{\mathcal{E}} = e\mathcal{E}\cdot x$, and within the above two–band description splits up into a diagonal part $H_{||}$ and into a non–diagonal part $H_\perp$. $H_{||}$ accelerates the carriers within the same band, and is responsible for the intracollisional field effect. This phenomenon is usually neglected in semiclassical transport. The action of $H_{||}$ can be simply described by its effect on the Bloch states

$$U_t(H_{||})\varphi_n(k_0,x) \equiv e^{-iH_{||}t/\hbar}\varphi_n(k_0,x) = \varphi_n(k_t,x)\,, \tag{3}$$

$$\varphi_n(k,x) = \frac{1}{\sqrt{V}}e^{i\,k\cdot x}u_n(k,x)\,, \quad \text{and} \quad k_t \equiv k_0 - \frac{e\mathcal{E}}{\hbar}t\,. \tag{4}$$

The non–diagonal part in the second–quantization scheme can be written as

$$H_\perp = \sum_k X_{vc}(k)\, c_k^+ d_{-k}^+ + \sum_k X_{vc}^*(k)\, d_{-k} c_k\,, \tag{5}$$

$$X_{vc}(k) = \frac{1}{V}\int d^3x\, u_c^*(k,x)\,(ie\mathcal{E}\frac{\partial}{\partial k})\, u_v(k,x)\,. \tag{6}$$

It induces "vertical" transitions (Zener transitions) from one band to the other and is of the form of the operator describing the action of a semiclassical light field within the dipole approximation [5,6].

The two–band model leads to a division of the usual electron–electron interaction into several conceptually different contributions. One contribution is given by impact ionization and its inverse, Auger recombination (for electrons)

$$H_{ii} = \sum_{k,k',q} M_q^e c_{k+q}^+ c_{k'-q}^+ d_{-k'}^+ c_k + \sum_{k,k',q} M_{-q}^{e*} c_{k+q}^+ d_{-(k'-q)} c_{k'} c_k\,, \quad M_q^e = \frac{e^2}{\epsilon\epsilon_0 V}\frac{F_{vc}F_{cc}}{q^2+\lambda^2} \tag{7}$$

where M_q^e (inverse screening length λ, Bloch overlap integrals F_{ij}, relative and absolute permittivity ϵ, ϵ_0 and crystal volume V) is the usual Debye screened Coulomb matrix element.

The most important effect of the electric field for transport properties of devices is, of course, the action of $H_{\|}$. For a space–independent electric field one can avoid to treat $H_{\|}$ as a perturbation. It turns out that in the accelerated frame described by the k_t the corresponding distribution function satisfies the following equation of motion without $H_{\|}$ in the Hamiltonian

$$\frac{d}{dt} f_n(k_t, t) = \frac{i}{\hbar} \left\langle [(H - H_{\|}), c^{+}_{k_t} c_{k_t}] \right\rangle , \qquad f_n(k_t, t) = \left\langle c^{+}_{k_t} c_{k_t} \right\rangle . \tag{8}$$

$\langle\rangle$ means the average with the statistical operator. So $H_{\|}$ does not contribute to the time evolution of $f_n(k_t, t)$. The effect of $H_{\|}$ enters only through k_t in analogy with the semiclassical Chambers picture [7].

ZENER TUNNELING

Neglecting for the moment impact ionization and Auger recombination we are left with the following Hamiltonian

$$H = H_0 + H_{\|} + H_{\perp} , \tag{9}$$

which, in the accelerated frame, gives a closed coupled system of differential equations for the two (electron and hole) distribution functions f_n and f_p and the interband polarisation equivalently to the well–known optical Bloch equations [6]. Integrating formally the equation for the polarisation one obtains the following equation for f_n

$$\left.\frac{\partial}{\partial t} f_n(k_t, t)\right|_Z = \frac{2\pi}{\hbar^2}\, Re\left[X_{vc}(k_t) \int_{t_0}^{t} dt' \, \exp\left\{\frac{i}{\hbar}\int_{t'}^{t} d\tau\, \Omega(\tau)\right\} X^{*}_{vc}(k_{t'})\, \mathcal{F}(k_{t'}, t')\right] , \tag{10}$$

with

$$\mathcal{F}(k) = 1 - f_n(k) - f_p(k) , \quad \Omega(t) = (E_c(k_t) + E_v(k_t))/\hbar . \tag{11}$$

It is assumed that at $t = t_0$ the polarisation is zero. In the Boltzmann limit transitions are not possible but Zener transitions occur because $\Omega(t)$ oscillates in time [8] due to the action of $H_{\|}$. The total Zener generation rate is obtained by summing over all k

$$\frac{dn}{dt} = \sum_k \left.\frac{\partial}{\partial t} f_n(k_t, t)\right|_Z . \tag{12}$$

We see that the Zener effect in transport is given by almost the same equations as in optics [6] but the effect of the electric field is different.

IMPACT IONIZATION

If we replace $H_{\perp}$ with H_{ii}, calculations in analogy with the previous section, although more tedious, lead to the following generation rate due to impact ionization [4]

$$\frac{dn(t)}{dt} = \sum_k \int_{t_0}^{t} dt' \, f_n(k_{t'}, t') \int_{\omega_{min}}^{\omega_c(k_{t+})} d\omega \; P_{ii}(\omega_c(k_{t+}), \omega)\, \frac{1}{\pi} \cos\{(\omega_{th}(\mathcal{E}, t_-) - \omega)t_-\} , \tag{13}$$

where $\hbar\omega_{th}(\mathcal{E}, t_-)$ is the sum of the semiclassical impact ionization threshold energy $\hbar\omega_{th}$ and a field–dependent term [4]. $P_{ii}(\omega_c(k), \omega_{th})$ is the usual semiclassical impact ionization scattering rate and $t_+ = 0.5(t+t')$, $t_- = t-t'$ are the absolute and relative scattering times. Switching off the electric field, one obtains in the limit $t_0 \to -\infty$ the semiclassical counterpart of (13)

$$\frac{\mathrm{d}n(t)}{\mathrm{d}t} = \sum_k f_n(k_t, t) \int_{\omega_{min}}^{\omega_c(k_t)} d\omega \; P_{ii}(\omega_c(k), \omega) \;\; \delta(\omega_{th} - \omega) \; . \tag{14}$$

The comparison of eq. (13) and eq. (14) shows that in the quantum mechanical case (as for eq. (10)), the scattering rate depends on the distribution function even at earlier times. Energy–nonconserving transitions are possible, the impact ionization threshold is not sharply defined as in the semiclassical limit but is smeared out. The ionized electron is allowed to make a transition to a final state in the conduction band by taking up an amount of energy smaller or greater than the real energy gap. Electrons near the bottom of the conduction band preferably make transitions to final states with a larger energy with respect to the semiclassical case. This phenomenon tends to increase the quantum mechanical rate.

A numerical anlysis was done using the parameter values $m_c = 0.067 m_0$, $m_v = 0.450 m_0$, spherical parabolic bands, $\mathcal{E}_{gap} = 1.422\, eV$ (corresponding to a simplified GaAs model). In this case the semiclassical impact ionization threshold energy is $1.606\, eV$. In Fig. 1 results for the quantum eq. (13) and the semiclassical eq. (14) are compared. In both cases we have calculated the lowest order correction $\Delta n(t)$ of the number of electrons at t, starting at $t_0 = 0$, with a time–dependent initial distribution function

$$f_n(k_t, t) := f_n^0(k_t, t) \tag{15}$$

given by a semiclassical generation rate due to a laser [6]. A constant electric field of $10^5 V/cm$ is applied. In Fig. 1 the full line represents the number of electrons $n_0(t)$ as generated by the laser. The two upper dotted and dash–dotted lines refer to the quantum mechanical Δn with and without electric field, all the lower curves refer to the semiclassical Δn. Due to energy–time uncertainty the quantum mechanical response is immediate and independent of the semiclassical threshold energy. The semiclassical counterparts start at much lower levels for both values of laser excess energies. Although not shown, the quantum mechanical Δn for the lower laser excess energy would almost coincide with the other quantum mechanical curve ($1.8\, eV$). This demonstrates that at short times the semiclassical threshold energy has no significance. Impact ionization has nearly the same influence independent of whether the carriers are mainly created below or above threshold. Similar enhancement of the generation rate at very short times is found for coherent optical excitation [6].

CONCLUSION

To summarize, we have tried to emphasize similarities between short–time laser experiments and semiconductor transport in small devices. As a first point we mentioned the creation of electron–hole pairs by a laser and the Zener effect in transport. As a second point from the analysis of laser experiments we showed that at short times and strong electric fields quantum corrections lead to an enhanced generation of carriers by impact ionization, as compared to the semiclassical theory.

The main conclusion is that the impact ionization threshold is not a fixed quantity as it would result from the lowest order perturbation theory (Fermi's golden rule) commonly used in semiclassical transport. As an application of our results to semiconductor microstructure devices, we want to stress that a carrier traveling from source to drain may not have the necessary energy to impact ionize under semiclassical energy conservation conditions, but may well be able to do this within the above more general approach.

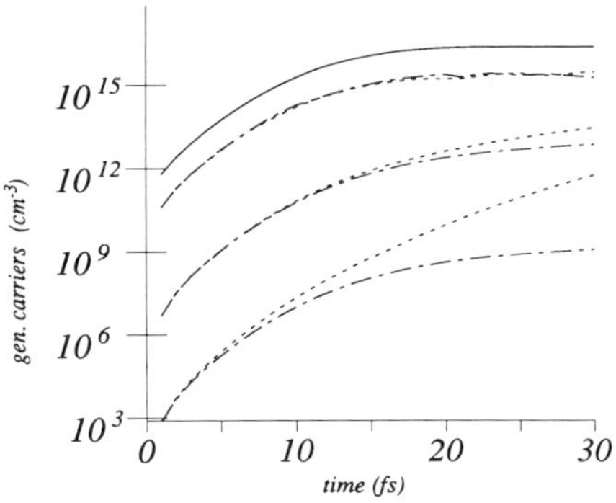

Figure 1. Number of generated carriers as a function of time. The full line represents $n_0(t)$ as generated by the laser. The upper dotted and dash–dotted lines represent the quantum mechanical Δn due to impact ionization with and without electric field for an excess energy of $1.8\,eV$. The dotted (with el. field) and dash–dotted (without el. field) lines in the middle represent the semiclassical Δn due to impact ionization for an excess energy of $1.8\,eV$. The lower dotted (with el. field) and dash–dotted (without el. field) lines represent the semiclassical Δn for an excess energy of $1.4\,eV$.

ACKNOWLEDGEMENT

This work was partially supported by Deutsche Forschungsgemeinschaft and by the European Communities.

REFERENCES

[1]R. Brunetti, C. Jacoboni, and F. Rossi, Phys. Rev. B **39**, 10781 (1989).
[2]C. Zener, Proc. R. Soc. A **145**, 523 (1934).
[3]A. R. Beattie and P. T. Landsberg, Proc. R. Soc. A **249**, 16 (1958).
[4]W. Quade, F. Rossi, C. Jacoboni, and E. Schöll, in *Proc. 21th int. Conf. Phys. Semicond.*, Beijing, Ed. by P. Jiang and H.–Z. Zheng, (World Scientific, Singapore 1992), p. 321.
[5]R. Zimmermann and M. Hartmann, Phys. Stat. Sol. (b) **150**, 365 (1988).
[6]T. Kuhn and F. Rossi, Phys. Rev. B **46**, 7496 (1992).
[7]R. Chambers, Proc. Phys. Soc. (London) **A65**, 458 (1952).
[8]W. Franz, Z. Naturforschg. **13 a**, 484 (1958).

OPTICAL GENERATION AND DETECTION OF CARRIERS IN ULTRAFAST PUMP-PROBE SPECTROSCOPY OF SEMICONDUCTORS

Christopher J. Stanton[1], Alex V. Kuznetsov[1], and Chang Sub Kim[2]

[1]Department of Physics
University of Florida
Gainesville, Florida 32611 USA

[2]Department of Physics
Chonnam National University
Kwangju, 500–757 Korea

INTRODUCTION

One of the most common experimental approaches to studying the relaxation of carriers in semiconductors on the femtosecond time scale is pump-and-probe spectroscopy. Essentially, it consists in exciting nonequilibrium carriers by an intense femtosecond pump pulse and then probing the density distributions with a weak time-delayed probe pulse. Comparison of such pump-and-probe spectra with theoretical simulations of carrier kinetics provides important information for understanding relaxation mechanisms that govern ultrafast carrier dynamics.[1]

Although at present there is quite a number of advanced Monte-Carlo simulations of carrier dynamics,[2–4] they all share the same problem of ambiguity in the generation of the initial distribution function, which obviously affects the subsequent evolution of carrier distribution function (DF). The simplest approach is to add carriers to the simulation monoenergetically[2] at t = 0. However, this violates the time-energy uncertainty principle which states that carriers created by a pump with the duration τ cannot have their energy defined better with an accuracy of the order $\hbar/\tau$. In an attempt to incorporate this restriction, the initial DF is often assumed to follow the power spectrum of the pump[4] or is chosen according to some other *ad hoc* model, and its time dependence is either ignored or oversimplified.

Recently[5,6] it was recognized that during and immediately after the optical excitation the carriers are in a coherent regime which cannot be treated classically. Therefore, the initial carrier distribution has to be determined by a full quantum mechanical calculation.

The purpose of the present paper is to explore the possible regimes of generation and also to study how the probe absorption is affected by the fast-changing nonequilibrium carrier distributions.

In this paper, we determine the dependence of the DF on time and energy during optical excitation by solving a coupled set of Bloch equations[7] for carrier density and interband polarization:

$$\frac{\partial p_{\mathrm{k}}}{\partial \mathrm{t}} = i(\varepsilon_{c\mathrm{k}}+\varepsilon_{h\mathrm{k}})p_{\mathrm{k}} - i\mu_{\mathrm{k}}E(t)(1-n_{c\mathrm{k}}-n_{h\mathrm{k}}) - \gamma p_{\mathrm{k}}\,; \tag{1}$$

$$\frac{\partial n_{c\mathrm{k}}}{\partial t} = \frac{\partial n_{h\mathrm{k}}}{\partial t} = 2\,\mathrm{Im}\{p_{\mathrm{k}}^{*}\mu_{\mathrm{k}}E\} \tag{2}$$

Here, $\varepsilon_{\alpha\mathrm{k}}$ and $n_{\alpha\mathrm{k}}$ are the energy dispersion and the carrier DF in the band α, the quantity p_{k} is the interband polarization[7], and *E(t)* is the electric field of the pump pulse (we use $\hbar = 1$ units).[7] The only relaxation mechanism included in (1) — (2) is the dipole dephasing characterized phenomenologically by the dephasing constant γ. We solve these equations numerically to obtain the time-dependent carrier distributions.

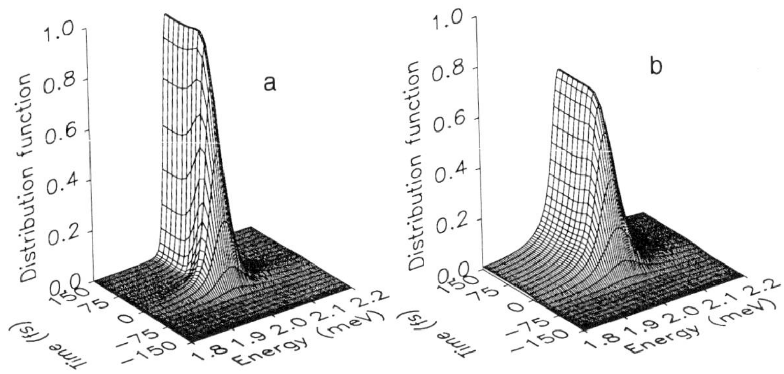

Figure 1. Carrier distribution functions generated with a 50 fs pulse centered at 2 eV for the parameters of GaAs. (a) For low dephasing ($\gamma^{-1} = 1$ ps) the pump creates transient population of virtual carriers that extends far from the central frequency of the pump; (b) At higher dephasing ($\gamma^{-1} = 100$ fs) these virtual carriers stay in the bands creating a very broad DF after the excitation

For a given pump pulse duration, the two most important parameters that control the photogenerated carrier distribution are the *pump intensity* and the *dephasing time*, γ^{-1}. We find that the only case where the photocarrier DF does approach the power spectrum of the pump pulse is for low intensity excitation and long dephasing times. However, this limit is not applicable to realistic experimental situations[8,9] where high intensities are used to achieve better signal to noise ratio, and the dephasing times are typically a few tens of femtoseconds.[10]

In Fig. 1(a) we show a typical time dependence of the DF of carriers excited by a pump pulse obtained by a direct numerical solution of the Bloch equations. The DF approaches the power spectrum of the pump at long times after the excitation. However, at intermediate times, during the pump pulse, the DF is much broader.

The reason the DF is broader while the pulse is on than after the excitation is that the pump creates virtual carriers in states far away from the excitation frequency. The density of virtual carriers falls off inversely proportional to the detuning $\Delta = \hbar\omega_o - \varepsilon$, so that at intermediate times the DF has power-law tails extending far away from the central frequency of the excitation.

Both high pump fields and the high dephasing rates encountered in realistic experiments will enhance this broadening effect because the height of the power-law shoulders is proportional to the pump intensity, and for dephasing times shorter than the excitation duration the carriers will stay in the bands because of relaxation processes (Fig. 1(b)). Therefore, at high dephasing rates a considerable fraction of the total number of photoexcited particles will be created far away from the central frequency of the pump.

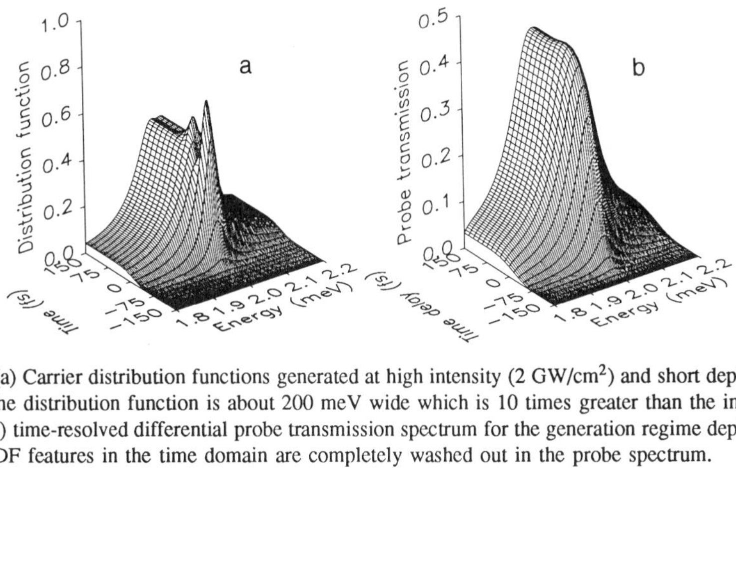

Figure 2. (a) Carrier distribution functions generated at high intensity (2 GW/cm^2) and short dephasing time of 25 fs. The distribution function is about 200 meV wide which is 10 times greater than the inverse pulse duration; (b) time-resolved differential probe transmission spectrum for the generation regime depicted in (a). The sharp DF features in the time domain are completely washed out in the probe spectrum.

This is illustrated by Fig. 2 where we plot the DF for a moderately strong pump field and a high dephasing rate. In this example, only 8 % of the total number of carriers is created within the energy interval $\hbar\omega_o \pm \Delta\omega$, ($\Delta\omega$ being the spectral width of the pump). Note also that the high dephasing suppresses the Rabi oscillations of carrier density around the center of the pulse, and the DF becomes "clipped" at $f_e = 0.5$ which corresponds to total absorption saturation. Let us note that in modulation-doped quantum wells[8] where high dephasing rates can be expected, these broadening effects may be even more pronounced and lead to almost structureless DF of photocarriers in a broad energy range.

Another important factor that leads to further broadening of the experimental spectra is that in pump-probe experiments the probe transmission is not a "snapshot" of the DF. In steady state (cw experiments) the probe absorption is related to the carrier populations by:

$$\alpha_{pr}(\omega) \propto \sum_{\mathbf{k}} |\mu_{\mathbf{k}}|^2 \delta(\varepsilon_{c\mathbf{k}} + \varepsilon_{h\mathbf{k}} - \hbar\omega)[1 - n_{c\mathbf{k}}(t) - n_{h\mathbf{k}}(t)] \quad (3)$$

However, in femtosecond spectroscopy the carrier populations change significantly *during* the probe pulse, and the probe has a finite spectral width. Therefore, for femtosecond excitation conditions the probe spectra are not described by (3) and also have to be determined by a full quantum mechanical calculation.

We have developed a procedure for calculating probe transmission that takes into account the quantum effects in pump-probe interaction.[11] To determine the probe transmission spectrum, we solve the Bloch equations for the polarization and the density generated by the probe:

$$\frac{\partial p_{\mathbf{k}}^{pr}}{\partial t} = i(\varepsilon_{c\mathbf{k}}+\varepsilon_{h\mathbf{k}})p_{\mathbf{k}}^{pr} - i\mu_{\mathbf{k}}E_{pr}(t-t_d)(1-n_{c\mathbf{k}}[t]-n_{h\mathbf{k}}[t]) - \gamma p_{\mathbf{k}}^{pr}\,, \tag{4}$$

$$\frac{\partial n_{c\mathbf{k}}^{pr}}{\partial t} = -2\,\mathrm{Im}\{p_{\mathbf{k}}^{pr}\mu_{\mathbf{k}}E_{pr}^{*}\}\,, \tag{5}$$

substituting the actual time and energy dependence of the DFs $n_{\alpha\mathbf{k}}(t)$ produced by the *pump* in the saturation term. The total number of carriers created by the probe pulse at $t \to \infty$, $\int d^3\mathrm{k}\cdot n_{c\mathbf{k}}^{pr}$, is a direct measure of the probe absorption. In this calculation, we neglect the coherent interactions between the pump and the probe[12] which would further complicate the results but nonetheless are present in some experimental conditions.

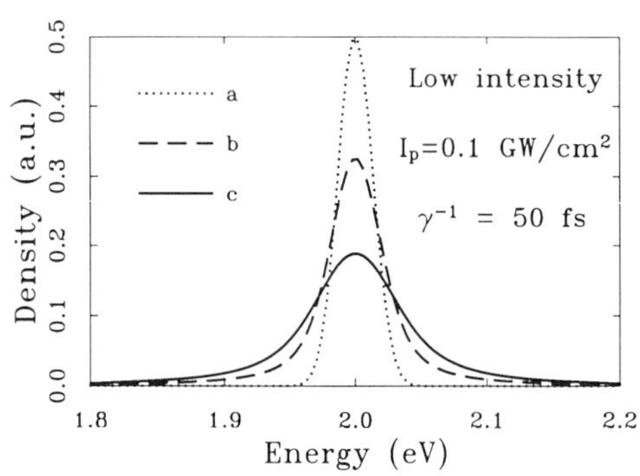

Figure 3. Comparison of calculated probe spectrum at 100 fs time delay (solid curve, c) with (a) pump power spectrum; (b) the carrier distribution function. This example is for low intensity ($I_p = 0.1\ GW/cm^2$) and moderate dephasing $\gamma^{-1} = 50\ fs$. Due to nonstationary effects, the probe spectrum is much broader than the distribution function.

Fig. 2(b) shows the calculated probe transmission as a function of the central frequency and the time delay of the probe for the pump-created carrier DF shown in Fig. 2(a). One sees that because the carrier density is spectrally and temporally averaged, the probe transmission does not have sharp temporal structures.

In Fig. 3 we compare the probe spectrum 100 fs after low-intensity excitation with the DF it is supposed to represent. The probe spectrum is significantly wider than the DF, which is a quite general result because the probe performs a spectral averaging of the DF. This may help explain why the probe usually does not "see" the sharp peaks in the DF that are often predicted by theory[3] at early times.

In conclusion, we demonstrate that the correct description of optical generation and detection of carrier distributions in pump-probe experiments leads to a significant spectral broadening of calculated probe spectra which makes them closer to experimental results.

Acknowledgments

This work was supported by U.S. Office of Naval Research through Grant N00091–JJ-1956 and by the NSF grant DMR 8957382. CJS gratefully acknowledges support from Alfred P. Sloan foundation.

REFERENCES

1. For reviews see W.H. Knox, in *Hot Carriers in Semiconductor Nanostructures: Physics and Applications* (ed. J. Shah, Academic Press, Boston, 1992), p 313; J. Shah, Solid State Electron. **32**, 1051 (1989).
2. S.M. Goodnick, and P. Lugli, Phys.Rev. **B 38**, 10135 (1988).
3. D.W. Bailey, C.J. Stanton and K. Hess, Phys. Rev. **B 42**, 3423 (1990).
4. M.A. Osman and D.K. Ferry, Phys.Rev. **B 36**, 6018 (1987).
5. A.V. Kuznetsov, Phys.Rev. **B 44**, 8813 (1991)
6. T. Kuhn and F. Rossi, Phys. Rev. Lett. **69**, 977 (1992); Phys. Rev. **B 46**, 7496 (1992).
7. M. Lindberg and S.W. Koch, Phys. Rev. **B 38**, 3342 (1988).
8. W.-Z. Lin, R.W. Schoenlein, J.G. Fujimoto and E.P. Ippen, IEEE J. Quantum Electron. **QE-24**, 267 (1988); W.-Z. Lin, J.G. Fujimoto, E.P. Ippen and R.A. Logan, Appl. Phys. Lett. **50**, 124 (1987); R.W. Schoenlein, W.-Z. Lin, E.P. Ippen and J.G. Fujimoto, *idem*, **51**, 1442 (1987).
9. W.H. Knox, C.A. Hirlimann, D.A.B. Miller, J. Shah, D.S. Chemla, and C.V. Shank, Phys. Rev. Lett. **56**, 1191 (1986); W.H. Knox, D.S. Chemla, G. Livescu, J. Cunningham, and J.E. Henry, *idem*, **62**, 1290 (1988).
10. P.C. Becker *et al*, Phys. Rev. Lett. **61**, 1647 (1988).
11. C.J. Stanton, A.V. Kuznetsov, and C.S. Kim, unpublished.
12. M. Lindberg and S.W. Koch, Phys. Rev. **B 38**, 7607 (1988).

STUDY OF EXCITON DEPHASING IN SUPERLATTICES USING RESONANT RAMAN SPECTROSCOPY

A. J. Shields[1] and M. Cardona

Max-Planck-Institut für Festkörperforschung
Heisenbergstr. 1
D-70569 Stuttgart
Federal Republic of Germany

INTRODUCTION

Inelastic scattering of light in semiconductors occurs via a process involving electronic excitations as intermediate states. This has facilitated the study of electronic properties through measurement of the Raman intensity as a function of the photon energy, known as the resonance Raman profile (RRP) [1]. Maxima in the RRP occur when either the incident or scattered photon energy equals that of an electronic transition, situations referred to as incoming and outgoing resonance, respectively. Resonances in the Raman scattering efficiency of semiconductor superlattices (SLs) have been observed due to transitions between the valence and conduction subbands [2,3].

In this paper we demonstrate how the LO Raman intensity can be used to investigate exciton dephasing in SLs. The method relies upon the fact that the Raman scattering efficiency, under resonant photoexcitation conditions, increases with the dephasing lifetime of the intermediate electronic states. We utilise this property to measure the variation of the homogeneous linewidth with energy across the ground state of the SL, which is inhomogeneously-broadened due to spatial fluctuations in the well width. The dephasing rate of excitons formed between different electron and hole subbands are compared. The effect of temperature and illumination density upon the dephasing rate are investigated also. Electric field dependent studies reveal sudden reductions in the dephasing lifetime of the ground state at the fields where its wavefunction hybridises with an excited state in an adjacent layer.

DEPENDENCE OF RAMAN INTENSITY ON DEPHASING TIME

Inelastic scattering of light occurs by the multistep process portrayed diagramatically in Fig. 1 for the case of phonon scattering. (Fig. 1 represents the simplest Raman scatter-

[1] present address : Toshiba Cambridge Research Centre
Science Park, Milton Road, Cambridge CB4 4WE. U.K.

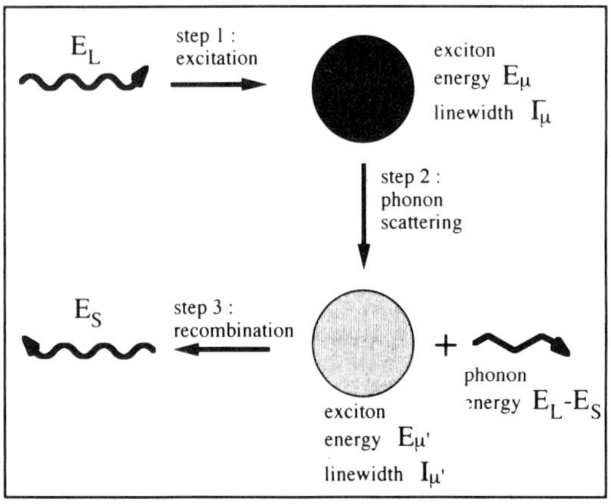

Figure 1. Schematic of the three-step process responsible for phonon Raman scattering. Its probability increases with the dephasing lifetime of the intermediate electronic states.

ing mechanism, although sometimes a higher-order process involving both elastic and phonon scattering is dominant [4,5].) Clearly the intensity of this process will increase with the dephasing lifetime of the real intermediate excitonic states, since the longer the state survives the more likely the following step becomes.

This intuitive argument can be substantiated using perturbation theory. The scattering amplitude of the process in Fig. 1 can be written,

$$W = \sum_{\mu,\mu'} \frac{\langle 0 \,| H_{pt} |\, \mu' \rangle \langle \mu' \,| H_{pn} |\, \mu \rangle \langle \mu \,| H_{pt} |\, 0 \rangle}{(E_S - E_{\mu'} + i\Gamma_{\mu'})(E_L - E_\mu + i\Gamma_\mu)} \tag{1}$$

where H_{pt} and H_{pn} are the Hamiltonians describing the interaction of the exciton with the photon and phonon, respectively. Here the probability of each of the three steps in Fig. 1 is described by a matrix element in the numerator. The denominator describes the 'detuning' of the exciton states from the incident (E_L) and scattered (E_S) photon energies. It is important to note that the broadening parameters in Eq. 1 are the *homogeneous* ones due to the finite dephasing lifetime of the states. It is evident from Eq. 1 that the scattering amplitude will have a local maximum when $E_L = E_\mu$ (incoming resonance) or $E_S = E_{\mu'}$ (outgoing resonance). Near an incoming or outgoing resonance, the photon energy dependence can be described by the resonating denominator alone, ignoring the other slowly varying terms, i.e. $W \sim (E_{L(S)} - E_{\mu(\mu')} + i\Gamma_{\mu(\mu')})^{-1}$.

The broadening of optical transitions in quantum wells is inhomogeneous in nature arising from spatial fluctuations in the well width [6]. Consequently the contribution of different well width regions to the scattering *intensity* ($\sim |W|^2$) must be summed. The inhomogeneous broadening is modelled by a function G, describing the density of excitons in the SL with energy E_h and homogeneous linewidth Γ_h. This distribution is centred around the energy E_i and has width Γ_i, corresponding to the inhomogeneous linewidth of the transition. Integrating the intensity due to the resonating denominator in Eq. 1 over the inhomogeneously broadened peak, yields the Raman intensity of the SL to be given by,

$$I_{RRP}(E) \sim \; K \int_0^\infty \frac{G(E_h, E_i, \Gamma_i)}{(E - E_h)^2 + \Gamma_h^2} \, dE_h \tag{2}$$

where $E = E_L$ for an incoming resonance and $E = E_S$ for an outgoing one. The factor K corrects for the change in the absorption depth across the exciton peak and is given by $K = (1 - e^{-(\alpha_L + \alpha_S)D})/(\alpha_L + \alpha_S)$, where $\alpha_{L(S)}$ is the absorption coefficient at $E_{L(S)}$ and D is the SL thickness. Although important in semi-infinite bulk samples, the factor K is almost energy-independent in the SLs studied here, since they are sufficiently thin that $(\alpha_L + \alpha_S)D << 1$ and hence $K \sim D$.

In structures were the well width is smaller than a few hundred Ånstroms, the inhomogeneous contribution to the broadening is typically much larger than the homogeneous one, i.e. $\Gamma_i >> \Gamma_h$, allowing Eq. 2 to be approximated,

$$I_{RRP}(E) \sim \; K \; \pi \; G(E, E_i, \Gamma_i) \; / \; \Gamma_h \tag{3}$$

This confirms that the resonance Raman intensity is inversely proportional to the homogeneous linewidth of the resonant electronic state, or, equivalently, that it is proportional to the dephasing lifetime.

In contrast the absorption spectrum is not sensitive to the homogeneous linewidth. This is essentially because absorption is only a one-step process and its intensity is therefore independent of the lifetime of the photoexcited state. The absorption strength of one homogeneously broadened oscillator is given by a Lorentzian whose integrated strength is independent of its width. Hence, we can write the total absorption strength of the SL to be given by,

$$\alpha(E) \sim K \int_0^\infty \frac{\pi^{-1}\Gamma_h G(E_h, E_i, \Gamma_i)}{(E - E_h)^2 + \Gamma_h^2} \, dE_h \;\; \sim \;\; K \; G(E, E_i, \Gamma_i) \tag{4}$$

The absorption spectrum therefore resembles the distribution of exciton energies in the SL and is not influenced by the value of the homogeneous linewidth, provided $\Gamma_h << \Gamma_i$.

PHOTON ENERGY DEPENDENCE

Figure 2 plots the photoluminescence (PL), PL excitation (PLE) and RRP measured on a GaAs(46 Å)/AlAs(41 Å) SL at 9 K. The PLE spectrum shows a large peak at 1.672 eV due to the $1s$ bound state of the $e1 - hh1$ exciton, while its higher bound and continuum states produce the weak peak $\sim$ 13.8 meV higher. The $1s$ bound state of $e1 - lh1$ can be seen as a peak around 1.708 eV. Notice the maximum of the PL lies $\sim$ 8.6 meV lower in energy than the bandedge peak of the excitation spectrum. This shift derives from the excitons relaxing at low temperature to the lowest energy states in the inhomogeneously broadened line before recombination.

The symbols in Fig. 2 plot the height of the Raman peak due to the LO_2 GaAs phonon as a function of the *scattered* photon energy. The Raman spectra were taken in the backscattering geometry, with the incident and backscattered light propagating normal to the surface and linearly polarised parallel to one of the crystal axes, $z(x,x)\bar{z}$. Maximum Raman intensity is seen when the scattered photon energy (ie. outgoing resonance) is close to that of the $e1 - hh1$ $(1s)$ exciton. The incoming resonance condition, where the *incident* photon energy nearly equals that of $e1 - hh1$ $(1s)$, produces the weaker feature one phonon energy ($\sim$ 295 cm^{-1}) to lower energy, near 1.629 eV. Another local maximum is discernible to higher energies, corresponding to outgoing resonance with the $e1 - lh1$ $(1s)$ exciton.

Bearing in mind that the Raman intensity is plotted on a logarithmic scale on Fig. 2, it can be seen that there is an enormous enhancement near outgoing resonance with $e1 - hh1(1s)$. This derives from the much longer exciton lifetime near the bandedge, resulting from the scarcity of states to dephase to. Notice that the $e1 - lh1$ feature in the Raman profile is about 6000 times weaker than that for $e1 - hh1$, whereas it is much more prominent in the PLE spectrum. This indicates that the $e1 - lh1(1s)$ exciton has a much larger homogeneous linewidth than $e1 - hh1(1s)$, due presumably to dephasing to the continuum of $e1 - hh1$. After allowing for the factor of 9 arising from the electron-photon matrix elements in Eq. 1, the RRP suggests the homogeneous linewidth of $e1 - lh1(1s)$ to be 2-3 orders of magnitude larger than that for $e1 - hh1$ The shorter dephasing lifetime of higher subbands also explains the much weaker (300 times) incoming resonance with $e2 - hh2$ compared to that for $e1 - hh1$, reported by Zucker et al [7] for their GaAs(96 Å)/$Al_{0.28}Ga_{0.72}As$(98 Å) SL.

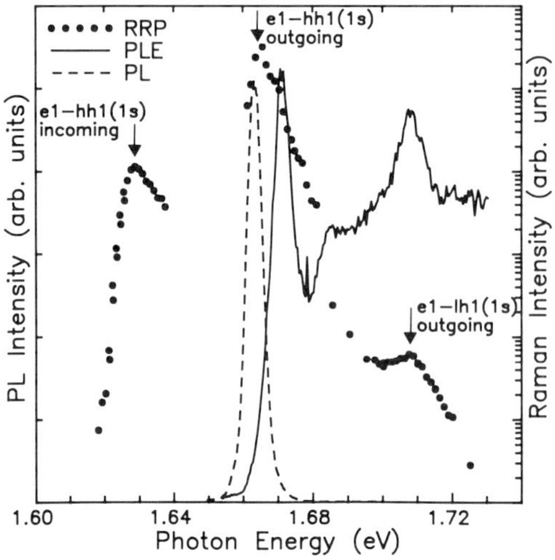

Figure 2. PL, PLE and resonance Raman profile measured on a GaAs(46 Å)/AlAs(41 Å) SL at 9K. The RRP is plotted as a function of the *scattered* photon energy. The much stronger outgoing resonance with $e1 - hh1$ compared to $e1 - lh1$ is due to its shorter dephasing time. The redshift of the RRP maximum with respect to the e1-hh1 PLE peak is also explained by a shorter dephasing lifetime in the lower-energy side of the exciton peak.

It can be seen in Fig. 2 that the maximum Raman efficiency for outgoing resonance does not correspond exactly with the $e1-hh1$ $(1s)$ peak of the PLE spectrum; the feature in the RRP lies several meV to lower energy and indeed closer to the peak of the PL. Similarly, that the incoming resonance also occurs at an *incident* photon energy below the PLE maximum is evident in Fig. 3a. This point is perhaps most clearly illustrated by the five Raman spectra plotted in Fig. 3b, which were recorded by scanning the laser energy at five different fixed spectrometer positions. The $e1 - hh1(1s)$ PLE can be seen as a background to one of these Raman spectra. Clearly the strongest Raman line is recorded when the laser energy lies in the low-energy tail of the PLE peak.

The redshift of the maximum of the RRP with respect to the PLE can be explained by an increase in the homogeneous linewidth with energy across the inhomogeneously broadened $e1 - hh1(1s)$ feature, due to the greater density of states for the exciton to dephase to. Hence lower-energy excitons in the inhomogeneously broadened peak have a narrower homogeneous linewidth and consequently make a stronger contribution to the Raman intensity, skewing the maximum of the RRP to lower energy. Degenerate four-wave mixing (DFWM) measurements on this sample (shown in Fig. 4), as well as on similar structures [8,9], have confirmed an increase in the homogeneous linewidth across the low-energy side of the exciton peak, as have calculations [10].

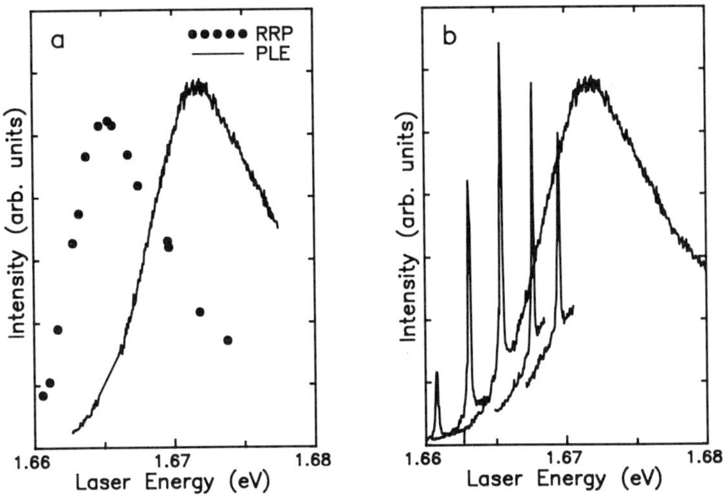

Figure 3. Raman data taken on GaAs(46 Å)/AlAs(41 Å) SL close to incoming resonance with $e1 - hh1(1s)$ at 9K. (a) PLE and RRP plotted as function of *incident* photon energy. Strongest Raman scattering is seen for photon energies in the low-energy side of the PLE peak where the homogeneous linewidth is narrowest. (b) Raman spectra recorded by fixing the spectrometer at five positions and scanning the laser energy. The PLE peak due to $e1 - hh1(1s)$ can be seen as a background to one of the spectra. Notice again that the strongest Raman line is seen in the low energy side of the PLE peak.

The energy dependence of the homogeneous linewidth can be deduced by taking the ratio of the PLE spectrum and the RRP using Eqs. 3 and 4. This ratio is plotted in Fig. 4 along with the homogeneous linewidth determined by DFWM measurements, further details of which are given in Ref. [11]. The curve determined from the RRP is scaled to match the DFWM measurement at one energy. The good agreement observed between the variation of the homogeneous linewidth deduced from the RRP and that measured more directly by DFWM, supports the explanation of the redshifts in Figs. 2 and 3 and highlights the potential of Raman spectroscopy for determining homogeneous linewidths.

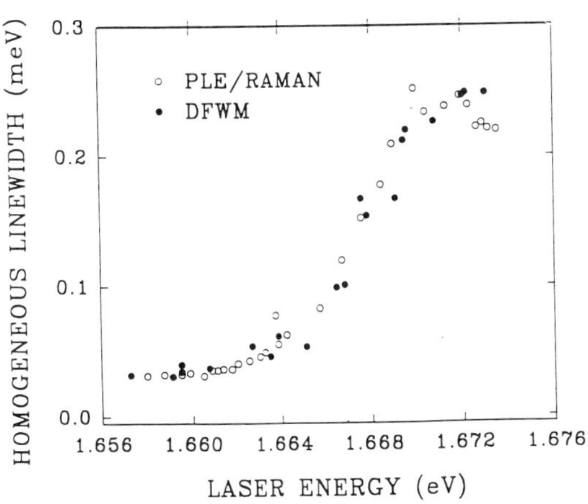

Figure 4. Comparison of energy-dependence of the homogeneous linewidth across the $e1 - hh1(1s)$ peak determined from Raman and DFWM measurements.

A similar red-shift was reported between the maximum of the Rayleigh scattering intensity with respect to the $e1 - hh1(1s)$ absorption peak in a GaAs(51 Å)/$Al_{0.3}Ga_{0.7}As$(200 Å) SL [12]. This could also be explained by the longer dephasing lifetime in the low-energy tail of the exciton peak. For the elastic scattering case both the incoming and outgoing photons are resonant with the transition, which should produce even greater sensitivity of the intensity to the dephasing lifetime. However, measurement of the Rayleigh signal is often hampered by background scattering from scratches and dirt on the sample surface.

TEMPERATURE AND ILLUMINATION INTENSITY DEPENDENCE

The sensitivity of the resonance Raman scattering intensity on Γ_h is further demonstrated by its temperature dependence, which is plotted in Fig. 5. Similar decreases in the resonance Raman [7] and Rayleigh [8] intensities with temperature have also been reported. The observed rapid decline of the scattering intensity can be explained by an increase of Γ_h with temperature. The homogeneous linewidth has been measured by DFWM [13] and calculated [10] to increase with temperature, due to the decrease in the exciton dephasing time caused by enhanced scattering with LA phonons.

Figure 6 shows the dependence of the resonance Raman intensity upon the power density of the incident laser beam, which can be seen to be linear over several orders of magnitude. At the highest illumination powers used there is a saturation in the Raman intensity. One cause of this behaviour could be an increase in the homogeneous linewidth due to exciton-exciton scattering at high illumination densities [14]. However, one cannot rule out the possibility that sample heating is producing this effect in our experiment. By using pulsed laser excitation this problem could be overcome, allowing the study of the influence of exciton-exciton scattering on the Raman intensity.

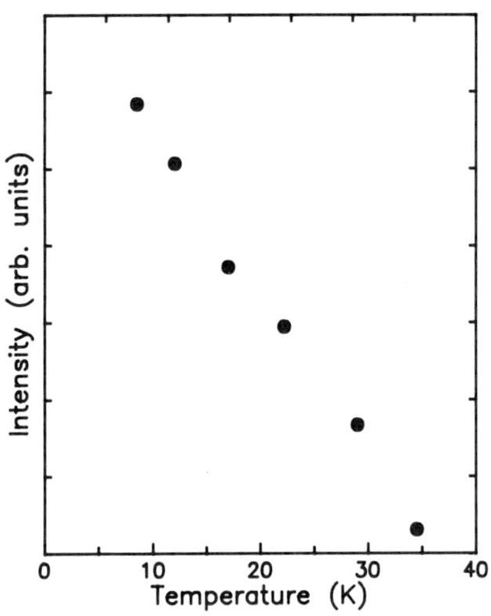

Figure 5. Temperature dependence of PL and Raman intensity for incoming resonance with $e1-hh1(1s)$ of GaAs(46 Å)/AlAs(41 Å) SL. The Raman intensity decreases sharply due to thermally activated dephasing by acoustic phonons.

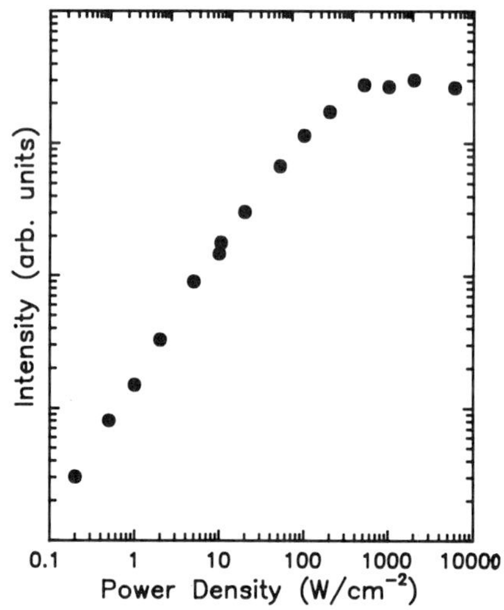

Figure 6. Illumination power density dependence of PL and Raman intensity for incoming resonance with $e1 - hh1(1s)$ of GaAs(46 Å)/AlAs(41 Å) SL at 9K.

ELECTRIC FIELD DEPENDENCE

We have also used resonance Raman spectroscopy to study exciton dephasing in SLs as a function of an electric field applied perpendicular to the layers. The sample for this experiment comprised 90 periods of GaAs(50 Å)/AlAs(22 Å) grown inside the intrinsic region of a $p^+ - i - n^+$ doping profile on a n^+-type (001) GaAs substrate. The electric field across the SL could then be varied by applying a bias between ohmic contacts made to the n^+- and p^+-regions. Biases of +1.5 V and -20 V correspond to uniform fields of 0 and 330 kV/cm across the SL, respectively.

The variation of the LO_2 Raman intensity with applied bias is shown in Fig. 7a, for a laser energy near resonance with the $e1 - hh1$ $(1s)$ exciton of the SL. It shows an initial rapid fall with increasing electric field, followed by a gradual recovery. Most striking is the sharp drop in Raman intensity close to -11 V, while a smaller feature around -3 V is also apparent. Figure 7b plots the variation in the PL and photocurrent (PC) intensity with applied bias. The PC was measured simultaneously with the Raman intensity in Fig. 7a, while the PL curve corresponds to the height of the $e1 - hh1$ $(1s)$ peak from the SL excited by a laser of higher energy. In the PC curve of Fig. 7b steps can be seen with thresholds close to -11 and -1.5 V.

By comparison with an effective-mass calculation, we ascribe these features to resonant tunnelling (or transfer) between subbands in adjacent wells (or adjacent well and barrier) [15]. The feature at -11 V is due to sequential resonant transfer of electrons to the level confined in the adjacent AlAs layers at the X-point of the bulk conduction band ($x1$), illustrated schematically in the inset to Fig 7a. Resonant tunnelling between $hh1$ and $lh1$ in adjacent wells is calculated to occur at a bias of -3.1 V, while LO-phonon-assisted tunnelling between the same subband in adjacent wells, would become resonant at -1.8 V, when the potential drop across one period equals the phonon energy. The weak features in Figs. 7a and b near -6 V agree with the bias calculated for resonant $hh1 \rightarrow hh2$ tunnelling.

The correspondence of the features in the Raman intensity versus bias curve of Fig. 7a to those in the PC and PL of Fig. 7b, suggests they also derive from resonant transfer of the carriers. The sharp drops in the Raman intensity can be explained by a decrease in the exciton dephasing lifetime caused by the transfer, as now discussed.

At the applied electric fields where an energy level aligns with an excited state in the adjacent well (or barrier), resonant tunnelling (or transfer) occurs. This can be described in terms of a hybridisation of the degenerate levels into bonding and antibonding states. In the limit which applies here, where the coupling is sufficiently weak that the splitting of the bonding and antibonding pairs is less than the inhomogeneous broadening, the perturbation on the exciton energy and envelope functions will be negligible compared to the effect on its dephasing lifetime. The lifetime of the hybridised state ($\tau_\pm$) can be written in terms of those of its pure components (a and b); $\tau_\pm^{-1} = (\tau_a^{-1} + \tau_b^{-1})/2$. Since the state close to the bandedge resonant with the photon in the Raman measurement will have a much longer dephasing lifetime (τ_a) than that of the state in the adjacent well (τ_b), the hybridisation produces a sharp drop in its lifetime ($\tau_\pm \sim 2\tau_b << \tau_a$) and hence the Raman intensity.

A similar phenomenon has been reported in four wave mixing studies on asymmetric double quantum wells [16]. At the electric field where the $e1$ levels in the two wells align, there was a drop in the dephasing lifetime of the narrow well, explained by hybridisation with the state in the wide well, which had a shorter lifetime due to the presence of free carriers in the wide well.

At a bias close to -11 V, the $e1$ level in the GaAs layer aligns with $x1$ in the AlAs (see inset Fig. 7a), producing resonant transfer. Since the hybridised $e1 - x1$ state has a shorter dephasing lifetime than the pure $e1$, due to LO-phonon-assisted relaxation of the $x1$ component to the (other) neighbouring GaAs layer at lower potential, this reduces the $e1 - hh1$ exciton dephasing lifetime and causes the sharp drop in the Raman intensity. The $\Gamma - X$ transfer rate in type-II GaAs/AlAs SLs has been measured to be < 1 ps [17], much less than typical $e1 - hh1$ (1s) dephasing lifetimes in type-I structures. The transfer between the Γ- and X-valleys of the conduction band requires scattering due to crystal imperfections, such as interface roughness, defects or impurities. Such a tunnelling process has previously been reported in transport studies of GaAs/AlAs double barrier structures [18]. The Raman intensity does not recover at biases beyond -11 V, where the $e1$ subbandedge lies above that of $x1$, as it is still possible for the Γ-electrons to scatter to states in the AlAs above the X-point minimum.

At low fields the $e1$ levels in adjacent wells interact [19,20]. The inhomogeneous broadening ($\sim$ 8 meV half-width-half-maximum) is comparable to the zero-field mini-bandwidth, calculated to be 6 meV, in this SL. At zero field the interaction produces strongest hybridisation in the regions where adjacent wells have the same width. However, as increasing electric field is applied, the $e1$ states resonantly excited by the photon energy will hybridise most strongly with e1 states from narrower regions. Since the ex-

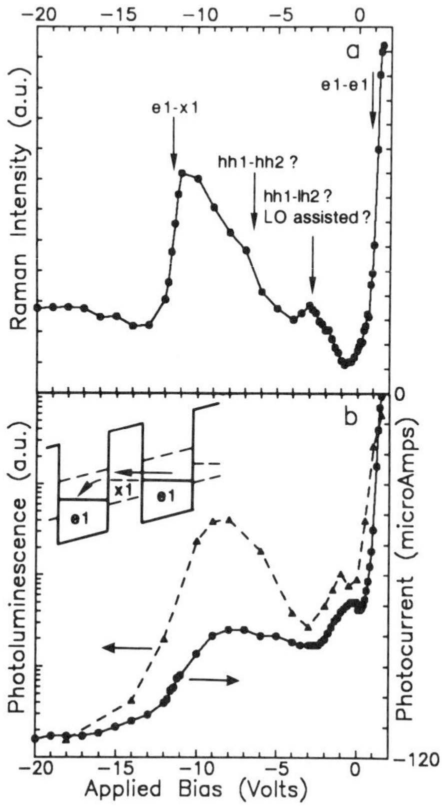

Figure 7 . Comparison of (a) Raman intensity and (b) PL and PC intensities measured for a GaAs(50 Å)/AlAs(22 Å) SL as a function of applied bias at 9K. Biases of +1.5 and -20 V, produce fields of 0 and 330 kV/cm, respectively. The arrows mark the calculated biases where levels in adjacent layers energetically align, as shown schematically in the inset for $e1 - x1$. These coincide with drops in the Raman intensity caused by a reduction of the dephasing lifetime of the ground state due to hybridisation with an excited state in an adjacent layer.

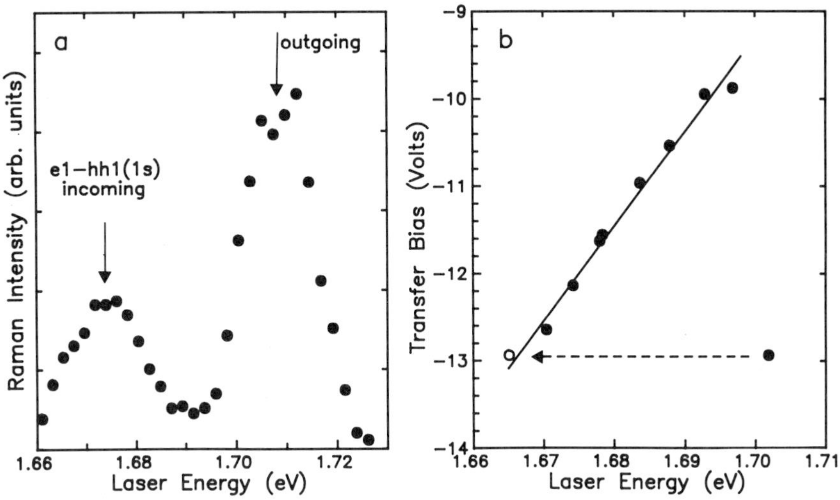

Figure 8. Laser energy dependence of the tunnelling effects upon the Raman intensity measured on GaAs(50 Å)/AlAs(22 Å) SL at 9K. (a) Resonance Raman profile at applied bias of -20V showing maxima due to incoming and outgoing resonance with $e1 - hh1(1s)$. (b) The applied bias required for the sharp drop in the Raman intensity due to $e1 \rightarrow x1$ tunnelling as a function of the laser energy. The linear shift is explained by inhomogeneous broadening of $e1$ as discussed in text.

citons in these narrower regions have a shorter dephasing lifetime, as demonstrated by Fig. 5, the lifetime of the hybridised state falls with increasing electric field, creating the drop in the Raman intensity between +1.5 and 0 V in Fig. 7a. The tunnelling should be inhibited when the potential drop across one period exceeds the sum of the inhomogeneous linewidth and the zero-field minibandwidth, calculated to occur at an applied bias of +0.2 V. With further increasing field the electron tunnelling is turned off, producing a recovery of the exciton lifetime and hence the Raman intensity (between 0 and -10 V).

Upon changing the laser energy within the incoming resonance (see Fig. 8a) there is a linear shift of the bias of the sharp drop in the Raman intensity ascribed to transfer between $e1$ and $x1$, which is also plotted in Fig. 8b. This linear shift of the $e1 \rightarrow x1$ transfer bias is explained by inhomogeneous broadening of the $e1 - hh1$ resonance. For some particular laser energy the Raman scattering will be perfectly resonant in only that part of the SL with the appropriate well width. Increasing the photon energy selects a higher energy $e1 - hh1$ exciton, deriving from a narrower GaAs region. Most of this exciton energy shift ($\sim$ 78 %) is manifested in the conduction band, since the electron has a smaller effective mass and hence larger confinement energy than the heavy hole. Since the electron confinement energy ($e1$) becomes larger, the reverse bias required to align with the $x1$-level in the barrier should be smaller. If the widths of adjacent GaAs and AlAs layers are spatially uncorrelated, this causes the critical bias to decrease at a rate of 0.073 Volt/meV. On the other hand, if we assume that wide GaAs regions neighbour narrow AlAs ones, we expect a maximum gradient of 0.139 Volt/meV. The slope of 0.108 Volt/meV in Fig. 8b suggests partial correlation.

When the laser energy is further increased to 1.7019 eV, the $e1 \rightarrow x1$ transfer is seen to return to a larger reverse bias of -12.9 V, see Fig. 8b. This behaviour is to be expected since this laser energy lies within the *outgoing* resonance (see Fig. 8a). Notice that after shifting this point one LO phonon (36.8 meV) lower in energy (plotted in Fig. 8b as open circle), it also lies close to the straight line through the other, incoming resonance, data.

CONCLUSIONS

The results presented here have demonstrated how resonance Raman spectroscopy can be used to study exciton dephasing in SLs. The technique has the advantage that it can be applied over wide ranges of photon energy, temperature, exciton density, electric field, as well as potentially many other experimental parameters. The sensitivity of the Raman intensity to the dephasing lifetime of resonant intermediate electronic states, derives from the fact that it is a multi-step process requiring coherence of the electronic state between subsequent stages. The huge enhancements in the Raman intensity for resonance with $e1 - hh1$ compared to higher subband transitions, are indicative of long dephasing times for excitons near the bandedge. Furthermore, strongest scattering is seen in the low-energy side of the inhomogeneously broadened $e1 - hh1$ peak, where the dephasing time is longest, due to the scarcity of states to dephase to. The variation in the homogeneous linewidth across the exciton peak deduced from the Raman intensity agrees well with that measured by DFWM. With increasing temperature the Raman intensity is quenched rapidly due to enhanced exciton dephasing by acoustic phonon scattering. Sharp drops are observed in the Raman intensity due to the bandedge exciton at the applied electric fields where the carriers can resonantly tunnel into the neighbouring well (or transfer to the barrier). These are explained by sudden reduction in the dephasing time caused by wavefunction hybridisation with excited levels in the adjacent layer.

Acknowledgements : We would like to thank many colleagues at the Max-Planck-Institut: R.Nötzel, K.Ploog and H.T.Grahn for providing the samples for these studies; G.O.Smith, E.Mayer, R.Eccleston and J.Kuhl for collaboration with the DFWM experiments and H.Hirt, M.Siemers, and P.Wurster for their technical assistance. A.J.S. acknowledges the financial support of the Royal Society under the European Science Exchange Programme.

References

[1] for a review of resonance Raman studies in bulk semiconductors, see M.Cardona, in *Light Scattering in Solids II*, edited by M.Cardona and G.Güntherodt (Springer, Berlin, 1982); resonance effects in SLs are covered by J.Menendez, J.Luminescence **44**, 285 (1989) and B.Jusserand and M.Cardona, in *Light Scattering in Solids V*, edited by M.Cardona and G.Güntherodt (Springer, Berlin, 1989)
[2] J.E.Zucker, A.Pinczuk, D.S.Chemla, A.Gossard and W.Wiegmann, Phys.Rev.Lett **51**, 1293 (1983)
[3] A.J.Shields, M.Cardona, R.Nötzel and K.Ploog, Phys. Rev. B **46**, 10490 (1992)
[4] A.J.Shields, C.Trallero-Giner, M.Cardona, H.T.Grahn, K.Ploog, V.A.Haisler, D.A.Tenne, N.T.Moshegov and A.I.Toropov, Phys. Rev. B **46**, 6990 (1992)
[5] A.J.Shields, V.A.Haisler, C.Trallero-Giner and M.Cardona, Proc. of NATO ARW on *Phonons in Heterostructures*, St Feliu de Guixols, Spain (15-18 Sept. 1992)
[6] C.Weisbuch, R.Dingle, A.C.Gossard and W.Weigmann, Solid State Comm. **38**, 709 (1981)
[7] J.E.Zucker, A.Pinczuk, D.S.Chemla and A.Gossard, Phys.Rev.B **35**, 2892 (1987)
[8] J.Hegarty, L.Goldner and M.D.Sturge, Phys.Rev.B **30**, 7346 (1984)
[9] M.B.Webb, S.T.Cundiff and D.G.Steel, Phys.Rev.B **43**, 12658 (1991)
[10] T.Takagahara, Phys.Rev.B **32**, 7013 (1985)
[11] A.J.Shields, G.O.Smith, E.Mayer, R.Eccleston, J.Kuhl, M.Cardona and K.Ploog, to be published in Phys. Rev. B
[12] J.Hegarty, M.D.Sturge, C.Weisbuch, A.C.Gossard and W.Wiegmann, Phys.Rev.Lett **49**, 930 (1982)
[13] L.Schultheis, J.Kuhl, A.Honold and C.W.Tu, Phys.Rev.B **34**, 9027 (1986)
[14] R.Eccleston, B.F.Feuerbacher, J.Kuhl, W.W.Rühle and K.Ploog, Phys.Rev.B **45**, 11403 (1992)
[15] A.J.Shields, M.Cardona, H.T.Grahn and K.Ploog, Phys. Rev. B **47**, 13922 (1993)

[16] K.Leo, J.Shah, E.O.Göbel and T.C.Damen, Appl. Phys. Lett. **56**, 2031 (1990)
[17] J.Feldmann, J.Nunnenkamp, G.Peter, E.Göbel, J.Kuhl, K.Ploog, P.Dawson and C.T.Foxon, Phys. Rev. B **42**, 5809 (1990)
[18] E.E.Mendez, W.I.Wang, E.Calleja and C.E.T.Goncalves da Silva, Appl. Phys. Lett. **50**, 1263 (1987)
[19] F.Capasso, K.Mohammed and A.Y.Cho, IEEE J. Quantum Electron. **QE-22**, 1853 (1986)
[20] J.Bleuse, G.Bastard and P.Voisin, Phys. Rev. Lett. **60**,220 (1988); E.E.Mendez, F.Agulló-Rueda and J.M.Hong, Phys. Rev. Lett. **60**, 2426 (1988)

BLOCH OSCILLATIONS IN SUPERLATTICES

P.Leisching[1], C.Waschke[1], P.Haring Bolivar[1], W.Beck[1], H.Roskos[1], K.Leo[1], H.Kurz[1], K.Köhler[2] and P.Ganser[2]

[1]Institut für Halbleitertechnik II, RWTH Aachen, Sommerfeldstrasse, 52056 Aachen, Germany
[2]Fraunhofer-Institut IAF, Tullastrasse 72, 79108 Freiburg, Germany

1. Introduction

Semiconductor superlattices (SL) have recently gained much attention as a model system for the investigation of the electronic states in a periodic potential (Bloch states) [1]. As an example, the Wannier-Stark ladder (WSL) [2] has been observed by cw optical techniques in semiconductor SL several years ago [3,4]: The extended Bloch states become localized due to the field F, which transforms the continuous energy miniband into evenly spaced ladder states with energies $E_n=E_0+n\Delta E$, where $\Delta E=eFd$ (d: lattice constant) [5]. This WSL behavior is, albeit with some qualitative changes, also observed when the excitonic correlation of photo-excited eh pairs is considered [6]. There is only one condition due to the excitonic behavior imposed to the miniband width: the bandwidth needs to be larger than the exciton binding energy (5-10 meV) [7]. Recently, the interest has focussed [8,9,10] on the time-domain equivalent of the WSL, the periodic motion of Bloch electrons in an homogeneous electric field (Bloch oscillations, BO [11]). A simple semiclassical description is frequently given in literature [12]: Without scattering, electrons will accelerate in the electric field till they reach the upper edge of the band at the end of the Brillouin zone. There, they will be reflected and return to the initial state. These oscillations in the Brillouin zone have a period $\tau_B=h/eFd$. More recently, several groups [5,13,14,15] have modeled the BO in a fully quantum-mechanical picture and shown that the essential features of the simple picture outlined above are preserved, e.g., the field-dependence of the oscillation frequency. Recent numerical calculations for realistic initial superposition of WSL states [14] predict the existence of an oscillating dipole moment, which is a prerequisite for the observation of strong THz emission.

2. Experimental

We here present results from five different samples grown by molecular beam epitaxy. Each of sample I-V contains 35 periods of GaAs wells and $Al_{0.3}Ga_{0.7}As$ barrier layers grown on n-doped substrate. The well width of the samples is 111/97/80/67/61 Å, respectively, while the barrier width is held constant at 17 Å. On top of the sample, a transparent Cr/Au Schottky contact is evaporated. A Kronig-Penney calculation gives electron miniband widths of 13/18/26/37/50 meV and heavy-hole miniband widths of 1-5 meV. The experimental configuration for the degenerate four-wave-mixing (FWM) experiments is a standard two-beam ($\mathbf{k}_1$,$\mathbf{k}_2$) set-up using a Kerr-lens modelocked Titanium-Sapphire laser emitting 130 fs (FWHM) pulses. Detected is the time-integrated FWM-signal in direction $\mathbf{k}_3=2\mathbf{k}_2-\mathbf{k}_1$ [9,16].

Coherent Optical Interactions in Semiconductors
Edited by R.T. Phillips, Plenum Press, New York, 1994

The decay of this signal as a function of the delay time τ between the two pulses reflects the dephasing of the coherent excitation with the phase relaxation time T_2. Oscillations in the FWM decay can be due to quantum-beat interference of a coherent superposition of electronic states (i.e. a many level system, as expected for BO) or polarization interference of independent quantum mechanical systems (i.e. two independent two level systems). There are two possibilities for an unambiguous experimental distinction: Time-resolving [17] or spectrally resolving [18] the diffracted signal. Here, we use a monochromator with 1 nm resolution to spectrally resolve the signal in direction $\mathbf{k}_3=2\mathbf{k}_2-\mathbf{k}_1$. The excitation density is below 1×10^9 cm^{-2} carriers per period. Despite this low excitation intensity, we observe strong quasi-static screening effects. The WSL is therefore measured by transmission spectroscopy with our broadband fs-pulse at the same excitation intensity. For this purpose we put the monochromator in direction $\mathbf{k}_2$.

The experimental setup for the time-resolved THz measurements is quite different [10]. One part of the laser pulse train excites the superlattice under 45^0. Corresponding to the generalized Fresnel equations, a THz wave is emitted in a direction nearly collinear with the reflected optical beam. The THz pulse leaves the cryostat through a silicon window and is focussed with a pair of off-axis paraboloidal mirrors onto a 50 μm subpicosecond dipole microantenna that is gated by the second time-delayed part of the femtosecond laser pulse. Tracing the antenna current as a function of delay time allows measurement of both the amplitude and the phase of the electromagnetic transients.

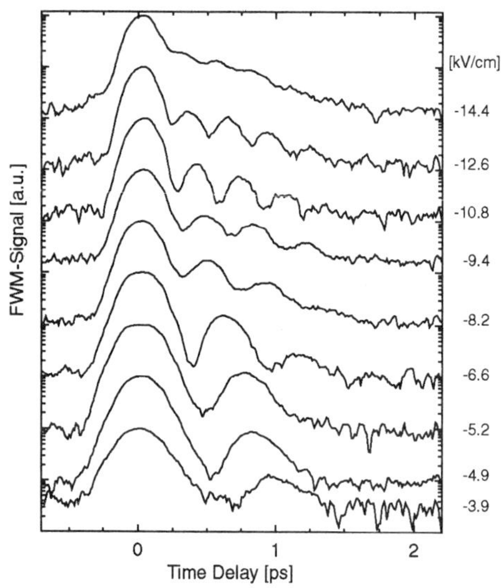

Figure 1. Semilogarithmic plot of the FWM traces of sample IV at 77 K as a function of the delay time for various electric fields (thickness of intrinsic region: 0.914 μm).

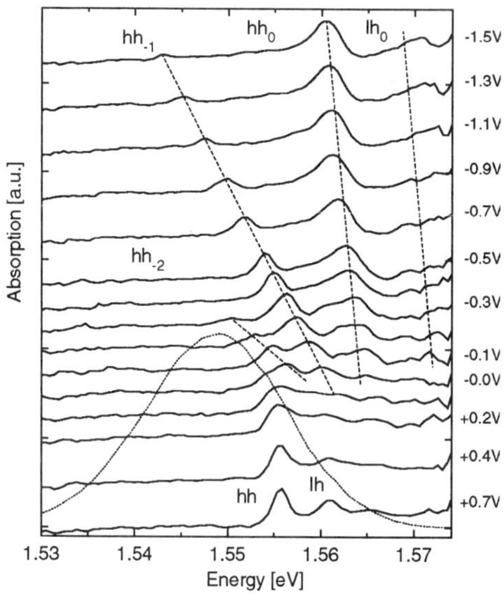

Figure 2. Wannier-Stark ladder of sample II at 10 K. Plotted is the absorption, calculated from the transmission spectra of the broadband fs-pulse, as a function of the reverse bias voltage.

3. Results and Discussion

Figure 1 shows transient FWM signals (sample IV at a lattice temperature of 77 K) as a function of the electric field. The excitation energy is peaked slightly below the heavy hole (hh) excitonic transitions of the superlattice. The bandwidth of the laser (about 20 meV) is large enough to cover several transitions of the WSL (see Fig. 2 and Fig. 4). The signals show a pronounced periodic modulation with a period which is strongly dependent on the electric field. Up to four oscillations are visible, even at 77 K. The absolute dephasing time of the signals is rather field-independent in the bias range studied here. For very high fields, when the oscillation frequency reaches about 5 THz, the amplitude ofthe oscillation starts to weaken due to the finite pulse duration of the laser. For even higher frequencies, we use the sample with the largest miniband width, since the maximum frequency is limited by the localization of the electronic wavefunction (which is dependent on the miniband width) in one well with increasing electric field. With sample V, we observe two oscillations at 200 K of frequencies of mor than 6 THz (T_2 at 200 K is about 400 fs).

The WSL of sample II measured by transmission spectroscopy at the same excitation density is depicted in Fig. 2. As guides for the eye, the dashed lines indicate some stairs of the hh WSL (lh_0 denotes the fundamental light hole transition), the dotted line indicates the spectrum of the laser pulse. The splitting of the WSL is clearly observed, with a static screening offset of about 0.7 V (flatband: +0.7 V). The screening offset is due to the accumulation of carriers using the high repetition rate (76 MHz) laser [19,20].

To compare the oscillation period measured in the time domain to the spectral data and to BO theory, we calculate the energies from the observed oscillation periods using the relation given above. Figure 3 shows the results for samples I and IV at 10 K and 77 K. The energy splitting reveals a linear dependence on the electric field. This is clear evidence that the oscillations originate from BO of coherently excited wavepackets. The proportionality constant is temperature-independent and in excellent agreement with that expected of BO theory for the parameters of the different superlattices (dashed lines).

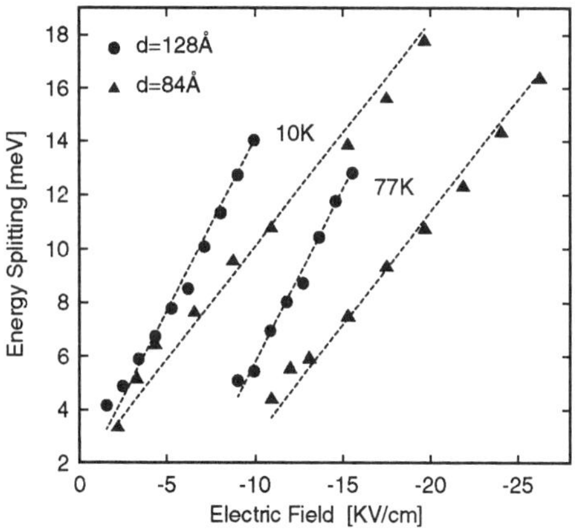

Figure 3. Energy splitting $\Delta E=eFd$ calculated from the BO period (using $\tau_B=h/\Delta E$). The dashed lines have slopes as expected for the different SL periods.

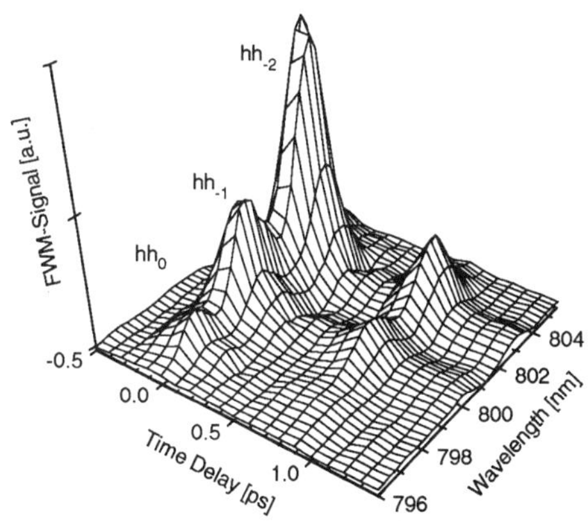

Figure 4. Spectrally resolved FWM-signal for sample II at 10 K. With this spectral position of the 20 meV fs-pulse we excite the hh_{-2}, hh_{-1} and hh_0 transitions.

Due to localization effects of the electronic wavefunction, the maximal *observable* energy splitting is different for both samples. The limit for the sample with the larger miniband width is caused by the spectral width of the laser pulse, for the small miniband it is caused by localization. Although the excitation conditions are the same, we observe that the field screening is *increasing* with increasing temperature. Currently the temperature dependence of screening is under detailed investigation, results will be published elsewhere [20].

We have spectrally resolved the FWM signal to unambiguously verify that we really observe quantum interference of coherently excited electron states and not polarization interference of WSL transitions. According to theory [18], one expects different beating behavior in the time integrated but spectrally resolved FWM signal for quantum beats and polarization interference: For quantum beats, the oscillations have the same phase for all frequencies, for polarization interference, the phase oscillates. Our experimental data correspond clearly to quantum interference (Fig. 4): the phase is in excellent agreement with the three level model - as expected for BO. The data are not corrected for the spectral intensity profile of the laser pulse, which is centered at 801 nm.

In the FWM experiments discussed above we trace the nonlinear *interband transition* of third order. With THz measurements, we detect directly the *intraband polarization*, associated with the dipole moment of the Bloch wave packet. We demonstrate the THz-wave emission of this coherently oscillating macroscopic dipole moment [10,19].

In Fig. 5, the detected THz transients of sample II at 10 K are shown as a function of bias voltage. In the low bias regime from +0.4 V to -1.0 V, the amplitude of the THz transients increases gradually with increasing bias and the waveform is only weakly structured. Above a reverse bias of -1.0 V, both the amplitude and the waveform change considerably with voltage. Several (up to six) oscillation periods are detectable. With increasing bias, the oscillation period decreases continuously. At a reverse bias of -1.8 V the amplitude of the measured transients passes through a maximum.

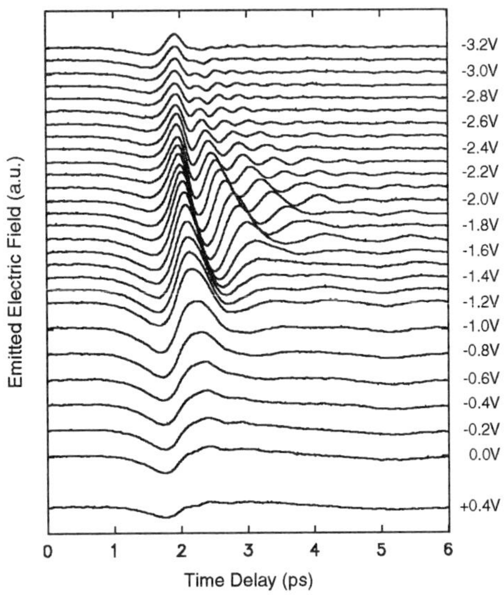

Figure 5. Measured coherent electromagnetic transients of sample II (T_L=10 K).

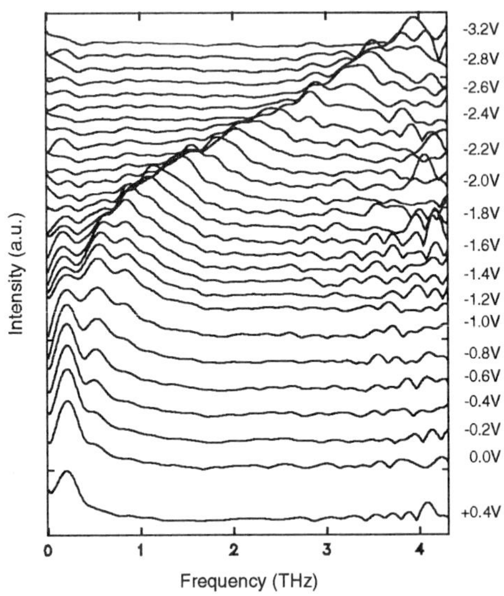

Figure 6. Fourier transforms of the time-domain data. The data are corrected for the spectral characteristics of the detection system.

Figure 6 shows the Fourier transforms of the THz data. These data are corrected for the spectral characteristics of the detection system (the absolute amplitude for frequencies above 3 THz is not reliable). From flatband up to a reverse bias of -1.0 V the Fourier spectra consists of two lines, a signal around 250 GHz (photon energy: 1 meV) and a signal around 600 GHz, whose exact origin is not clear yet. At higher reverse bias (-1.8 V), a signal at several hundred GHz emerges and shifts to higher frequencies. We can trace the signal up to nearly 4 THz before the limited bandwith of the measurement system cuts off detection. In this bias regime, the emission frequency depends linearly on the applied voltage, as expected for BO. A weak structure on the signal in the low bias regime has a frequency around 1 THz and is assigned to hh/lh beats. The signal disappears at higher voltages, which corresponds to the FWM measurements (not shown here [19]). Due to the quasi-static field screening, we have compared the THz data with *in-situ* reflection measurements [10].

The data show that the emission frequencies as a function of field correspond to the energy splitting of the WSL.

Comparing the FWM and the THz-emission results, we find general agreement between the data. In the miniband regime, we observe with both techniques a signal around 1 THz that we associate with hh/lh beats. In the transition regime from the miniband to the WSL regime, both experiments reveal a pronounced oscillating signal with bias-tunable frequency. We assign this signal to BO of coherently excited wave packets, because the dependence of the observed energy splitting (calculated from the oscillation period) on the applied electric field is linear and the slope is in excellent agreement with BO theory.

4. Conclusion

We have demonstrated the possibility of a Bloch oscillator emitting bias tunable submillimeter-wave radiation. The tuning range covers about one order of magnitude from 0.5 up to 5 THz, with the dependence on the applied electric field in excellent agreement with theory. For the first time this could be demonstrated even at the technologically important temperature of 77 K and up to 200 K. It is interesting to speculate whether BO can be observed at room temperature. T_2 at room temperature should be on the order of 250 fs [21]. Using shorter pulses and samples with even larger miniband widths, we expect that BO with oscillation frequencies in excess of 10 THz are possible at room temperature.

Acknowledgment: We thank J. Shah and R. Schwedler for stimulating discussions, and the Deutsche Forschungsgemeinschaft (contract Le747/2) and the Volkswagen-Stiftung for financial support.

References

1. L. Esaki and R. Tsu, IBM J. Res. Dev. **14**, 61 (1970).
2. G.H. Wannier, Rev. Mod. Phys. **34**, 645 (1962).
3. E.E. Mendez, F. Agulló-Rueda, and J.M. Hong, Phys. Rev. Lett. **60**, 2426 (1988).
4. P. Voisin, J. Bleuse, C. Bouche, S. Gaillard, C. Alibert, and A. Regreny, Phys. Rev. Lett. **61**, 1639 (1988).
5. G. Bastard and R. Ferreira, in *Spectroscopy of Semiconductor Microstructures*, NATO ASI Series (Plenum Press, New York 1989), p.333.
6. M.M. Dignam and J.E. Sipe, Phys. Rev. Lett. **64**, 1797 (1990).
7. A.M. Fox, D.A.B. Miller, G. Livescu, J.E. Cunningham, W.Y. Jan, Phys. Rev. B **44**, 6231 (1991).
8. J. Feldmann, K. Leo, J. Shah, D.A.B. Miller, J.E. Cunningham, S. Schmitt-Rink, T. Meier, G. von Plessen, A. Schulze, and P. Thomas, Phys. Rev. B **46**, 7252 (1992).
9. K. Leo, P. Haring Bolivar, F. Brüggemann, R. Schwedler, and K. Köhler, Solid State Commun. **84**, 943 (1992).
10. C. Waschke, H.G. Roskos, R.Schwedler, K.Leo, H.Kurz, and K. Köhler, Phys.Rev.Lett. **70**, 3319 (1993).
11. F. Bloch, Z. Phys. **52**, 555 (1928); C. Zener, Proc. R. Soc. London **A 145**, 523 (1934).
12. N.W. Ashroft and N.D. Mermin, Solid State Physics 1st, edition, CBS Publishing Asia LTD. (1988),p. 224 and 225.
13. J.B. Krieger and G.J. Iafrate, Phys.Rev. B **33**, 5494 (1986).
14. A.M. Bouchard and M.Luban, Phys. Rev. B **47**, 6815 (1993).
15. G. von Plessen and P. Thomas, Phys.Rev. B **45**, 9185 (1992).
16. T. Yajima and Y. Taira, J. Opt. Soc. Jpn. **47**, 1620 (1979).
17. M. Koch, J. Feldmann, G. von Plessen, E.O. Göbel and P. Thomas, Phys.Rev.Lett. **69**, 3631 (1992).
18. T. Tokizaki, A. Nakamura, Y. Ishida, T. Yaijma, I. Akai and T. Karasawa, Springer Series in Chemical Physics 53, Ultrafast Phenomena VII, Springer Verlag 1990, p. 253.
19. C. Wasche, P. Leisching, P. Haring Boliviar, R. Schwedler, F. Brüggeman, H.R. Roskos, K. Leo, H. Kurz and K. Köhler, to be published in Solid State Electronics.
20. P.Leisching and C. Waschke, et al, in preparation.
21. K. Leo, P. Haring, H.J. Bakker, G. Maidorn, H. Kurz and K. Köhler, XXIth Int. Conf. on the Phys. of Semic., Beijing, 1992, p. 983.

VIOLATION OF TIME-REVERSIBILITY IN SEMICONDUCTORS BY TRANSIENT OPTICAL EXCITATION

Nikolay I. Zheludev

Department of Physics
University of Southampton
Southampton S09 1ST
England

INTRODUCTION

The problem of time non-reversibility in solids has attracted growing attention during the last few years. Here first of all we mean "natural" symmetry-breaking phenomena in superconducting cuprate materials associated with the anyon superconductivity model. In this paper we investigate a different problem of loss of time reversal symmetry and related light-polarization affecting phenomena in semiconducting materials resulting from intense optical excitation.

Optical manifestation of broken time-reversibility may be formulated in terms of optical susceptibilities which may be derived from the more general principle of symmetry of kinetic coefficients, where we consider optical susceptibility as a particular case of the kinetic coefficients. If a System is affected by N external forces $f_s(t)$, (s = 1,2, ... N), the energy variation resulting from this excitation may be presented in the form:

$$V = x_s \, f_s(t) \tag{1}$$

where x_s are the internal coordinates corresponding to the external forces $f_s(t)$. The general form of the linear solutions to the corresponding dynamic problem in spectral presentation has the following structure: $x_s(\omega) = \alpha_{sp}(\omega) f_p(\omega)$. The so-called kinetic coefficients, $\alpha_{sp}(\omega)$ (s, p = 1,2,... N) should obey the following relationship: $\alpha_{sp}(\omega) = \pm\alpha_{ps}(\omega)$. Here, the (+) sign is appropriate if both the force $f_s(\omega)$ and the corresponding coordinate $x_s(\omega)$ have the same symmetry with respect to time inversion ($t \Rightarrow -t$), and the (-) sign if they have opposite symmetry. This is the principle of symmetry of the kinetic coefficients and the systems obeying it are said to be time-reversible. If this concept is applied to optics it can be shown that with a material equation in the form

$$D_i^{lin} = \varepsilon_{ij}E_j + \gamma_{ijl}\nabla_l E_j \qquad (2)$$

the material tensors describing a time-reversible crystal should obey the following remarkable symmetry relations:

$$\varepsilon_{ij} = \varepsilon_{ji} \qquad (3)$$

$$\gamma_{ijl} = -\gamma_{jil} \qquad (4)$$

These restrictions supplement the symmetry restrictions on the internal symmetry of the ε_{ij} and γ_{ijk} tensors appearing from the point group of the crystal that they are destined to describe (see, for example[1]).

In the time-non-reversible state this additional symmetry is broken. There are several possible reasons for violation of time reversibility: the most obvious is the presence of an external magnetic field or internal magnetic moment; some theoretical models of high-T_c superconductivity predict broken time-reversibility in cuprate materials; there is evidence that time-reversibility does not hold in semiconductors with a strong contribution of time-nonreversible spin-orbit coupling terms in the band structure.

TWO MECHANISMS

Here we study optical excitation as the reason for violation of time-reversibility. There are two main mechanisms of optical excitation induced time-nonreversible behaviour of a crystal: the first one is the optical magnetization of the medium with an intense circularly polarized pump; the second one is due to creation of a non-equilibrium state via a transient excitation. Observation of these optical effects is one of the most direct methods of detecting time non-reversibility. Optical methods with short-pulse lasers allow very sensitive, high-speed detection and simultaneous use of light both for monitoring and creation of time-nonreversible behaviour. Below we present the results of the experiments where polarization-sensitive time-resolved optical detection of a reflected probe wave was used for monitoring of time non-reversible state in GaAs and InSb which appeared as the result of transient optical excitation, and we investigate the most evident manifestations of these two main mechanisms.

Optical magnetization

Optical magnetization with a circularly polarized intense electromagnetic wave as detected by a probe wave is known as the Inverse Faraday Effect. In fact it is light-induced circular birefringence and dichroism (see fig.1). In the configuration "on reflection", when the polarization state of the reflected light is used for monitoring of the result of the excitation, this effect is called the Specular Inverse Faraday Effect. In the most convenient configuration the probe wave is linearly polarized. In terms of the material equation (2) "magnetization" results in appearance of $\varepsilon_{xy}^{a}=-\varepsilon_{yx}^{a}$, i.e. in violation of the restriction (3) and rotation $\delta\alpha$ of the polarization state of the reflected probe wave governed by the equation[1]

$$\delta\alpha = \frac{1}{n(n-1)(1+n)} \mathrm{Re}\{\varepsilon_{xy}^{a,\,induced}\} \qquad (5)$$

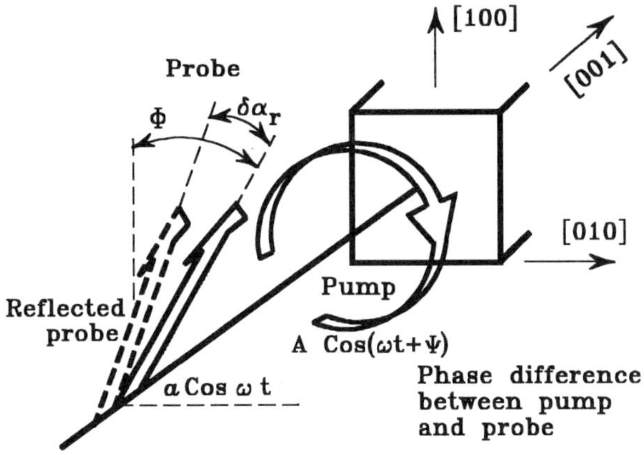

Figure 1. Observation of the Specular Induced Faraday Effect in cubic crystal

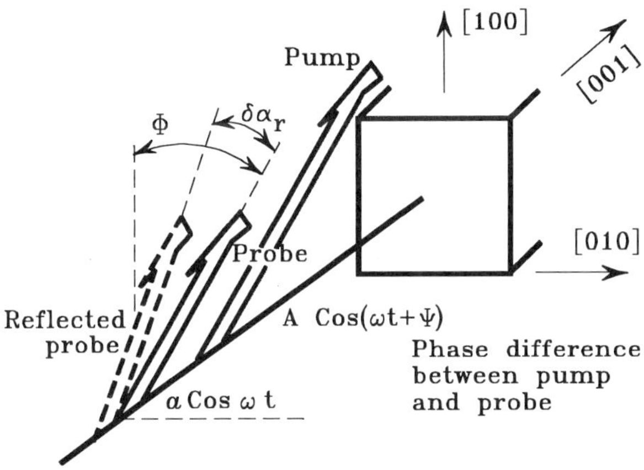

Figure 2. Observation of the Specular Induced Gyrotropic Linear Dichroism in zinc-blend crystal

Violation of thermodynamic equilibrium

The additional symmetry requirements (3-4) are derived as a direct result of the fluctuation-dissipation theorem, which is however, established for an equilibrium system and consequently they should not necessary hold for a non-equilibrium state. A non-equilibrium state may be created by a transient optical excitation. Zinc-blende materials ($\bar{4}3m$ point group) such as $A^{III}B^{V}$ semiconductors provide a nearly ideal testing ground for observation of time non-reversible state due to transient excitation creating a non-equilibrium state This is because as the result of excitation, the tensor ε_{ij} remains isotropic and time-reversible but the tensor γ_{ijk} may acquire a "symmetric" component $\gamma_{xyz}^{s} = \gamma_{yzx}^{s} = \gamma_{zxy}^{s} = \gamma_{xzy}^{s} = \gamma_{yxz}^{s} = \gamma_{zyx}^{s}$. The presence of γ^{s} leads in particular to the polarization-affecting phenomena of specular gyrotropic linear birefringence and dichroism (SIGLD). The polarization state of the reflected light alters due to γ^{s} ($\mathbf{k}\|[001]$)[1]:

$$\delta\alpha = \frac{4\pi \cos 2\Phi}{\lambda(n-1)(n+1)} \operatorname{Im}\{\gamma_{xyz}^{s,induced}\} \quad \textbf{(6)}$$

Here Φ is the initial orientation of the probe wave polarization with respect to the [100] direction of the crystal (see fig.2).

Coherent contributions

The main shortcoming of the above presentation of the light-induced effects via material equation (2) is loss of the **coherent** contributions to the effects, i.e. those contributions to time-nonreversible behaviour which depend not only on **intensity** of excitation, but also on the relative **phase** Ψ between the probe and the pump waves. These coherent contributions may be described via a more general, nonlinear material equation

$$D_i = D_i^{lin} + \chi_{ijlm} E_j E_l E_m^* + \Gamma_{ijlmk} \nabla_k E_j E_l E_m^* \quad \textbf{(7)}$$

They appear as additional coherent terms $\delta\alpha^{coh}$ to the total rotation $\alpha^{total}=\delta\alpha+\delta\alpha^{coh}$ and may be detected with mutually coherent probe and pump if the relative phase Ψ between them varies monotonicaly.

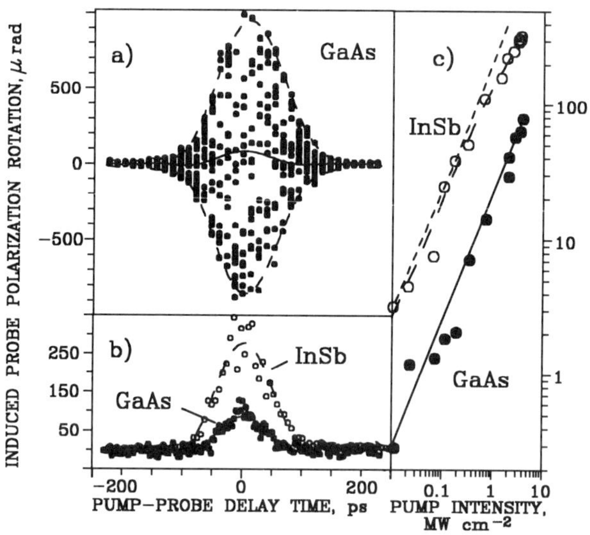

Figure 3. Variation of the SIFE in InSb and GaAs with pump-probe delay: (a) the total effect; (b) the incoherent component. Curve (c) represent the intensity dependence of the incoherent effect.

These coherent contributions are given by the following formulae: in SIFE

$$\delta\alpha_r^{coh} \propto (\operatorname{Re}\{\chi_{xxyy}\}\sin 2\Psi - \operatorname{Im}\{\chi_{xxyy}\}\cos 2\Psi) I_{p+} \quad \textbf{(8)}$$

and in SIGLD

$$\delta\alpha^{coh} \propto (\mathrm{Re}\{\Delta\Gamma\}\sin 2\Psi + \mathrm{Im}\{\Delta\Gamma\}\cos 2\Psi) I_p$$

$$\Delta\Gamma = 2\Gamma_{yxxxz} - \Gamma_{xyxxz} + \Gamma_{xxxyz} \qquad (9)$$

Where we introduce the incident pump beam intensity with linear polarization I_p for SIGLD and with circular polarization I_{p+} for SIFE.

EXPERIMENTAL

Below we present some recent results on observation of light-induced SIFE[2] and SIGLD[3] in bulk crystals of GaAs and InSb. The experiments were performed with a CW mode-locked frequency-doubled YAG laser on the wave-length of 532nm. i.e. in the highly absorbing range of the crystals (see fig.3 and 4).

Using the formulae (5-6) and the SIFE data one can calculate that circularly polarized excitation with intensity 1MW cm^{-2} created the time-nonreversible component of the dielectric tensor of $\varepsilon_{xy}{}^{a}$=6 10^{-4} in GaAs and $\varepsilon_{xy}{}^{a}$=7 10^{-3} in InSb. From the SIGLD data one can calculate that linearly polarized excitation with intensity 1MW cm^{-2} creates a time-nonreversible component of the nonlocality tensor of $\gamma_{xyz}{}^{s}$= 4 10^{-10}esu in GaAs and $\gamma_{xyz}{}^{s}$ = 1.6 10^{-9}esu in InSb.

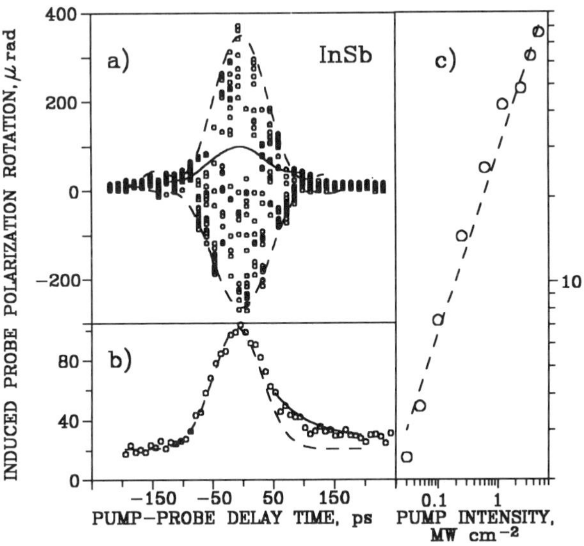

Figure 4. The SIGLD in InSb: (a) the total effect; (b) the incoherent component (notice the "tail" of 60ps). Curve (c) is the intensity dependence of the incoherent effect.

Acknowledgement

Author is grateful to S.V.Popov for assistance with preparation of this manuscript. This work is supported by the RW Paul Instrument Fund.

REFERENCES

1. A.R.Bungay, Yu.P.Svirko, N.I.Zheludev, Broken symmetry of the kinetic coefficients and specular polarization phenomena, *Phys.Rev.B*. 47:16141 (1993); A.R.Bungay, S.V.Popov, Yu.P.Svirko, N.I.Zheludev, Time-non-invariant linear birefringence and dichroism due to spin-orbit interaction. *Chem. Phys. Lett.* (1994, in press).
2. S.V.Popov, N.I.Zheludev, Yu.P.Svirko, Coherent and incoherent specular inverse Faraday effect: $\chi^{(3)}$measurements in opaque materials. *Optics Letters*. 19: 1 (1994); N.I.Zheludev, M.A.Brummell, R.T.Harley, A.Malinowski, S.V.Popov, D.E.Ashenford, B.Lunn. Giant specular inverse Faraday effect in $Gd_{0.6}Mn_{0.4}Te$. *Solid State Comm.* (1994, in press).
3. A.Malinowski, S.V.Popov, N.I.Zheludev, D.Yu.Paraschuk. Transient optical excitation breaks time-reversibility in GaAs and InSb crystals. *J. of Lumin.* (1994, in press).

ULTRAFAST OPTICAL GENERATION OF CARRIERS IN A DC ELECTRIC FIELD

Alex V. Kuznetsov and Christopher J. Stanton

Department of Physics
University of Florida
Gainesville, Florida 32611 USA

INTRODUCTION

Recently, several experimental groups have demonstrated that the ultrafast optical excitation of bulk semiconductors and semiconductor microstructures subjected to a DC electric field produces submillimeter electromagnetic radiation.[1–5] The fast-changing photocurrent caused by the ultrafast excitation has been identified as a source of such electromagnetic transients.[1] However, the theoretical understanding of the photocurrent on femtosecond time scale is still incomplete.

The principal difficulty in the description of current during and immediately after ultrafast photoexcitation consists in the presence of two competing mechanisms of charge transport. On one hand, the photoexcited particles accelerate in the DC electric field and produce a *transport* current (Fig. 1). On the other hand, each electron-hole pair acquires a dipole moment in the DC field. Optical generation of such "polarized pairs"[2,4] creates a time-dependent dipole moment in the sample, whose time derivative is the *displacement* current (Fig. 1).

The displacement current is the only radiation source possible in quantum well structures[2] where the transport current across the structure is suppressed. In the bulk samples, however, both mechanisms should be operative, and the theory of the transient photocurrent should be able to describe the translational motion of carriers and their polarization by the DC field at the same time. The primary goal of this paper is to present a fully quantum mechanical description of ultrafast carrier dynamics that can treat both transport and displacement contributions to the total photocurrent on the same footing.

The DC electric field distorts the electronic eigenstates in the field direction. In a bulk sample, the eigenstates can be expressed in terms of the Airy functions:

$$\psi_\nu^\alpha(x) = \sqrt{\frac{2}{F}}\,\beta_\alpha \frac{1}{l_o} Ai[\pm\beta_\alpha(x+\nu)] \;, \tag{1}$$

Coherent Optical Interactions in Semiconductors
Edited by R.T. Phillips, Plenum Press, New York, 1994

Here, F is the DC field, ν is the energy eigenvalue in units of $\varepsilon_o = Fl_o$, $l_o \equiv (\hbar^2/2m_r F)^{\frac{1}{3}}$ is a characteristic length, and $\beta_\alpha = (m_\alpha/m_r)^{1/3}$, where m_r is the reduced mass, m_α is the band masses for the conduction (α = c) and valence (α = v) bands.

We use the Airy eigenstates to construct the density matrix (DM)[6]. Let $a^\dagger_{\alpha\nu}(t)$, $a_{\alpha\nu}(t)$ denote the Heisenberg operators that create and annihilate electrons in an eigenstate ν of the band α. The density matrix (DM) is defined as:

$$N^{\alpha\beta}_{\nu\lambda}(t) \equiv \langle a^\dagger_{\alpha\nu}(t)\, a_{\beta\lambda}(t)\rangle = \begin{pmatrix} \langle a^\dagger_{c\nu}\, a_{c\lambda}\rangle & \langle a^\dagger_{c\nu}\, a_{v\lambda}\rangle \\ \langle a^\dagger_{v\nu}\, a_{c\lambda}\rangle & \langle a^\dagger_{v\nu}\, a_{v\lambda}\rangle \end{pmatrix}_{\alpha\beta}$$
$$= \begin{pmatrix} n^c_{\nu\lambda} & p_{\nu\lambda} \\ p^*_{\lambda\nu} & n^v_{\nu\lambda} \end{pmatrix}_{\alpha\beta}, \tag{2}$$

where <...> denotes the statistical average over the nonequilibrium state of the system. The DM gives a complete description of the electronic system: its diagonal elements give the number of particles in each state, while the non-diagonal elements describe quantum-mechanical correlations between different states.

For ultrashort excitation pulses there is no strict energy conservation for optical transitions,[7] and the presence of the DC field also lifts the momentum conservation requirements. As a result, an electron in the valence band can be optically excited into a number of states in the conduction band. The final state is then a *coherent superposition* of conduction band states which is described by the non-diagonal elements of the DM.

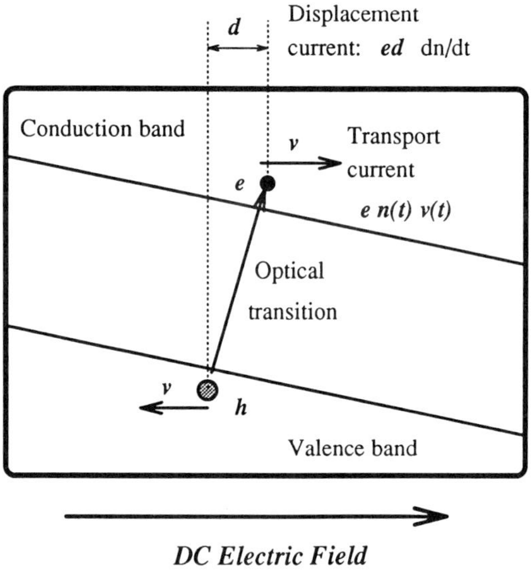

Figure 1. Two sources of photocurrent in a sample excited by a femtosecond optical pulse. The *transport* current is caused by translational motion of the carriers in a DC field, while the *displacement* current is produced by changes in the number of polarized electron-hole pairs

To describe the process of the optical excitation, we derive the equations of motion for the DM. These equations are a direct generalization of the optical Bloch equations[7,8] for the case with a DC field:

$$\frac{\partial p_{\nu\lambda}}{\partial t} = i(\varepsilon_{c\nu} - \varepsilon_{v\lambda})p_{\nu\lambda} + i\mu_{cv}E(t)\int d\rho\{n^v_{\rho\lambda}\chi^{cv}_{\nu\rho} - n^c_{\nu\rho}\chi^{cv}_{\rho\lambda}\} \tag{3}$$

$$\frac{\partial n^c_{\nu\lambda}}{\partial t} = i(\nu - \lambda)n^c_{\nu\lambda} - i\int d\rho\{p_{\nu\rho}\chi^{vc}_{\rho\lambda}E^* - \chi^{cv}_{\nu\rho}p^*_{\rho\lambda}E\} \ . \tag{4}$$

Here, χ is the overlap integral between conduction and valence band eigenstates (Eq. (1)), μ is the interband matrix element, and $E(t)$ is the external optical field. These equations can be solved numerically to obtain the time-dependent DM.[9]

Once the DM is known, we can calculate the time-dependent charge density for electrons and holes. The total charge density for the band α can be expressed as:

$$P^\alpha(x) = \sum_{\lambda\nu} n^\alpha(t, \lambda - \nu)\cdot\psi^\alpha_\nu(x)\psi^\alpha_\lambda(x) \equiv \sum_\lambda \rho^\alpha_\lambda(x,t) \ , \tag{5}$$

where we have introduced the Wigner function:

$$\rho^\alpha_\lambda(x,t) \equiv \sum_\eta n^\alpha(t,\eta)\,\psi^\alpha_\lambda(x)\cdot\frac{1}{2}\{\psi^\alpha_{\lambda+\eta}(x) + \psi^\alpha_{\lambda-\eta}(x)\} \ . \tag{6}$$

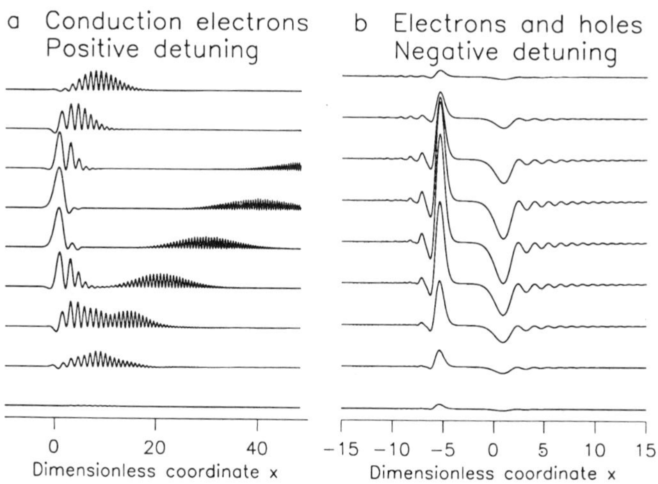

Figure 2. (a) Sequence plot of the Wigner function for conduction electrons created by a Gaussian excitation pulse with central frequency above the band gap. Time increases from bottom to top from —1 to 5 pulse durations. Initially an electron is created as a localized wavepacket, which later splits into two "particles" that go along the field (right) and against the field (left). Translational motion of the wavepackets gives rise to the transport current. (b) Charge density for virtual carriers (excitation below the band edge). Time ranges from —1.5 to 2 pulse durations. Virtual electrons (right) and holes (left) disappear after the excitation. However, each virtual pair possesses a considerable dipole moment, so that changes in their density during the excitation produce displacement current.

Because the final states are superpositions of delocalized eigenstates, the Wigner function and the charge density $\rho_\lambda(x) = \rho_\lambda^h(x) - \rho_\lambda^c(x)$ are no longer uniform. We have found that the charge distribution of photoexcited carriers substantially depends on the excitation energy. For excitation above the band gap (Fig. 2(a)) the electrons and holes behave as classical particles. They are created as localized wavepackets formed by coherent superposition of delocalized Airy eigenstates. These wavepackets later accelerate in the DC field according to the classical laws of motion producing the transport current. The holes are created around the same point in space as the electrons, so the initial pair dipole moment is small.

For the case of excitation below the band gap, however, the situation is completely different. Such excitation produces "virtual carriers"[10] that disappear after the excitation is turned off (Fig. 2(b)). The virtual carriers show no signs of classical translational motion in the DC field and do not contribute to the transport current. Instead, they are strongly polarized by the DC field and carry a considerable dipole moment per pair. As they are created and annihilated in the course of the optical excitation process, the medium acquires dielectric polarization that adiabatically follows the excitation intensity, and the time derivative of this polarization gives the displacement current.

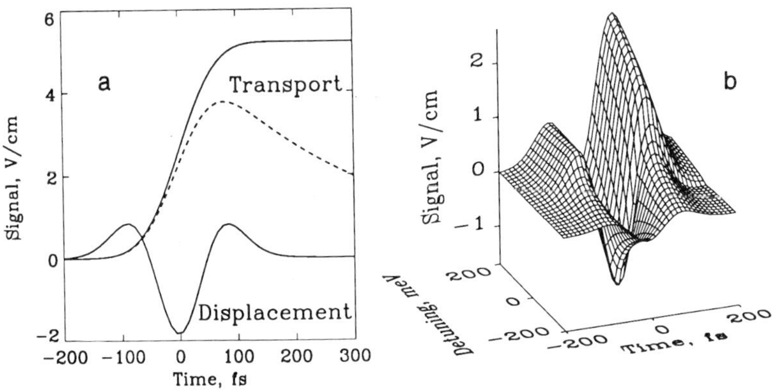

Figure 3. (a) Transport and displacement contribution to the radiated electromagnetic signal. The transport contribution follows the number of real carriers and is positive (taking into account scattering causes this contribution to decay (dashed curve) but does not change its sign). Displacement signal is a second derivative of the excitation envelope and has a negative main peak. (b) Sign reversal of the signal with increasing detuning. Below the band gap the negative displacement signal dominates, while above the gap the positive transport contribution is more important

Using the time-dependent charge densities, we can calculate the photocurrent. The experimentally measured radiated field[1–3,5] is proportional to the time derivative of the photocurrent. Therefore, the transport contribution to the overall signal from the ballistically accelerating real carriers should be positive and proportional to the number of real carriers (Fig. 3(a)). Note that the inclusion of scattering processes will lead to the decay of the transport signal as the current relaxes to its stationary value. On the other hand, the displacement contribution from virtual carriers follows the second derivative of the excitation intensity and therefore has a negative main peak.

Generally, excitation above the band gap will produce real carriers in the states with low perpendicular momentum $k_\perp^2 < 2m_r(\hbar\omega - E_g)/\hbar^2$, while virtually populating states with higher perpendicular momentum, so that *both real and virtual carriers* contribute to the observed signal. The relative contributions of real and virtual carriers to the total photocurrent depend on the pulse duration, DC field, and the central frequency of the excitation.

Fig. 3(b) illustrates the competition of transport and displacement processes. At negative and small positive detunings only virtual carriers are present, which leads to radiated waveform with a negative main peak. As the detuning increases, the positive transport contribution from real carriers starts to offset the displacement signal, which results in sign reversal of the signal at sufficiently high detunings. Such sign reversal has been observed experimentally[5] and was found qualitatively consistent with the above described picture of competing transport and displacement currents.[11]

In conclusion, we have developed a fully quantum mechanical treatment of femtosecond carrier dynamics in a DC electric field. We find that the carriers are optically generated as localized wavepackets that can move and carry the transport current while having certain initial dipole moment that is responsible for the displacement current. The transport and displacement contributions to the THz emission have different dependence on the parameters of the problem so either one can be made dominant. Although our theory does not yet include scattering and excitonic effects, it is qualitatively consistent with the experimental findings.[5,11]

Acknowledgments

This work was supported by U.S. Office of Naval Research through Grant N00091–JJ-1956 and by the NSF grant DMR 8957382. CJS gratefully acknowledges support from Alfred P. Sloan foundation.

REFERENCES

[1] X.-C. Zhang, B.B. Hu, J.T. Darrow, and D.H. Auston, Appl. Phys. Lett. **56**, 1011 (1990); B.B. Hu, X.-C. Zhang, and D.H. Auston, Phys. Rev. Lett. **67**, 2709 (1991).

[2] P.C.M. Planken, M.C. Nuss, W.H. Knox, D.A.B. Miller, and K.W. Goossen, Appl.Phys. Lett. **61**, 2009 (1992); P.C.M. Planken, M.C. Nuss, I. Brener, K.W. Goossen, M.S.C. Luo, S.L. Chuang, and L. Pfeiffer, Phys. Rev. Lett. **69**, 3800 (1992).

[3] N. Katznellenbogen and D. Grischkowsky, Appl. Phys. Lett. **58**, 222 (1991).

[4] S.L. Chueng, S. Schmitt-Rink, B.I. Greene, P.N. Saeta, and A.F.J. Levi, Phys. Rev. Lett. **68**, 102 (1992).

[5] X.-C. Zhang, Y. Jin, K. Yang, and L.J. Schowalter, Phys. Rev. Lett. **69**, 2303 (1992)

[6] R. Bertoncini and A.P. Jauho, Phys. Rev. Lett. **68**, 2826 (1992).

[7] A. V. Kuznetsov, Phys. Rev. **B 44**, 8721 (1991).

[8] M. Lindberg and S.W. Koch, Phys. Rev. **B 38**, 3342 (1988).

[9] A.V. Kuznetsov and C.J. Stanton, Phys. Rev. **B**, to be published.

[10] For a review, see S. Schmitt-Rink, D.S. Chemla, and D.A.B. Miller, Adv. Physics **38**, 89 (1989).

[11] B.B. Hu, A.S. Weling, D.H. Auston, and A.V. Kuznetsov, *Proceedings of the Quantum Electronics and Laser Science Conference, Baltimore, MD, 1993*, Postdeadline papers, p. QPD16–1/35

MANY-BODY EFFECTS AT THE FERMI EDGE OF MODULATION DOPED SEMICONDUCTORS: A NUMERICAL STUDY

Wilfried Schäfer[1], Igal Brener[2], and Wayne Knox[2]

[1]Forschungszentrum Jülich, HLRZ, 52425 Jülich, FRG
[2]AT&T Bell Laboratories, Holmdel, New Jersey 07733

INTRODUCTION

Optical properties near the Fermi-edge of one-component Coulomb systems have been studied for more than half a century (for recent reviews see [1,2]). Originally the interest was focussed on x-ray spectra of metals. The development of modern crystal-growth technique allowed to realize similar systems in semiconductor materials. In these doped quantum-wells many-body effects which are responsible for the lineshape of optical spectra can be studied over a large range of densities applying the various tools of ultra-short time spectroscopy. The current interest in these systems results as well from the technological importance of them as from conceptual reasons, concerning basic aspects of quantum mechanics and manybody physics. Especially the study of nonlinear optical properties on an ultrashort timescale provides sensitive information about the relevant scattering processes which determine the dephasing and thus the lineshape of nonlinear spectra.

Experimental investigations of the ac Stark-effect in a doped quantum-well show strong deviations in comparison with the undoped case, even if the excitation conditions are equivalent. The striking difference is a nearly complete missing of bleaching of the oscillator strength which does not fit into the common interpretation of the ac Stark-effect in undoped systems [3]. In that case there are several competing effects which in summary are responsible for the details of the lineshape of the excitonic resonance: Pauli-blocking, exciton-exciton and exciton-photon interaction together with broadening effects due to the spectral width of the pump pulse result for strictly off-resonant ultra-short excitation in a moderate blue-shift and a bleaching of the excitonic resonance. This kind of analysis is based on the semiconductor Bloch-equations (SBE) in which relaxation and dephasing are described by phenomenological time constants. As was demonstrated in several recent investigations [4] this approach describes the most important features of coherent phenomena in semiconductors correctly, as long as induced scattering and dephasing processes may be neglected. Application of the SBE's to the ac Stark-effect in doped materials under comparable excitation conditions yields for shift and bleaching of the excitonic resonance very similar results as in the undoped case and fails to explain experimental data. This is not at all surprising as already a correct description of linear optical absorption of doped systems must allow for the dynamical response of the Fermi sea to the optically generated holes. Consequently nonlinear optical properties are not only determined by optically excited carriers and their mutual interaction but even more by their scattering with the Fermi sea.

THEORY

Most approaches, which include the dynamical response of the Fermi sea into the calculation of optical spectra are based on the following assumptions:

i) There is only one infinitely heavy hole, which acts as structureless scattering center for conduction electrons

ii) Electron-electron interaction is assumed to cause near the Fermi surface only renormalized quasiparticle properties and is therefore neglected.

iii) The interaction potential is assumed to be separable and in most calculations replaced by a constant.

The assumption of an infinite hole mass is well justified in the case of deep core-hole of metals but fails in semiconductors. As was shown in [5] indirect transitions become possible due to the finite hole-mass. Electrons and holes with different **k**-vectors are excited under simultaneous scattering with an excitation of the Fermi sea, which ensures energy and momentum conservation. The energy spacing between direct and indirect thresholds is $m_e/m_h\, \varepsilon_F$ and a singularity at the Fermi-edge is smeared over a corresponding energy range. A description of these processes is only possible if electron-electron interaction is taken into account, i.e. the second assumption has to drop out, too. The third assumption seems to be merely technical. It has, however, important consequences for the interpretation of the excitonic resonance occuring near the Fermi-edge, as will be discussed briefly in the following.

In the simplest description dynamical correlations of the Fermi-sea are neglected. This may be justified in the limit of small doping, i. e. if the Fermi-energy is small in comparison with the excitonic Rydberg energy. With increasing concentrations of cold carriers the exciton energy shifts above the renormalized band-gap i.e. in an energy-range, where boundstates cease to be stable (Mott-transition). Nevertheless as long as dynamical correlations of the Fermi-sea are neglected the theory predicts a bound-state below the Fermi-edge, which is present as long as the Fermi-edge is sufficiently sharp. Technically spoken this bound state results from a ladder-approximation for the polarization, which accounts for the electron-hole attraction as result of repeated scattering, whereas crossed diagrams are neglected. The latter describe backward scattering of electrons, resulting from the excitation of the Fermi-sea due to the action of the holes. These processes partially compensate the electron-attraction, i. e. to a n-th order ladder diagram, there is a corresponding diagram with maximal crossing, which reduce the ladder contribution. The magnitude of this reduction depends on details of the electron-hole interaction. If the screened Coulomb interaction can be approximated by a contact potential - this is justified in the high doping limit - the balance between ladder diagrams and crossing diagrams converts the excitonic resonance, into a powerlaw singularity [2]. For low and moderate doping concentrations this balance is modified if the **k**-vector dependence of the screened interaction is considered. For the ratio of the second order ladder and crossed diagrams one obtains e. g. for a contact interaction 1/3 [2]. In the case of a statically screened Coulomb interaction this ratio becomes dependent on density and frequency. In the vincinity of the resonance it is roughly 1/8, when r_s is of order one. This indicates that the contribution of maximal crossed diagrams compared to ladder diagrams, is considerably reduced for not too high doping concentrations. In the following we therefore take no diagrams with more than two crossing Coulomb lines into account. A systematic improvement of the present approach, which is discussed in the following, would require the application of parquet theory [6].

Our approach is based on the quantum-kinetic equations for diagonal and off-diagonal elements of the one-particle propagators, which reduce in the equal-time limit to the one-particle distribution and the polarization [7]. A well known special case of these equations are the already mentioned SBE's, where scattering contributions are replaced by phenomenological relaxation rates. In the following we will treat scattering processes microscopically, however, in contrast to [7] within a Marcovian approximation. Thus only propagators, which are diagonal in time need to be considered and the scattering contribution has the form.

$$\frac{\partial}{\partial t} G^{<}_{cv}(t,t)\,|_{scatt} = \sum_{a=c,v} \int_{-\infty}^{t} dt' [(\Sigma^{>}_{ca}(t,t') G^{<}_{av}(t',t) - G^{>}_{ca}(t,t')\Sigma^{<}_{av}(t',t)) + (\,> \rightleftarrows <)] \qquad [1]$$

(1) treats self-energy contributions (diagonal elements of the self-energy with respect to band-indices c and v) and vertex-corrections (off-diagonal elements of the self-energy) on the same footing. In a second order Born approximation with a statically screened interaction w, the self-energy is given by

$$\Sigma^{<}_{cv}(\mathbf{k}tt') = \sum_{\mathbf{k}'\mathbf{k}''ab} \{2w^2(\mathbf{k}') G^{<}_{cv}(\mathbf{k}+\mathbf{k}', tt') G^{>}_{ab}(\mathbf{k}'+\mathbf{k}'', t't) G^{<}_{ba}(\mathbf{k}'', tt')$$

$$- w(\mathbf{k}') w(\mathbf{k} - \mathbf{k}'') G^{<}_{ca}(\mathbf{k}+\mathbf{k}', tt') G^{>}_{ab}(\mathbf{k}'+\mathbf{k}'', t't) G^{<}_{bv}(\mathbf{k}'', tt')\} \qquad [2]$$

To proceed further, we factorize the propagators in a slowly varying contribution and a rapidly varying phasefactor, which represents the lowest order solution, i. e. for the case without interaction (for details compare [7]). The time integration in (1) is then only performed with respect to the phasefactors, whereas the slowly varying contributions are taken at the upper limit of the integral. This procedure corresponds to a common Marcovian approximation and yields e.g. in the kinetic equations for distribution functions the usual Boltzmann type scattering terms. For the dephasing we obtain in terms of the polarization $P_{\mathbf{k}}$ and the distribution functions $f_{\mathbf{k}}$:

$$\frac{\partial}{\partial t} P_{\mathbf{k}}(t)\,|_{scatt} = -\{ \sum_{\mathbf{k}'\mathbf{k}''a=e,h} 2w^2(\mathbf{k}')\{\{g(\varepsilon^a_{\mathbf{k}} - \varepsilon^a_{\mathbf{k}+\mathbf{k}'} + \varepsilon^e_{\mathbf{k}'+\mathbf{k}''} - \varepsilon^e_{\mathbf{k}''})$$

$$[P_{\mathbf{k}}(1 - f^a_{\mathbf{k}+\mathbf{k}'}) f^e_{\mathbf{k}'+\mathbf{k}''}(1 - f^e_{\mathbf{k}''})] + [f \mathrel{\substack{\leftarrow \\ \rightarrow}} (1-f)] - P_{\mathbf{k}} P_{\mathbf{k}'+\mathbf{k}''} P^{\times}_{\mathbf{k}''}\} - \{\mathbf{k} \mathrel{\substack{\leftarrow \\ \rightarrow}} (\mathbf{k}+\mathbf{k}')\}\}$$

$$- \sum_{\mathbf{k}'\mathbf{k}''} w(\mathbf{k}') w(\mathbf{k} - \mathbf{k}'')\{\{g(\varepsilon^e_{\mathbf{k}} - \varepsilon^e_{\mathbf{k}+\mathbf{k}'} + \varepsilon^e_{\mathbf{k}'+\mathbf{k}''} - \varepsilon^e_{\mathbf{k}''}) \qquad [3]$$

$$[P_{\mathbf{k}}(1 - f^e_{\mathbf{k}+\mathbf{k}'}) f^e_{\mathbf{k}'+\mathbf{k}''}(1 - f^e_{\mathbf{k}''})] + [f \mathrel{\substack{\leftarrow \\ \rightarrow}} (1-f)] - P_{\mathbf{k}} P_{\mathbf{k}'+\mathbf{k}''} P^{\times}_{\mathbf{k}''}\} - \{\mathbf{k} \mathrel{\substack{\leftarrow \\ \rightarrow}} (\mathbf{k}+\mathbf{k}'), \mathbf{k}'' \mathrel{\substack{\leftarrow \\ \rightarrow}} (\mathbf{k}''+\mathbf{k}')\}$$

$$+ g(\varepsilon^h_{\mathbf{k}+\mathbf{k}'} - \varepsilon^h_{\mathbf{k}} + \varepsilon^e_{\mathbf{k}'+\mathbf{k}''} - \varepsilon^e_{\mathbf{k}''})\{[(1 - f^h_{\mathbf{k}})(f^h_{\mathbf{k}+\mathbf{k}'} P_{\mathbf{k}'+\mathbf{k}''} - P_{\mathbf{k}+\mathbf{k}'} f^e_{\mathbf{k}'+\mathbf{k}''})(1 - f^e_{\mathbf{k}''})] + [f \mathrel{\substack{\leftarrow \\ \rightarrow}} (1-f)]\}$$

$$+ g(\varepsilon^h_{\mathbf{k}} - \varepsilon^h_{\mathbf{k}+\mathbf{k}'} - \varepsilon^e_{\mathbf{k}'+\mathbf{k}''} + \varepsilon^e_{\mathbf{k}''}) P_{\mathbf{k}}(P^{\times}_{\mathbf{k}+\mathbf{k}'} - P^{\times}_{\mathbf{k}'+\mathbf{k}''}) P_{\mathbf{k}''}\}\} - \{e \rightleftarrows h\}$$

where we have switched to the e-h picture and introduced the definition

$$g(\omega) = P\frac{i}{\omega} + \pi\delta(\omega) \qquad [4]$$

(3) and corresponding contributions for the relaxation replace the phenomenological relaxation rates in the SBE's. For the RPA contributions in (3) (terms~ $w^2(\mathbf{k})$) a non-Marcovian approximation has been proposed in [8].

To apply these results to a two beam experiment in which two pulses E and δE with wave-vectors q_1 and q_2 interact with the sample, one has to perform a Fourier analysis, which was introduced earlier for the SBE's [9,4]. The resulting expressions are rather lengthy and for lack of space we only illustrate the structure of the dephasing processes of the polarization δP, induced by the probe pulse. Replacements of the following type have to be performed e.g. in the first contribution to (3):

$$P_{\mathbf{k}}(1 - f^e_{\mathbf{k}+\mathbf{k}'})f^e_{\mathbf{k}'+\mathbf{k}''}(1 - f^e_{\mathbf{k}''}) \rightarrow (\delta P_{\mathbf{k}}(1 - f^e_{\mathbf{k}+\mathbf{k}'}) - P_{\mathbf{k}}\delta f^e_{\mathbf{k}+\mathbf{k}'})f^e_{\mathbf{k}'+\mathbf{k}''}(1 - f^e_{\mathbf{k}''})$$

$$+ P_{\mathbf{k}}(1 - f^e_{\mathbf{k}+\mathbf{k}'})(\delta f^e_{\mathbf{k}'+\mathbf{k}''}(1 - f^e_{\mathbf{k}''}) - f^e_{\mathbf{k}'+\mathbf{k}''}\delta f^e_{\mathbf{k}''}) + o(\delta f^2) \quad [5]$$

where δf denotes the occupation modulation, which results from the combined action of pump and probe pulse. The contributions, depending on δf describe induced dephasing effects, which are characterized by scattering processes of electrons of the Fermi sea with electrons or hole which are distributed according to the occupation modulation. The numerical studies discussed in the following have shown that just these induced dephasing effects are responsible for the lack of bleaching found experimentally.

RESULTS AND DISCUSSION

Numerical studies of the modified SBE's using the description of dephasing discussed in the last section together with corresponding expressions for the energy-relaxation have been performed for bulk GaAs. Fig. 1 shows the absorption of a doped system with and without pump-pulse in comparison with the undoped case for equivalent excitation conditions. In both cases the 100fs pump pulse is centered 6 Ryd below the excitonic resonance, the Rabi-frequency is 1.5 Ryd and the delay with respect to the probe is zero. The broadening of the excitonic resonance in the doped system results mainly from e-e interaction. A second broadening effect is due to the finite temperature of 10K . Additionally we have added a small background damping of 0.2 Ryd. The linear spectrum of the undoped case is broadened only due to the phenomenological dephasing rate of 0.4 Ryd in the SBE's. For the doped system the nonlinear spectrum shows a small blue-shift and a small increase of the line-width, whereas the maximum of the resonance remains essentially unchanged. This is in in contrast to the undoped case in which the blue-shift is accompanied by a strong bleaching of the pulse. This different behaviour is in good agreement with experimental data

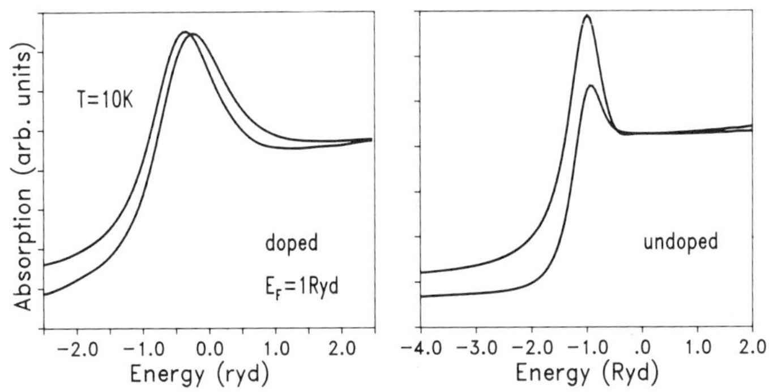

Figure 1. Calculated linear and nonlinear absorption spectra for a doped and a undoped system. For details see text.

shown in Fig.2, where the ac-Stark effect in a modulation n-doped quantum-well system is compared with the undoped case. In both cases the detuning is 50meV below the excitonic resonance and the excitation-energy is about 1GW/cm^2. The Fermi-energy in the n-doped sample is 15 meV corresponding to a doping density of $3 \times 10^{11} cm^{-2}$.

From a number of numerical tests the different behaviour found as well in the calculations as in experiments, can be unambigously traced back to induced scattering processes between the electrons of the Fermi sea and particles generated by the combined action of excite and probe pulse. This is a new type of many-body effect in nonlinear optics of semiconductors and should be observable even more pronounced in frequency or time-resolved four-wave-mixing signals.

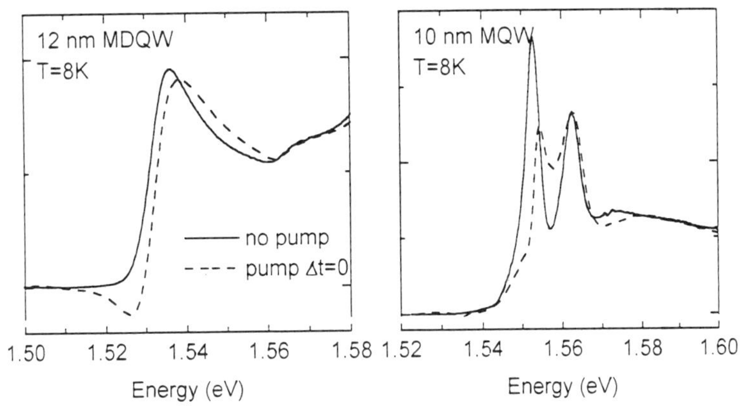

Figure 2. Experimental excite and probe spectra in comparison with the linear absorption for doped and undoped quantum-well systems.

REFERENCES

(1) For a recent reviews, see e.g. K. Ohtaka, Y. Tanabe, Rev. Mod. Phys. 62, 929 (1990)
(2) G. Mahan, Many-Particle Physics (Plenum, New York, 1990), Sec. 8.3
(3) R. Binder, S. W. Koch, M. Lindberg, W. Schäfer, F. Jahnke, Phys. Rev. B 43, 6520 (1991) and references therein
(4) W. Schäfer, in:'Optics of Semiconductor Nanostructures', F. Henneberger, S. Schmitt-Rink, E. O. Göbel (eds.), Akademie Verlag, Berlin 1993
(5) A.E. Ruckenstein, S. Schmitt-Rink, Phys. Rev. B 35, 7551 (1987)
(6) B. Roulet, J. Gavoret, P. Nozieres, Phys. Rev. 178, 1072 (1969); P. Nozieres, J. Gavoret, B. Roulet, Phys. Rev. 178, 1084 (1969)
(7) M. Hartmann, W. Schäfer, phys. stat.sol. (b) 173, 165 (1992)
(8) D.B. Tran Thoai, H. Haug, Z. Phys. B 91, 199 (1993)
(9) M. Lindberg, R. Binder, S. W. Koch, Phys. Rev A 45, 1865 (1992)

THEORY OF POLARIZATION-DEPENDENT FOUR-WAVE MIXING IN QUANTUM WELLS USING THE OPTICAL BLOCH EQUATIONS

[1]T. Meier, D. Bennhardt, and P. Thomas
[2]Y.Z. Hu, R. Binder, and S.W. Koch

[1]Department of Physics and Materials Sciences Center
Philipps University, Renthof 5, D-35032 Marburg, Germany
[2]Optical Sciences Center and Department of Physics
University of Arizona, Tucson 85721, USA

1. INTRODUCTION

Transient four-wave mixing (TFWM) experiments have been widely used to study the excitation dynamics of excitons in quantum wells [1]. Recently is has been observed that TFWM signals, especially quantum beats between heavy-hole (hh) and light-hole (lh) excitons, strongly depend on the polarization of the incident laser pulses [2-6]. This dependence is presently being investigated.

In the theoretical treatment presented here we calculate the TFWM signal emitted in the direction $2\vec{k}_2 - \vec{k}_1$ for incident pulses with wavevectors $\vec{k}_1$ and $\vec{k}_2$ in third order perturbation theory. In Section 2 this signal is calculated for different angles between linearly polarized incident pulses using a degenerate-three level model, which is based on the optical Bloch equations (OBE). These equations arise as an approximation to the semiconductor Bloch equations, which have been used to study TFWM and photon echo signals within the Hartree-Fock approximation [7,8]. Within the simplified model used in Section 2 we can explain some general characteristics of the experiments. Afterwards in Section 3 we extend this model and phenomenologically include a disorder induced coupling of excitonic transitions [5], Coulomb-effects as local-fields [9] and excitation induced dephasing processes [10,11].

2. FIRST MODEL

The simplest model that can be used to explain the dependence of the hh - lh quantum beat signal on the polarization of the incident pulses is a degenerate three-level system, as proposed by Schmitt-Rink et al. [3], Fig. 1(a). Each of these three-level

Coherent Optical Interactions in Semiconductors
Edited by R.T. Phillips, Plenum Press, New York, 1994

systems includes oppositely circular polarized optical dipole transitions between the conduction band states ($m_j = \pm\frac{1}{2}$) and the hh- ($m_j = \pm\frac{3}{2}$) and lh- ($m_j = \mp\frac{1}{2}$) states. The nonlinear optical response of this system is described by the OBE for the polarization $p_{mn} =< v_m^\dagger c_n >$, conduction electron population $c_{nn'} =< c_n^\dagger c_{n'} >$, and valence electron population $v_{mm'} =< v_m^\dagger v_{m'} >$. Here, c_n destroys a conduction electron with moment $m_j = \pm\frac{1}{2}$ and $v_m^\dagger$ creates a valence electron with moment $m_j = \pm\frac{3}{2}$ or $m_j = \pm\frac{1}{2}$. Introducing the Rabi frequency $R_{mn}(t) = \vec{\mu}_{nm}^* \cdot \vec{E}(t)$, where $\vec{E}(t)$ is the laser field, these equations read within the rotating-wave approximation

$$\begin{aligned} \partial_t p_{mn} + i(\varepsilon_n^c - \varepsilon_m^v)p_{mn} + \gamma_{nm}^{cv} p_{mn} &= i\sum_{m'} v_{mm'} R_{m'n}(t) - i\sum_{n'} R_{mn'}(t) c_{n'n} \\ \partial_t v_{mm'} + i(\varepsilon_{m'}^v - \varepsilon_m^v) v_{mm'} &= i\sum_n (p_{mn} R_{m'n}^*(t) - R_{mn}(t) p_{m'n}^*) \\ \partial_t c_{n'n} + i(\varepsilon_n^c - \varepsilon_{n'}^c) c_{n'n} &= -i\sum_m (R_{mn'}^*(t) p_{mn} - p_{mn'}^* R_{mn}(t)) \quad . \end{aligned} \tag{1}$$

Here, $\varepsilon_{n(m)}^{c(v)}$ is the conduction (valence) electron band energy for band n (m) and we have added phenomenological dephasing rates γ to the equations. The selection rules are included in the dipole matrix element $\vec{\mu}_{nm} =< v, m|e\vec{r}|c, n >$. The laser field consists

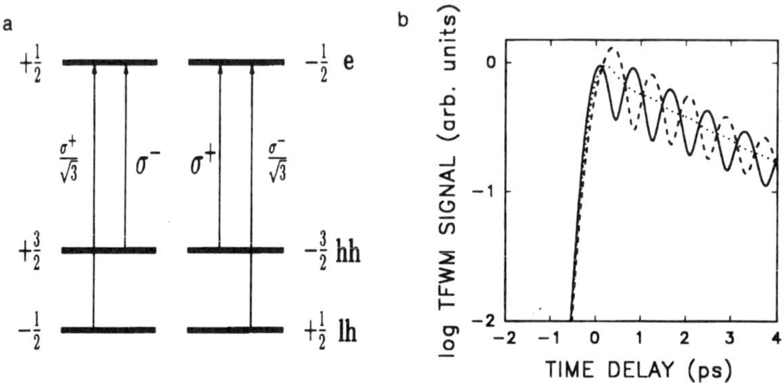

Figure 1. (a) hh and lh exciton circularly polarized transitions. (b) Calculated TI TFWM signal versus time delay for different orientations of the excitation pulses, $\Theta_{12} = 0^o$ (solid), 45° (dotted) and 90° (dashed). For further parameters see text.

of two pulses with wavevectors $\vec{k}_1$ and $\vec{k}_2$, with a time delay τ between them. Positive τ means that the pulse with wavevector $\vec{k}_1$ acts first. We calculate numerically the third order polarization $\vec{P}^{(3)}(t,\tau) = \sum_{nm} \vec{\mu}_{nm} p_{mn}^{(3)}$ which radiates in the direction $2\vec{k}_2 - \vec{k}_1$. For time-resolved (TR) and time-integrated (TI) measurements the signals are

$$I_{TR}(t,\tau) \propto |\vec{P}^{(3)}(t,\tau)|^2 \qquad \text{and} \qquad I_{TI}(\tau) \propto \int dt |\vec{P}^{(3)}(t,\tau)|^2 \quad . \tag{2}$$

The TI-signal is calculated for various angles, Θ_{12}, between the linearly polarized incident pulses. The following parameters were used for the calculations $\varepsilon_{hh}^v - \varepsilon_{lh}^v = 5meV$, $\gamma_{hh}^{cv} = 5ps^{-1}$, $\gamma_{lh}^{cv} = 3ps^{-1}$, a laser pulse width of $300fs$ (FWHM) and the central frequency of the laser pulse was tuned in the middle between the hh and the lh transitions.

Fig. 1(b) shows that the quantum beats in the TFWM signals for parallel ($\Theta_{12} = 0^0$) and perpendicular ($\Theta_{12} = 90^0$) incident polarizations are 180^0 out of phase. This is a consequence of the shift in the hh and lh transition phases induced by the relative polarizations of the laser pulses. This antiphase beating was observed experimentally [3].

For an angle of $\Theta_{12} = 45^0$ this model predicts that the beating should be absent, in contrast to experiments where in this case a beating in phase with the signal for $\Theta_{12} = 0^0$ is observed [4,5].

In addition to this the experiments show distinct signal decay times for $\Theta_{12} = 0^0$ and $\Theta_{12} = 90^0$. Depending on the sample quality the signal for the perpendicular case decays faster than the one for the parallel case [2-6].

There are at least two further observations which can also not be explained within the present model. First, the intensities of the signals for the parallel and the perpendicular case at zero time delay can differ from each other more than one order of magnitude. This difference was shown to depend on the excitation density [2,4]. Second, the observed signals for negative time delay are not present in this model [3-6].

3. EXTENDED MODEL

We now phenomenologically extend the degenerate three-level model discussed above in three ways in order to show which mechanisms can explain the experimental findings.

First of all we include a disorder induced coupling of the σ^+ and σ^- exciton transitions in the model. This coupling, which is inhomogeneous in strength, was proposed by Bennhardt et al. [5]. Possible microscopic mechanisms for this coupling are interface roughness, dislocations, stress effects, impurities in the well and compositional disorder in the barrier material. This disorder interaction constitutes a local symmetry breaking and lifts the degeneracy of the twofold degenerate hh and lh transitions. Calculations have shown that even a small inhomogeneous coupling leads to remarkable changes of the TI and TR four-wave mixing signals [5]. In agreement with experimental observations, the TI signals for $\Theta_{12} = 90^0$ decays faster than that for $\Theta_{12} = 0^0$ and the signal for $\Theta_{12} = 45^0$ shows a beating in phase with the signal for the parallel case. Also the TR signals are changed through this coupling. The signal for the parallel case is, depending on the strength of the coupling, partially emitted as an echo, while the signal for the perpendicular case shows always a free decay [2,4].

Since it is well known that Coulomb-effects have a strong influence on the nonlinear optical response of semiconductors we include them into our model in the well known local-field approximation [9]. These local-fields are the dominant effects caused by coulomb interaction for very low excitation densities. We replace the Rabi frequency $R_{mn}(t)$ in the OBE by $\vec{\mu}_{nm} \cdot \vec{E}(t) + Lp_{nm}$, where L is a parameter that determines the strength of the local-fields. The local-fields lead to contributions to the four-wave mixing signal which are known as polarization scattering. These terms increase the signal for positive time delay τ and give rise to signals for negative τ. The beating and the polarization dependence of the signal is not distorted by the inclusion of local-fields in the calculations. This is reasonable since these terms do not change the symmetries of the system.

Recently, we have shown that the dependence of the dephasing rate on the excited density of excitons leads to another contribution to the four-wave mixing signal [10,11]. The simplest approach including this contribution in the OBE is to linearize this dependence around a certain average density which is determined by the experimental conditions. The dephasing rate γ is then replaced by $\gamma_0 + \gamma' \, \delta N$, where γ' is the derivative of γ evaluated at the average density N and $N + \delta N$ is the density excited by the four-wave mixing pulses. γ' is a parameter which determines the strength of the excitation induced dephasing process. The inclusion of this term in the OBE results in different signal intensities for the parallel and the perpendicular case, since at $\Theta_{12} = 0^0$ the density grating due to N are created in phase giving a finite value, while for $\Theta_{12} = 90^0$ they are out of phase and exactly cancel. Even if γ' is very small and $\gamma' N$, the change of the dephasing rate, is only about $(10^4 ps)^{-1}$ the intensities of the two signals differ by about one order of magnitude, as we will show below.

We have calculated the four-wave mixing signals for three values of Θ_{12}. All three effects we have discussed are included in the calculation. The parameters used are the following: $L = 0.2meV$, $\gamma' N = (10^4 ps)^{-1}$ and $g(J) = exp(-(\frac{J}{0.4meV})^2)$, where $g(J)$ is the distribution function of the inhomogeneous coupling. The features we have discussed above are clearly displayed in Fig. 2(a).

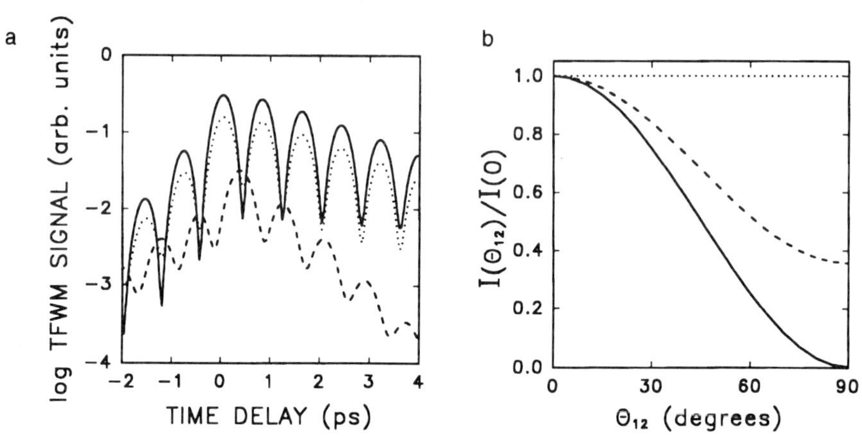

Figure 2. (a) Calculated TI TFWM signal versus time delay for different orientations of the excitation pulses, $\Theta_{12} = 0^0$ (solid), 45^0 (dotted) and 90^0 (dashed). For further parameters see text. (b) Intensity of the TFWM signal versus Θ_{12}. $L = 0.2meV$, for further parameters see text.

It is now useful to look more closely at the dependence of the signal intensity on the angle Θ_{12}. Therefore we take τ equal to $1ps$ and consider only the hh transitions. Fig. 2(b) shows this dependence for three sets of parameters. If we neglect the disorder induced coupling and the excitation induced dephasing process there is no dependence at all (dotted line). For a disorder coupling as in Fig. 2(a) a dependence is seen, which is due to the different decay times of the signals for the parallel and the perpendicular cases (dashed line). If we then set $\gamma' N$ to $(10^3 ps)^{-1}$ (solid line) a strong dependence is found. The different curves for $I(\Theta_{12})/I(0)$ can all be fit with $\alpha \cos^2(\Theta_{12}) + \beta$ where α and β depend on the disorder coupling and γ'. A comparison of these curves to experimental data from Eccleston et al. [4] suggests that γ' decreases with increasing excitation density. This means that the dephasing rate varies strongly with the excitation density for small densities. At high densities the increase saturates.

4. CONCLUSIONS

We have presented a model that reproduces some experimental observations on the polarization dependence of TFWM signals and we have shown physical mechanisms, which can explain the observed features. In future work we have to address the relative importance of the physical processes discussed above, as well as biexciton effects [12,13]. The basis for such an analysis will be a detailed microscopic description of both the disorder and the excitation induced dephasing processes.

Acknowledgements

We wish to acknowledge J. Kuhl, S.T. Cundiff, D.G. Steel, H. Wang, M. Lindberg, K. Bott, and O. Heller for many stimulating and fruitful discussions.

REFERENCES

1. see for example, F. Henneberger, S. Schmitt-Rink, and E.O. Göbel eds., "Optics of Semiconductor Nanostructures", Akademie Verlag, Berlin (1993).
2. S.T. Cundiff, H. Wang, and D.G. Steel, "Polarization-dependent picosecond excitonic nonlinearities and complexities of disorder", *Phys. Rev. B* 46:7248 (1992).
3. S. Schmitt-Rink, D. Bennhardt, V. Heuckeroth, P. Thomas, P. Haring, G. Maidorn, H. Bakker, K. Leo, D.-S. Kim, J. Shah, and K. Köhler, "Polarization dependence of heavy- and light-hole quantum beats", *Phys. Rev. B* 46:10460 (1992).
4. R. Eccleston, J. Kuhl, D. Bennhardt, and P. Thomas, "Intensity Dependent Four-Wave-Mixing Rules in Quantum Wells", *Sol. Stat. Comm.* 86:93 (1993).
5. D. Bennhardt, P. Thomas, R. Eccleston, E.J. Mayer, and J. Kuhl, "Polarization Dependence of Four-Wave-Mixing Signals in Quantum Wells", *Phys. Rev. B* 47:13485 (1993).
6. H.H. Yaffe, Y. Prior, J.P. Harbison, and L.T. Florez, "Polarization dependence and selection rules of transient four-wave mixing in GaAs quantum-well excitons", *J. Opt. Soc. Am. B* 10:578 (1993).
7. M. Lindberg, R. Binder, and S.W. Koch, "Theory of the semiconductor photon echo", *Phys. Rev. A* 45:1865 (1992).
8. Y.Z. Hu, R. Binder, and S.W. Koch, "Photon echo and valence-band mixing in semiconductor quantum wells", *Phys. Rev. B* 47:15679 (1993).
9. M. Wegener, D.S. Chemla, S. Schmitt-Rink, and W. Schäfer, "Line shape of time-resolved four-wave mixing", *Phys. Rev. A* 24:5675 (1990).
10. Y.Z. Hu, R. Binder, and S.W. Koch, "Valence-band mixing and photon echo in semiconductor quantum wells", *Quantum Electronics and Laser Science Conference*, Technical Digest Series, Vol. 12:182 (1993).
11. H. Wang, K. Ferrio, D.G. Steel, Y.Z. Hu, R. Binder, and S.W. Koch, *Phys. Rev. Lett.*, accepted for publication.
12. T. Saiki, M. Kuwata-Gonokami, T. Matsusue, and H. Saiki, "Elementary excitation model of the excitonic resonant nonlinearity in a GaAs quantum well", *Quantum Electronics and Laser Science Conference*, Technical Digest Series, Vol. 12:180 (1993).
13. K. Bott, O. Heller, D. Bennhardt, S.T. Cundiff, P. Thomas, E.J. Mayer, G.O. Smith, R. Eccleston, J. Kuhl, to be published.

POLARIZATION STUDIES OF FOUR-WAVE MIXING IN GaAs QUANTUM WELLS

G.J. Denton and R.T. Phillips

Cavendish Laboratory
Madingley Road
Cambridge
CB3 9JQ
U.K.

INTRODUCTION

Recently, polarization-dependent four-wave mixing experiments have demonstrated the need to account for the angular-momentum states of excitons when modelling their nonlinear optical response in GaAs quantum wells[1]. If exciton-exciton interactions are neglected, the appropriate model of the heavy-hole (hh) exciton is a pair of independent two-level systems which interact with left and right circularly polarized photons respectively. Abella et al.[2] invoked this model to explain the results of their photon-echo experiments on ruby in the 1960s. If the light-hole (lh) and heavy-hole exciton transitions are excited simultaneously, the six-level model illustrated in Figure 1 is required.

If the non-interacting model of the hh exciton is applied to self-diffraction experiments using linearly polarized light, the signal intensity is predicted to be independent of the polarization of the incident beams. Recently, injection-dependent deviations from this behaviour have been reported[3], and attributed to exciton-exciton interactions[4]. One potential consequence of these interactions is the introduction of biexcitonic effects. We have therefore measured the polarization dependence of four-wave mixing at higher injection levels than those previously reported, since this should lead to a clear biexcitonic contribution to the signal.

EXPERIMENT

A typical two-beam self-diffraction geometry was used as illustrated in Figure 2. The two beams, derived from the same laser system, have wavevectors $\mathbf{k_1}$ and $\mathbf{k_2}$ and may be separated by a delay ΔT, although for all the experiments described here $\Delta T=0$. The polarization of beam 2 was rotated using a Soleil-Babinet compensator; this was used rather than a mica waveplate to prevent etalon effects. The laser was a synchronously pumped styryl-9 dye-laser producing pulses of $\approx$ 1.8meV FWHM linewidth. The pump beams were focussed onto the sample and the self-diffracted signals were detected close to the directions $\mathbf{2k_1\text{-}k_2}$ and $\mathbf{2k_2\text{-}k_1}$ by silicon photodiodes giving the time-integrated signal.

Coherent Optical Interactions in Semiconductors
Edited by R.T. Phillips, Plenum Press, New York, 1994

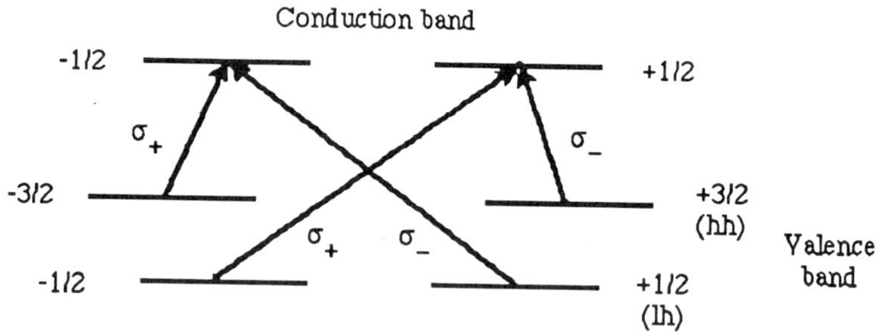

Figure 1 The six-level model employed in Reference 1 to describe the interaction of light with the light- and heavy-hole excitons in GaAs quantum wells. σ_+ and σ_- represent right and left circularly polarized photons respectively. In the case of heavy-hole excitation only, the model reduces to a pair of independent two-level systems. This model does not account for exciton-exciton interactions.

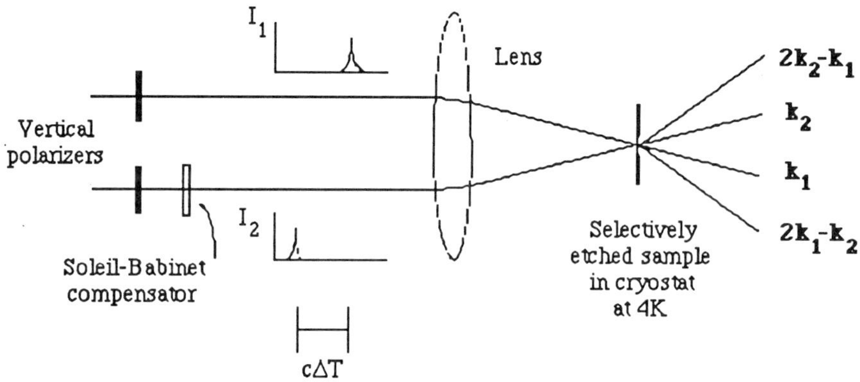

Figure 2 Schematic illustration of the self-diffraction geometry used in these experiments.

A window was etched in the GaAs substrate to allow transmission of the self-diffraction. This geometry leads to scattering from the pump beams into the detected directions so a background-free measurement was produced by chopping both incident beams and using phase-sensitive detection.

Two samples were investigated: a GaAs multiple quantum well (MQW) structure consisting of 100 repeats of a 102Å QW with 150 Å $Al_{0.33}Ga_{0.66}As$ barriers, and a 102Å GaAs single quantum well (SQW) structure. The MQW showed inhomogeneously broadened heavy-hole exciton photoluminescence (PL) with a full width at half maximum of 2.3meV and signs of well-width fluctuations. The SQW had a hh exciton PL linewidth of 0.7meV and showed no signs of well-width fluctuations or a Stokes shift.

RESULTS

Polarization of self-diffracted beams

Figure 3 shows the polarizations of both self-diffracted signals, θ_{sd}, versus the angle between the electric fields of the pump beams, θ_i. The experiment was performed on the MQW with excitation at the low-energy side of the heavy-hole exciton at 5×10^{11} cm^{-2} (as shown by line 1 in Figure 4).

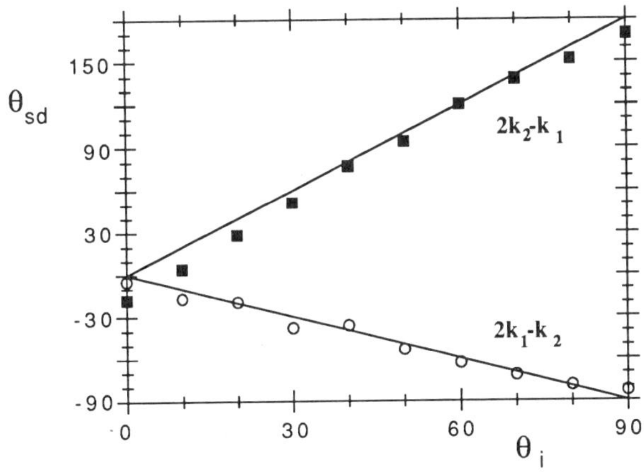

Figure 3 Polarization of self-diffracted beams, θ_{sd}, as a function of the angle separating the pump beams' polarizations, θ_i. The solid lines through data labelled **$2k_2$-k_1** and **$2k_1$-k_2** are $\theta_{sd} = 2\theta_i$ and $-\theta_i$ respectively and are the results expected for two independent two-level systems.

The lines through **$2k_2$-k_1** and **$2k_1$-k_2** are $\theta_{sd} = 2\theta_i$ and $\theta_{sd} = -\theta_i$ respectively which describe the results of Abella et al.[2] and the predictions of Schmitt-Rink et al.[1] for the hh exciton alone; these polarization directions and those of the transmitted beams are illustrated in Figure 5. We see that even at this high-injection level, the non-interacting model describes the experimental results. We repeated the experiment with laser excitation resonant with the hh exciton peak and with the peak marked A, shown by lines 2 and 3 respectively in Figure 4; in both cases we obtained results identical to those illustrated in Figure 3.

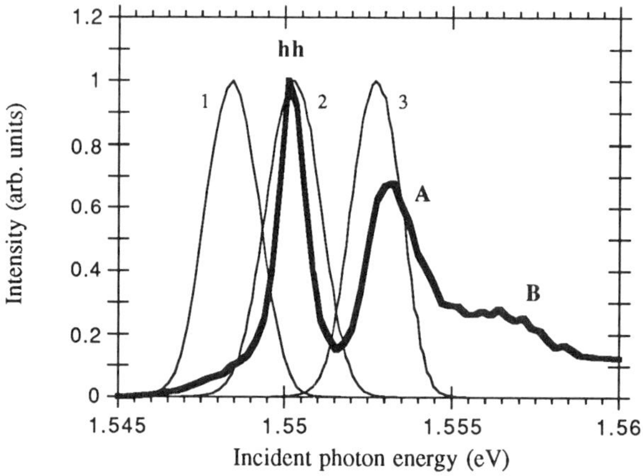

Figure 4 (Bold line) Self-diffraction intensity versus incident photon energy for the MQW sample excited at 8×10^{11} cm^{-2} at the hh exciton. The lines labelled 1...3 are laser spectra showing different excitation tunings used on this sample. The peaks marked A and B are believed to be due to well-width fluctuations.

Similar results were obtained by Eccleston et al.[3] for a single QW at 4×10^9 cm^{-2}, but at 4×10^8 cm^{-2} they found that $\theta_{sd} \approx 0°$ for $\theta_i = 0$ to 60°; for larger θ_i the predicted behaviour returned. These authors obtained a good fit to the low-injection data using an exciton-exciton interaction model[4] and explained the absence of its effect at 4×10^9 cm^{-2} by invoking screening of the interaction.

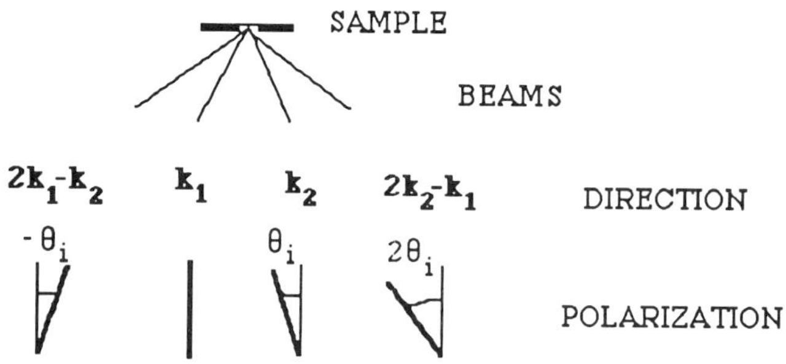

Figure 5 The polarization of the pump beams and the self-diffracted beams; the latter are shown by the solid lines in Figure 3.

Dependence of self-diffracted intensity on the incident polarizations

Figure 6 shows the time-integrated self-diffraction intensity as a function of θ_i for excitation of lh excitons in the SQW (open circles) and at the peak of the hh exciton transition in the MQW (solid circles). The solid lines represent $\cos^2\theta_i$ and $\sin^2\theta_i$ fits to the data.

For excitation of a single transition, the non-interacting model[1,2] predicts no variation of intensity with θ_i. A $\cos^2\theta_i$ dependence was reported by Eccleston et al.[3] for a hh exciton injection of 4×10^8 cm^{-2} in a GaAs SQW, but to our knowledge, there have been no investigations where the signal intensity for $\theta_i = 90°$ (perpendicular-polarization) is greater than that for $\theta_i = 0°$ (parallel-polarization). We find that for excitation of the hh exciton in the MQW, perpendicular-polarization produces a signal which is about 3 times greater than that from parallel-polarization and the intensity follows a $\sin^2\theta_i$ dependence. The same behaviour was also found for excitation at the low-energy side of the hh exciton in the SQW, thus eliminating the step-splitting or inhomogeneous broadening in the MQW as possible causes of this phenomenon.

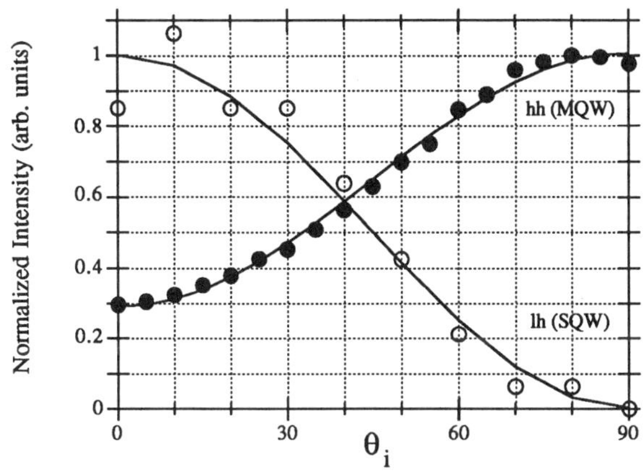

Figure 6 Time-integrated self-diffraction intensity as a function of angle between the pump-beam polarizations, θ_i. Excitation was at the light-hole exciton in the single quantum well (open circles) and at the peak of the heavy-hole exciton in the MQW (solid circles). The solid lines represent $\cos^2\theta_i$ and $\sin^2\theta_i$ fits to the data.

To determine whether the behaviour displayed in Figure 6 is affected by the excitation conditions, we measured the ratio of the signal intensities obtained with parallel- and perpendicular-polarizations (I_0/I_{90}) at a series of excitation energies across the hh transition in the SQW. This was performed at several injection levels ranging from 1×10^{10} to 1×10^{11} cm^{-2}, all measured when excitation was at the centre of the transition.

For the highest injection level, I_0/I_{90} was about 0.3 at the low energy side of the transition. This ratio increased to be 0.8 at the centre of the hh exciton and reached approximately 1 at the high-energy side. The cause of this behaviour is not clear. It has been reported that the exciton dephasing rate is energy dependent in samples with well-width fluctuations[5, 6, 7] and the parallel- and perpendicular-polarized cases may be affected differently by this. However, the SQW shows no evidence of such inhomogeneity.

A comparison of I_0/I_{90} at the line centre for the different injection levels is shown in Figure 7. The ratio is seen to be less than 1 for the highest level but it increases to about 2.6 for the lowest-injection case. $I_0/I_{90} = 1$ indicates the behaviour predicted by the non-interacting model, that is the signal intensity is independent of θ_i. When the ratio is greater or less than 1 we have an intensity variation following $\cos^2\theta_i$ or $\sin^2\theta_i$ respectively. Figure 7 clearly shows that the ratio is injection-level dependent and is only less than 1 for the highest level. This indicates that a density-dependent phenomenon, such as exciton-exciton interaction, influences the behaviour shown in Figure 6.

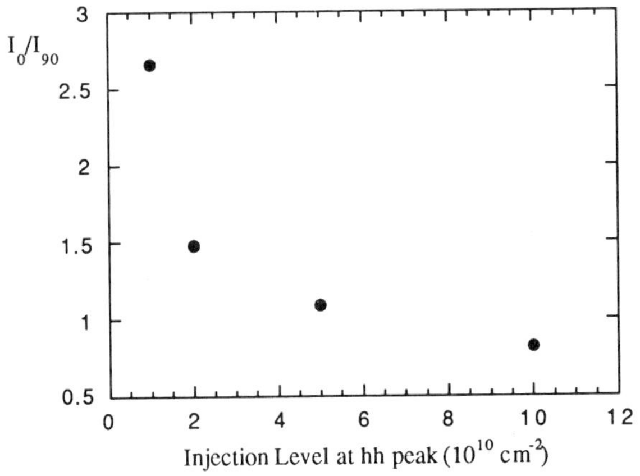

Figure 7. Ratio of the self-diffracted signals obtained with parallel- and perpendicular-polarization, I_0/I_{90}, for a range of injection levels, all measured at the peak of the hh exciton transition.

Spectral dependence of self-diffraction for parallel- and perpendicular-polarized incident beams

Figures 8 and 9 show the time-integrated self-diffraction intensity from the SQW as a function of incident photon energy at four different injection levels for parallel- and perpendicular-polarizations respectively.

At 1×10^{10} cm^{-2} the signal for both polarization cases has a single peak at 1.5478eV, attributed to the hh exciton, with a small low-energy shoulder. The Figures clearly show that the signal from the lh exciton is polarization dependent. For higher injection, the $\theta_i = 0°$ lineshape remains constant except for broadening, however, the $\theta_i = 90°$ case shows a rapid increase in the intensity of the low-energy shoulder until it dominates over the hh exciton signal.

The shoulder is about 1meV below the heavy-hole exciton, a splitting close to the biexciton binding energy in this sample as determined by low-temperature PL[8]. Yaffe et al.[5] attributed the signal they obtained with perpendicular-polarization to the biexciton. Their experiment was carried out at less than 3×10^9 cm^{-2} and the signal obtained with perpendicular-polarization was two orders of magnitude smaller than the parallel-polarization case. The greater relative intensity of the perpendicular-polarization signal in our work may be due to the larger injection levels.

DISCUSSION

Despite the high injection levels used in our experiments, we find that the dependence of θ_i on θ_{sd} follows the predictions of the non-interacting model. The results of Kuhl et al.[3] for an injection of 4×10^8 cm^{-2} disagree with the non-interacting model, but at 4×10^9 cm^{-2} the predicted behaviour returns and they attribute this to screening of the exciton-exciton interaction[4]. Our experiments were performed at about an order of magnitude higher injection than those reported by Kuhl and co-workers.

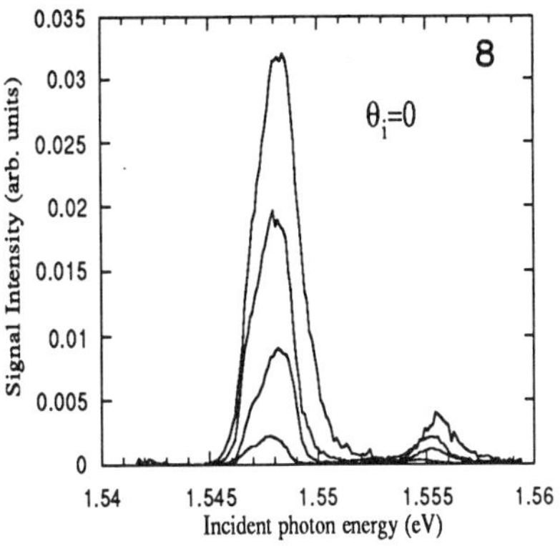

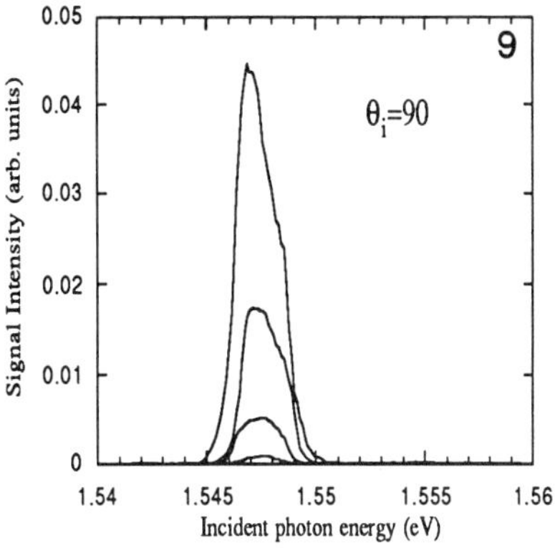

Figures 8 & 9 Self-diffraction intensity from the SQW versus incident photon energy for four different injection levels: 1×10^{11}, 5×10^{10}, 2×10^{10} and 1×10^{10} cm^{-2} measured at the hh peak. Figure 8 was measured with $\theta_i = 0°$ and Figure 9 with $\theta_i = 90°$. Although the intensity units are arbitrary, both Figures use the same scale to allow comparison of their relative values.

Although our measurements of θ_{sd} as a function of θ_i do not show any exciton-interaction effects, the dependence of the signal intensity on θ_i suggests that excitons are interacting to form biexcitons in these experiments. This can be understood with reference to Figures 8 and 9. If excitation occurs at the low-energy side of the hh transition, we see that for low injection where the shoulder is small, there will be little signal variation with θ_i. However, for larger injection levels the shoulder increases and causes the signal for perpendicular-polarization to exceed that for parallel-polarization. Therefore, we will see an increase of the signal on going from θ_i=0° to 90° as shown in Figure 6. We attribute the shoulder causing this behaviour to the biexciton, because of its spectral position[8] and dependence on polarization[5] and injection[8].

REFERENCES

1. S. Schmitt-Rink, D. Bennhardt, V. Heuckeroth, P. Thomas, P. Haring, G. Maidorn, H. Bakker, K. Leo, D-S. Kim, J. Shah, and K. Köhler, Phys. Rev. B **46**, 10460 (1992)

2 I.D. Abella, N.A. Kurnit, and S.R. Hartmann, Phys. Rev. **141**, 391 (1966); N.A. Kurnit, I.D. Abella, and S.R. Hartmann, Phys. Rev. Lett. **13**, 567 (1964)

3. R. Eccleston, J. Kuhl, D. Bennhardt, and P. Thomas, Solid State Commun. **86**, 93 (1993)

4 J. Kuhl, E.J. Mayer, G. Smith, R. Eccleston, D. Bennhardt, P. Thomas, K. Bott, and O. Heller, *Optical dephasing of excitons in III-V semiconductors*, in this volume.

5. H.H. Yaffe, Y. Prior, J.P. Harbison, and L.T. Florez, J. Opt. Soc. Am. B **10**, 578 (1993)

6. J. Hegarty, and M.D. Sturge, J. Opt. Soc. Am. B **2**, 1143 (1985)

7. T. Takagahara, Phys. Rev. B **32**, 7013 (1985)

8. R.T. Phillips, D.J. Lovering, G.J. Denton, and G.W. Smith, Phys. Rev. B **45**, 4308 (1992)

9. R.C. Miller, D.A. Kleinman, A.C. Gossard, and O. Munteanu, Phys. Rev. B **25**, 6545 (1982)

PARTICIPANTS

Dr I Abram
CNET
196 Avenue Henri Ravera,
92220 Bagneux
France

Dr J Allam
Hitachi Cambridge Laboratory
Madingley Road
Cambridge CB3 0HE
UK

Dr J-Y Bigot
Group d'Optique Nonlinéaire et d'Optoélectronique
5, rue de l'Université
67084 Strasbourg Cedex
France

Mr I Blewett
Physics Department, Heriot-Watt University
Riccarton
Edinburgh EH14 4AS
UK

Dr W Brocklesby
Department of Physics
University of Southampton
Highfield
Southampton SO9 5NH
UK

Prof W Bron
University of California
Department of Physics
Irvine
California 92717
USA

Dr M Burt
British Telecom Research Laboratories
Martlesham Heath
Ipswich IP5 7RE

Mr G Denton
Cavendish Laboratory
Madingley Road
Cambridge CB3 0HE
UK

Prof B Deveaud
Ecole Polytechnique Fédérale de Lausanne
Institut de Micro-optoélectronique
CE Ecublens
CH-1015 Lausanne
Switzerland

Dr K Duppen
Chemische Fysica
Rijksuniversiteit Groningen
Nijenborgh 4
9747 AG Groningen
Netherlands

Dr J Feldman
Fachbereich Physik- AG Halbleiterphysik
Renthof 5
D-35032 Marburg
Germany

Dr M Fox
Clarendon Laboratory
Parks Road
Oxford OX1 3PU
UK

Mr N Gallaher
Department of Physics
Heriot-Watt University
Riccarton
Edinburgh EH14 4AS
UK

Prof. Dr. Göbel
Fachbereich Physik- AG Halbleiterphysik
Renthof 5
D-35032 Marburg
Germany

Mr D Guerrier
Department of Physics
University of Southampton
Southampton SO9 5NH
UK

Dr R Harley
Department of Physics
University of Southampton
Southampton SO9 5NH
UK

Prof J Hvam
Mikroelectronik Centret
Technical University of Denmark
Building 345 East
DK-2800 Lyngby
Denmark

Mr M Jütte
Universität GH Paderborn
Warburgerstrasse 100A
Fachbereich 6, A2.201
D-33095 Paderborn
Germany

Dr A Kar
Department of Physics
Heriot-Watt University
Riccarton
Edinburgh EH14 4AS
UK

Prof C S Kim
Department of Physics
Chonnam National University
Kwangju 500-757
South Korea

Prof S Koch
Fachbereich Physik
Renthof 5
D-35032 Marburg
Germany

Dr Jürgen Kuhl
Max-Planck-Institut für Festkörperforschung
Heisenbergstrasse 1
D-70506 Stuttgart
Germany

Mr O Kuhn
CNRS Service National des Champs Intenses
25 Avenue des Martyrs
F-38042 Grenoble Cedex 9
France

Prof T Kuhn
Institut für Theoretische Physik und Synergetik
Universität Stuttgart
Pfaffenwaldring 57/IV
D-70550 Stuttgart
Germany

Dr A Kuznetsov
Department of Physics
University of Florida
215 Williamson Hall
Gainesville, FL 32611
USA

Mr P Leisching
Institut für Halbleitertechnik II
RWTH Aachen
Sommerfeldstrasse 24
D-52056 Aachen
Germany

Mr J-P Likforman
ENSTA Centre d'Yvette
F-91120 Palaiseau
France

Dr David Lovering
Cavendish Laboratory
Madingley Road
Cambridge CB3 0HE
UK

Mr T Meier
Fachbereich Physik
Renthof 5
D-35032 Marburg
Germany

Dr R Phillips
Cavendish Laboratory
Madingley Road
Cambridge CB3 0HE
UK

Mr M Portuné
Fachbereich Physik
Universität, Erwin-Schrödinger-Strasse 1
D-67663 Kaiserslautern
Germany

Dr W Quade
Institut für Theoretische Physik
TU Berlin
Hardenbergstrasse 36
D-1 Berlin 12
Germany

Dr F Rossi
Fachbereich Physik
Renthof 5
D-35032 Marburg
Germany

Dr W Schäfer
HLRZ
Forschungszentrum Jülich GmbH
D-52425 Jülich
Germany

Dr Jagdeep Shah
AT&T Bell Laboratories
Holmdel
New Jersey 07733
USA

Dr A Shields
Toshiba Cambridge Research Ltd.
260 Cambridge Science Park
Milton Road
Cambridge CB4 4WE
UK

Dr N Shigekawa
Department of Physics
University of Nottingham
University Park
Nottingham NG7 2RD
UK

Prof A Smirl
Centre for Laser Science & Engineering
University of Iowa
124 AMRF
Oakdale
Iowa City, IA 52242
USA

Prof C Stanton
Department of Physics
University of Florida
215 Williamson Hall
Gainesville, FL 32611
USA

Prof Duncan G Steel
Harrison M. Randall Laboratory of Physics
University of Michigan
Ann Arbor, Michigan 48109
USA

Mr N Traynor
Department of Physics
University of Southampton
Southampton SO9 5NH
UK

Ms A Uhrig
Fachbereich Physik
Universität, Erwin-Schrödinger-Strasse 1
D-67663 Kaiserslautern
Germany

Prof. W. von der Osten
Fachbereich Physik
Universität-Gesamthochschule
Warburger Strasse 100A
D-4790 Paderborn
Germany

Mr G von Plessen
Fachbereich Physik
Renthof 5
D-35032 Marburg
Germany

Prof Dr M Wegener
Institut für Physik
Universität Dortmund
D-44221 Dortmund
Germany

Dr S Weiss
Lawrence Berkeley Laboratory
MS 2-300
1 Cyclotron Road
Berkeley, CA 94720
USA

Mr R Worsley
Department of Physics
University of Southampton
Southampton SO9 5NH
UK

Dr N Zheludev
Department of Physics
University of Southampton
Southampton SO9 5NH
UK

INDEX